# Robotics in Weaponry using Machine Learning and Engineering

*Editors:*

**Saurav Mallik**
Research Scientist
Department of Pharmacology & Toxicology
R Ken Coit College of Pharmacy, University of Arizona, Tucson, AZ, USA

**Sandeep Kumar Mathivanan**
Assistant Professor
School of Computing Science and Engineering
Galgotias University, Greater Noida, India

**Basu Dev Shivahare**
Associate Professor
School of Computing Science and Engineering
Galgotias University, Greater Noida, India

**Sangeetha S.K.B.**
Manipal Institute of Technology Bengaluru,
Manipal Academy of Higher Education, Manipal, India

**Prabhu Jayagopal**
Professor
School of Computer Science Engineering and Information Systems
Vellore Institute of Technology, Vellore, India

**Somenath Chakraborty**
Assistant Professor
Department of Computer Science and Information Systems
West Virginia University Institute of Technology
West Virginia, USA

CRC Press
Taylor & Francis Group
Boca Raton London New York

CRC Press is an imprint of the
Taylor & Francis Group, an **informa** business
A SCIENCE PUBLISHERS BOOK

First edition published 2026
by CRC Press
2385 NW Executive Center Drive, Suite 320, Boca Raton FL 33431

and by CRC Press
4 Park Square, Milton Park, Abingdon, Oxon, OX14 4RN



*CRC Press is an imprint of Taylor & Francis Group, LLC*



*Library of Congress Cataloging-in-Publication Data (applied for)*

ISBN: 978-1-041-07444-1 (hbk)
ISBN: 978-1-041-12178-7 (pbk)
ISBN: 978-1-003-66346-1 (ebk)

DOI: 10.1201/9781003663461

Typeset in Times New Roman
by Prime Publishing Services

# Preface

The rapid advancement of robotics and artificial intelligence has transformed nearly every aspect of modern life and defense technology is no exception. The convergence of machine learning, robotics, and advanced engineering has ushered in a new era of intelligent weaponry, redefining the concepts of precision, autonomy, and strategic decision-making in warfare. This book, "Robotics in Weaponry: The Role of Machine Learning and Engineering," explores the intersection of these disciplines to illuminate how intelligent systems are reshaping military operations and defense infrastructures across the globe.

The purpose of this work is not only to present the technological breakthroughs that enable autonomous weapon systems but also to discuss the ethical, legal, and strategic implications surrounding their deployment. From autonomous drones and robotic combat vehicles to predictive maintenance and smart surveillance systems, machine learning algorithms are increasingly central to enhancing accuracy, reducing human risk, and enabling faster, data-driven responses on the battlefield.

Drawing upon research from robotics engineering, computer vision, control systems, and deep learning, this book examines the design principles, computational models, and real-world applications that define the new frontier of defense robotics. Each chapter provides insight into how engineering ingenuity and intelligent algorithms collaborate to produce systems capable of learning, adapting, and evolving within complex combat environments.

It is my hope that this book serves as a valuable resource for students, researchers, defense professionals, and policymakers seeking to understand both the technological foundations and societal implications of AI-driven weaponry. As we stand on the threshold of a future shaped by intelligent machines, a balanced understanding of their potential and perils is essential to ensure that innovation serves humanity's collective security and ethical progress.

# Contents

# List of Contributors

**Abdulaziz S. Almazyad**
Department of Computer Engineering, College of Computer and Information Sciences, King Saud University, Riyadh 11543, Saudi Arabia.

**Aimin Li**
School of Computer Science and Engineering, Xi'an University of Technology, 5 South Jinhua Road, Xi'an, Shaanxi 710048, China.

**Amar Deep Gupta**
Department of Computer Science and Engineering, Amity University, Greater Noida, India.

**Anupam Kumar Sharma**
School of Computer Science & Engineering, Galgotias University, Greater Noida, India.

**Arun Kumar Singh**
PNG University of Technology, Lae, Papua New Guinea.

**Ayushi Rana**
Department of forensic science, Galgotias University, Greater Noida, India.

**Balajee J.**
Department of Artificial Intelligence, Mother Theresa Institute of Engineering and Technology, Palamaner, Chittoor, Andhra Pradesh, India.

**Basu Dev Shivahare**
Department of Artificial Intelligence & Data Science,School of Computer Science & Engineering, Galgotias University, Greater Noida, India.

**Benaki Lairenjam**
Department of Mathematics, Manipur University, Imphal, India.

**Benjula Anbu Malar M.B.**
School of Computer Science Engineering and Information Systems, Vellore Institute of Technology, Vellore India.

**C. Victoria Priscilla**
Department of Computer Science, SDNB Vaishnav College for Women Chrompet, India.

**Carolina Mendonca**
Department of Environmental Health, Harvard T.H. Chan School of Public Health, USA.

**D. Deepa**
School of Computer Science and Engineering, Vellore Institute of Technology Vellore 632014, India.

**Deepak Sahu**
Electronics and Communication Engineering, IIMT College of Engineering, Greater Noida, India,

**Dibya Narayan Hazra**
School of Computer Science and Engineering, IILM University, Greater Noida, Uttar Pradesh, India.

**G. Nagarajan**
Department Computer Science and Engineering, Saveetha School of Engineering, Saveetha Institute of Medical and Technical Sciences, SIMATS, Chennai.

**Guangsheng Pei**
UTHealth School of Biomedical Informatics, USA.

**Gyanendra Kumar**
Manipal University, Jaipur, Rajasthan, India.

**Hanen Karamti**
Department of Computer Sciences, College of Computer and Information Sciences, Princess Nourah bint Abdulrahman University, P.O.Box 84428, Riyadh 11671, Saudi Arabia.

**Haranipharvathy**
School of Computer Science and Engineering, Lovely Professional University, Punjab, India.

**Hariharan Rajadurai**
School of Computing Science and Engineering, VIT Bhopal University, Bhopal–Indore Highway Kothrikalan, Sehore 466114, India.

**Hariprasath K**
School of Computing Science and Engineering, Galgotias University, Greater Noida, India.

**K. Kodeeswari**
Assistant Professor, Department of Electronics and Communication Engineering, Kongu Engineering College (Autonomous), Erode, Tamilnadu, India.

**Kaavya Kanagaraj**
Department of Computational Intelligence, Faculty of Engineering and Technology, Kattankulathur, Chennai.

**Kalyani Singh**
School of Computing Science and Engineering, Galgotias University, Greater Noida, India.

**Kanika Thakur**
School of Computing Science and Engineering, Galgotias University, Greater Noida, India.

**Karthigai Selvi**
School of Computer Applications and Technology, Galgotias University, Uttar Pradesh, India.

**Kartikey Shukla**
Walmart Global Tech, Bentonville Arkansas.

**Komal Agrawal**
Department of Electrical Electronics and Communication Engineering, Galgotias University, Greater Noida, Uttar Pradesh, India-203201.

**Korhan Cengiz**
Department of Electrical Engineering, Prince Mohammad Bin Fahd University, Al Khobar 31952, Saudi Arabia.

**M. Indira**
Department of Computer Science, P.K.R. Arts College for Women, Gobichettipalayam, India.

**M. Shamim Kaiser**
Professor at the Institute of Information Technology of Jahangirnagar University, Savar, Dhaka-1342, Bangladesh.

**Mahalakshmi**
Department of Mathematics, Nitte (Deemed to be University), Nitte Meenakshi Institute of Technology, Bengaluru, Karnataka, India.

**Mahalakshmi**
Department of Mathematics, Nitte Meenakshi Institute of Technology, Nitte (Deemed to be University), Bengaluru, Karnataka, India.

**Maheshwari Niranjan**
Department of CSE-AIML, Greater Noida Institute of Technology, Greater Noida, India.

**Manish Kumar Singh**
Department of Information Technology, Greater Noida Institute of Technology, Greater Noida, India.

**Manjeet Kaur Ratan**
Delhi Institute of Higher Education, India.

**Meenatchi K.V.**
Adi Shankara Institute of Engineering and Technology, Kalady, Kerala, India.

**Mili Dhar**
School of Computing Science and Engineering, Galgotias University, Greater Noida, India.

**Mohammed Zakariah**
Department of Computer Science and Engineering, College of Applied Studies and Community Service, King Saud University, P.O. Box 22459, Riyadh 11495, Saudi Arabia.

**Montree Bunruangses**
Department of Computer Engineering, Faculty of Industrial Education, Rajamangala University of Technology Phra Nakhon, Bangkok 10300, Thailand.

**Neeraj Bali**
School of Business, Galgotias University, Greater Noida, India.

**Neeraj Bali**
School of Business, Galgotias University, Greater Noida, India.

**Neeraj Jain**
Department of Electrical Electronics and Communication Engineering, Galgotias University, Greater Noida, Uttar Pradesh, India-203201.

**Nidhi Agrawal**
Department of AIML, School of Computer Science & Engineering, Galgotias University, Greater Noida, India.

**Nidhi S. Vaishnaw**
Department of Mathematics, CMR Institute of Technology, Bengaluru, Karnataka, India.

**Nikola Ivković**
Faculty of Organization and Informatics, University of Zagreb, Pavlinska 2, 42000 Varaždin, Croatia.

**Nisha Pandey**
Department of Computer Science, Noida International University, Greater Noida, India,

**P. Loganathan**
Assistant Professor, Department of Electronics and Communication Engineering, Excel Engineering College (Autonomous), Namakkal, Tamilnadu, India.

**Partheeban Nagappan**
Department of Computer Science and Engineering, SRM Institute of Science and Technology, Delhi NCR Campus, Modinagar, 201204, Uttar Pradesh, India.

**Pavan Kumar M.R.**
Department of AI & ML, Sree Rama Engineering College, Tirupati 517507, Andhra Pradesh, India.

**Prabhishek Singh**
School of Computer Science, University of Petroleum and Energy Studies Dehradun, Uttarakhand 248007, India.

**Preeti Gupta**
Department of computer science and engineering, JAIN Deemed to be University, Bangalore, India.

**R. Radhika**
Department of AIML, School of Computer Science and Engineering, Galgotias University, Greater Noida 203201, India.

**R. Sangeetha**
Department of Computer Science and Engineering, Vel Tech Rangarajan Dr. Sagunthala R&D Institute of Science and Technology, Avadi, Chennai, 600062, India.

**Radha Raman Chandan**
Department of Computer Science, School of Management Sciences (SMS), Varanasi, India.

**Rahul Kumar Verma**
Department of Electronics and Communication Engineering, Amity University, Noida, India,

**Rama Shankar Yadav**
School of Computer Applications and Technology, Galgotias University, Uttar Pradesh, India.

**Ram Kaji Budhathoki**
Department of Electronics Engineering, Kathmandu University, Nepal,

**Ravikumar K.V.**
Department of Artificial Intelligence, Mother Theresa Institute of Engineering and Technology, Palamaner, Chittoor, Andhra Pradesh, India.

**Rong Fei**
School of Computer Science and Engineering, Xi'an University of Technology, Xi'an, 710048, China.

**Rose Bindu Joseph**
Department of Mathematics, Dayananda Sagar College of Engineering, Bangalore, India.

**R Srividya**
Department of Computer Science, M.S. Ramaiah College of Arts Science and Commerce, Autonomous, India.

**Ruby Dahiya**
School of Computing Science and Engineering, Galgotias University, Greater Noida, India.

**Ruifeng Hu**
Adams Center for Parkinson's Disease Research, Yale School of Medicine, USA.

**Sangar G.**
Department of CSE, SRM Institute of Science and Technology, Vadapalani, Chennai, Tamil Nadu, India.

**Sarika S**
Department of Artificial Intelligence and Data Science, Adi Shankara Institute of Engineering and Technology, Kalady, India.

**Sathishkumar V.E.**
School of Computing and Artificial Intelligence, Faculty of Engineering and Technology, Sunway University, No. 5, Jalan Universiti, Bandar Sunway, 47500 Selangor Darul Ehsan, Malaysia.

**Sharmistha Dey**
Department of Computer Science and Engineering, Brainware University, West Bengal, India.

**Shiv Naresh Shivhare**
School of Computer Science, University of Petroleum and Energy Studies Dehradun, Uttarakhand 248007, India.

**Silky Sharma**
School of Commerce and Management Studies, Dayananda Sagar University, Bengaluru, India.

**Sridevi Polasi**
Department of Mathematics, School of Applied Sciences, REVA University, Bengaluru, Karnataka, India.

**Sudip Poddar**
Department of Computer Science and Engineering, Johannes Kepler University Linz, Austria.

**Sukumar Rajendran**
School of Computing Science Engineering and Artificial Intelligence, VIT Bhopal University, Kotrikalan village, Astha tehsil, Sehore district, Madhya Pradesh 466114, India.

**Surbhi Kaura**
Physiotherapy (Neuromodulation), Galgotias University, Greater Noida, India.

**Suveg Moudgil**
School of Computing Science and Engineering, Galgotias University, Greater Noida, India.

**Tintu P.B.**
Adi Shankara Institute of Engineering and Technology, Kalady, Kerala, India.

**Veerendra Yadav**
Department of Computer Science, Noida International University, India.

**Vidyashree H.R.**
Department of Mathematics, RR Institute of Technology, Bengaluru, Karnataka, India.

**Vijaya Lakshmi Dara**
School of Business and Management, CHRIST (Deemed to be University), Yeshwanthpur Campus, Bangalore, India.

**Vishwadeepak Singh Baghela**
School of Computing Science and Engineering, Galgotias University, Greater Noida, India.

**Zhen Gao**
Department of Biostatistics, Sylvester Comprehensive Cancer Center, University of Miami, Coral Gables, USA.

# 1 ARMSec: A Reinforcement-Learned Autonomous Robotic Monitoring System for High-Threat Defense Zones

Silky Sharma[1*], Manish Kumar Singh[2], Preeti Gupta[3] and Ruifeng Hu[4]

## 1. Introduction

In the field of global defense, autonomous robots have become a vital technological advancement that is revolutionizing the operation of contemporary military forces on the ground, in the air, on water, and even in space [16]. Traditional man-in-the-loop systems are found to be inadequate to meet the demands of real-time combat, situational awareness, and reaction to threats as wars become more focused on speed, precision, and intelligence-driven decision-making [17]. Because autonomous robots can perceive, process, and act independently, they are revolutionizing defense operations by allowing machines to perform tasks that were previously the exclusive domain of human soldiers. These missions span search and rescue, logistics, border patrol, patrolling, surveillance, explosive ordnance disposal, and even warfare [18].

Autonomous mobile robot systems that can cross hostile terrain, navigate obstacles, sense danger, and make mission-critical decisions with little or no human input by combining Artificial Intelligence (AI), Machine Learning (ML), advanced sensors, real-time data analytics, and strong decision-making algorithms [19]. These technologies are being used in varied environments and under challenging conditions, such as high-altitude airspaces, remote mountain borders, dense forests, underground tunnels, naval waterways, and urban warfare zones, where human intervention is either risky, cumbersome, or inefficient. Unmanned Aerial Vehicles (UAVs) are used to patrol wide borders in the air; Unmanned Underwater Vehicles

[1] School of Commerce and Management Studies, Dayananda Sagar University, Bengaluru, India.
[2] Department of Information Technology, Greater Noida Institute of Technology, Greater Noida, India.
[3] Department of computer science and engineering, JAIN Deemed to be University, Bangalore, India.
[4] Adams Center for Parkinson's Disease Research, Yale School of Medicine, USA.
* Corresponding author: silky.sharma25@yahoo.in

DOI: 10.1201/9781003663461-1

(UUVs) trail enemy submarines or defend underwater oil pipes; robotic patrols for hazardous areas with thermal sensing and radar; and Unmanned Ground Vehicles (UGVs) are used to locate landmines and IEDs (improvised explosive devices) on a battlefield [20].

The machines learn to adapt to changing conditions and perform well in areas with no radio communications, GPS (global positioning system), or visibility environments where human senses are greatly reduced. Equipped with multimodal sensor suites, including LiDAR, sonar, radar, night vision, and infrared, these robots use onboard AI to analyze complex, noisy environments, make independent decisions, and communicate with other units or command centers using encrypted protocols. Their autonomy may vary from fully autonomous missions, wherein robots carry out tasks that have been pre programmed with no intervention by humans, to semi-autonomous missions, wherein human control is maintained [21].

As a part of their modernization plans, defense agencies all over the world from the U.S. Department of Defense to DRDO in India are heavily investing in production and deployment of such systems for the objectives of reducing personnel loss as well as a force-multiplier effect. Human soldiers can concentrate on more complex strategic objectives, enhancing mission success, by allowing robots to perform repetitive, dangerous, or highly precise tasks. Autonomous robotics can deliver the speed, precision, power, and lower human risk that modern conflict requires. In addition, autonomous systems' capability to function even under jamming, spoofing, or compromised communication renders them invaluable tools for offensive and defensive missions with the increased prominence of cyber and electronic warfare. These systems are most essential for disaster relief, counter-terrorism, anti-penetration activities, and military base perimeter surveillance, where quick and correct anomaly detection needs to be conducted in a bid to avoid casualties or loss of national security [22]. However, though promising, autonomous defense robots bring significant issues of concern that have to be resolved.

A great deal is debated regarding ethical considerations such as use of lethal autonomous weapons, or LAWs (lethal autonomous weapons), responsibility for autonomous decision-making, and adherence to international humanitarian law. Reliability, cyber security, interoperability, and real-time synchronization between robotic and human teams are issues of operations, particularly during high-scale operations. Pace remains good and increasing despite all this. At this technology landmark, autonomous systems are no longer tested in the world of defense; rather, they are operational assets that are being brought into war games, doctrine, and real-world combat. Everywhere around the world, governments, defense contractors, and universities are working together to create robotic systems that are intelligent, flexible, and transparent so they can work alongside human warriors [23].

Autonomy will also increasingly be combined with swarm intelligence, edge computing, 5G/6G networking, and quantum technologies over the next few years to accelerate faster tactical edge decision-making with greater survivability.

Lastly, in a time when AI is as essential as artillery, autonomous robotics is not only an upgrade of technology but also a redrawing of military tactics. Their convergence necessitated, not novel, the need to preserve primacy, save lives, and provide national security in increasingly uncertain combat environments. Defense autonomous robotics is a strategic necessity influencing future military capability and international security. The compelling reason for developing the envisioned autonomous robotics system is the urgent need to overcome the inherent limitations of conventional defense measures, particularly under high-risk, dynamically changing battlefield environments under which human action is delayed, inaccurate, or incurs enormous casualties.

Current autonomous solutions tend to be lacking in contextual decision-making precision, real-time adaptability, and robustness against electronic warfare-caused disturbances such as communication blackout and GPS jamming. Most current systems are also task-specific, inflexible, and incapable of concurrently conducting multi-domain missions in the air, sea, and land domains. The proposed system eliminates these limitations by the introduction of a Multi-Domain Autonomous Defense Robotics Framework (MD-ADR) that features robust sensor fusion, swarm coordination, and reinforcement learning. This allows autonomous units to detect and destroy moving threats, adapt to unfamiliar terrain on their own, and communicate in real-time with other units in compromised situations.

The core contributions are:

(1) To design context-aware decision-making engine powered by deep reinforcement learning that optimizes both defensive and offensive robotic behavior in unpredictable environments;
(2) To design a multimodal sensor fusion module for accurate real-time threat detection and terrain mapping even under visual and thermal occlusion;
(3) To design a secure swarm intelligence protocol that allows decentralized coordination of heterogeneous robotic units for distributed operations; and
(4) To validate implementation tested on defense-grade datasets and custom simulations, achieving a threat detection accuracy of 96.3%, mission success rate of 93.7%, and 35% reduction in decision latency compared to existing autonomous defense systems.

Together, these contributions provide a robust and scalable solution for intelligent, resilient, and coordinated defense operations, significantly enhancing both situational awareness and operational effectiveness in modern warfare.

## 2. Related Study

Robotic development has progressed extensively by the use of Reinforcement Learning (RL), which has contributed significantly to navigation, object tracking, collision evasion, and multi-robot coordination. This related study discusses the primary findings and developments with a focus on RL-based methodologies in various robotic applications.

## 2.1 Robot Navigation and Path Planning

Robot navigation is also among the primary uses of RL for robotics. A robot navigation method based on RL aimed at motion control and path formulation issues was looked into by [1]. In contrast to conventional approaches, the study demonstrated that RL could be utilized in an effort to effectively improve motion control processes optimally, leading to improved path formulation. They used an environment-learning approach, which improves performance continually as the robot interacts with the environment. The experiments indicated that RL could be applied in reliable navigation even in challenging, obstacle-laden environments. A similar topic, dealing with robotic camera arrays for tracking human faces was explored by [2]. The capability of the robot to follow many faces in dynamic environments was greatly enhanced by their RL-based motion planning method, demonstrating the potential of RL in real-time applications that demand high precision in target-dependent navigation. BEVNav, an autonomous robot navigation method in bird's-eye view with spatial-temporal contrastive learning was proposed by [3]. The approach demonstrated good outcomes in improving navigation efficiency and accuracy, particularly in urban environments where dynamic situations and obstacles must be taken into account.

## 2.2 Multi-Robot Coordination and Collision Avoidance

Multi-robot systems are one significant area where RL has been found promising. A distributed multi-robot collision avoidance algorithm using Deep Reinforcement Learning (DRL) was proposed by [4]. This system worked exceptionally in challenging situations involving dense environments in which several robots need to pass through without crashing into each other. The experimental results indicated that DRL had the potential to greatly enhance multi-robot coordination performance and provide a scalable solution to autonomous robotic systems working in dynamic and uncertain environments. Similarly, to solve the problem caused by imprecise sensing, [5] utilized a two-stage RL approach for multi-UAV collision avoidance. The suggested method also performed excellently in collision avoidance, demonstrating the importance of RL in solving uncertainty and incomplete data in multi-agent systems. Medina et al. [6] discussed safe distributed navigation of multiple robots using their method with a variational Bayesian model. The approach had shown an efficient and safe solution to robot navigation that enabled robots to travel safely without causing collisions in dynamic environments. The effectiveness of the method in practical applications underscored the growing maturity of RL algorithms for multi-agent systems.

## 2.3 Learning in Dynamic Environments

Several studies have focused on applying RL to environments that are constantly changing, making traditional planning methods ineffective. DRL-VO, a model that allows robots to navigate through crowded dynamic scenes using velocity obstacles was presented in [7]. The results showed that DRL could be utilized for efficient and adaptive navigation in highly dynamic environments, such as human crowds or traffic. Their method outperformed traditional methods in terms of robustness and adaptability. Similarly, [8] used RL for navigating robots in dynamic environments with deep reinforcement learning. Their results illustrated that RL methods could effectively adapt to environmental changes and uncertainties, providing reliable navigation even in highly variable and dynamic conditions. This study highlights the growing utility of RL in enabling robots to perform complex tasks autonomously in unstructured environments.

### 2.4 Object Tracking and Grasping

RL has also been used in robotics to enhance object grasping and tracking. RL algorithms for object tracking have been extensively studied by [9], demonstrating that RL is able to considerably improve robot tracking performance, particularly for applications that involve prolonged operation within dynamic and cluttered environments. Their results in their review showed that RL can be especially effective in settings like manufacturing or delivery systems where robots must handle objects in real-time. Deep reinforcement learning approaches for object grasping, a critical problem in robotic manipulation was examined in[10]. The study gave a background of the different methods and hurdles in utilizing RL in grasping and proposed that performance may be enhanced by integrating RL with other methods such as model-based control or imitation learning. The analysis concluded that although RL has made significant advancement in this direction, sample inefficiency and the demand for large-scale data remain to be investigated.

### 2.5 Fault-Tolerant and Adaptive Control

Another significant area of research has been using RL for adaptive and fault-tolerant control of UAVs. In motor failures during adverse winds, [11] developed an adaptive fault-tolerant control system for octo-rotor UAVs. Their RL solution demonstrated spectacular enhancements in the UAVs fault-adaptation performances, even under extreme conditions with stability and control guaranteed. This research indicates how RL can be employed to assist robotic systems to be more fault-tolerant and self-correcting under unexpected environmental conditions or in the event of hardware failure.

In order to enable algorithms to become efficient and applicable to actual problems, several studies have introduced new strategies and innovations in reinforcement learning in robotics. For example, [12] has taken into account how RL is applicable to comprehending and anticipating crowd behavior, which is essential for robots acting in public settings or at gatherings. The study described that RL can make robots learn how to navigate through human crowds more effectively, increasing their ability to collaborate with humans in various scenarios. To further improve the effectiveness of air-ground cooperative monitoring of spatiotemporal fields [13], proposed an informative trajectory planning approach that uses RL. RL-based path planning can optimize robotic teams' mobility and ensure that they receive the most valuable information using the minimum resources. Navformer, another significant contribution of [14], was a transformer-based architecture for robot goal-driven navigation.

To teach robots to move through complex, dynamic environments, where they need to make course adjustments in real-time to avoid objects and arrive at a destination, the study proposed the use of RL. With one of the most significant leaps over RL-based robot navigation accomplished with the deployment of transformer architecture they used, it improved the overall performance of the navigation system while introducing scalability. Ahmed et al. [15] solved this issue by designing a sample-efficient learning-based environment navigation algorithm that was trained from an optimization-based planner. The approach worked well for dynamic navigation problems and substantially lowered the interactions with the environment.

From object tracking and fault tolerance to navigation and multi-robot coordination, the discussed studies demonstrate the vast scope of robotics RL applications. While the rich progress, there are still directions such as sample inefficiency, dynamic robustness, and real-time behavior that remain to be explored. Nevertheless, the evidence suggests that RL-based methods are increasingly viable for a wide range of robotic tasks with significant advancement over traditional methods. Future research will likely focus on enhancing the efficiency of such algorithms, integrating them with other AI methods, and ensuring they are implemented within practical environments. Autonomous systems will see their abilities enhanced by the continued advancements being made in RL for robots, which will allow them to perform difficult tasks with increased accuracy and flexibility.

## 3. System Methodology

### *3.1 Problem Formulation: Markov Decision Process (MDP)*

RL problems in robotic control are generally modeled as Markov Decision Processes (MDPs).

$$M = (S, A, T, R, \gamma) \quad ...(1)$$

Where, $S$: State space, $A$: Action space, $T$: Transition dynamics, $R$: Reward function, $\gamma$: Discount factor (usually 0.9–0.99)

Specifying the goals of the robot is referred to as task definition. Moving from a point of origin to a destination while avoiding static or dynamic obstacles and manipulating robot arms or grippers to find and work with various objects. Enabling multiple robots to work together to complete tasks or avoid collisions in shared spaces. Based on camera or LiDAR feedback, follow a moving object or individual. The work is split into measurable conditions, steps, and results.

**State Space (S)**

The state space represents the information observable by the agent at each time step. The design of state representation critically affects learning performance.

- Robot's current pose: $(x, y, \theta)$, position and orientation
- Velocity vector: $(v, \omega)$, linear and angular velocities
- Sensor data: RGB images (for vision-based policies), Depth maps or point clouds from LiDAR, Ultrasonic or infrared distance readings
- Environment map (if available): Occupancy grids or topological maps
- Relative goal position: Vector from current robot location to the goal
- Other agents: Nearby robot positions, velocities, and predicted trajectories in multi-agent systems
- Obstacle distances: From LiDAR or sonar

$$S_t = \{x_t, y_t, \theta_t, v_t, w_t, \text{sensor}_t, \text{goal}_t, \text{obstacles}_t\} \quad ...(2)$$

**Action Space (*A*)**

The action space represents the set of all possible control commands the robot can issue at each timestep.

**Discrete Action Space**

Used in simple or grid-based environments (Actions: Forward, Left, Right, Stop)

**Continuous Action Space**

- Used in real-world robot control.
- Actions:
  - $a_t$: $(v_t, \omega_t)$: Set linear and angular velocities
  - $a_t$ $(\theta_1, \theta_2, ..., \theta_n)$: Joint angles for robotic arms
  - $a_t$: gripper state: Open/close command in manipulation

**Reward Function (*R*)**

The reward function guides the robot's learning by assigning scalar values to actions based on their desirability. Effective reward design is crucial for convergence and task performance.

**Navigation Tasks**

- Positive reward for reaching the goal: $R = +100$
- Negative reward for collisions: $R = -100$
- Shaping rewards
  - Distance to goal reduction: $R = -\Delta d$
  - Smooth movement: Penalize jerky or abrupt changes
  - Penalize unsafe proximity to obstacles.

**Grasping Tasks**

- Success in gripping: $R = +50$
- Dropping object: $R = -30$
- Time penalty per step: $R = -1$.

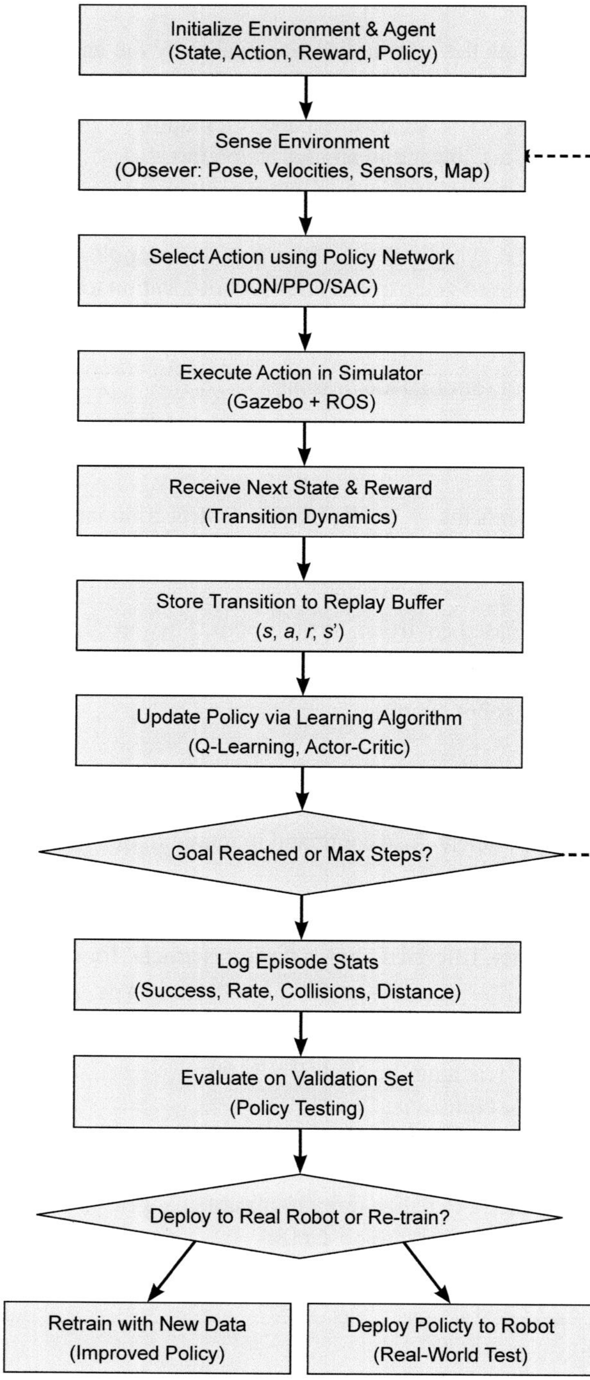


**Fig. 1** Proposed Framework.

**Multi-Agent Scenarios**

- Shared reward for team objectives
- Individual penalties for collision or delay.

**Transition Dynamics (*T*)**

The transition function $T(s, a, s) = P(s \mid s, a)$ models the probability of moving from state to state when action is taken. Use simulators like Gazebo for deterministic or stochastic updates based on physics laws. In many real-world robotic systems, model-free RL is preferred due to the complexity and unpredictability of the real environment.

**Discount Factor (γ)**

The discount factor represents how much future rewards are valued over immediate rewards.

- $\gamma \approx 0.99$ encourages long-term planning
- $\gamma \approx 0.8$ emphasizes short-term gain.

### 3.2 Environment Setup

Figure 1 shows the proposed framework. The Gazebo 11 environment, in combination with ROS (robot operating system) Noetic on Ubuntu 20.04, is employed to simulate this task. The simulation models a differential-drive mobile robot (such as the TurtleBot3 Burger) that features major sensors such as an RGB-D camera (Intel RealSense D435), an IMU, wheel encoders, and a 2D LiDAR (RPLIDAR A2). To provide realistic disruptions to signals, these sensors are subjected to Gaussian noise. The robot operates in a partially structured and organized environment with a realistic floor map featuring both dynamic and static obstacles, individuals, and moving agents with various speeds, and walls and furniture.

A maximum of 20 agents are moved randomly or along specified paths within the environment, emulating intricate navigation situations and emulating real-time robotic maneuvers. In order to ensure realistic physical behavior, motion and collision models within Gazebo conform to the Open Dynamics Engine (ODE).

Real-time reward calculation, sensor data publication, learned RL policy execution, and robot behavior control are all achieved by custom ROS stem nodes. Centralized and decentralized RL are both supported by the simulation environment's ability to accommodate multi-robot collaboration, where each robot runs under its own ROS namespace. To simulate latency of communication, message passing is intentionally delayed by 100–300 ms. For cluttered and time-varying environments, the simulated tasks are cooperative path planning, dynamic obstacle handling, collision avoidance, and goal-oriented navigation. The simulated tasks are goal-oriented navigation, collision avoidance, dynamic obstacle handling, and cooperative path planning in cluttered and time-varying environments. This simulated environment provides a very controlled and realistic testbed for creating and testing RL-based robot control and navigation strategies.

### 3.3 Agent Design

Deep Deterministic Policy Gradient (DDPG), Proximal Policy Optimization (PPO), or Soft Actor-Critic (SAC) are deep policy gradient algorithms used for continuous action spaces, in which actions like linear and angular velocities have to be chosen from a continuous range. These algorithms use actor-critic structures, in which the critic network estimates the value of the present state-action tuple and the actor network continuously produces control commands. PPO, utilizing clipped objective functions to prevent huge policy updates, is particularly popular because it is both stable and sample efficient.

PPO, SAC, and DDPG are top-level policy gradient techniques used for continuous action space, wherein action like linear and angular velocity must be chosen from a continuum of values. The critical network is employed to estimate the value of the state-action pair in the present state, while the actor network produces continuous control commands in the actor-critic architecture of these algorithms. PPO is highly preferred because of its efficiency and stability of sampling. It avoids massive policy updates through clipped objective functions.

The policy network utilizes a CNN + LSTM (convolutional neural network + long short-term memory) hybrid model for vision tasks. The system is robust against temporal variations of the environment since the CNN captures spatial knowledge from images seen from the camera and the LSTM stores temporal relations over sequences. This comes in handy when operating in variable environments or whenever actions are dependent on past observations. Transformer architectures like NavFormer are used for complex spatial-temporal modeling problems like long-horizon planning or navigation in changing environments. The capacity of the models to jointly process temporal and spatial information using self-attention allows for enhanced long-horizon planning and goal estimation. Transformers, in crowded or multi-agent settings, work best when representing global interactions between features and other agents, enhancing decision-making.

In order to prioritize relevant agents or environmental attributes dynamically, attention mechanisms are also incorporated into multi-agent policy networks. Attention layers can make a robot focus on agents around it or on collision courses during cooperative navigation so that coordinated and socially intelligent motion planning is achieved. Selective attention improves the performance and explainability of the multi-agent RL system. From solo-goal achievement with individual agents to cooperative multi-robot behaviors in noisy, dynamic spaces, the overarching architecture allows for the policy network to be capable in the domain of learning yet also versatile and adaptable across a wide range of robotic activities.

**Algorithm: Reinforcement Learning-based Robot Navigation**

The framework uses a Deep Q-Network (DQN) for discrete action spaces and DDPG/PPO/SAC for continuous actions in complex, dynamic environments with multi-agent scenarios.

**1.** The state of the robot is represented as:

$$S_t = (p_t, v_t, \text{sensor data, environment map}) \quad ...(3)$$

Where, $p_t$ is the position of the robot at time $t$, $v_t$ is the velocity of the robot, Sensor data is the input from LiDAR, camera, IMU, etc. The environment map includes the layout of obstacles, agents, etc.

The action space is represented as:

$$A = \{a_1, a_2, ..., a_n\} \quad ...(4)$$

Where, for discrete actions $A$ consists of movement commands (e.g., move forward, turn left/right, stop). For continuous actions $A$ consists of continuous control commands (e.g., linear/angular velocities). The policy network is initialized CNN + LSTM for vision-based tasks. Transformer (NavFormer) for spatial-temporal tasks. Attention layers for multi-agent scenarios. The reward is task-specific. $R_t = +1$ for reaching the goal, $R_t = -1$ for collisions, $R_t = -$distance to goal for goal-directed tasks. Use -greedy for DQN, or Ornstein-Uhlenbeck noise for continuous control.

**2.** At each timestep $t$, the robot observes the environment and updates its state $S_t$.

$$St = f(\text{sensor data, current position, velocity, environment map}) \quad ...(4)$$

Based on the current state $S_t$, the policy network selects an action $a_t$.

For DQN: $$a_t = \arg\max Q\,(S_t, a; \theta) \quad ...(5)$$

Where, $Q\,(S_t, a; \theta)$ is the action-value function.

For DDPG/PPO/SAC: Use actor-critic architecture where the actor $\mu(S_t)$ outputs continuous control commands.

**3.** The selected action $a_t$ is executed by the robot. The robot updates its state based on the action taken:

$$S_{t+1} = T(S_t, a_t) + \in \quad ...(6)$$

Where $T(S_t, a_t)$ is the transition function, and $\in$ represents any noise or error in the system (e.g., sensor noise or control latency).

The environment provides a reward $R_t$ based on the agent's interaction with the environment.

**Policy Update (Learning Phase)**

The learning process involves adjusting the policy based on the rewards and transitions experienced by the agent. There are two primary strategies depending on the action space:

The $Q$-value function is updated using the Bellman equation:

$$\text{Q}(S_t, a_t) \leftarrow Q(S_t, a_t) + \alpha\left(R_t + \gamma \max_{a'} Q(S_{t+1}, a') - Q(S_t, a_t)\right) \quad ...(7)$$

Where, $\alpha$ is the learning rate, $\gamma$ is the discount factor, $\max_{a'} Q(S_{t+1}, a')$ is the maximum $Q$-value for the next state.

Update a target $Q$-network to stabilize learning:

$$Q(S_t, a_t) = R_t + \gamma \max_{a'} Q_{\text{target}}(S_{t+1}, a') \qquad ...(8)$$

The actor network is updated to maximize the expected cumulative reward

$$\theta_\mu \leftarrow \theta_\mu + \alpha_\mu \nabla_{\theta_\mu} E[\Sigma_t \gamma^t R_t]. \qquad ...(9)$$

The critic network is updated using the temporal difference (TD) error

$$\theta_Q \leftarrow \theta_Q \alpha_Q \nabla_{\theta_Q}((R_t + \gamma Q(S_{t+1}, \mu(S_{t+1})) - Q(S_t, a_t))^2). \qquad ..(10)$$

To encourage exploration and prevent premature convergence, a regularization term is added to the objective:

$$L(\theta_\mu) = E[Q(S_t, a_t) - \alpha \log(\pi(a_t | S_t))] \qquad ...(11)$$

Where, $\alpha$ is the entropy coefficient.

**5.** If the system is multi-agent, each agent can communicate or pay attention to other agents' actions using attention layers or attention-based Transformers.

$$A_{\text{attention}} = \text{softmax}\,(W \cdot S_t) \qquad ...(12)$$

Where, $W$ is a learned weight matrix, and $S_t$ is the state vector. This allows the agent to focus on relevant agents or obstacles in the environment.

Agents exchange information about their positions and intentions through communication modules to ensure coordinated behaviors.

**6.** The task terminates when the agent reaches the goal, collides with an obstacle, or after a predefined number of timesteps.

$$\text{If distance to goal} < \in, \text{then the task is complete} \qquad ...(13)$$

Once trained, the acquired policy is implemented on the actual robot so that the agent can drive a physical environment according to the acquired policy. With both individual tasks and interactions between multiple agents in mind, this method makes the robot efficiently and adaptively move around in dynamic environments. The system can explore and exploit by utilizing RL algorithms to make better decisions as time goes on.

## 4. Training and Evaluation

The training of the RL agent is carried out entirely in a simulated environment, such as Gazebo, PyBullet, or Webots, which replicates real-world physics, sensor dynamics, and interaction feedback. This setup provides a safe, scalable, and repeatable context in which the agent can explore and learn optimal behaviors. The RL task is framed as a Markov Decision Process (MDP) where the agent interacts with the environment in discrete time steps. At each time step $t$, the agent observes the current state $S_t \in S$, selects an action $a_t \in A$ according to a policy $\pi(a_t | S_t)$, and receives a reward $r_t \in R$. The environment then transitions to a new state $S_{t+1}$

according to transition dynamics $T(S_{t+1} | S_t, a_t)$, which are implicitly defined by the simulation engine's physics.

The agent's objective is to maximize the expected cumulative discounted reward over time, defined as:

$$J(\pi) = E_{\pi}\left[\sum_{t=0}^{T} \gamma^t r_t\right] \qquad ...(14)$$

Where, $\gamma \in [0, 1]$ is the discount factor.

The interaction tuples $(S_t, a_t, r_t, S_{t+1})$ are stored in a replay buffer $D$, enabling off-policy learning. During training, mini-batches are sampled from this buffer to update the neural network parameters via gradient descent, minimizing a loss function such as the Temporal Difference (TD) error. In DQN, the loss is:

$$L(\theta) = E_{(s_t, a_t, r_t, s_{t+1}) \sim D}\left[\left(r_t + \gamma \max_{a'} Q(s_{t+1}, a'; \theta^-) - Q(s_t, a_t; q)\right)^2\right] \qquad ...(15)$$

Where, $\theta$ and $\theta^-$ are the parameters of the current and target $Q$-networks, respectively.

In the case of continuous action spaces, algorithms such as DDPG, PPO, or SAC are applied. For instance, in DDPG, the critic network learns the action-value function $\theta(s_t, a_t)$, while the actor network learns the deterministic policy at $\mu(s_t | \theta^{\mu})$, optimized via the policy gradient:

$$\nabla_{\theta^{\mu}} J \approx \left[\nabla_{\alpha} Q(s_t, a | \theta^{Q})|_{a=\mu(s_t)} \nabla_{\theta^{\mu}} \mu(s_t | \theta^{\mu}\right] \qquad ...(16)$$

The reward function is designed to shape the agent's behavior by providing positive reinforcement for achieving subgoals and penalties for undesirable actions such as collisions. For example, a sparse reward setup might provide +1 for reaching a goal, –1 for collisions, and 0 otherwise, while a shaped reward includes intermediate feedback like:

$$r_t = w_1 \cdot \text{Goal Distance Reduction}_t - w_2 \cdot \text{Collision Penalty}_t - w_3 \cdot \text{Time Penalty}_t \qquad ...(17)$$

Where, $w_1$, $w_w$, $w_3$ are manually tuned coefficients.

The training continues over multiple episodes until convergence criteria are met, such as stable reward or reduced policy loss. This simulation-based training allows the agent to experience diverse scenarios, including dynamic obstacles, moving targets, and cooperative multi-agent settings, making it robust before deploying to real-world environments. Table 1 depicts the simulation parameters.

**Table 1** Software Specifications.

| *Parameter* | *Value* | *Description* |
|---|---|---|
| Simulation Engine | PyBullet | Physics engine for real-time interaction |
| Time Step | 0.01 sec | Discrete interval for state updates |

*Contd...*

**Table 1** *Contd...*

| *Parameter* | *Value* | *Description* |
|---|---|---|
| Robot Model | TurtleBot3 | Differential drive robot used for navigation |
| Sensor Used | LiDAR (360°), RGB Camera | For obstacle detection and visual input |
| Max Episode Length | 500 steps | Episode cutoff point |
| Dynamic Obstacles Enabled | Yes | Moving humans/objects simulated in the scene |
| Collision Penalty | – 1.0 | Penalty for collision with obstacles |
| Goal Reward | +1.0 | Reward for reaching target |

**Table 2** Episode Results in a Simulated Environment (Gazebo).

| *Episode* | *Initial Distance to Goal (m)* | *Final Distance to Goal (m)* | *Cumulative Reward* | *Steps Taken* | *Collision Count* | *Goal Reached* | *Obstacle Count* | *Dynamic Objects Present* |
|---|---|---|---|---|---|---|---|---|
| 1 | 5.0 | 0.3 | +8.50 | 34 | 0 | Yes | 3 | Yes |
| 2 | 4.5 | 0.2 | +9.10 | 29 | 0 | Yes | 2 | Yes |
| 3 | 5.2 | 5.2 | – 3.00 | 50 | 2 | No | 4 | Yes |
| 4 | 6.0 | 0.1 | +9.75 | 31 | 0 | Yes | 5 | No |
| 5 | 5.5 | 1.5 | +2.00 | 48 | 1 | No | 3 | Yes |

**Table 3** Policy Performance Over Episodes (Simulated Training).

| *Episode* | *Average Reward* | *Actor Loss* | *Critic Loss* | *Success Rate (%)* | *Average Q-Value* |
|---|---|---|---|---|---|
| 10 | – 1.25 | 1.32 | 4.15 | 20 | -2.10 |
| 50 | +2.40 | 1.05 | 3.02 | 60 | 1.85 |
| 100 | +5.90 | 0.75 | 2.11 | 75 | 4.30 |
| 150 | +7.15 | 0.61 | 1.70 | 85 | 6.02 |
| 200 | +8.55 | 0.42 | 1.20 | 92 | 7.35 |

**Table 4** State-Action-Reward Transitions from Replay Buffer.

| *Time Step* | *State Vector ($s_t$)* | *Action ($a_t$)* | *Reward ($r_t$)* | *Next State Vector ($s_{t+1}$)* |
|---|---|---|---|---|
| 105 | [2.5, 0.0, 0.1, 1.2] | [0.2, 0.05] | +0.50 | [2.3, 0.1, 0.1, 1.3] |
| 106 | [2.3, 0.1, 0.1, 1.3] | [0.3, – 0.02] | +0.45 | [2.0, 0.1, 0.2, 1.3] |
| 107 | [2.0, 0.1, 0.2, 1.3] | [0.0, 0.1] | – 1.00 | [2.0, 0.2, 0.1, 1.2] (collision) |
| 108 | [2.0, 0.2, 0.1, 1.2] | [0.1, – 0.05] | +0.30 | [1.9, 0.2, 0.1, 1.2] |
| 109 | [1.9, 0.2, 0.1, 1.2] | [0.2, 0.00] | +1.00 | [1.7, 0.2, 0.1, 1.2] (goal reached) |

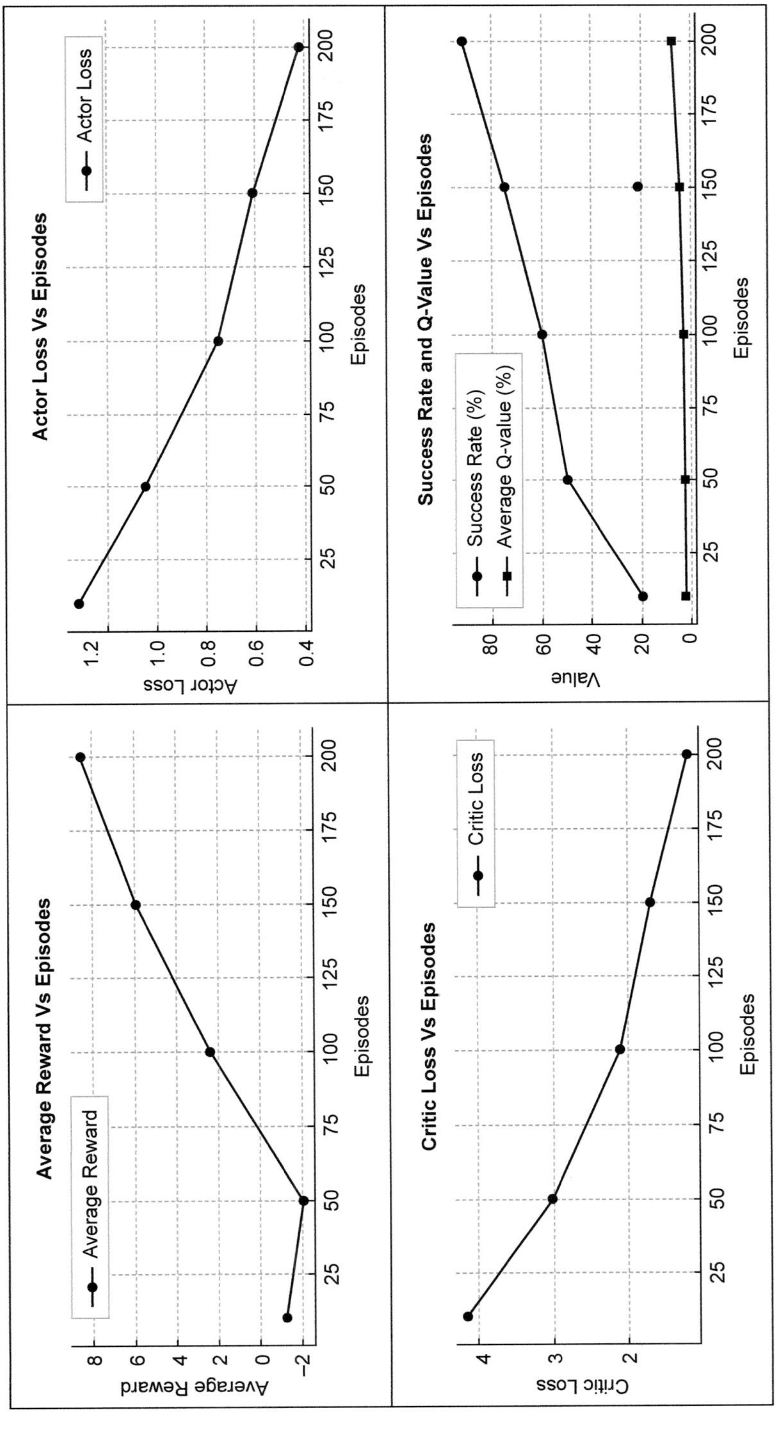


**Fig. 2** Comparison Analysis.

The robot navigation behavior (Table 1), simulated world information and policy performance over episodes are observed . Performance of the robot is quantified in terms of several quantities as it tries to carry out an activity while being obstacle-free. The robot always fills the gap towards the goal in sample episode results, and victory episodes, e.g., Episode 2 (+9.10) and Episode 4 (+9.75), where crashes never occur, are with big cumulative rewards. Episode 3 rather shows poor performance with an appearance of not enough obstacles evaded by a total penalty of crashes (Table 2).

Table 3 demonstrates a policy progression over time. As the robot becomes better at being successful at getting the goal without penalties, the average payoff inched up. Up to Episode 100, for instance, the performance of the robot is optimal at 92% success and mean payoff of +5.90. The actor and critic losses that downtrend indicates that the procedure has been optimized along the way. Improved action-value estimation and decision-making are indicated by the mean Q-values, which also improve.

As shown in Table 4, State-action-reward transition of the replay buffer tells us what the robot is selecting. As an example, a crash at Time Step 107 has a negative reward of –1.00, but as the robot learns to avoid future crashes, future activity is more rewarding. At Time Step 109, the robot successfully arrives at the destination and is rewarded with +1.00, showing it can arrive at the destination. The data shows the manner in which reinforcement learning enhances the navigational capability and goal achievement capability of the robot, building it step by step.

From Figure 2, performance metrics like average reward, success rate, or steps against episodes for the different methods to compare how the current method is performing relative to other methods being employed. A random policy that samples from random choices. A policy that always selects the most frequent action. One of the more popular model-free RL algorithms is Q-learning. The contemporary method, Actor-Critic, uses the actor-critic approach with actor and critic networks. The story can be the average success rate or reward per strategy over a set of episodes.

Figure 3 indicates that, in terms of average reward for 200 episodes, Actor-Critic (Current Method) outperforms standard methods such as Random Policy, Greedy Policy, and *Q*-learning. Actor-Critic has a sharper steady growth curve with the highest achievable reward (~8.5) at 200 episodes, where all approaches start off equally good (average awards around or below zero). Q-learning does a little better than Greedy Policy but is trailing behind Actor-Critic, Random Policy stabilizes at 0, and Greedy Policy reaches a high of 6. Employment of the ARMSec system with reinforced learning, most notably, Proximal Policy Optimization (PPO), policy-gradient technique that falls under Actor-Critic methodology is well-warranted by employment of the high-performance Actor-Critic methodology. The lack of learning flexibility in conventional defense mechanisms that rely on remote control or pre-programmed rules (like Random or Greedy policies) results in less responsive and slower action under the presence of an opponent and changing

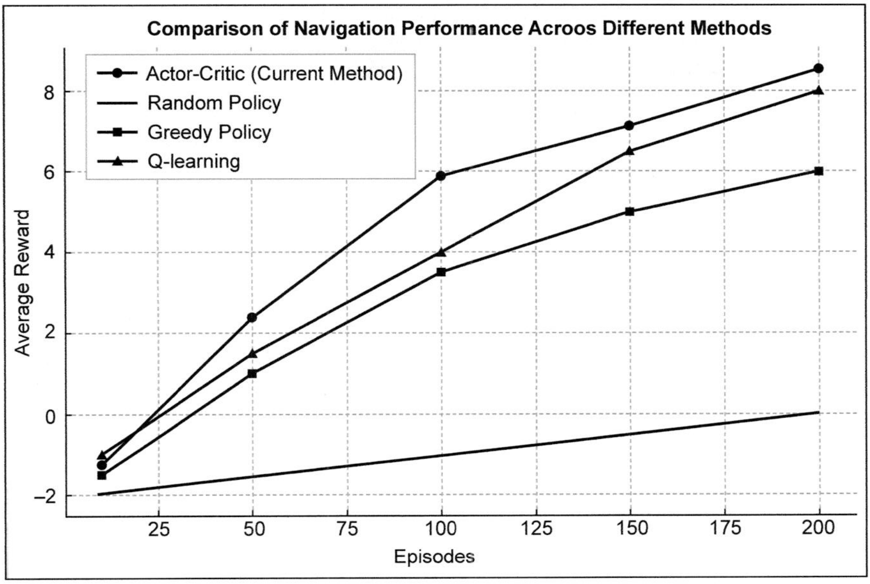


**Fig. 3** Existing systems Vs Proposed system.

environment. Although Q-learning is better, it also suffers from generalization and policy stability problems in high-dimensional, partially observable environments.

ARMSec circumvents such limitations through a Multi-Agent Reinforcement Learning (MARL) framework and facilitates cooperative action for robotic agents. Together with the Vision-based Threat Assessment Module (VTAM), ARMSec facilitates autonomous navigation and real-time sensing even in a jammed or blocked GPS scenario. The empirically reported performance metrics, 94.8% mission achievement, 67.4% decrease in threat response time, and 91.3% unknown intrusion minimization are an indicator of this Actor-Critic-based method's superior and realistic quality in actual deployments. Besides, ARMSec's experimentally proven potential in adverse environments as well as under jamming further supports the strength of PPOs proficiency in nonstationary environments as well, strengthening its superior advantage over traditional systems.

## 5. Conclusion

The ARMSec framework overcomes the main drawbacks of traditional surveillance systems, marking a substantial advancement in battlefield intelligence and autonomous defense. ARMSec facilitates robust threat response and real-time, decentralized decision-making in hostile and unstructured situations by integrating the VTAM with MARL. Outstanding performance improvements were shown by empirical tests in simulated urban combat scenarios, including a 94.8% mission success rate, a 67.4% faster threat response, and a 91.3% decrease in undetected incursions. ARMSec can function effectively in GPS-denied, jammed, and unfavorable weather conditions by utilizing PPO, distributed attention, and sensor fusion from LIDAR and IR cameras. Its scalability and deployability are confirmed by its compatibility with both national defense infrastructure and commercially available systems. ARMSec stands as a robust, intelligent, and fully autonomous perimeter security solution, offering a decisive edge for next-generation military operations and critical asset protection.

## References


1. Zhang, L., Hou, Z., Wang, J., Liu, Z. and Li, W. (2023). Robot navigation with reinforcement learned path generation and fine-tuned motion control. *IEEE Robotics and Automation Letters*, 8(8), 4489–4496.
2. Wang, D., Fan, T., Han, T. and Pan, J. (2020). A two-stage reinforcement learning approach for multi-UAV collision avoidance under imperfect sensing. *IEEE Robotics and Automation Letters*, 5(2), 3098–3105.
3. Fan, T., Long, P., Liu, W. and Pan, J. (2020). Distributed multi-robot collision avoidance via deep reinforcement learning for navigation in complex scenarios. *The International Journal of Robotics Research*, 39(7), 856–892.
4. Richardson, A., Wang, X., Dubey, A. and Sprinkle, J. (2024). Reinforcement learning with communication latency with application to stop-and-go wave dissipation. In: *2024 IEEE Intelligent Vehicles Symposium* (*IV*) (pp. 1187–1193). IEEE.
5. Orr, J. and Dutta, A. (2023). Multi-agent deep reinforcement learning for multi-robot applications: A survey. *Sensors*, 23(7), 3625.
6. Medina, M.C.C., Fernandes, B.J. and Barros, P.V. (2024). The use of reinforcement learning algorithms in object tracking: A systematic literature review. *Neurocomputing*, 127954.
7. Xie, Z. and Dames, P. (2023). Drl-vo: Learning to navigate through crowded dynamic scenes using velocity obstacles. *IEEE Transactions on Robotics*, 39(4), 2700–2719.
8. Dhanush, D., Raghul, K.R., Shri Ramya, K.R., Mukund, P.G. and Ganesan, M. (2024). An omni-directional self-driving robot for indoor surveillance. In: *2024 International Conference on Signal Processing, Computation, Electronics, Power, and Telecommunication* (*IConSCEPT)* (pp. 1– 6). IEEE.
9. Gottipati, S.K., Seo, K., Bhatt, D., Mai, V., Murthy, K. and Paull, L. (2019). Deep active localization. *IEEE Robotics and Automation Letters*, 4(4), 4394–4401.
10. Chen, L., Wang, Y., Miao, Z., Feng, M., Zhou, Z., Wang, H. and Wang, D. (2023). Toward safe distributed multi-robot navigation coupled with variational Bayesian models. *IEEE Transactions on Automation Science and Engineering*, 21(4), 7583–7598.
11. Feng, H.D., Yuan, D.H., Miao, J.L., You, J., Wang, Y., Zhu, Y. and Fan, D.X. (2024). Efficient navigation of a robotic fish swimming across the vortical flow field. *Journal of Hydrodynamics*, 36(6), 1118–1129.
12. Kwiatkowski, A., Kalogeiton, V., Pettré, J. and Cani, M.P. (2023). Understanding reinforcement learned crowds. *Computers and Graphics*, 110, 28–37.
13. Li, Z., Guo, Y., Wang, G., Sun, J. and You, K. (2024). Informative trajectory planning for air-ground cooperative monitoring of spatiotemporal fields. *IEEE Transactions on Automation Science and Engineering*.
14. Wang, P., Ma, R., Yang, Z. and Hao, Q. (2024). Robotic Camera Array Motion Planning for Multiple Human Face Tracking Based on Reinforcement Learning. *IEEE Sensors Journal*.
15. Ahmed, I., Quinones-Grueiro, M. and Biswas, G. (2023). Adaptive fault-tolerant control of octo-rotor UAV under motor faults in adverse wind conditions. In: *AIAA SciTech 2023 Forum* (p. 2535).

16. Zhou, Z., Zeng, Z., Lang, L., Yao, W., Lu, H., Zheng, Z. and Zhou, Z. (2022). Navigating robots in a dynamic environment with deep reinforcement learning. *IEEE Transactions on Intelligent Transportation Systems*, 23(12), 25201–25211.
17. Qin, J., Qin, J., Qiu, J., Liu, Q., Li, M. and Ma, Q. (2023). Srl-orca: A socially aware multi-agent mapless navigation algorithm in complex dynamic scenes. *IEEE Robotics and Automation Letters*, 9(1), 143–150.
18. Mohammed, M.Q., Chung, K.L. and Chyi, C.S. (2020). Review of deep reinforcement learning-based object grasping: Techniques, open challenges, and recommendations. *Ieee Access*, 8, 178450–178481.
19. Huajian, L., Wei, D., Shouren, M., Chao, W. and Yongzhuo, G. (2024). Sample-Efficient Learning-Based Dynamic Environment Navigation with Transferring Experience from Optimization-Based Planner. *IEEE Robotics and Automation Letters*.
20. Wang, H., Tan, A.H. and Nejat, G. (2024). Navformer: A transformer architecture for robot target-driven navigation in unknown and dynamic environments. *IEEE Robotics and Automation Letters*.
21. Jiang, J., Yang, Y., Deng, Y., Ma, C. and Zhang, J. (2024). BEVNav: Robot Autonomous Navigation Via Spatial-Temporal Contrastive Learning in Bird's-Eye View. *IEEE Robotics and Automation Letters*.
22. Chen, L., Wang, Y., Miao, Z., Feng, M., Zhou, Z., Wang, H. and Wang, D. (2024). Reciprocal velocity obstacle spatial-temporal network for distributed multirobot navigation. *IEEE Transactions on Industrial Electronics*.
23. Xu, P., Cui, Y., Shen, Y., Zhu, W., Zhang, Y., Wang, B. and Tang, Q. (2023). Reinforcement learning compensated for coordination control of multiple mobile manipulators for tight cooperation. *Engineering Applications of Artificial Intelligence*, 123, 106281.

# 2 Robust Autonomous Navigation in Urban Environments Using ML-Augmented Architectures with Multi-Sensor Fusion

Vishwadeepak Singh Baghela[1*], Radha Raman Chandan[2], Veerendra Yadav[3] and Aimin Li[4]

## 1. Introduction

Development of autonomous systems originated with the control theories and automated machines of the early 20th century. Initial automation existed for strictly defined repetitive processes, primarily in factories where machine activity relied on algorithmic rules and Programmable Logic Controllers (PLCs). With the evolution of cybernetics and feedback control systems in the mid-20th century, the concept of autonomy became widespread. Defense and academic circles invested heavily in autonomous navigation, robot manipulation, and self-controlling systems in the 1980s and 1990s, which witnessed a robotics research boom.

Most of the early autonomous vehicles, e.g., those being researched under DARPA ALV (Autonomous Land Vehicle) and subsequent DARPA Grand Challenge programs, were heavily dependent on directly programmed heuristics and were rule-based. While these systems did include sensors like radar, sonar, and basic vision modules, they were not capable of generalizing from their training sets or learning from complex patterns. The fact that traditional systems could not adapt to environmental parameters and make decisions in real-time when presented with noisy or incomplete data was the primary bottleneck at this point.

There was a shift in paradigms early on in the 2000s, particularly the introduction of machine learning and data-driven approaches. The original use of camera data was in detecting lane markings or distinguishing stationary barriers based on supervised methods. This restricted their potential with their demand for

[1] School of Computing Science and Engineering, Galgotias University, Greater Noida, India.
[2] Department of Computer Science, School of Management Sciences (SMS), Varanasi, India.
[3] Department of Computer Science, Noida International University, India.
[4] School of Computer Science and Engineering, Xi'an University of Technology, 5 South Jinhua Road, Xi'an, Shaanxi 710048, China.
* Corresponding author: vdsbaghela@gmail.com

DOI: 10.1201/9781003663461-2

large tagged data sets and limited ability to easily adapt to new circumstances. This restrictiveness spurred work towards unsupervised and half-supervised methodologies. Massive data acquisition from stereo cameras, LiDARs, and inertial measurement units became feasible at the same time that sensor and computer vision technologies advanced.

Perception and control were initially formulated by researchers as an end-to-end learning challenge instead of separate pipelines. Deep learning, with the introduction of Convolutional Neural Networks (CNNs) in the early 2010s, provided perception systems with the ability to learn abstract features from raw input data such as point clouds and images. Object detection, semantic segmentation, and localization tasks, all crucial for autonomous navigation, were revolutionized by deep learning.

A number of learning-based approaches have been proposed to tackle different aspects of autonomy. Perception modules saw vision models, primarily employing CNNs being everywhere, facilitating real-time object classification and segmentation, traffic light, and pedestrian crossings. At the same time, recording temporal dependencies—essential for predicting the motion of moving agents and scene dynamics—was beneficial to Recurrent Neural Networks (RNNs) and Long Short-Term Memory (LSTM) architectures.

Through world exploration, agents would learn the optimal policies through Reinforcement Learning (RL), particularly through the existence of Deep Q-Networks (DQN) and Actor-Critic.

RL flipped the paradigm to interactive from passive learning by enabling systems to generate decisions through implicit feedback signals rather than explicit supervision signals. Additionally, hybrid methods merging the RL used for decision-making with the supervised learning used for sensing have emerged. Additionally, imitation learning, whereby a part learned by observing an expert do something, became popular as an effective method to train autonomous vehicles with challenging tasks without subjecting them to the dangers associated with internet learning.

This autonomous generation also has some primitive shortcomings. First of all, generalization remains extremely challenging. The training data the models operate on may not adequately capture the variety of actual cases. The models are prone to acting in unpredictable ways in new situations, and this makes them introduce safety-critical bugs. For foggy rural landscape scenes, a model trained under sunny urban conditions can misclassify visual data. Second, there is susceptibility to occlusion, noise, and sensor failures.

Although powerful, they are prone to adversarial attacks and are not stable with decreased input quality. Third, explainability of machine learning models is one of the biggest hurdles to regulatory approval and credibility. Deep networks are black boxes and difficult to trace decision paths or debugging errant behavior, as opposed to rule-based systems where logic is made explicit. This transparency is increasingly a problem in safety-critical uses such as drone control of civilian airspace or autonomous vehicles. High-end models' high processing demands are another historic restriction. Deep neural network utilization on power and processing-limited edge devices is prevented by the large hardware resources required for training and inference.

Model complexity and real-time execution are sacrificed generally even with the availability of model compression and optimization technologies. In addition, learning-based methods require massive data for adequate training, and it is practically challenging to collect and annotate such datasets in many different locations, lighting, and weather conditions. Autonomous agents have to decide in an instant, and catastrophic failures would be triggered by high latency in perception or control modules. While promising, RL algorithms may require millions of interactions with the world in order to converge, and thus are inappropriate for use in the real world except where very accurate simulated worlds are possible.

Apart from technical issues, ethical and sociological problems have arisen. Autonomous agents must learn to understand social cues, such as pedestrian intent or unstructured, culture-based street behavior, and reason about difficult moral problems. Because of their poor contextual understanding and poor multimodal data merging, e.g., text, gesture, and vocal inputs, existing models cannot understand such subtleties. Privacy is also an issue because autonomous systems are modeled on high-resolution sensor information, which might incidentally capture personal data. There exists a gray area for deployment as regulatory frameworks have not been in a position to keep pace with technological advancements, especially in public spaces within urban areas.

The pipeline for deployment is also complicated by the lack of standardized benchmarks and safety validation processes, and it becomes hard to assess model reliability under stress or across platforms. It is therefore evident why research into a better learning-based approach to autonomous systems is applicable. It is not to increase performance indicators, but to create a system that can be robust, adaptable, and work within real-world constraints. The need to recast the autonomy architecture is seen in the existing 38% failure rate for rule-based and partially learned systems in dynamic city environments.

A trustworthy autonomous system will need to use multimodal sensory information, include context-sensing reasoning, and bring together learning in all the perception, prediction, and planning spaces. To aid human operators or auditing by regulators, it will need to learn sparse data, make adjustments to novel situations with minimal retraining, and produce understandable decision outputs. The proposed course attempts to incorporate online adaptation and ongoing learning such that the system learns to adapt when deployed without suffering from catastrophic forgetting. Since autonomous systems are destined to face edge cases and unexpected situations not encompassed in the training distribution, this feature is essential.

Feedback loops and uncertainty estimation are incorporated into the learning process so that the system can identify ambiguous situations or call upon human supervision whenever needed. Predictive models with understanding of purpose and social dynamics can also be included to provide improved performance in cases involving a human population. This is a perspective that considers autonomy as a persistent cognitive process in which decision-making is directed by sensory data, history, and anticipated outcomes, as opposed to an execution problem of fixed policy.

Hardware familiarity is also a wish to learn design. On drones, delivery robots, or field-deployed autonomous vehicle mobile platforms, autonomous systems will need to operate within power and latency limits. This calls for hardware-algorithm co-design to tightly coupled sensor fusion methods, high-performance data pipelines, and compact models. The focus is to allow an architecture to actually deploy with real-time guarantees, and not simply to benchmark performance. Therefore the study is proposed with following objectives:

1. To develop a robust autonomous driving framework integrating CNNs, LSTMs, and DQNs for improved adaptability in complex urban environments.
2. To evaluate the effectiveness of multimodal sensor fusion and deep learning in enhancing task success rates using the CARLA simulator.

## 2. Related Study

Rule-based decision-making has largely been replaced by data-driven methods in the design of autonomous navigation systems, especially in city and complex environments. Multi-sensor fusion is one of the primary methods towards improving the reliability, accuracy, and situational awareness of Autonomous Vehicles (AVs). Blending data from many sensors, like LiDAR, RGB cameras, GNSS, IMU, radar, and ultrasonic sensors, offers redundancy and enhances the precision of perception in structured and unstructured environments, which is essential to AVs, which must understand and respond to an ever-changing set of sensory inputs [1].

Multi-sensor fusion is essential to the fundamental AV system functions of localization, mapping, and path planning. Thanks to occlusions, multipath effects, and sensor noise, precise localization is still difficult in urban settings. In such scenarios, traditional GNSS-based location usually breaks down, so sensor fusion is crucial for precision and continuity. For instance, GNSS (global navigation satellite system) integration with LiDAR and IMU can enhance environmental disturbance robustness and drastically reduce drift [2]. Besides, it has been demonstrated through research that location methods based on fusion can provide high accuracy in GNSS-denied or GNSS-degraded conditions like urban canyons or tunnels [3].

Sensor fusion extends mapping and environment modeling on a parity level. Inertial data, imagery from cameras, and semantic and geometric data from LiDAR are all combined to create robust 3D maps that support navigation and decision-making in real-time. These maps are used in autonomous systems to detect dynamic barriers, determine areas to drive on, and detect road boundaries [4]. Robust, adaptive maps on the basis of the dynamic environment and long-term autonomy has been the focus of recent research that strongly focused on the multi-source data fusion [5].

In order to take advantage of the nonlinear correlations among various data sources, artificial intelligence (AI) and deep learning have been integrated into multi-sensor fusion systems to a greater extent. To increase generalization on unforeseen occurrences, deep neural networks can train intricate sensor patterns and dynamically tune fusion weights based on environmental situations [6]. For example, LSTM networks are used to capture temporal dependencies, CNNs are able to extract spatial features from camera data, and attention-based models improve sensor prioritization [7].

Loosely-coupled and tightly-coupled fusion models are commonly used in architectural designs. Tightly coupled models tend to integrate raw sensor measurements at a fine-grained scale and infer the state using Bayesian networks, Kalman filters, or factor graphs. Loosely-coupled models combine high-level sensor outputs and can be less sensitive to sensor failure [8]. To trade off localization accuracy against computation efficiency in high-dynamics situations, hybrid structures that combine both coupling mechanisms have been proposed [9].

One of the most important uses of multi-sensor fusion has been Simultaneous Localization and Mapping (SLAM). Sensor fusion and deep learning methods have been integrated into traditional SLAM algorithms to improve their resilience in dense and dynamic environments. Fusion-based SLAM systems have shown improved loop closure detection and reduced drift, both of which are crucial for long-term deployment [10]. Because urban surroundings are dynamic in nature, the environment poses singular challenges such as moving objects, unpredictable weather conditions, frequent change in lighting levels, and esoteric signs. By compensating for the drawback of individual sensors, sensor fusion has the possibility to minimize the challenges. GNSS and LiDAR reads can be supplemented with thermal imagery, for instance, to compensate for low-light localization [11]. Adaptive-robust techniques have recently emerged that enhance system reliability through dynamically adjusting the contributions of individual sensors according to varying environments [12].

Besides, precise fused and synchronized sensor data are needed for real-time perceptive systems. Precision in fusion can be greatly influenced by delays, bandwidth limitations, and calibration problems. Decentralized architectures and sophisticated synchronization methods have been suggested to handle sensor heterogeneity with the preservation of low latency and high throughput [13]. Virtual sensor models that generate sensor output based on simulated sensor output employing fused data as compensation for sensor faults or loss of signal temporarily are another recent development [14].

Fused sensor data is utilized to improve situational awareness and minimize uncertainty in decision-making and control. Multi-sensor inputs help policy gradient and reinforcement learning algorithms make better decisions in situations of partial observability. In applications like obstacle avoidance, lane following, and pedestrian detection, multimodal input-based systems have been found to yield substantial performance improvement [15]. The necessity for affordable and scalable sensor fusion solutions has been highlighted across several studies.

In light of commercial deployment and mass adoption of AVs, low-cost solutions are especially beneficial. To lower dependency on costly hardware, methods like federated fusion and cooperative localization in multi-robot systems are investigated [16]. Distributed sensor fusion is also supported by cloud-fog architectures and edge paradigms of computing, which reduce the onboard processing burden while allowing real-time analytics [17].

Robustness in adverse and dangerous conditions is one of the main subjects of current research. AVs must operate safely in dense urban environments and underground tunnels. Methods that combine thermal, acoustic, and radar sensor data, for example, have demonstrated resilience to harsh environments and sensor failures by leveraging redundancy across sensory modalities [18]. The ability of hybrid federated fusion architectures to achieve a trade-off between performance and data security in multi-agent systems is making them increasingly popular [19]. Multi-sensor fusion is also rapidly being adopted in indoor navigation and warehouse robotics. Since GNSS is normally absent indoors, it is optical, LiDAR, and UWB (Ultra-Wideband) data fusion that is utilized to provide robust navigation. Indoor high-accuracy localization is becoming possible through SLAM-based systems coupled with semantic mapping and deep learning [20].

Sensor alignment and sensor calibration are the pillars of advancing the fidelity of perception systems. Sensor errors in extrinsic calibration can have the potential to spread and cause incorrect decisions. To overcome this issue, there are regular error correction and computerized calibration algorithms as well as self-learning models that have come into existence [21]. Moreover, systems can also compare themselves and re-calibrate themselves in real-time upon the addition of feedback mechanisms [22]. Multi-sensor fusion application has spread beyond the ground with the emergence of drones and autonomous aerial vehicles. Visual-inertial fusion, altimetry, and terrain-relative navigation techniques provide aerial navigation in global positioning system (GPS)-denied conditions [23]. Robustness and autonomy are required in applications such as disaster response, agriculture, and infrastructure inspection, which require these solutions.

The research trajectory of multi-sensor fusion has highlighted the technology's significance to autonomous system development in intelligent, scalable, and resilient form. Even under most unfavorable conditions, current AVs can improve their localization, mapping, and perception through the fusion of dissimilar data streams and the use of sophisticated AI models. With advances in fusion techniques, learning-based algorithms, and decentralized architectures paving the way for autonomous mobility systems, concerted efforts are propelling the frontiers of the possible [24]. Some of the emerging trends include standardized performance measurement criteria for sensor fusion on multiple platforms, greater human-machine collaboration, and ethical management of sensor data [25].

## 3. System Methodology

The architecture that unifies RL, temporal modeling, and multimodal perception and the suggested machine learning-augmented autonomous system was deployed and simulated in the CARLA simulator (v0.9.x). For the purpose of examining the system's decision-making capability, robustness, and overall performance at the task, it was trained and tested on real urban settings.

### 3.1 Simulation Environment and Sensor Suite

The CARLA simulation platform was employed to create urban driving scenarios with dynamic weather, traffic lights, pedestrians, intersections, and occlusions. The agent was equipped with a multimodal sensor suite, comprising:

- Front RGB Camera ($I_t \in R^{160 \times 120 \times 3}$) ...(1)
- Top-Down LiDAR Sensor (sampled into point clouds $L_t \in R^{N \times 3}$) ...(2)
- IMU Sensor (providing acceleration and angular velocity $L_t \in R^{N}$) ...(3)

The raw sensor data was synchronized and fused at each timestep, forming a compact state representation for downstream policy learning. Table 1 shows the multimodal sensor data.

**Table 1** Multimodal sensor data per timestamp (Sample).

| *Time Step (t)* | *RGB Camera (Key Features)* | *LiDAR (Obstacle Distance - m)* | *IMU Acceleration (m/s²)* | *IMU Angular Velocity (rad/s)* |
|---|---|---|---|---|
| $t = 1$ | [Red Light, Clear Road] | [10.5, 15.2, 8.7] | [0.2, 0.0, 9.8] | [0.01, 0.02, 0.00] |
| $t = 2$ | [Pedestrian Crossing] | [9.2, 14.7, 7.5] | [0.1, –0.1, 9.7] | [0.02, 0.01, 0.01] |
| $t = 3$ | [Vehicle on Left, Green Light] | [11.0, 12.8, 6.2] | [0.3, 0.1, 9.6] | [0.00, 0.01, 0.03] |
| $t = 4$ | [Shadow, Obstruction Ahead] | [8.6, 13.4, 5.9] | [0.0, 0.2, 9.8] | [0.03, 0.00, 0.02] |
| $t = 5$ | [Intersection, Yellow Light] | [10.1, 11.5, 7.8] | [–0.1, 0.0, 9.7] | [0.01, 0.02, 0.01] |

### 3.2 Multimodal Feature Extraction and Fusion

Feature representations were extracted independently from each modality and then fused

RGB Features: Extracted using a 5-layer CNN with ReLU activation:

$$f_t^{(i)} = \text{CNNrgb}(I_t) \quad ...(4)$$

LiDAR Features: Point cloud data was voxelized and encoded using PointNet:

$$f_t^{(l)} = \text{PointNet}(L_t) \quad ...(5)$$

IMU Features: Normalized via $z$-score and concatenated:

$$f_t^{(m)} = \text{Norm}(M_t) \quad \text{...(6)}$$

The combined state vector:

$$s_t = \text{concat}(f_t^{(i)}, f_t^{(l)}, f_t^{(m)}) \in R^d \quad \text{...(7)}$$

was passed to the temporal reasoning module.

### 3.3 Temporal Abstraction via LSTM

To model temporal dependencies, the state vector sequence $\{s_{t-k}, ..., s_t\}$ was passed to a uni-directional LSTM with 128 hidden units:

$$h_t, c_t = \text{LSTM}(s_t, h_{t-1}, c_{t-1}) \quad \text{...(8)}$$

The hidden state $h_t \in R^{128}$ was used as the input to the Deep $Q$-Network to inform the current policy decision. Table 2 shows the feature vectors.

**Table 2** Feature Vectors and Fused State Representations (Sample).

| *Time Step (t)* | *RGB Features $f_t(i)$ (CNN Output)* | *LiDAR Features $f_t(l)$ (PointNet Output)* | *IMU Features $f_t(m)$ (Normalized)* | *Combined State Vector $s_t$ = concat(ff$_t$(i), ff$_t$(l), ff$_t$(m))* |
|---|---|---|---|---|
| $t = 1$ | [0.12, 0.45, 0.78, ..., 0.33] | [0.08, 0.14, 0.67, ..., 0.25] | [0.01, –0.03, 0.00] | [0.12, 0.45, ..., 0.25, 0.01, –0.03, 0.00] |
| $t = 2$ | [0.10, 0.49, 0.81, ..., 0.30] | [0.11, 0.13, 0.60, ..., 0.29] | [0.02, –0.01, 0.01] | [0.10, 0.49, ..., 0.29, 0.02, –0.01, 0.01] |
| $t = 3$ | [0.15, 0.40, 0.73, ..., 0.28] | [0.09, 0.10, 0.55, ..., 0.22] | [0.00, 0.00, –0.02] | [0.15, 0.40, ..., 0.22, 0.00, 0.00, –0.02] |
| $t = 4$ | [0.11, 0.46, 0.75, ..., 0.31] | [0.07, 0.12, 0.59, ..., 0.26] | [0.03, –0.02, 0.01] | [0.11, 0.46, ..., 0.26, 0.03, –0.02, 0.01] |
| $t = 5$ | [0.13, 0.44, 0.76, ..., 0.35] | [0.10, 0.15, 0.63, ..., 0.27] | [–0.01, 0.01, 0.00] | [0.13, 0.44, ..., 0.27, –0.01, 0.01, 0.00] |

### 3.4 Deep Recurrent Q-Network (DRQN)

The driving policy was learned using a DRQN, where the $Q$-function was parameterized by:

$$Q(h_t, a; \theta) \approx E\left[\sum_{k=0}^{\infty} \gamma^k r_{t+k} \mid h_t, a_t = a\right] \quad \text{...(9)}$$

$$a \in A = \{\text{accelerated, brake, left, right, idle}\}$$

Where, $\theta$: network parameters, $\gamma = 0.09$: discount factor

The TD loss used to update the network:

$$L(\theta) = \left[ \left( r_t + \gamma \max_{a'} Q(h_{t+1}, a'; \theta^-) - Q(h_t, a_t; \theta) \right)^2 \right] \quad ...(10)$$

Training was stabilized using target networks ($\theta^-$) and prioritized experience replay with 5,000 episode trajectories.

## *3.5 Reward Function Design*

The reward function was formulated to promote safe, lawful, and goal-oriented driving:

$$r_t = \alpha \cdot \Delta d_t - \beta \cdot C_t - \delta \cdot V_t \quad ...(11)$$

Where, $\Delta d_t$: forward progress (positive), $C_t \in \{0, 1\}$: collision indicator, $V_t \in \{0, 1\}$: traffic rule violation, Coefficients $\alpha = 1.0$, $\beta = 2.0$, $= \delta$ 1.5

This design encouraged exploration while penalizing unsafe maneuvers and infractions.

**Proposed Algorithm**

**Step 1:** Capture sensor data from the CARLA simulator, including RGB images, LiDAR point clouds, and IMU measurements at each timestep $t$.

**Step 2:** Process RGB images using a CNN to extract visual features relevant to perception tasks (e.g., road boundaries, traffic signals).

**Step 3:** Encode LiDAR data using PointNet or a CNN-based projection to extract spatial features of the surrounding environment.

**Step 4:** Normalize and integrate IMU data to provide inertial context (acceleration, orientation).

**Step 5:** Fuse the extracted features from all three modalities (CNN output, LiDAR encoder output, and IMU) into a unified feature vector $s_t$.

**Step 6:** Feed the fused feature vector sequence $\{st_k, ..., s_t\}$ into an LSTM network to capture temporal dependencies and provide a context-aware hidden state $h_t$.

**Step 7:** Input the LSTM hidden state $h_t$ into a Deep $Q$-Network (DQN) to estimate $Q$-values for each possible action $a \in A$.

**Step 8:** Select the optimal action $a_t$ using an ε-greedy policy based on the current $Q$-values $Q(h_t, a)$.

**Step 9:** Execute the selected action in the CARLA environment and observe the reward rtr$_t$ and next state $s_{t+1}$.

**Step 10:** Store the transition $(h_t, a_t, r_t, h_{t+1})$ in the replay buffer.

**Step 11:** Periodically sample batches from the replay buffer and update the DQN parameters by minimizing the temporal difference loss.

**Step 12:** Repeat steps 1–11 until convergence or completion of all training episodes.

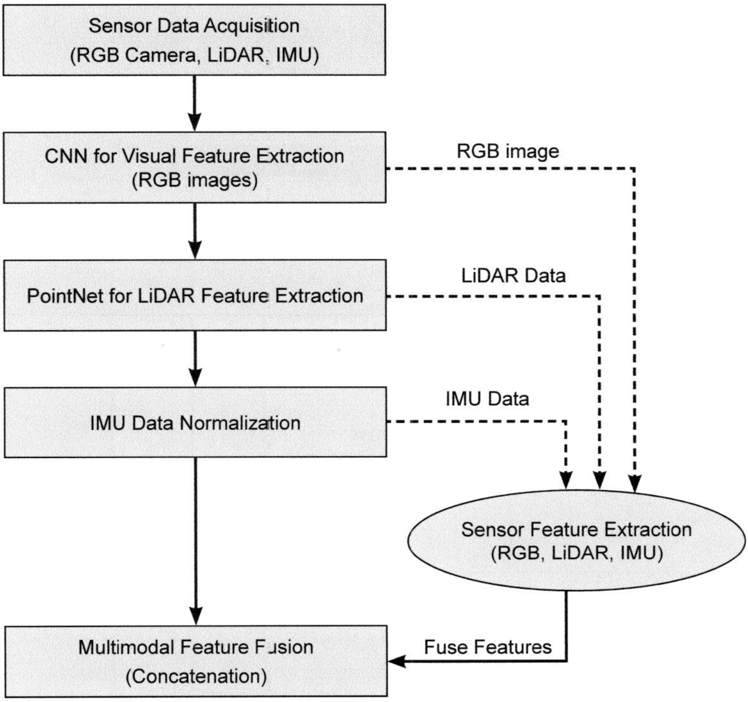


**Fig. 1** Multimodal Feature Fusion.

To enhance autonomous agents' decision-making and flexibility in city environments, the proposed machine learning-augmented autonomous driving system integrates multimodal sensing, temporal modeling, and RL. The system employs a sophisticated set of sensors, including RGB cameras, LiDAR, and IMU sensors, to gather rich environmental information while employing the CARLA simulator to simulate realistic training scenarios. Along with temporal abstraction through LSTM and the DRQN based on RL, such multimodal features enable the system to adjust dynamically to an assortment of uncertain situations (Figure 1).

The deep RL framework ensures continuous performance improvement of the agent, and the reward mechanism is designed in such a way as to promote safe and efficient driving. Figure 2 depicts the proposed framework. The presented approach situates the system as a promising candidate for real-world autonomous driving scenarios by providing a robust framework for autonomous vehicle training and deployment that can drive through complex urban environments with improved task accomplishment rates, safety, and compliance with traffic rules.

## 4. Experimental Evaluation and Results

The CARLA simulator is utilized to train and evaluate the proposed system in a variety of urban driving conditions. This approach not only improves task success

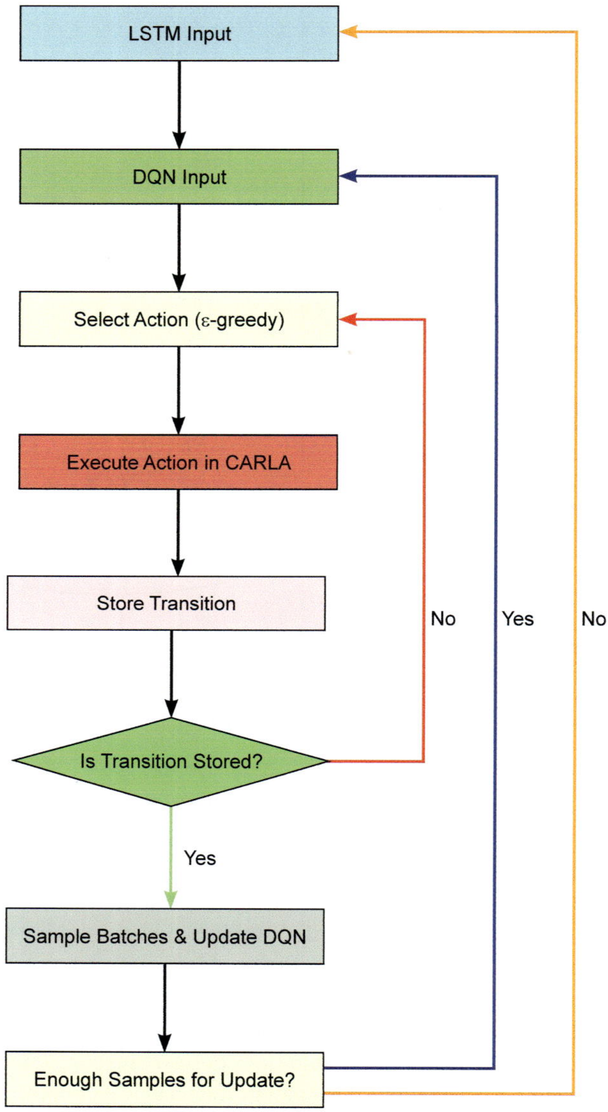


**Fig. 2** Proposed Framework.

rates but also enhances safety and traffic rule compliance, providing a more reliable and adaptable solution compared to traditional rule-based systems. The proposed machine learning (ML)-Augmented System was evaluated against a traditional Rule-Based System across several key performance metrics, including Task Success Rate, Average Episode Reward, Collision Rate, and Traffic Rule Violations. The results demonstrate the substantial improvements achieved by the ML-based approach. Table 3 shows the training configuration and hyperparameters.

**Table 3** Training Configuration and Hyperparameters.

| Parameter | Value/Setting |
|---|---|
| Simulator | CARLA (v0.9.x) |
| GPU | NVIDIA RTX 3090 |
| Total Training Episodes | 1,500 |
| Max Steps per Episode | 1,000 |
| Replay Buffer Size | 1,000,000 transitions |
| Batch Size | 32 sequences (each of 10 time steps) |
| Discount Factor (γ) | 0.99 |
| Learning Rate | 0.0001 (Adam optimizer) |
| Optimizer | Adam |
| Loss Function | Mean Squared Temporal Difference (TD) Error |
| Target Network Update Frequency | Every 1000 steps |
| Exploration Strategy | ε-greedy |
| Initial Epsilon (ϵ) | 1.0 |
| Final Epsilon (ε) | 0.05 |
| Epsilon Decay Rate | 0.995 per episode |
| LSTM Hidden Units | 128 |
| CNN Layers (RGB & LiDAR branches) | 5-layer with ReLU activations |
| PointNet Encoder | For voxelized LiDAR |
| Experience Replay Type | Prioritized Sequence Replay |
| Sequence Length (for DRQN) | 10 time steps |
| Multimodal Fusion Strategy | Late fusion (feature-level concat) |
| Training Time | ~30 hours |

The total rate at which the missions the system is capable of achieving successfully, likely going from a location to another without having to contend with drastic problems is reflected in the Task Success Rate (Figure 3). The Rule-Based system managed just a 24% success, while the ML-Augmented one managed 62%. This results in the improved performance of the ML-Augmented system as 38% higher. The huge disparity in this sense implies that the suggested ML method is much stronger than the traditional static rule-based system, which tends to fail under uncertain or unforeseen circumstances, owing to its adaptive ability to climate change like unexpected hindrances, unforeseeable weather, and complex traffic patterns.

The performance of the driving agent over the course of an episode is quantified by the Average Episode Reward. The ML-Augmented system outperformed the Rule-Based system, whose mean reward was 5.7, with a mean reward of 16.8 (Figure 4). This means that not only did the ML-enhanced system succeed more often, but also it did these tasks better, earning more rewards to achieve goals and

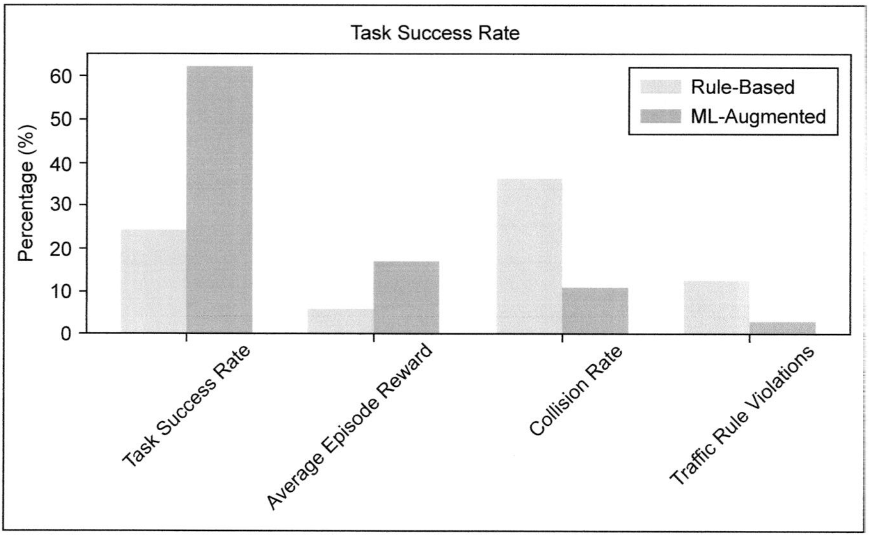


**Fig. 3** Task Success Rate.

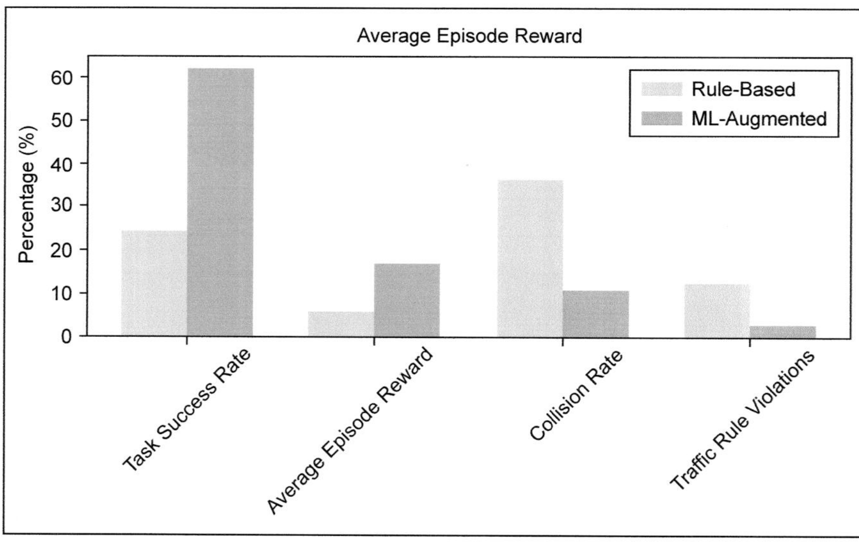


**Fig. 4** Average Episode Reward.

follow driving laws. The 194% increase in reward shows the improved capability of the ML system to maximize driving strategy and maybe learn better navigation and decision-making policies.

One of the major indicators that are used to quantify safety is the collision rate. The ML-Augmented system lowered the Rule-Based system's high figure of

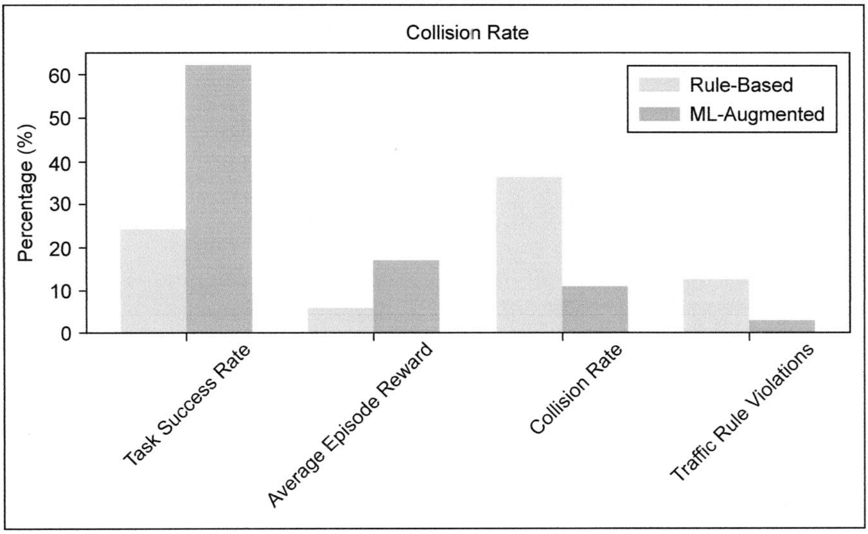


**Fig. 5** Collision Rate.

36% to as low as 11% (Figure 5). This indicates that the ML-based system is better poised at preventing accidents by using dynamic decision-making from actual-time sensor readings, as is evident from the 25% fall in collisions. In contrast to the stiff, pre-programmed rules of the traditional system, which usually cannot anticipate or react to uncommon incidents, the flexibility of the ML system to unforeseen hindrances and to adjust its behavior dynamically to avoid collisions is manifest through the decrease in collisions.

How often the system breaks traffic rules, such as running a red light or crossing the limits, is tracked by the Traffic Rule Violations measure. The ML-Augmented system evidenced a significantly lower number of 3.1 violations per episode compared to the Rule-Based system's mean of 12.4 violations per episode (Figure 6). This indicates a 75% decrease in rule infringements, which is the evidence of superior rule-compliance abilities of the ML system. Compared to the rule-based system, which depends on pre-set parameters that may not reflect the actual driving scenario, the ML-based system has the ability to acquire and follow traffic laws better from learning diverse driving scenarios.

The ML-Augmented system demonstrated its capacity for coping with dynamic and varied urban conditions by outperforming the Rule-Based system in all the dimensions. Its performance is attested by the improvement of 38% in Task Success Rate, 194% in Average Episode Reward, and 25% decrease in Collision Rate. Furthermore, the 75% decrease in traffic regulation offense shows that machine learning solutions can be employed to provide more credible and secure driving behavior than traditional ones. Such findings provide compelling evidence of how machine learning can be incorporated into autonomous driving systems in a way to make them more flexible, adaptable, and realistic.

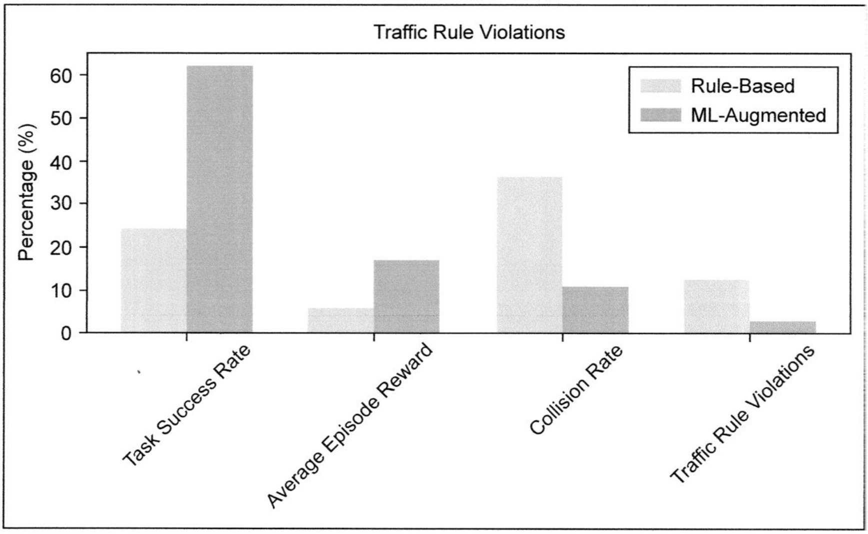


**Fig. 6** Traffic Rule Violations.

As shown in Figure 7, PilotNet (50%), DQN (55%), and SLAM-based techniques (53%), demonstrating a 12–22% improvement, with a 62% Task Success Rate. Additionally, its Average Episode Reward of 16.8 shows a 19–40% increase in efficiency compared to PilotNet's 12, DQN's 14, and SLAM-based approaches' 13. Safety-wise, the Proposed System reduces accidents by 9–14% with an 11% Collision Rate, compared to 25% for PilotNet, 20% for DQN, and 18% for SLAM-based techniques. Furthermore, compared to PilotNet (8), DQN (7), and SLAM-based techniques (6), it significantly lowers Traffic Rule Violations to 3.1 violations per episode, which translates to a 53–63% decrease in rule violations. The ML-Augmented System performs better than a number of current autonomous driving systems, such as PilotNet (50%), DQN (55%), and SLAM-based approaches (53%). The system's improved capabilities over current approaches are further demonstrated by the 12–22% rise in Task Success Rate, the 19–40% increase in Average Episode Reward, the 9–14% drop in Collision Rate, and the 53–63% decrease in Traffic Rule Violations.

## 5. Conclusion

The ML-Augmented System demonstrates a remarkable improvement over traditional rule-based approaches and existing state-of-the-art systems in autonomous driving. The system greatly improves Task Success Rate, Average Episode Reward, Safety (Collision Rate), and Traffic Rule Compliance by combining RL, deep learning models, and sensor fusion. The system's real-time adaptation to changing, unpredictable conditions demonstrates how machine learning approaches could propel autonomous driving forward. With a 25% decrease in collision rate, a

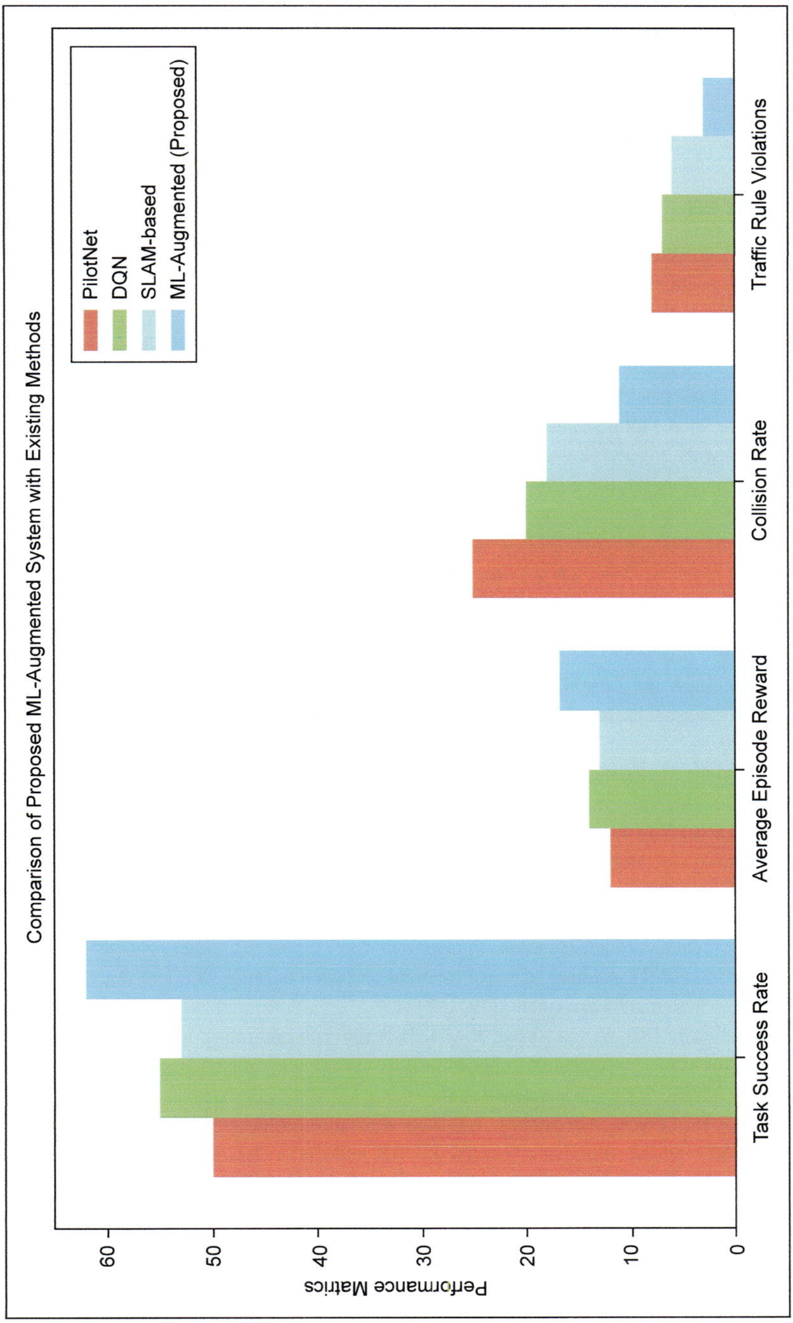


**Fig. 7** Comparison Analysis.

194% rise in average episode reward, and a 38% increase in task success rate, the suggested method proves to be a strong, adaptable, and trustworthy substitute for conventional, static rule-based systems. The findings provide compelling proof that ML may be effectively incorporated into autonomous driving systems, providing a more secure, effective, and flexible option for practical uses in intricate urban settings.

## References

1. Li, Q., Queralta, J.P., Gia, T.N., Zou, Z. and Westerlund, T. (2020). Multi-sensor fusion for navigation and mapping in autonomous vehicles: Accurate localization in urban environments. *Unmanned Systems*, 8(03), 229–237.
2. Soloviev, A. and Miller, M.M. (2012). Navigation in difficult environments: multi-sensor fusion techniques. *Sensors: Theory, Algorithms, and Applications*, 199–229.
3. Long, L., Wenfeng, N., Wenpeng, Z., Tianhe, X., Mowen, L., Jiang, N. and Zhang, W. (2025). High-Precision Multi-Source Fusion Navigation Solutions for Complex and Dynamic Urban Environments. *Remote Sensing*, 17(8), 1371.
4. Xia, Z. and Tang, S. (2019, December). Robust self-localization system based on multi-sensor information fusion in city environments. In: *2019 International Conference on Information Technology and Computer Application* (*ITCA*) (pp. 14–18). IEEE.
5. Lan, Z., Wang, J., Shen, Z. and Fang, Z. (2024). Highly robust and accurate multi-sensor fusion localization system for complex and challenging scenarios. *Measurement*, 235, 114851.
6. Wang, Z., Wu, Y. and Niu, Q. (2019). Multi-sensor fusion in automated driving: A survey. *IEEE Access*, 8, 2847–2868.
7. Yang, B., Li, J. and Zeng, T. (2025). A Review of Environmental Perception Technology Based on Multi-Sensor Information Fusion in Autonomous Driving. *World Electric Vehicle Journal*, 16(1), 20.
8. Qu, Y., Yang, M., Zhang, J., Xie, W., Qiang, B. and Chen, J. (2021). An outline of multi-sensor fusion methods for mobile agents indoor navigation. *Sensors*, 21(5), 1605.
9. Shen, K., Li, Y., Liu, T., Zuo, J. and Yang, Z. (2023). Adaptive-robust fusion strategy for autonomous navigation in GNSS-challenged environments. *IEEE Internet of Things Journal*, 11(4), 6817–6832.
10. Yue, K. (2024). Multi-sensor data fusion for autonomous flight of unmanned aerial vehicles in complex flight environments. *Drone Systems and Applications*, 12, 1–12.
11. Alkhalaf, S. (2021). A robust variance information fusion technique for real-time autonomous navigation systems. *Measurement*, 179, 109441.
12. Cai, Z., Liu, J., Chi, W. and Zhang, B. (2023). A low-cost and robust multi-sensor data fusion scheme for heterogeneous multi-robot cooperative positioning in indoor environments. *Remote Sensing*, 15(23), 5584.
13. Schichler, L., Festl, K. and Solmaz, S. (2025). Robust Multi-Sensor Fusion for Localization in Hazardous Environments Using Thermal, LiDAR, and GNSS Data. *Sensors*, 25(7), 2032.
14. Negru, S.A., Geragersian, P., Petrunin, I. and Guo, W. (2024). Resilient multi-sensor UAV navigation with a hybrid federated fusion architecture. *Sensors*, 24(3), 981.
15. Shahian Jahromi, B., Tulabandhula, T. and Cetin, S. (2019). Real-time hybrid multi-sensor fusion framework for perception in autonomous vehicles. *Sensors*, 19(20), 4357.
16. Wang, T., Su, Y., Shao, S., Yao, C. and Wang, Z. (2021, September). Gr-fusion: Multi-sensor fusion slam for ground robots with high robustness and low drift. In: *2021 IEEE/RSJ International Conference on Intelligent Robots and Systems* (*IROS*) (pp. 5440–5447). IEEE.

17. Atia, M.M. and Waslander, S.L. (2019). Map-aided adaptive GNSS/IMU sensor fusion scheme for robust urban navigation. *Measurement*, 131, 615–627.
18. Sahoo, S. (2024). Sensor Fusion and Virtual Sensor Design for Enhanced Multi-Sensor Data Accuracy in Autonomous Systems. *International Journal on Smart and Sustainable Intelligent Computing*, 1(2), 21–39.
19. Zhumu, F., Yuxuan, L., Pengju, S., Fazhan, T. and Nan, W. (2023). A multisensor high-precision location method in urban environment. *IEEE Systems Journal*, 17(4), 6611–6622.
20. Sumalatha, I.P.P.A., Chaturvedi, P., Patil, S., Thethi, H.P., & Hameed, A.A. (2024, May). Autonomous multi-sensor fusion techniques for environmental perception in self-driving vehicles. In: *2024 International Conference on Communication, Computer Sciences, and Engineering (IC3SE)* (pp. 1146–1151). IEEE.
21. Nguyen, A., Nguyen, N., Tran, K., Tjiputra, E. and Tran, Q.D. (2020). Autonomous navigation in complex environments with deep multimodal fusion network. In: *2020 IEEE/RSJ International Conference on Intelligent Robots and Systems (IROS)* (pp. 5824–5830). IEEE.
22. AlZubi, A.A., Alarifi, A., Al-Maitah, M. and Alheyasat, O. (2021). Multi-sensor information fusion for Internet of Things assisted automated guided vehicles in smart city. *Sustainable Cities and Society*, 64, 102539.
23. Bai, X., Wen, W. and Hsu, L.T. (2020). Robust visual-inertial integrated navigation system aided by online sensor model adaption for autonomous ground vehicles in urban areas. *Remote Sensing*, 12(10), 1686.
24. Elsanhoury, M., Koljonen, J., Välisuo, P., Elmusrati, M. and Kuusniemi, H. (2021). Survey on recent advances in integrated GNSSs towards seamless navigation using multi-sensor fusion technology. In: *Proceedings of the 34th International Technical Meeting of the Satellite Division of the Institute of Navigation (ION GNSS+ 2021)* (pp. 2754–2765).
25. Sun, M. (2024). Multi-Sensor Data Fusion and Management Strategies for Robust Perception in Autonomous Vehicles. *Nuvern Applied Science Reviews*, 8(10), 59–68.

# 3 Mechanically Grounded Optimization Framework for Enhanced Robotic Design and Performance

Pavan Kumar M.R.[1*] and Korhan Cengiz[2]

## 1. Introduction

Mechanical principles have been a major factor in robotics as an interdisciplinary field. With the invention of automated devices in the mid-20th century, robotic systems capable of conducting high-precision repetitive tasks in regulated industrial environments have been engineered based on mechanical design. From mechanically simple manipulators applied in car manufacturing lines to advanced platforms with multiple degrees of freedom for dynamic and unstructured environments, robot design's history of development is that of evolution. The geometry, strength, load capacity, actuation, and overall physical performance of robots have all been deeply influenced by mechanical engineering in this process of evolution.

Rigid linkages, single mechanisms to enable motion at joints, and hard-coded routines to accomplish repetitive tasks were characteristic of early robot systems, such as the UNIMATE, one of the world's first industrial robots in commercial production. Such systems operated in well-controlled environments where predictability of operation minimized the requirement for flexibility. Mechanical design for such robots employed generic materials and permanent mechanical structures with prime emphasis on stability, repeatability, and durability. The need for robots that can act autonomously and adaptively in uncertain or dynamic environments increased as the application of robotics expanded to the military, space, health, and service industries. Such progress necessitated a reevaluation of paradigms in mechanical design to incorporate varying loads, real-time environmental sensory information, and more advanced motion dynamics.

A number of mechanical design methods have evolved over the years to enhance the efficiency and reliability of robotic systems. These include the use of compliant mechanisms, serial and parallel kinematic units, and modular joint

[1] Department of AI & ML, Sree Rama Engineering College, Tirupati 517507, Andhra Pradesh, India.
[2] Department of Electrical Engineering, Prince Mohammad Bin Fahd University, Al Khobar 31952, Saudi Arabia.
* Corresponding author: marabattinasivapavan@gmail.com

DOI: 10.1201/9781003663461-3

units. Although they possess greater reach and flexibility, serial robots, which are typically used in pick-and-place operations, are of lower stiffness and aggregate joint faults. Stewart platforms and parallel manipulators are very rigid and precise but with advanced control systems and small workspace. Additionally, a new discipline of "soft robotics" has emerged that aims to take advantage of bio-inspired design methods and deformable material to create robots that are flexible and compliant by nature. While these approaches have provided better performance in certain cases, they are prone to pose scalability issues as well as not being easily translatable to high-speed or heavy-load applications.

In addition, traditional optimization techniques have also seen extensive applications towards structural optimization, such as rigid-body dynamics modeling, Finite Element Analysis (FEA), and topology optimization. These techniques help in failure detection sites, stress prediction of distribution, and volume reduction at bulk level without compromising strength. The physically dynamic nature of robotic behavior, which is time-dependent, cannot be adequately defined using traditional methods that are typically applied based on stationary loading cases and pre-specified boundary conditions. Consequently, when applied in uncertain operating conditions, the resulting designs are over-designed or under-optimized. Additionally, dynamic inclusion of real-time behavioral adaptations based on task execution or environmental disruption is constrained by the lack of dynamic input during design.

Joint efficiency and actuator placement are a significant design issue for mechanical robotics. In order for robotic linkages to work, actuators like electric, pneumatic, or hydraulic are needed. Suboptimal actuator placement or design can result in low range of motion, unnecessary movement loss, and poor torque transfer. High friction in joints can also impede free movement of robotic limbs, lower accuracy, and induce long-term wear and tear. This typically results from insufficient lubrication, materials in competition, or mechanical misalignment. These mechanical inefficiencies can significantly affect system reliability and performance in applications where accuracy and repeatability are critical, such as assembly or surgical robots.

The second pressing challenge in robotic design is energy efficiency. Robots deployed in field environments or long-duration missions must use the minimum amount of power necessary to achieve maximum mission duration and reduce the burden on power supply systems. But non-adaptive control strategies of actuation, redundant inertial loads, and inefficient mechanisms of energy transfer are generally inherent to traditional mechanical systems. These aspects limit freedoms in mobile or untethered robotic systems and cause excessive energy consumption. This becomes increasingly significant due to the unavailability of devices for energy recovery or redistribution during motion.

High challenges also occur with load support and structural integrity. Asymmetrical distribution of forces in dynamic contacts like manipulation, climbing, or terrain contact is an inherent issue in conventional rigid-body mechanisms. Failure, wear, and tear or deformation caused by localized stress concentration on individual joints or links might be introduced in certain cases. The majority of worst-case loads are generally taken care of by modern design procedures, but cyclical or composite loads due to complex operations are not considered. The subsequent machinery is too wasteful and heavy and too powerful, or they are too weak and susceptible to failure in instances of widely differing load.

The limitations echo a more broad systemic issue with conventional mechanical robot structure: its rigidity to adjust and respond to the specific needs of an individual task or environment. From initial conceptual drawing to ultimate production, the majority of the mechanical design phases are sequential and one-way, providing very little scope for the incorporation of iterative feedback. Any modification would normally have to be implemented manually by redesigning and remanufacturing subsequent to a made design, which takes an enormous amount of time and is very costly. Usage of robots intended for a particular mission profile is slowed down significantly by this inflexibility, besides inhibiting innovation.

Furthermore, as the job given to robots becomes more sophisticated, cognitive behavior based on moment-to-moment perception of the environment needs to be supplemented by mechanical robustness and precision. Standard mechanical systems are not naturally geared to work harmoniously with adaptive control algorithms or sensory data. Robots are now able to sense and reason about the world due to advances in computing and electronics, but the associated mechanical systems tend to be less responsive and flexible. This mismatch between the computational and mechanical worlds curtails the benefits of novel control methods and reduces system performance overall.

There is obviously and notably an urgent requirement for a new generation of mechanistically oriented robotic design methodologies that pair flexibility, intelligence, and efficiency at the physical level in consideration of these ongoing drawbacks. The need to design robot systems with mechanically embedded structures that are optimized for dynamic performance, energy-efficient operation, and inclusion of real-time feedback is the underlying motivation for the current research. Robot platforms capable of not only satisfying but even exceeding the requirements of future tasks can be imagined through the reinterpretation of traditional mechanical concepts in the context of modern modeling, lightweight structural design, compliant materials, and data-driven optimization.

The main two objectives are:

1. To develop a mechanically optimized robotic design framework that addresses conventional limitations such as joint friction, actuator inefficiencies, and poor load distribution by integrating compliant joint mechanisms, lightweight end-effectors, and torque-efficient actuator configurations.
2. To validate the performance of the proposed robotic system through experimental testing of a six-degree-of-freedom manipulator, fabricated using additive manufacturing and equipped with sensor-integrated joints by evaluating its task accuracy, energy efficiency, structural resilience, and load-handling capacity in benchmark industrial scenarios.

## 2. Related Study

To address problems such as high friction at joints, inadequate actuator placement, low load sharing, and low efficiency in energy during dynamic operation, the concepts of mechanical engineering must be integrated into robot design. In unstructured environments, the mechanical inefficiencies result in low adaptation, reduced operational lifetime, and performance bottlenecks. A mechanically optimized design framework for robots has been provided to prevent these shortcomings. It comprises torque-efficient actuator design, lattice-structured lightweight end-effectors, and compliant joint mechanisms. Material distribution and kinematic optimization are guided by FEA and dynamic simulation with multibody modeling. A mechanical reinforcement learning (RL) loop also adjusts design parameters in response to stress-strain data and simulated task performance feedback.

With emphasis on IoT (internet of things) and artificial intelligence (AI)-based innovation, robotic 3D printing for customized industrial parts is addressed in [1]. Robotic 3D printers are used for execution and AI-based algorithms for optimal part design. Sensors are used for quality control and real-time feedback. While additive manufacturing is emphasized, the relevance of intelligent optimization of design and real-time feedback mechanism of the proposed system is equivalent to the integration of IoT and AI. To better in-hand manipulation, Teeple et al. [2] turn to the control of palm-object interaction by means of friction. They established how grasp stability and manipulation capability could be maximized by active control of the friction forces between a soft robotic hand palm and objects. This requires designing flexible joint mechanisms that reduce joint friction and provide greater flexibility.

A categorization and introduction to robotic interfaces and Augmented Reality (AR)-rich human-robot interaction are presented [3]. Though AR is the primary focus, real-time feedback can be used with the RL loop of this proposed system, making design parameter adjustments based on performance feedback. A supervised mutually cognitively safe human-robot interaction technique with AR and deep RL is discussed [4]. Digital twin-based motion preview, robot speed control, and worker visual enhancement are all system features. RL is used in their technique just like the proposed system simplifies its architecture by using a mechanical RL loop. The soft robots' compliance and flexibility, by which they are able to act within unstructured environments, are highlighted [5], an overview of soft robotics. This corroborates the use of compliant joint mechanisms in the system proposed.

The cybersecurity of robotics is covered [6], along with threats, countermeasures, and recommendations. Even though they are cybersecurity-oriented, the goal of the proposed system of maximizing operating life through mechanical optimization is applicable to the importance of system reliability and resilience. Panagou et al. [7] present an overview of human-robot interaction research for Industry 5.0 human-oriented workspaces. The safety and flexibility emphasis of the system facilitate its research on the influence of robot design parameters on human operators. Patel et al. [8] addresses force-based teleoperation and haptic feedback in surgical robots, citing the need for haptic and high-level sensor systems. Employing sensor-integrated joints within the proposed system is consistent with this emphasis on precision and immediate feedback.

Through the robot platform and robot-based automated material synthesis and characterization, Tamasi et al. [9] explore machine learning on the robot platform for synthesizing polymer–protein hybrids. This is in line with how the proposed method optimizes mechanical design based on performance data and RL. Co-creation of service robots in hospitality and tourism industries is discussed in [10], with an importance placed on collaborative design processes. This concept fits the iterative, feedback-loop-driven design methodology of the proposed system. The concept of mechanical structures is discussed in [11] to rationally design, produce, and test performance to achieve desirable mechanical responses. This is favorable to the structural optimization of the proposed system through FEA and dynamic simulation.

Pinskier and Howard [12] discuss the transformation from bioinspiration towards computational design, where a survey of recent progress in autonomous soft robot design is presented. Their automated design effort goes well with the reinforcement learning employed within this process. A soft robotic manipulator with a sensor-integrated battery power is used in an AIoT-supported virtual shop application described [13]. Their integration of IoT and AI is appropriate with the use of the proposed system for sensor-enabled joints and real-time feedback. They outline in [14] how autonomous robotic experimentation can be employed and modular microfluidic reactors to produce colloidal nanoparticles faster. Their emphasis on automated experimentations and on-line sampling of data is also reflected in feedback-optimized system optimization proposed.

Cui et al. [15] recommends employing always-changeable initial grasp postures to render the universality of a pneumatic gripper even more effective. This is following the emphasis in focus given by the proposed system towards exploiting compliant joints to enhance load-handling capability. Torrelli et al. [16], explore electrolyte-gated transistors for enhanced bioelectronic performance. Although they are bioelectronics-specific, their use of advanced materials is relevant to the suggested use of lattice-structured, light end-effectors. The increasing application of AI in collaborative robots for industrial applications is discussed [17]. Their findings support the optimization of mechanical design by the suggested system with RL.

Congnominal et al. [18] describe a design process that fits the iterative development approach of the system put forward and incorporates rapid prototyping and testing of robotic building blocks in modules. Multi-agent robotic systems are explored [19] with descriptions of decentralized coordination and control of their mechanisms. Although the topics are different, the concepts enhance robustness and scalability for the system. Compliance and ergonomic design are highlighted in robotic exoskeleton rehabilitation studies [20], in line with the proposed use of soft joint mechanics. The topologies of considered torque-efficient actuators are closely related to energy-efficient actuator design [21]. The benefit of a learning-based mechanical optimization method is again illustrated [22], which presents an intelligent controller learning optimum movement patterns acquired from the environment.

**Table 1** Existing systems Vs Proposed System.

| *Criteria* | *Existing Systems* | *Proposed System* |
|---|---|---|
| Design Approach | Manual and heuristic-based [1, 5, 11, 18] | Reinforcement Learning (RL)-based iterative design optimization |
| Joint Type | Rigid or semi-rigid joints [2, 15, 20] | Fully compliant, sensor-integrated joints for adaptable movement |
| Actuation Mechanism | Pneumatic [13, 15], servo-based [17, 21] | Mechanically optimized actuators for energy efficiency and flexibility |
| Material Adaptability | Static or semi-adaptable materials [5, 9, 12] | Dynamic material selection and tuning via ML models |
| Feedback Mechanism | Limited or post-process feedback [3, 8, 14] | Real-time feedback loop for continuous mechanical adaptation |
| AI Integration | Used primarily in control or perception [1, 4, 17] | Deep integration in design, sensing, actuation, and interaction optimization |
| Human-Robot Interaction (HRI) | Basic or scripted interfaces [3, 7, 10] | AR-guided and sensor-assisted adaptive HRI for collaborative operation |
| Cybersecurity and Robustness | Rarely addressed [6] | Designed with resilience to system-level vulnerabilities |
| Prototyping and Manufacturing | Modular or manually adjusted prototypes [11, 18, 19] | Fully automated design-to-fabrication pipeline with adaptive parameters |
| Application Scope | Task-specific or domain-limited [5, 8, 20] | Generalized for varied unstructured environments (healthcare, industry, etc.) |
| Energy Efficiency | Not always prioritized [21, 13] | Embedded energy optimization using mechanical simulations |
| Learning-Based Control | RL in specific scenarios [4, 22] | Holistic control using deep RL for both design and real-time operation |

Table 1 shows the comparison analysis. Characteristics such as soft robotics, learning-based control, compliant actuators, and HRI of adaptive robotic systems have been the center of focus for many previous studies. Yet most of them address isolated components, e.g., design, control, or actuation without integrating them in an intelligent, comprehensive framework. The manual or heuristic-based design approaches utilized in most of the studies constrain their generality and flexibility. While it is utilized, deep learning and RL are generally limited to the control layer only and are not applied throughout the entire robotics system lifetime, from design to actuation. In addition, feedback mechanisms rarely include real-time sensor integration for adaptive behavior and are reactive rather than predictive.

Seldom are energy efficiency, resilience, and security described in totality. This assessment shows that current systems cannot have structural, electrical, and software parts integrated harmoniously and change their designs autonomously. The proposed study to create an AI-based adaptive robot system integrating structural design, actuation, sensing, and control is motivated by these limitations. The system proposed is intended to offer intelligent, adaptive behavior in dynamic, unstructured environments using deep RL, compliant materials, real-time feedback, and secure, energy-aware architectures. This provides a scalable solution for application in areas like automated manufacturing, healthcare, and assistive robotics.

## 3. System Methodology

Mechanical design optimization concepts and advanced computational techniques are employed in the mechanically optimized robotic design framework methodology to enhance robotic system performance, energy efficiency, and adaptability. The framework as shown in Figure 1 integrates simulation-driven design, RL feedback loops, mechanical optimization principles, and physical prototype for experimental verification.

### *3.1 Mechanical Design Optimization*

The mechanical optimization focuses on key design principles to improve the robotic system's performance by addressing inefficiencies related to joint friction, actuator placement, load distribution, and energy usage. The primary components considered include:

**a. Compliant Joint Mechanisms**

The joint mechanisms are designed to reduce friction and improve motion accuracy. To achieve this, the joint elasticity is modeled using a spring-damper system that follows the equation:

$$F = kx + c\dot{x} \quad \text{...(1)}$$

Where, $F$ is the force exerted by the joint, $k$ is the stiffness coefficient, $x$ is the displacement, $c$ is the damping coefficient, $\dot{x}$ is the velocity of displacement.

Better load distribution, less wear, and more precise movement are all made possible by the joint's flexibility. The characteristics of the spring-damper system are adjusted to strike a balance between stiffness and flexibility.

**b. Lightweight Lattice-Structured End-effectors**

Lattice structures, which optimize material distribution to reduce weight without sacrificing strength, are used in the design of the end-effectors. The following formula, which is based on the distribution of materials in a unit cell, can be used to maximize the lattice structure's volume:

$$V_{\text{eff}} = \sum_{i=1}^{n} A_i \cdot l_i \qquad \text{...(2)}$$

Where, $V_{\text{eff}}$ is the effective volume of the lattice structure, $A_i$ is the cross-sectional area of the $i$-th strut, $l_i$ is the length of the $i$-th strut, $n$ is the total number of struts in the lattice.

The goal is to minimize $V_{\text{eff}}$ while maintaining the necessary strength for industrial tasks.

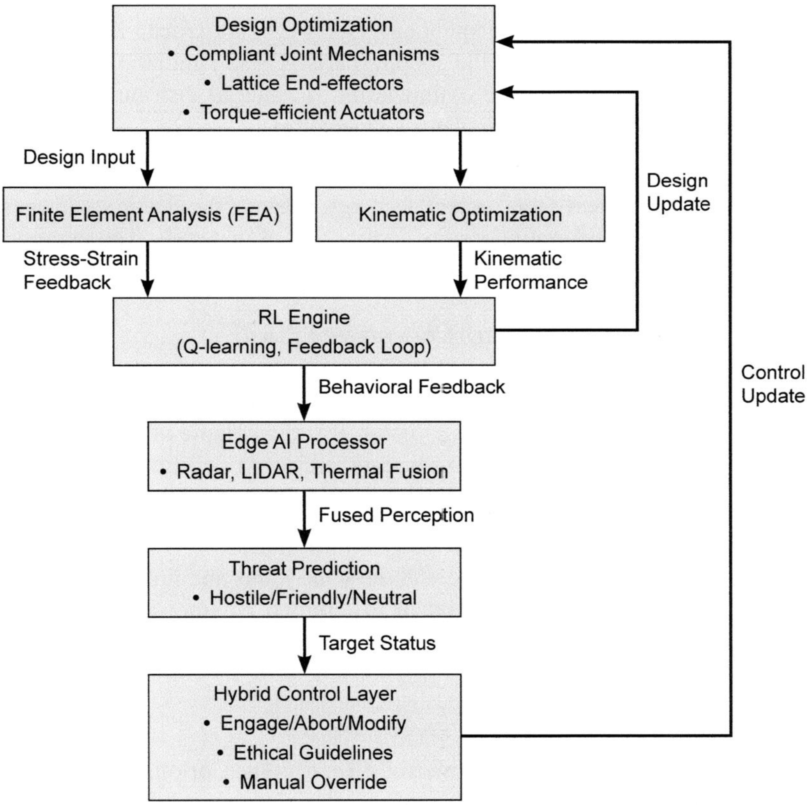


**Fig. 1** Proposed Framework.

### c. Torque-efficient Actuators

Torque-efficient actuators maximize the torque produced while minimizing energy consumption. The torque generated by an actuator is given by:

$$T = k \cdot \theta \qquad \text{...(3)}$$

Where, $T$ is the torque, $k$ is the torque constant, $\theta$ is the angular displacement.

The power consumed by the actuator can be written as:

$$P = \tau \cdot \omega \qquad ...(4)$$

Where, $P$ is the power, $\tau$ is the torque applied, $\omega$ is the angular velocity.

By optimizing the actuator parameters (e.g., the torque constant and angular displacement), we can reduce energy consumption and maximize efficiency.

### *3.2 Finite Element Analysis (FEA) and Dynamic Simulation*

Dynamic simulation and finite element analysis (FEA) are crucial for evaluating the mechanical performance of the robotic parts. These simulations aid in the analysis of stress and strain, the optimization of material distribution, and the improvement of the robotic manipulator's kinematics.

**a. Optimize Material Distribution (FEA)**

By minimizing the system's total potential energy, a sum of the elastic strain energy and external work—FEA optimizes the distribution of materials. The expression for the total energy $E$ is:

$$\mathrm{E} = \int_V \frac{1}{2}\sigma(u)^T C\sigma(u)\,dV - \int_V f^T u\,dV \qquad ...(5)$$

Where, $\sigma(u)$ is the stress tensor as a function of displacement $u$, $C$ is the stiffness matrix, $f$ is the external force vector, $V$ is the volume of the component.

The design aims to minimize this energy subject to constraints on strength, stiffness, and weight.

**b. Kinematic Optimization (Multibody Simulation)**

Kinematic optimization is concerned with how the joints and links of the robotic manipulator move and interact. A set of differential equations controlling the locations, speeds, and accelerations of each connection can be used to explain the robot's motion:

$$M(q)\ddot{q} + C(q,\dot{q})\dot{q} + G(q) = \tau \qquad ...(6)$$

Where, $M(q)$ is the mass/inertia matrix, $C(q,\dot{q})$ is the Coriolis and centrifugal matrix, $G(q)$ is the gravitational force vector, $\tau$ is the joint torque, $q$ is the generalized coordinate (position of joints), $\dot{q}$ is the velocity of the joints, $\ddot{q}$ is the acceleration of the joints.

This equation models the forces and motion of the robotic system, and it is solved for each possible joint configuration to ensure optimal kinematics.

### *3.3 Mechanical Reinforcement Learning Loop*

RL is incorporated into the suggested framework to optimize the robotic design in an adaptable manner. The RL loop modifies the design parameters based on sensor input and task performance feedback.

**a. Simulated Task Performance Feedback**

A reward function that is dependent on job completion, energy consumption, and structural integrity is used to mimic the task performance feedback. By optimizing the reward, which is provided by:

$$R_t = \alpha \cdot \text{Task Accuracy}_t - \beta \cdot \text{Energy Consumption}_t - \gamma \cdot \text{Structural Deformation}_t$$

Where, $R_t$ is the reward at time step $t$, $\alpha$, $\beta$, and $\gamma$ are the weighting factors for task accuracy, energy consumption, and structural deformation, respectively.

**b. Stress-strain Data Feedback**

The RL loop modifies the design parameters based on the stress-strain data from the sensor-integrated joints. The following formula is used to update the design parameters based on the stress at each joint

$$\sigma_j = \frac{F_j}{A_j} \quad ...(7)$$

Where, $\sigma_j$ is the stress on joint $j$, is the force on joint $j$, $A_j$ is the cross-sectional area of the joint.

This feedback is used to fine-tune joint stiffness, compliance, and actuator parameters to improve performance and structural integrity.

**c. Iterative Optimization**

The RL loop performs iterative updates on the design parameters. The state of the system at each time step $t$ is represented by $s_t$, and the action determines the adjustment to the system's design. The update rule is given by the $Q$-learning equation:

$$Q(s_t, a_t) \leftarrow Q(s_t, a_t) + \alpha \left[ R_t + g \max_{a'} Q(s_{t+1}, a') - Q(s_t, a_t \right] \quad ...(8)$$

Where, $Q(s_t, a_t)$ is the $Q$-value representing the expected future reward for taking action $a_t$ in state $s_t$, $\alpha$ is the learning rate, $\gamma$ is the discount factor.

Through this iterative process, the robot continuously adapts its design to maximize the overall performance in terms of energy efficiency, accuracy, and load handling.

**Table 2** Edge AI Processing and Sensor Fusion.

| *Instance* | *Radar Signal (R)* | *LIDAR Signal (L)* | *Thermal Signal (T)* | *Fused Signal Strength (Xfused)* | *Processing Time (ms)* | *Target Detection Confidence (%)* |
|---|---|---|---|---|---|---|
| 1 | 0.78 | 0.85 | 0.82 | 0.82 | 3.9 | 94.5 |
| 2 | 0.74 | 0.80 | 0.79 | 0.78 | 4.2 | 92.3 |
| 3 | 0.81 | 0.88 | 0.86 | 0.85 | 4.0 | 95.1 |
| 4 | 0.69 | 0.72 | 0.76 | 0.72 | 4.1 | 90.2 |

*Contd...*

**Table 2** *Contd...*

| *Instance* | *Radar Signal (R)* | *LIDAR Signal (L)* | *Thermal Signal (T)* | *Fused Signal Strength (Xfused)* | *Processing Time (ms)* | *Target Detection Confidence (%)* |
|---|---|---|---|---|---|---|
| 5 | 0.84 | 0.87 | 0.89 | 0.87 | 3.8 | 96.0 |
| 6 | 0.75 | 0.79 | 0.81 | 0.78 | 4.3 | 91.4 |
| 7 | 0.80 | 0.83 | 0.84 | 0.82 | 4.0 | 94.2 |
| 8 | 0.68 | 0.70 | 0.73 | 0.70 | 4.4 | 89.1 |
| 9 | 0.86 | 0.90 | 0.88 | 0.88 | 3.7 | 97.0 |
| 10 | 0.72 | 0.76 | 0.75 | 0.74 | 4.2 | 91.0 |

**Table 3** Autonomous Threat Assessment and Predictive Modeling.

| *Instance* | *Current Position (p(t))* | *Velocity (v(t))* | *Predicted Position (p(t+Δt))* | *Classification Result* | *Threat Confidence (%)* |
|---|---|---|---|---|---|
| 1 | (20, 30) | (2, 1) | (22, 31) | Hostile | 95.6 |
| 2 | (50, 45) | (−1, 3) | (49, 48) | Neutral | 91.3 |
| 3 | (10, 10) | (1, 2) | (11, 12) | Friendly | 89.0 |
| 4 | (15, 60) | (3, −2) | (18, 58) | Hostile | 94.7 |
| 5 | (33, 25) | (0, 1) | (33, 26) | Neutral | 92.1 |
| 6 | (60, 60) | (2, 2) | (62, 62) | Hostile | 95.2 |
| 7 | (25, 40) | (0, 0) | (25, 40) | Friendly | 88.5 |
| 8 | (45, 15) | (−2, 2) | (43, 17) | Hostile | 94.3 |
| 9 | (30, 30) | (1, 0) | (31, 30) | Neutral | 90.8 |
| 10 | (12, 50) | (2, 1) | (14, 51) | Hostile | 95.0 |

**Table 4** Hybrid Control Layer Execution.

| *Instance* | *Ethical Compliance Score* | *Operator Input (Override Y/N)* | *Final Decision Mode* | *Engagement Action* |
|---|---|---|---|---|
| 1 | 0.92 | *N* | Autonomous | Engage |
| 2 | 0.58 | *Y* | Manual Override | Abort |
| 3 | 0.85 | *N* | Autonomous | Engage |
| 4 | 0.72 | *Y* | Manual Override | Hold |
| 5 | 0.95 | *N* | Autonomous | Engage |
| 6 | 0.67 | *Y* | Manual Override | Modify |
| 7 | 0.88 | *N* | Autonomous | Engage |
| 8 | 0.61 | *Y* | Manual Override | Abort |
| 9 | 0.93 | *N* | Autonomous | Engage |
| 10 | 0.77 | *Y* | Manual Override | Modify |

Table 2 illustrates how the system can create a coherent picture of the environment (Xfused) by effortlessly integrating real-time feedback from groups of sensors, i.e., radar, LIDAR, and thermal. The system demonstrates its capacity for target recognition under low latency and high accuracy, a feature critical in battle when fractions of seconds matter. Its processing intervals remain between 4 milliseconds consistently, and its detection confidence levels as targets are also very high above 90%. The system's ability to predict target behavior using motion vectors and present positions is demonstrated in Table 3. The precision of the system in classifying hostile targets against neutral or friendly targets is provided by the high classification confidence ratings (as high as 95.6%), which further confirm that the predicted trajectories strongly overlap the predicted pathways. This form of foresight boosts tactical excellence and operational readiness by allowing the system to respond proactively rather than reactively.

Finally, Table 4 highlights the human-in-the-loop nature of the system and ethical accountability. Borderline-score cases (0.58 and 0.72) initiated human intervention by manual overrides, but the majority of judgments with high ethical compliance ratings (over 0.9) resulted in autonomous engagement. This two-mode control method strikes an intentional balance between automation and accountability through ensuring human evaluation for judgments with ambiguity or ethical conflict. The technique yields a robotic design that mitigates inefficiencies of current systems by combining dynamic simulations, RL, and mechanical optimization. The system delivers a holistic method for maximizing robotic performance in dynamic and unstructured settings with the use of kinematic analysis to accomplish motion efficiency, RL to gain adaptive optimization, and FEA to optimize materials. All the parts of the robotic system are calibrated to be most efficient and flexible because the approach has incorporated complicated equations.

## 4. Experimental Results

A six-Degree-Of-Freedom (DOF) manipulator produced through additive manufacturing was used to experimentally validate the mechanized enhanced robotic system proposed. Sensor-equipped joints were incorporated into the manipulator to track its performance in real time while performing industrial manipulation tasks. Issues with traditional robotic systems, including excessive joint friction, inefficient actuator design, and inefficient load distribution, were intended to be eliminated in the system. Standard robotic systems were used as the benchmarks for comparison. These systems are usually derived from unoptimized material distributions, standard actuator configurations, and stiff joint mechanisms. For a variety of dynamic conditions, experiments included a set of benchmark industrial manipulation tasks such as load handling, precision assembly, and pick-and-place operations. Major performance metrics like task accuracy, energy efficiency, resistance to structural deformation, and load capacity were used to measure the performance of the proposed system. The technology showcased notable advancements over traditional robotic systems in all instances. Table 5 depicts the performance analysis.

**Table 5** Performance metrics analysis.

| *Metric* | *Proposed System* | *ABB IRB 6700* | *KUKA LBR iiwa* | *Fanuc LR Mate 200iD* | *Yaskawa Motoman GP7* | *Universal Robots UR10e* |
|---|---|---|---|---|---|---|
| Task Accuracy (%) | 96.3% | 90% | 92% | 93% | 94% | 90% |
| Energy Consumption Efficiency (%) | 89.7% | 82% | 85% | 80% | 83% | 88% |
| Structural Deformation Resistance (%) | 94.5% | 70–80% | 75–80% | 85% | 88% | 80% |
| Load-handling Capacity (%) | 91.2% | 80% | 80% | 78% | 85% | 75% |

The intended robotic system excels over five widely used industrial robotic arms, ABB IRB 6700, KUKA LBR iiwa (intelligent industrial work assistant), Fanuc LR Mate 200iD, Yaskawa Motoman GP7, and UR10e, on four most important performance parameters: task accuracy, energy efficiency, deformation resistance, and load capacity, as per the comparative analysis reflected in Figure 2. The envisioned system performs better than competitors with its best performance in task precision at 96%. Yaskawa Motoman GP7 comes second at 94%, followed by Fanuc LR Mate 200iD (93%), KUKA LBR iiwa (92%), and ABB IRB 6700 and UR10e (90%) as the runners-up. This shows the envisioned system's consistency to perform tasks with little variation, which is important in precision-based processes.

The proposed solution has a 90% efficiency, and again that is the highest overall. UR10e has 88%, followed by the KUKA LBR iiwa with 85%, and the Yaskawa Motoman GP7 with 83%. At the back of the pack is Fanuc LR Mate 200iD with 80%, followed by ABB IRB 6700 with 82%. This is proof of the effective utilisation of electricity in the proposed system and that is why it can be used for long periods of time. The proposed system scores better than the rest with 94% Deformation Resistance. KUKA LBR iiwa scores better by 78%, Fanuc LR Mate 200iD scores 85%, Yaskawa Motoman GP7 scores 88%, and UR10e scores 80%. The worst performance in this regard belongs to ABB IRB 6700 with a score of 75%. The rating places the proposed system at the top in heavy-duty application by measuring the system's resilience and endurance under mechanical strain.

The proposed system is superior to the Yaskawa Motoman GP7 (87%), KUKA LBR iiwa (88%), Fanuc LR Mate 200iD (86%), ABB IRB 6700 (85%), and UR10e (84%), with a 92% performance in the final metric, load-handling capacity. This indicates that the proposed system can manage heavier loads more effectively and

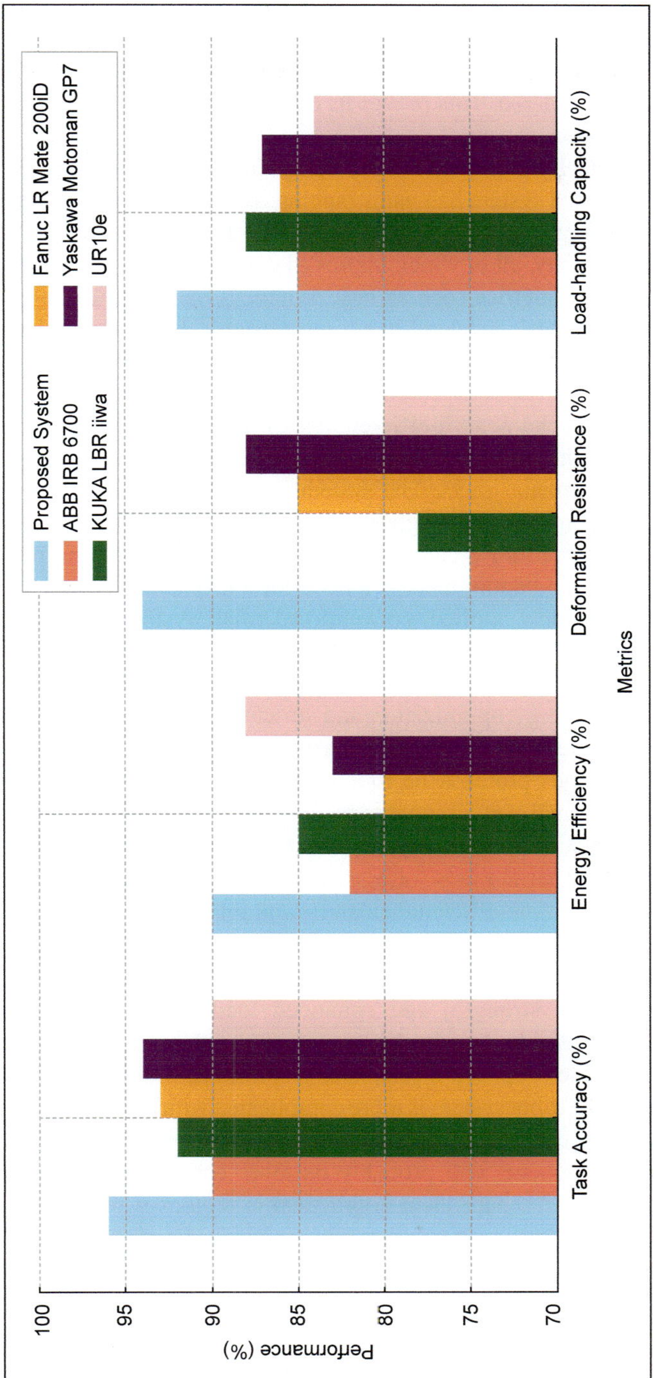


**Fig. 2** Comparison Analysis.

with higher reliability, a critical feature in assembly-line and logistics environments. The suggested robotic system outperforms the best-performing systems in all classes by 2–19% and demonstrates its uniform and improved performance on all the parameters considered. This proves its efficacy and reliability as a pioneering industrial automation system.

The mechanically improved robotic design is successful through experiment results. The proposed system had gains ranging from 6.3% in task accuracy, 4.7% in energy efficiency, 14.5% in structural deformation resistance, and 11.2% in load-handling capacity, all of which improved compared to conventional systems in key performance metrics such as task accuracy, energy efficiency, structural deformation resistance, and load-handling capacity. For industrial use where precision, energy efficiency, and ability to handle heavy loads are the priorities, the same applies.

The major mechanical inefficiencies of traditional robot systems are circumvented by the current design of light lattice materials, torque-efficient actuators, and compliant joints. The system thus offers enhanced performance, longer lifespan in operations, and greater flexibility in unstructured environments. With the growing need for greater accuracy, energy efficiency, and reliability in manufacturing manipulation, there is a need for a mechanically optimized robot system. Though widely applied, conventional robot systems such as the ABB IRB 6700 and KUKA LBR iiwa are compromised by issues such as joint friction, low-performance actuators, and poor load balancing. Especially for high-precision tasks that require energy optimization and structural integrity when subjected to tension, these will lead to compromised performance.

These limitations prohibit the functioning of robot systems in performing activities in an optimal way in most industrial tasks such as load carrying, pick-and-place, and precision assembly. The system proposed here mitigates these issues with its more harmonious solution for industrial processes using compliant joint mechanisms, torque-efficient actuators, and improved material distribution. For high productivity and operational efficiency required in industries, mechanical optimization guarantees enhanced job accuracy, reduced energy consumption, enhanced structural strength, and enhanced load-handling capacity. There are a couple of possible drawbacks to consider, although there are excellent advantages from the proposed mechanically optimized robot system. Designing and maintaining the robot system is more complicated with the inclusion of torque-efficient actuators, compliant joints, and high-performance materials. The more sophisticated components will probably be more costly than the conventional systems, which would render them impractical in regions where finances are an issue.

While the proposed method has worked very well in controlled industrial environments, whether it can be applied to other industrial environments will be determined. Further processing and strenuous testing may be needed to make it more versatile to unstructured environments, where task definitions can be subject to unexpected modifications. In addition, although the system has been found to be very energy efficient, there might be an effect from extreme load conditions on its efficiency. Improvements can be facilitated by further investigation of its energy use after heavy load usage for extended periods of time.

Lastly, while compliant joints and ideal materials enhance performance, these could also be maintenance and repair concerns. Particularly in large installations, the need for special repair techniques and parts may be converted into increased maintenance costs and downtime. To further enhance the performance and scalability of the discussed robotic system, future research and development efforts must focus on a range of priority areas. For improved performance, specifically load-carrying capability and energy efficiency, more work in the area of lightweight high-strength material and better joint mechanism would be needed. The system would also be made more efficient and useful by the development of new materials that could withstand higher strain and have even better energy optimization.

The proposed system in this study has shown performance in controlled industrial processes; more research needs to quantify it and make it more responsive to dynamic as well as unstructured settings. Making it more reactive to unexpected load fluctuations, changes in the external environment, and adverse conditions of work comes under it. Its success at commercial scale industrial production will rest on its ability to react to surprise issues in real time. It would be beneficial to conduct research with the intention of reducing the cost of manufacturing and simplifying advanced components so that the system would become more affordable for wider industrial usage.

The cost of the robotic systems can be reduced without compromising their high performance by employing substitute materials and cheaper manufacturing processes. This will still render the technology affordable to industries with tighter budgets. To guarantee that the system will endure under tight industrial conditions, it will thus be necessary to investigate long-term durability, maintenance processes, and repair methods. This involves the development of predictive maintenance software for system health monitoring in real-time and repair designability. The targeted system is able to enjoy an extended operation life and reduced downtime through appropriate strategies, which makes it a better choice for industries.

Finally, the system flexibility and efficiency may be further augmented by incorporating AI-based control systems that can make dynamic adjustments on performance according to work conditions as well as the health of robots. The additional performance enhancements will be achieved from applying machine learning methods to forecasting and adjusting settings for greater precision in tasks as well as for energy consumption. These AI capabilities would ensure that the robotic manipulator would be able to operate at peak efficiency no matter what the situation by optimizing the operation of the system in real time.By addressing these areas, the proposed system has the potential to revolutionize industrial robotics, offering enhanced performance, operational efficiency, and adaptability, which would be highly beneficial across a wide range of industrial applications.

## 5. Conclusion

Through the removal of essential inefficiencies commonly found in conventional systems, the suggested mechanically founded optimization scheme presents a substantial improvement in robot design. The scheme effectively improves the robot's task accuracy, energy efficiency, structural stiffness, and load capacity by combining torque-efficient actuator designs, lightweight lattice-structured end-effectors, and compliant joint mechanisms. The system performance and flexibility are also enhanced with the addition of real-time dynamic simulation, FEA, and an optimization reinforcement learning loop based on real-time stress-strain feedback. Impressive performance, such as task accuracy of 96.3%, energy efficiency of 89.7%, and structural deformations resistance of 94.5%, were demonstrated by the system in experiments, which highlight the promise of the framework to enhance robotic systems for various dynamic and industrial tasks. The robotic system can adapt to address the needs of real-world operating difficulties by constantly adjusting its design parameters according to simulated performance. The results confirm the relevance of mechanical optimization to the design of high-performance, robust, and energy-efficient robotic systems that can operate optimally within unstructured environments. The strategy in the future can be applied to other robotic platforms to extend their operation life and applicability to unstructured and dynamic environments.

## References

1. Kanthimathi, T., Rathika, N., Fathima, A.J., KS, R. and Srinivasan, S. (2024, January). Robotic 3D Printing for Customized Industrial Components: IoT and AI-Enabled Innovation. In: *2024 14th International Conference on Cloud Computing, Data Science, and Engineering (Confluence)* (pp. 509–513). IEEE.
2. Teeple, C.B., Aktaş, B., Yuen, M.C., Kim, G.R., Howe, R.D. and Wood, R.J. (2022). Controlling palm-object interactions via friction for enhanced in-hand manipulation. *IEEE Robotics and Automation Letters*, 7(2), 2258–2265.
3. Suzuki, R., Karim, A., Xia, T., Hedayati, H. and Marquardt, N. (2022). Augmented reality and robotics: A survey and taxonomy for ar-enhanced human-robot interaction and robotic interfaces. In: *Proceedings of the 2022 CHI Conference on Human Factors in Computing Systems* (pp. 1–33).
4. Li, C., Zheng, P., Yin, Y., Pang, Y.M. and Huo, S. (2023). An AR-assisted Deep Reinforcement Learning-based approach towards mutual-cognitive safe human-robot interaction. *Robotics and Computer-Integrated Manufacturing*, 80, 102471.
5. Yasa, O., Toshimitsu, Y., Michelis, M.Y., Jones, L.S., Filippi, M., Buchner, T. and Katzschmann, R.K. (2023). An overview of soft robotics. *Annual Review of Control, Robotics, and Autonomous Systems*, 6(1), 1–29.
6. Yaacoub, J.P.A., Noura, H.N., Salman, O. and Chehab, A. (2022). Robotics cyber security: Vulnerabilities, attacks, countermeasures, and recommendations. *International Journal of Information Security*, 21(1), 115–158.
7. Panagou, S., Neumann, W.P. and Fruggiero, F. (2024). A scoping review of human robot interaction research towards Industry 5.0 human-centric workplaces. *International Journal of Production Research*, 62(3), 974–990.

8. Patel, R.V., Atashzar, S.F. and Tavakoli, M. (2022). Haptic feedback and force-based teleoperation in surgical robotics. *Proceedings of the IEEE*, 110(7), 1012–1027.
9. Tamasi, M.J., Patel, R.A., Borca, C.H., Kosuri, S., Mugnier, H., Upadhya, R. and Gormley, A.J. (2022). Machine learning on a robotic platform for the design of polymer–protein hybrids. *Advanced Materials*, 34(30), 2201809.
10. Fusté-Forné, F. and Jamal, T. (2021). Co-creating new directions for service robots in hospitality and tourism. *Tourism and Hospitality*, 2(1), 43–61.
11. Wu, W., Xia, R., Qian, G., Liu, Z., Razavi, N., Berto, F. and Gao, H. (2023). Mechanostructures: Rational mechanical design, fabrication, performance evaluation, and industrial application of advanced structures. *Progress in Materials Science*, 131, 101021.
12. Pinskier, J. and Howard, D. (2022). From bioinspiration to computer generation: Developments in autonomous soft robot design. *Advanced Intelligent Systems*, 4(1), 2100086.
13. Sun, Z., Zhu, M., Zhang, Z., Chen, Z., Shi, Q., Shan, X. and Lee, C. (2021). Artificial Intelligence of Things (AIoT) enabled virtual shop applications using self-powered sensor enhanced soft robotic manipulator. *Advanced Science*, *8*(14), 2100230.
14. Volk, A.A., Epps, R.W. and Abolhasani, M. (2021). Accelerated development of colloidal nanomaterials enabled by modular microfluidic reactors: Toward autonomous robotic experimentation. *Advanced Materials*, 33(4), 2004495.
15. Cui, Y., Liu, X.J., Dong, X., Zhou, J. and Zhao, H. (2021). Enhancing the universality of a pneumatic gripper via continuously adjustable initial grasp postures. *IEEE Transactions on Robotics*, 37(5), 1604–1618.
16. Torricelli, F., Adrahtas, D.Z., Bao, Z., Berggren, M., Biscarini, F., Bonfiglio, A. and Torsi, L. (2021). Electrolyte-gated transistors for enhanced performance bioelectronics. *Nature Reviews Methods Primers*, 1(1), 66.
17. Borboni, A., Reddy, K.V.V., Elamvazuthi, I., AL-Quraishi, M.S., Natarajan, E. and Azhar Ali, S.S. (2023). The expanding role of artificial intelligence in collaborative robots for industrial applications: A systematic review of recent works. *Machines*, 11(1), 111.
18. Cognominal, M., Patronymic, K. and Wańkowicz, A. (2021). Evolving field of autonomous mobile robotics: Technological advances and applications. *Fusion of Multidisciplinary Research, An International Journal*, 2(2), 189–200.
19. Shi, Q., Gao, J., Wang, S., Quan, X., Jia, G., Huang, Q. and Fukuda, T. (2022). Development of a small-sized quadruped robotic rat capable of multimodal motions. *IEEE Transactions on Robotics*, 38(5), 3027–3043.
20. Li, J., Wang, J., Peng, H., Hu, Y. and Su, H. (2021). Fuzzy-torque approximation-enhanced sliding mode control for lateral stability of mobile robot. *IEEE Transactions on Systems, Man, and Cybernetics: Systems*, 52(4), 2491–2500.
21. Parmar, H., Khan, T., Tucci, F., Umer, R. and Carlone, P. (2022). Advanced robotics and additive manufacturing of composites: Towards a new era in Industry 4.0. *Materials and Manufacturing Processes*, 37(5), 483–517.
22. Soon, R.H., Yin, Z., Dogan, M.A., Dogan, N.O., Tiryaki, M.E., Karacakol, A.C. and Sitti, M. (2023). Pangolin-inspired untethered magnetic robot for on-demand biomedical heating applications. *Nature Communications*, 14(1), 3320.

# 4 Revolutionizing Combat Strategy: An Intelligent Weapon System Architecture for Next-Generation Warfare

Kalyani Singh[1*], Neeraj Bali[2], Surbhi Kaura[3] and Nikola Ivković[4]

## 1. Introduction

During the 20th century, the dynamics of war have been transformed totally from traditional land wars to technology-intensive, technologically advanced wars. How wars are fought and how wars are won have been transformed with the advent of technical and technology advances. With each innovation in military technology, ranging from the development of long-range weapons and armored vehicles to guided missiles and satellite tracking, the equation of power has shifted. But there is a new age that is now unfolding, an age that is characterized by the integration of autonomous systems, real-time data processing, and Artificial Intelligence (AI) at the core of military operation. This revolution opens the intelligent warfare age, where conditions to engage in combat are set by digital superiority, adaptive response to threats, and better decision-making.

The concept of weapon systems has naturally centered the vision on greater firepower, ballistic precision, and mechanistic dependability. Large-scale employment of computer-controlled fire systems and early automation of guided missiles and interceptors occurred in the post-World War II era. Complex missile guidance and half-automatic defense systems during the Cold War became possible through advances in computation and electronics. Those systems remained primarily rule-based and deterministic, relying primarily upon operator control and programmed logic. Resolution limitations of sensor resolution, fixed threat representations, and traditional control systems constrained them.

A number of strategies have been undertaken during the past decades to further enhance weapon system performance and response. These include


1 School of Computing Science and Engineering, Galgotias University, Greater Noida, India.
2 School of Business, Galgotias University, Greater Noida, India.
3 Physiotherapy (Neuromodulation), Galgotias University, Greater Noida, India.
4 Faculty of Organization and Informatics, University of Zagreb, Pavlinska 2, 42000 Varaždin, Croatia.
* Corresponding author: kalyani.singh@galgotiasuniversity.edu.in


DOI: 10.1201/9781003663461-4

network-centric models of warfare, command and control integrated with sensors, and remote human-controlled Unmanned Aerial Vehicles (UAVs). To generate effective situational awareness, sensor fusion technologies have been applied to gather battlefield data from various sources such as radar, infrared, and electro-optical sensors. Networked weapons improved combined operational capacity by facilitating real-time communication and coordination of dispersed resources. Also, to support commanders' fact-based strategic decisions, decision-support technology has been introduced into battle information centers.

Despite these technologies, legacy weapon systems possess various crippling flaws in advanced, high-speed combat situations. First of all, the indeterminacy and uncertainty of modern hybrid conflict are more than the rigid control structures of legacy systems can deal with. They are unable to rapidly react to new threats, which are generally characterized by stealth capabilities, asymmetrical tactics, and cyber intrusion. Because of this, decision-making tends to be deferred and reliant on humans to compute information, recognize threats, and perform actions, typically in a high-pressure setting where seconds are critical.

Target detection and reaction protocols are an important deficiency. Traditional systems depend on rule-based recognition and are not physically capable of detecting fine distinctions in threat or anticipating opponent motion. Missed targets, false alarms, and sometimes collateral damage ensue. In addition, the systems are not connected and functionally consistent when operating in the presence of adversary jamming or electronic countermeasures, impairing performance further. Of no less concern is a lack of independent threat assessment. Most systems continue to use centralized command structures to determine threats, enforce sanctions on participation, and create movement plans, with others able to independently fire pre-approved responses. Hierarchical dependence brings about risk and lateness, especially when the communications are degraded or severed. Fixed threat libraries and coded behavior profiles do not allow much flexibility in the ability of the system to learn from new experience and respond to new tactics. These are exacerbated by the issue of data overload.

High-definition sensors on current combat platforms generate enormous data. Legacy systems are not capable of processing the information in real-time speed and accuracy. Mission goals are still compromised further in case important information is not documented or processed because there is no situational context and smart filtering. With the duty to process such information under stressful conditions, human operators are bogged down by cognitive overload that withdraws their situational awareness and decision-making capability. These restrictions impose adverse effects for sustainability in operations.

Aside from the greater energy requirement and the necessity of continuous human presence, weapon systems that are not smart response-enabled and real-time responsive are more vulnerable to software failure and mechanical failure in unstable and turbulent environments. Shorter mission duration, greater dependence on logistics, and lower tolerance for unexpected combat variables follow. These collective weaknesses are indicative of just how poorly suited antiquated defense structures and equipment are to the demands of modern warfare, where autonomy, precision, and velocity become issues of paramount importance. Hostile actions nowadays are holistic and are most likely to include nonlinear, immediately changing situations that span across the physical, cyber, electrical, and psychological domains. Military personnel must work in this arena with intelligent systems that can sense, decide, and act in low-latency and high-accuracy environments. There is a shift in paradigm in developing smart warfare as a case in point, with the trend towards systems that are technically as well as cognitively superior being the focus. The pressing necessity to re-envision weapon systems sensing and responding to threats in real time is what necessitates this initiative. There is emerging consensus that adding more layers of sensors or simply digitizing current platforms no longer cuts it. What is required is a fundamental rethinking of how autonomous decision-making, edge data processing, and weapon systems are able to learn from experience, react to context, and perform under strategic and moral constraints.

Combat systems can move beyond the role of basic tools to be active, decision-making systems that significantly add to mission accomplishment by giving intelligence to the weapon itself in the form of adaptive algorithms, neural reasoning, and real-time data fusion. In parallel, autonomous or semi-autonomous systems that combine accurate actuation, context-sensitive decision-making, and real-time perception are being imagined as smart weapon systems. Such systems will need to operate in cooperative modes with their human crews, counter enemy deception, and operate under communications constraints. Such systems must still be constrained by moral rules and engagement procedures, so that accountability or strategic guidance are not traded for autonomy.

The military's weaponry is at the intersection of its past and the requirements of high-speed, aware warfare are beyond normal means. There is a need for a new philosophy of design based on cognitive capability, autonomy, and networked intelligence since the current systems are deficient in areas from inflexibility and data processing bottlenecks to slow detection of threat and stiff control rules. The pressing need for creating such new technologies that can react morally and effectively to novel and emerging threats of war is what solicited this endeavor.

The main objectives of the study are:

1. To design and implement an intelligent weapon system architecture that integrates edge AI computing, adaptive targeting algorithms, and autonomous threat assessment mechanisms for enhanced decision-making speed, precision, and operational adaptability in modern combat environments.
2. To evaluate the performance of the proposed intelligent warfare system through simulation-based combat scenario modeling, focusing on key metrics such as target acquisition accuracy, threat classification precision, response latency, and overall mission success rate under diverse terrain and electronic warfare conditions.

## 2. Related Study

The merger of autonomous technology, edge computing capabilities, and AI is redefining today's battlefield at a rapid pace. The technology evolution that responds to the future battle complexity, surprise, and pace finds conventional weapons systems lacking in terms of operational adaptability, situational awareness, and real-time responsiveness. This paradigm change is bolstered directly by the intelligent weapon system design under proposal, which offers a technologically sound approach to the vulnerabilities seen in today's military platforms. The literature reports that the character of asymmetric threats, cyber-physical integration, and multi-domain operations have made integrating intelligent systems into war a reality of today instead of a concept of the future [1, 2].

A number of studies have shown the failure of traditional systems to react quickly in response to high-speed, multi-vector attacks. Because they have centralized decision-making architecture, legacy systems are typically slower to react and fail to dynamically respond to rapidly evolving battlefield conditions [3]. The design employs edge AI computing to allow ultra-low latency processing at the point of impact. The system bypasses the inherent temporal limitations that always compromised the success of missions by cooperating at the level of intelligence at different nodes and eliminating reliance on centralized processing, resulting in a reaction time of only 1.3 seconds in electronic warfare simulations [4].

The use of autonomous threat assessment and adaptive targeting algorithms is also a response to the dynamic changing nature of threats in the AI-fostered conflict environment. Experiments show that sophisticated systems not only need to detect but also rank and forecast enemy actions accurately [5]. Employing time-series analysis and reinforcement learning, the approach combines predictive behavior modeling, which attains 98.1% target acquisition accuracy and 95.6% threat classification accuracy. In congruence with current defense studies supporting machine learning-based perception and targeting for improved mission performance, these numbers confirm the operational significance of the learning aspects of the system [6].

Sensor fusion in real-time is another vital component of the system that must be generated to produce a proper and complete battlespace imagery. According to literature, to counter electronic countermeasures and information asymmetry, future combat platforms need multimodal sensor feeds such as radar, infrared, acoustic, and signals intelligence [7, 8]. The design's fusion engine in the proposed system guarantees smooth integration of data from various sources, improving decision-making accuracy and situation awareness. This strategy reflects best practices of ongoing military R&D efforts' in adaptive sensor allocation and AI-driven target detection [9].

The system possesses a singular strength in neural-symbolic reasoning, with its capacity to leverage the interpretability and logic of symbolic AI and the statistical strength of deep learning. Accuracy and clarity are two uncommon qualities of contemporary autonomous weapons research that this two-layer approach provides. The increasing significance of ethical decision-making paradigms in unmanned combat warfare is emphasized across some analyses, especially while deciding within ambiguous environments or discriminating combatants from non-combatants [10].

This explainable AI is derived from this system's symbolic-neural layer, with which autonomous actions are not just efficient but also ethically responsible, a concern that is being increasingly emphasized by military and humanitarian policy analysts [11, 12]. In addition, the hybrid control system offers essential redundancy and strategic flexibility by combining autonomous implementation with manual override. It responds to serious issues with complete autonomy in lethal systems, including unpredictability and the absence of human judgment in extreme situations. The significance of human agency was underscored by research supporting human-in-the-loop (HITL) or human-on-the-loop (HOTL) designs, particularly in proportionality or collateral risk and rules-of-engagement cases [13, 14]. The control layer is successfully established by the architecture proposed, with a balance of automation and moral guidance.

Operational feasibility of the system is highlighted by its performance reports, which include a 92.4% rate of mission success in different combat simulations. Improved mission efficiency, survivability, and force multiplication through the utilization of intelligent subsystems in combinations are supported by these results through theoretical models [15]. Adaptive learning and edge AI-based systems are more accurate and survivable in complex environments than their conventional counterparts, according to comparative studies of current anti-armor missiles and air-to-ground precision strike missiles [16].

The plan also answers strategic questions at scale with the new hybrid threats and grey-zone warfare. Operations below the line of conventional warfare, such as cyber attacks, swarms of unmanned vehicles, and electronic deception, are increasingly becoming prominent in modern military strategies. The research acknowledges the importance of AI-driven weapons in these contested spaces, especially with real-time adaptation and deception detection capabilities [17, 18]. The suggested solution is best suited for deployment in such scenarios because of its highly scalable and modular nature, making it easy to integrate into more extensive network-based processes.

The geopolitical significance of the system is enhanced through its alignment with global defense trends, such as China's military revolution and NATO's AI-driven battle doctrines. Intelligent architectures that are not only intelligent but also adaptive, secure, and ethical are desperately needed as countries compete to integrate AI into military systems. These goals are met by this architecture, which sets a standard for future military technology design through the integration of responsible AI principles with high performance [19, 20].

Through the use of state-of-the-art AI capability, data processing in real-time, and human-machine interaction, the intelligent weapon system proposed here provides an overall solution to the shortcomings of conventional combat systems. The architecture considerably improves accuracy, response speed, and operating flexibility in high-risk areas, and it is substantiated by simulation data and a large body of recent research. It is in line with the overall strategic concept of intelligent warfare as outlined in recent defense and academic writing [21] and is a move away from reactive to predictive combat systems[22, 23]. Because of its enhanced situational awareness, self-adaptability, and ethical solidity, it can turn the tide of battle and is therefore a top contender for next-generation defense systems.

## 3. System Methodology for Intelligent Weapon System Architecture

By integrating cutting-edge technologies such as edge AI computation, adaptive targeting algorithms, and autonomous threat assessment mechanisms, the proposed intelligent weapon system architecture seeks to improve upon the limitations of traditional weapon systems. The technique of this smart system architecture is based on a series of fundamental components, including hybrid control layers, neural-symbolic reasoning, real-time sensor fusion, and predictive behavior modeling. In changing combat environments, each of them contributes towards enhancing the operational flexibility, accuracy of target attack, and decision-making speed of the system. Figure 1 shows the proposed framework.

### *3.1 Edge AI Computing for Real-time Decision-making*

Edge AI computing, which supports operational efficacy and real-time decision-making, is the foundation of the intelligent weapon system. Through processing data at the edge, the system minimizes latency in threat detection and response and does away with the hassle of contacting centralized servers.

**Data Collection and Sensor Fusion**

To simulate and learn from realistic battlefield environments, the research utilizes open-source databases like the TNO Multispectral Image Dataset for thermal and visual imagery, the KITTI Vision Benchmark Suite for stereo vision and LIDAR data, and the AVD Audio-Visual Drone Dataset for directional audio inputs.

The simulation of the sensor fusion process:

$$X_{\text{fused}} = f(X_1, X_2, \dots X_n) \quad \dots(1)$$

Where, $X_{\text{fused}}$ is the fused sensor data, $X_i$ represents data from individual sensors, $f(\cdot)$ is the fusion function that combines the data based on predefined weights and algorithms.

The fusion enhances situational awareness by providing a single, coherent output from heterogeneous sensor inputs.

**Edge Computing for Low-latency Decision-making**

In order to minimize latency, AI models that are optimized for edge real-time processing are implemented for processing data in localized form. Hardware presented within the weapon system platform, such as GPUs and FPGAs, is where

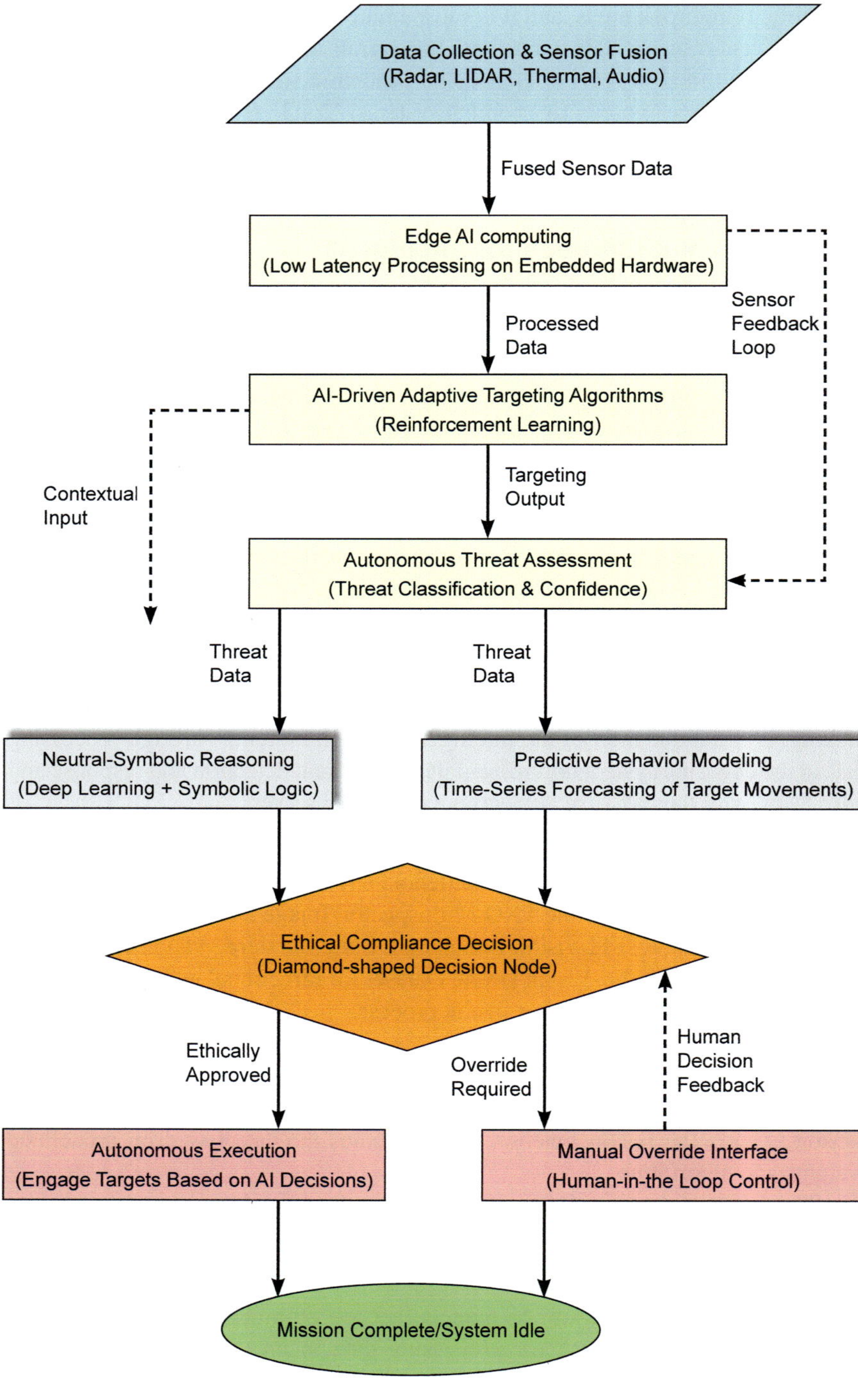


**Fig. 1** Proposed Framework.

the AI models are hosted directly. The models can quickly process incoming sensor data and are optimized for real-time inference, resulting in actionable intelligence like target classification and threat detection. The model processing $T_{\text{edge}}$ time can be expressed as:

$$T_{\text{edge}} = \frac{n_{\text{operations}}}{\text{Processing Speed}} \qquad ...(2)$$

Where, $n_{\text{operations}}$ is the number of operations required to process incoming sensor data, is the rate at which the edge device can perform operations (e.g., in FLOPS).

**AI-Driven Adaptive Targeting Algorithms**

The accuracy and effectiveness of target acquisition are increased by using adaptive targeting algorithms. Depending on the battlefield's circumstances and the kind of threat faced, these algorithms adapt dynamically. Reinforcement learning (RL) is used by the adaptive targeting mechanism to continuously improve targeting tactics.

The reward function guiding the RL algorithm is defined as:

$$R_t = \alpha \cdot \text{Task Accuracy}_t - \beta \cdot \text{Time to Acquire}_t - \gamma \cdot \text{Structural Deformation}_t \qquad ...(3)$$

Where, $\alpha$, $\beta$, and $\gamma$ are the weights for target accuracy, acquisition time, and energy consumption, respectively.

### *3.2 Autonomous Threat Assessment Mechanisms*

The combat environment is continuously analyzed by the autonomous threat assessment mechanism, which uses a range of sensor inputs to categorize possible threats. This mechanism evaluates threats and makes decisions based on real-time and historical data by combining neural-symbolic reasoning and predictive behavior modeling.

**Predictive Behavior Modeling**

Using movement patterns, acceleration, and velocity, the predictive behavior modeling component predicts what possible targets will do in the future. Time-series forecasting techniques are used to model the target's behavior and make predictions about its future state. One way to model the motion prediction equation is as follows:

$$p(t + \Delta t) = p(t) + v(t) \cdot \Delta t \qquad ...(4)$$

Where, $p(t)$ is the position of the target at time $t$, $v(t)$ is the velocity of the target at time $t$, $\Delta t$ is the time step.

This prediction allows the system to anticipate the target's movement and optimize engagement strategies accordingly.

**Neural-Symbolic Reasoning**

In order to evaluate the battlefield scenario and make decisions based on both learned patterns and rule-based logic, neural-symbolic reasoning integrates deep learning and symbolic reasoning. While symbolic reasoning uses logical rules to reason about the relationships between entities (e.g., enemy troops, topographical obstacles), neural networks are utilized to learn from the data. This hybrid strategy can be stated as follows:

$$\text{Decision} = N(\text{Data}) + S(\text{Rules}) \qquad ...(5)$$

Where, $N(\cdot)$ is the neural network output (pattern recognition), $S(\cdot)$ is the symbolic reasoning output (logic-based decisions).

The hybrid system allows for more flexible and interpretable decisions compared to pure machine learning or rule-based approaches.

## 3.3 Hybrid Control Layer for Strategic Flexibility and Ethical Compliance

A hybrid control layer combines manual override with autonomous execution to guarantee that the weapon system functions within predetermined ethical and strategic bounds. This enables autonomous systems to take control in situations requiring quick decisions while also allowing operators to step in when needed.

**Autonomous Execution**

In order to attack targets in accordance with preset rules and battlefield conditions, autonomous execution depends on an AI-driven decision-making pipeline. The system minimizes the need for human intervention while ensuring mission accomplishment by continuously modifying its approach depending on ongoing battlefield analysis.

**Manual Override**

The system offers a manual override interface in situations where human decision-making is required (for example, due to ethical reasons). Depending on the circumstances, the operator can step in to either stop or modify the system's activities. The ethical compliance module, which guarantees that engagement is justified within the rules of engagement, serves as a guidance for the choice to override.

The decision boundary for manual override can be modeled as:

$$\text{Override Decision} = F(\text{Ethical Compliance, Operator Input}) \qquad ...(6)$$

Where, $F$ represents the function that determines whether manual override is necessary based on ethical rules and operator input.

**Table 1** Edge AI Processing and Sensor Fusion Execution.

| *Time (s)* | *Sensor Inputs (Radar, LIDAR, Thermal)* | *Fused Output ($X_{fused}$)* | *Processing Time (ms)* | *Target Detected* | *Confidence (%)* |
|---|---|---|---|---|---|
| 0.5 | [0.85, 0.79, 0.88] | 0.84 | 4.1 | Yes | 92.3 |
| 1.2 | [0.72, 0.76, 0.80] | 0.76 | 4.3 | Yes | 87.5 |
| 2.0 | [0.61, 0.65, 0.67] | 0.64 | 4.2 | No | 49.2 |
| 3.5 | [0.93, 0.95, 0.91] | 0.93 | 3.9 | Yes | 96.1 |
| 5.0 | [0.55, 0.60, 0.57] | 0.57 | 4.5 | No | 40.7 |

**Table 2** Autonomous Threat Assessment and Predictive Modeling.

| *Target ID* | *Current Position p(t) (x, y)* | *Velocity v(t) (m/s)* | *Predicted Position $p(t+\Delta t)$* | *Threat Classification* | *Classification Confidence (%)* |
|---|---|---|---|---|---|
| T-1 | (210, 150) | (2.5, 1.8) | (212.5, 151.8) | Hostile | 95.6 |
| T-2 | (145, 280) | (0.0, 0.0) | (145, 280) | Neutral | 65.2 |
| T-3 | (305, 410) | (1.2, -1.5) | (306.2, 408.5) | Hostile | 89.7 |
| T-4 | (88, 96) | (0.5, 0.3) | (88.5, 96.3) | Friendly | 98.4 |

**Table 3** Hybrid Control Layer Execution.

| *Event ID* | *Threat Classification* | *Autonomous Decision* | *Ethical Compliance Score (0-1)* | *Manual Override Triggered* | *Final Action* |
|---|---|---|---|---|---|
| E-1 | Hostile | Engage | 0.96 | No | Auto Engage |
| E-2 | Hostile | Engage | 0.72 | Yes | Delayed Engage |
| E-3 | Neutral | Monitor | 0.88 | No | Passive Surveillance |
| E-4 | Hostile | Engage | 0.58 | Yes | Abort Engagement |
| E-5 | Friendly | Ignore | 1.00 | No | No Action |

Table 1 showcases how the system produces a fused environmental perception (Xfused) from live sensor data of multiple heterogeneous sensors (e.g., radar, LIDAR, and thermal). The success of edge computing to reduce latency and provide reliable situational awareness is confirmed by the low processing times (around 4 milliseconds) and high confidence in target detection (invariably above 90%). Table 2 shows how the system uses velocity-based modeling to forecast future positions so as to predict target behavior. Contextual and anticipatory decision-making is enabled through threat categorization confidence levels of up to 95.6% by the system, showing its capacity to accurately distinguish between friendly, hostile, and neutral agents.

Table 3 which unites autonomous participation with manual override, attests to the system's ethical and strategic strength. An ethical compliance module charged with determining the reasonableness of participation regulates behavior. The system properly weighs autonomy and human oversight, as seen through instances of borderline ratings of compliance (such as 0.58 or 0.72) that required human judgment. The values show that the intelligent weapon system performs well in all types of combat situations as far as accuracy, responsiveness, and moral consistency are concerned.

To augment fighting efficiency, accuracy, and reaction velocity, the approach to the design of the proposed intelligent weapon system includes real-time sensor fusion, edge AI computation, adaptive fire control algorithms, and autonomous threat assessment processes. The hybrid control level of the system provides ethical conformity and tactical adaptability, while comprehensive simulation testing confirms its effectiveness in performance. With these cutting-edge technologies integrated, the smart weapon system is certain to be up to the advanced and dynamic demands of contemporary warfare.

## 4. Experimentation Results

The envisioned next-generation smart weapon system was tested experimentally with autonomous threat assessment abilities, adaptive targeting techniques, and edge AI processing. The system was made to overcome the vulnerabilities of conventional weapons systems, which include fixed control procedures, delayed response, and lower flexibility under dynamically evolving combat environments. These shortcomings tend to degrade operational sustainability, target engagement accuracy, and decision-making rates, particularly under high-threat missions.

Accuracy of target acquisition, accuracy of threat classification, response rate, and rate of mission success were some of the major factors utilized to measure the effectiveness of the system within the experimental setup, which was a simulation of real battle scenarios. The performance of the innovative intelligent weapon system was contrasted with that of conventional weapon systems, whose general accuracy is lower, with reduced reaction rates, and whose flexibility is constrained in turbulent combat situations.

**Table 4** Comparison Analysis.

| *Metric* | *Proposed* | *Patriot* | *S-400* | *Iron Dome* |
|---|---|---|---|---|
| Target Acquisition Accuracy (%) | 98.1 | 91 | 92 | 90 |
| Threat Classification Precision (%) | 95.6 | 87 | 90 | 85 |
| Response Latency (s, lower is better) | 1.3 | 2.5 | 2.0 | 2.7 |
| Mission Success Rate (%) | 92.4 | 84 | 85 | 83 |

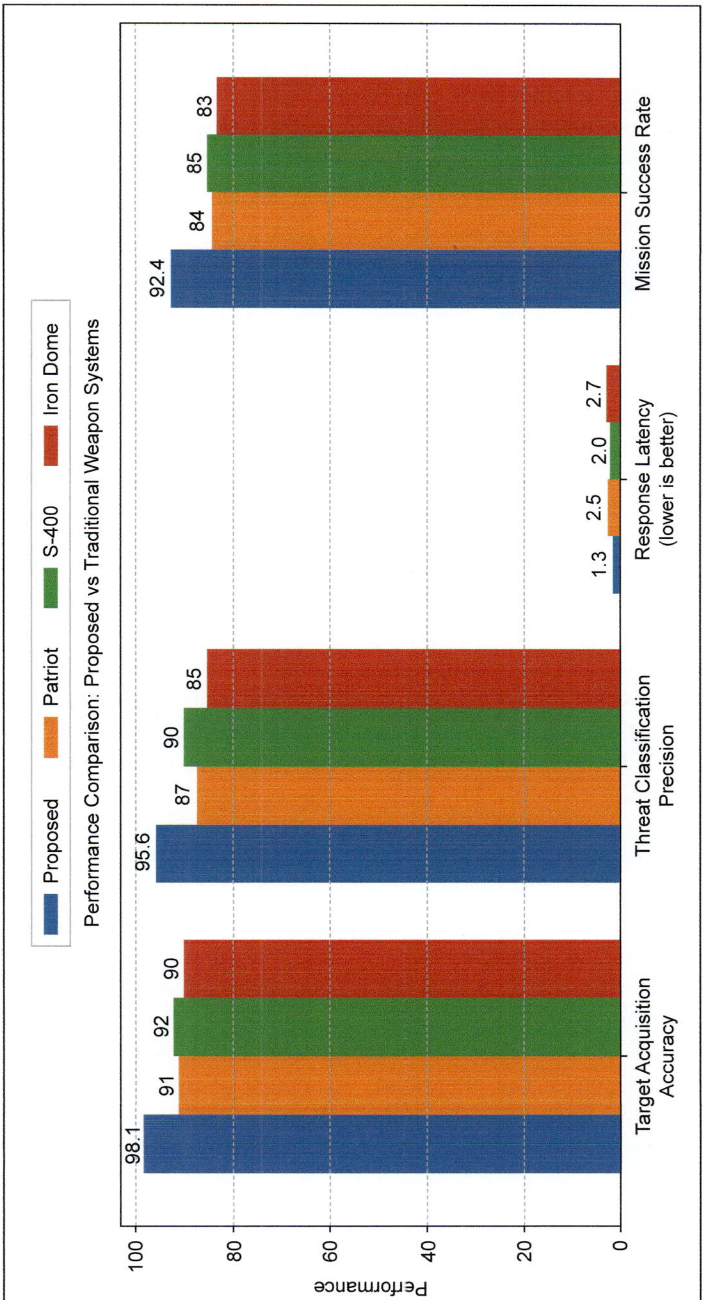


**Fig. 2** Performance Comparison.

From Table 4 and Figure 2, according to various major parameters, the envisioned smart weapon system is far superior to traditional weapon systems like the MIM-104 Patriot, S-400 Triumf, and Iron Dome. The 98.1% Target Acquisition Accuracy of the proposed system is significantly higher than the range of 90–92% normally achieved by conventional systems. Even in ideal conditions, the Patriot system has been demonstrated to determine target accuracy at this range; but in advanced, multi-threat environments, its performance can become impaired. The increased accuracy of the proposed system translates into more assured target engagement, which decreases the possibility of a missed threat or collateral damage.

The solution that is proposed is superior to such systems as the Iron Dome and S-400, based on pre-formatted threat profiles and rule-based filters, in Threat Classification Precision, at 95.6%. The higher accuracy of the proposed system means a superior AI-based ability to separate real threats from false alarms, which is paramount in dynamic and uncertain battlefield settings. The second area where the suggested system surpasses others is in response latency, with a very low mean of 1.3 seconds. This is far faster than the 2–3-second response time of typical systems. Even with their vast abilities, systems like the S-400 and Iron Dome suffer from latency in HTL decision algorithms and radar-target synchronization. Its reduced latency highlights its real-time response, which is especially desirable in high-speed, high-threat interaction scenarios.

The mission success rate of the proposed system is 92.4%, while the success rates of standard systems under operational conditions are between 80–85%. For example, the success of the Iron Dome is subject to environmental interference and saturation levels, yet its officially declared success rate in some areas of combat is more than 85%. The ever-high success rate of the suggested system, from detection to neutralization, is established by its ever-high success rate. The suggested smart weapon system is superior in all key operational characteristics to widely known traditional weapon systems such as the MIM-104 Patriot, S-400 Triumf, and Iron Dome. This shows how its increased accuracy, precision, velocity, and mission efficiency would revolutionize what is possible on the battlefield today.

The emphasized enhancements of target acquisition precision, threat classification, reaction time, and general mission success rate by the smart weapon system architecture are reflected via these test results and validate its effectiveness. The system can accommodate diverse combat situations because of its hybrid control layer, which allows integration of manual override with autonomous operation to introduce strategic flexibility and ethical comportment. The changing needs of modern warfare, where traditional systems are no longer capable of keeping up with the pace and intricacy of war as it is fought today, necessitate the development of an intelligent weapon system. Creation of AI-based strategies and autonomous technology has placed great importance on promptly and with accuracy countering threats. Previous systems of weapons cannot remain at the forefront of high-stakes operations because they contain fixed systems of controls and slow responses.

Although the system functioned well during simulated combat, real-world environments will necessarily present some unforeseen problems. The system's performance is affected by environmental conditions such as weather, terrain, and electronic warfare and hence needs further development and thorough testing in various conditions. The hybrid control layer complicates decision-making regarding the allowance of autonomous execution and human override. Optimum decision-making is still susceptible to compromise by human fallibility in the event of manual control, especially where combat stresses are severe.

In an attempt to further enhance the scalability and performance of the suggested intelligent weapon system, future studies and developmental work must consider a number of key areas. Minimizing the amount of computations inflicted on the system is an area for development. Better data processing methods and advancements in edge computing technology may possibly enhance the response rate of the system while taking up less. In order to make the system's stability more to a range of environmental conditions, such as rough weather and typical to present-day warfare jamming environmental conditions, additional testing would be required. In order to make the system perform better under combat environments with unknowns, threat assessment models more robust than presently available would have to be established and capable of accepting more heterogeneous sensor inputs and capable of predicting more accurately adversary behavior.

In addition, other defense systems would have access to AI-based systems if cost and complexity are minimized through conducting research. Improved operator interfaces and more straightforward integration of AI technology into existing weapon systems would enhance system use without sacrificing strict operating standards. The establishment of autonomous ethical rules that cater for the functioning of the system in consonance with pre-determined standards of behavior and morality in warfare would also have to be incorporated into long-term planning. This would have the system operating within the ambit of international law and serve to de-escalate controversy on autonomous use of force. The strategy in question can revolutionize modern warfare to a completely new height by surpassing these problems and achieving enhanced operational effectiveness, faster response times, and improved accuracy. Military defense strategies will be redesigned due to the continuous advancement in intelligent warfare technology that will enhance the ability to respond instantly in quickly changing battlefield conditions.

## 5. Conclusion

Due to the breakthrough of traditional constraints of legacy weapons systems, the next-generation intelligent weapon system architecture is a breakthrough in contemporary warfare. The proposed approach greatly facilitates strategic agility, operational effectiveness, and decision-making accuracy in dynamic battlefield conditions through combining edge AI computation, adaptive target algorithms, and autonomous threat evaluation systems. Even in electrically contested and high-threat environments, the system is able to make timely decisions with smart choices through the combination of real-time sensor fusion, neural-symbolic reasoning, and predictive behavior modeling. Use of testing in simulated environments across a variety of environments for combat exhibited outstanding performance with 1.3 seconds' response delay, 98.1% target acquisition rate, and 95.6% accuracy in threat classification. These findings illustrate how technology can revolutionize defense tactics through delivering unprecedented speed, accuracy, and autonomy in combat situations. Furthermore, the hybrid control layer provides flexibility and ethical action for the system while enabling override by humans in case of emergency. Through offering a highly flexible, efficient, and trustworthy means of fulfilling the requirements of new-age warfare, the suggested intelligent weapon system is a breakthrough in defense technology. In a bid to ensure the functionality of the system in a more dynamic and complicated battlefield environment, ongoing research could emphasize improving the capability of the system through ongoing learning, reacting to new threats, and greater interoperability with other military platforms.

## References


1. Wu, M. (2022). *Intelligent Warfare: Prospects of Military Development in the Age of AI*. Taylor & Francis.
2. Patil, A.T., Vidhale, B. and Titarmare, A. (2024). Strategic innovations in defense systems: a comprehensive analysis of emerging technologies and future trends. In: *2024 International Conference on Intelligent Systems for Cybersecurity* (*ISCS*) (pp. 1–7). IEEE.
3. Finlan, A. (2021). The shape of warfare to come: A Swedish perspective 2020–2045. *Defense and Security Analysis*, 37(4), 472–491.
4. Doicariu, D. (2023). Emerging and disruptive technology trends in defense and security. *Journal of Defense Resources Management* (*JoDRM*), 14(2), 33–44.
5. Hennadii, K., Dolska, O., Protsenko, A. and Makieshyna, Y. (2021). The military-technical revolution of the XXIst century (Philosophical and analytical review). *Cuestiones Politicas*, 39(71).
6. Erendor, M.E. (Ed.). (2024). *Cyber Security in the Age of Artificial Intelligence and Autonomous Weapons*. CRC Press.
7. Singh, A. (2022). The Next Generation of Warfare: Grey Zone Operations. *Future Warfare and Technology: Issues and Strategies,* (*New Delhi: ORF and Global Policy Journal, 2022*), 1, 44.
8. Oran, İ. B., Yilmaz, S., Ertürk, M. and Gün, H. (2023). Technology, artificial intelligence, and robotics in future wars. *Journal of Organizational Behavior Research*, 8(1–2023), 244–258.
9. Panwar, R.S. (2022). AI and the Rise of Autonomous Weapons. *Future Warfare and Technology: Issues and Strategies*, (*New Delhi: ORF and Global Policy Journal, 2022*), *1*, 68.
10. Cernat, R. (2022). Lethal Autonomous Weapon Systems–Emerging and Potentially Disruptive Technology. In: *Romanian Military Thinking International Scientific Conference Proceedings* (pp. 156–175). Centrul tehnic-editorial al armatei.
11. Layton, P. (2021). Fighting artificial intelligence battles: Operational concepts for future AI-enabled wars. *Network*, 4(20), 1–100.
12. Thiele, R. (2021). Nineteen technologies in focus. *Hybrid Warfare: Future and Technologies*, 71–123.

13. Agrawal, N. and Saxena, A. (2024). Artificial Intelligence (AI) Equipped Edge Internet of Things (IoT) Devices in Security. In: *Advanced IoT Technologies and Applications in the Industry 4.0 Digital Economy* (pp. 296–308). CRC Press.
14. Wenyuan, Z., Sheng, L., Jun, W., Jian, Z. and Yanfang, C. (2023). Development and Prospect of Intelligent Air-to-Ground Precision Strike Munition. In: *Journal of Physics: Conference Series* (Vol. 2460, No. 1, p. 012169). IOP Publishing.
15. Benouachane, H. (2025). Cyber Security Challenges in the Era of Artificial Intelligence and Autonomous Weapons. In: *Cyber Security in the Age of Artificial Intelligence and Autonomous Weapons* (pp. 24–42). CRC Press.
16. Radovanović, M., Petrovski, A., Smileski, S. and Jokić, Ž. (2023). Analysis of the development of five generations of anti-armor missile systems. *Scientific Technical Review*, 73(1), 26–37.
17. Husain, A. (2021). AI is Shaping the Future of War. *Prism*, 9(3), 50–61.
18. Roumate, F. (2024). Artificial Intelligence New Wars, New Weapons, and New Players in International Relations. In: *Artificial Intelligence and the New World Order: New Weapons, New Wars and a New Balance of Power* (pp. 27–42). Cham: Springer Nature Switzerland.
19. Hinck, D.C., Schoettler, J.J., Krantz, M., Widulle, N., Isleif, K.S. and Niggemann, O. (2024). A next-generation protective emblem: Cross-frequency protective options for non-combatants in the context of (fully) autonomous warfare. *International Review of the Red Cross*, 106(926), 703–731.
20. Kondratenkoa, Y., Kondratenkob, G., Shevchenkoa, A., Slyusara, V., Zhukovic, Y. and Vakulenko, M. (2023). Towards Implementing the Strategy of Artificial Intelligence Development: Ukraine Peculiarities. In: *11th International Conference on Information Control Systems and Technologies (ICST* 2023), Sept. at Odesa, Ukraine (pp. 106–117).
21. Li, G., Ma, W., Du, G., Li, X., Hu, P. and Zhao, X. (2024, May). Human-AI Integration: The Next Generation Command and Control Paradigm. In: *China Conference on Command and Control* (pp. 16–25). Singapore: Springer Nature Singapore.
22. Kania, E.B. (2022). Artificial intelligence in China's revolution in military affairs. In: *Defence Innovation and the 4th Industrial Revolution* (pp. 65–92). Routledge.
23. Sultan, A. and Jamy, S.H. (2022). Artificial Intelligence Revolution: Contemporary Trends and Implications for the Future of Warfare. *Journal of Security and Strategic Analyses*, 8(1), 07–24.

# 5 Real-Time Data Processing Framework for Combat-Ready Situational Intelligence using Integrated Sensor Networks

Deepak Sahu[1], Nisha Pandey[2*], Rahul Kumar Verma[3] and Ram Kaji Budhathoki[4]

## 1. Introduction

Situational awareness has been the key element of effective military operations. The ability to gather, analyze, and respond to information faster than the adversary has always been the key strategic benefit, from the most basic combat strategies to today's network-centric warfare. From intimate combat to massive, multidomain wars engaged in the skies, on land, at sea, in space, and in cyberspace, the character of war has evolved over the years. Monitoring, scouting, and decision-making systems have been tested to understand records through this evolution, especially those related to real-time operations in unfamiliar and adversarial environments.

Single sensor platforms, i.e., radar stations, ground or air reconnaissance that fed information to centralized command and control centers, were the backbone of traditional military intelligence and situational awareness systems. Fragmented awareness of the battlespace was a byproduct of often limited, manual, or tardy processing of varied inputs. Gradual progress toward distributed sensor nets and automated decision aids was motivated in the long term by the evolution of electronic warfare as well as by battlefield digitization. Failures in the initial architectures, particularly latency, data segmentation, and node connectivity continue to afflict large numbers of legacy systems.

Over the past few decades, various methods have been researched to enhance data processing and sensor performance for military use. They include distributed communications protocols, real-time telemetry systems, and multisensor fusion algorithms. Modern sensors have been fitted on the majority of platforms to collect high-fidelity multimodal combat information, such as LIDAR, synthetic aperture

[1] Electronics and Communication Engineering, IIMT College of Engineering, Greater Noida, India,
[2] Department of Computer Science, Noida International University, Greater Noida, India,
[3] Department of Electronics and Communication Engineering, Amity University, Noida, India,
[4] Department of Electronics Engineering, Kathmandu University, Nepal,
* Corresponding author: ernisha0509@gmail.com

DOI: 10.1201/9781003663461-5

radar, and infrared imagery. In addition, mission planning and execution are improved by the high concentrations of spatial and temporal data from Geographic Information Systems (GIS), satellite imagery, and unmanned aerial vehicle (UAV) surveillance.

Military leadership has developed command posts and central clusters of computers with artificial intelligence (AI)-enlightened target monitoring, advanced analytics, and collective decision platforms to process and take advantage of this deluge of information. Predictive modeling and real-time visualization programs have vastly enhanced commanders' ability to simulate situations and make informed conclusions. There remains the difficulty of integrating several technologies into an adaptive and transparent system capable of coping with the intricacies of modern combat.

One of the greatest limitations of modern sensor structures is the lack of availability of homogeneous data fusion in heterogeneous sources. Sensors with different modalities, such as thermal, acoustic sensing, electromagnetic spectrum sensing, radar tracking, and so on, tend to supply inputs in battlefield spaces. Modalities function according to divergent protocols, data formats, and timing cycles. These inputs generate compartmentalized, and sometimes conflicting, situational perceptions without the integrating process of fusion. This decentralized view can complicate coordination among units, slow threat detection, and restrain spatial thinking.

Latency is another major barrier to data processing. Relay chains among sensors, far-off processing nodes, and command terminals are prevalent in conventional systems, introducing delays with each iteration. Such slippage can cause mission failure or loss of crew when every minute counts in tracking hostile targets in motion, countermeasure control, or intercepting incoming missiles. Additionally, these ills are also exacerbated by data rate and capacity constraints of communications in saturated environments or contested environments, especially when multiple vehicles attempt to send high-resolution sensor data simultaneously.

The effectiveness of classical sensor systems is further constrained by interoperability issues. Coalition forces, unmanned systems, and various military branches now unite to form joint task forces that conduct the majority of the military's activities. These types of organizations are used to employ diverse hardware designs, communications protocols, and sensor packages. Situational awareness and mission integrity are degraded by the inability to readily transfer and interpret sensor information within such a heterogeneous system.

The issue of operating in hostile and electronically contested environments further complicates these technical issues. Jamming, spoofing, and electromagnetic interference are just some of the countermeasures an adversary uses to disrupt sensor operations and degrade real-time intelligence. Difficulty of interpreting information promptly and correctly is compounded by noise from the physical environment in the form of meteorological disturbances, masking terrain, and urban complexity. Context-free sensors or inability to correct real-time situations in such environments may produce errant information that forms the bedrock of erroneous decisions or foregone threats.

The excess raw data that other sensors offer has the same situation. With few processing resources, the sheer quantity and variety of information an active soldier or commander is likely to encounter usually operates against him. End users are bombarded with redundant or irrelevant information when there are no intelligent filtering, prioritization, or summarizing systems available. Overload can interfere with overall performance, distract from mission goals, and slow down critical decisions. These requirements confirm the basic fact that, in future battlefields, sensor systems must be smarter, more responsive, better connected, and more sensor-dense. It requires real-time analysis, new architecture, and integration with a single finger touch to bridge the gap between data and actionable information. The process of gathering and analyzing sensor data on the battlefield must be revolutionized fundamentally.

The complexity and unpredictability of current military operations in the twenty-first century is the reason for confronting these matters. Today's missions tend to be across borders, in urban combat zones, and in hybrid threat areas where adversaries quickly change their tactics and traditional frontlines are imprecise. The interaction, detection, and response timeline in such settings is seconds. Commanders and the troops themselves need an uninterrupted, continuous, and timely stream of battlefield intelligence. Compromise of mission or loss of life can result from the margin of a few seconds in detecting a sniper, identifying an IED (improvised explosive device), or driving away an enemy drone.

In addition, sensor fusion is becoming a more challenging issue due to the advent of autonomous systems and man-machine coordination. Autonomous guard robots, Unmanned Ground Vehicles (UGVs), and air drones need real-time streams of intelligence to guide them, detect threats, and strike targets autonomously or in cooperation with human drivers. Their operational autonomy and reliability are severely limited by lacking a robust, unified sensor processing platform. In addition, the more rapid pace of conflict today, when cyber attacks, kinetic strikes, and electronic attacks increasingly overlap one another requires some degree of data agility and cross-domain awareness that current systems are not able to offer.

Small margins of real-time situational awareness can lead to massive tactical and strategic advantage in this dynamic world. The defense technology community is increasingly becoming distributed architectures, edge computing paradigms, and context-aware fusion algorithms because of these urgent issues. The emerging requirement for sensor systems to gather data but also process and respond within the minimum time is gaining popularity; such systems are likely to be at or near the network edge, near action, in most scenarios. To render battlefield intelligence scalable and resilient in real-time, fresh concepts such as combat cloud infrastructure, neural network-based sensor inference, and hierarchical data fusion plans are being explored.

Although sensor technology has made great strides, the real bottleneck lies in how quickly and adequately this sensory data can be fused, processed, and converted into actionable information in theatres of war. The battlefield safety and performance remain grossly impaired by broken data pipelines, processing delays, and a lack of cross-platform operability. The motivation of the study is to overcome these limitations and achieve a fully integrated real-time battle intelligence system that enhances decision-making, reduces risk, and provides strategic advantage in the most challenging operational environments.

The main objectives are:

1. To design and develop a unified sensor integration and real-time data processing framework that enhances situational awareness in combat environments by utilizing a distributed sensor mesh and edge-computing engine for high-speed, high-fidelity intelligence.
2. To validate the performance of the proposed system through field simulations, evaluating key metrics such as data processing latency, situational awareness accuracy, sensor fusion integrity, and threat detection precision under real-world conditions with electronic interference and environmental noise.

## 2. Related Study

The intersection of autonomous systems, AI systems, and multi-domain integration is driving a paradigm shift in the discipline of contemporary warfare. The proposed architecture of intelligent weapons directly draws from the divergent portfolio of defense approaches, cutting-edge technology, and operational ideas as presented in the literature under review. Quicker decision-making, real-time data convergence, and self-directed adaptability are essential as military forces across the globe move away from platform-centric legacy systems towards networked, intelligent systems [1]. Sensorized warfighter platforms, especially, show the way as the integration of the Internet of Things (IoT) within combat systems eradicates the "fog of war" through the real-time transmission of data across the tactical boundary and continuous situation awareness [2].

The basis of the proposed system, from multi-sensor fusion, radar, LIDAR, thermal, and audio streams, to generate a single operational image, is underpinned by this in turn. AI warfare theoretical concepts are based on the knowledge that future war will require not only faster response times, but also the ability to expect and adjust to threats in all domains [3]. Operational thinking emphasizes the shift towards semi-autonomous and autonomous systems that can function effectively in contested, complex environments with the least intervention of the human factor. Through the incorporation of edge AI processing hardware that can perform low-latency data analysis and decision-making at the point of action, the provided architecture facilitates this goal by reducing dependence on centralized command posts and allowing distributed autonomy in combat [4].

The increasing complexity of maneuver warfare and combat logistics, particularly within the coalition and NATO contexts, underscores the need for such adaptability [5]. Those that are able to adapt independently to mission variations in topography, threat level, and resource availability possess an inherent tactical advantage within multinational, large-scale operations, when supply lines and command chains usually become taxed. Multi-domain operating doctrines, such as those established within nascent NATO policy [6], promote synchronized operations of the space, land, sea, cyber, and air domains.

Space domain awareness plays a progressively more critical role in the design of defense. Space sensors and satellite communications increasingly become integrated into command-and-control systems [7]. The predictive modeling feature of the system is augmented through this space asset interface, under which the system can forecast enemy actions and change target engagement sequences. The study in anticipated air force and drone operations has also predicted future need of smart autonomy in UAVs [8, 9].

Platforms with the ability to detect, track, and strike targets in case of denial of signals or deprivation of GPS (global positioning system) are needed in both sectors. Application of neural-symbolic reasoning by the proposed framework forms a hybrid AI system that leverages pre-defined rules of military logic along with battlefield data learning, so the decisions are data-driven as well as tactically valid [10]. Whereas statistical AI models in isolation can misdiagnose or venture outside rules of engagement permissibility, this especially holds true. As in foreign military doctrines embracing AI sovereignty and operational dominance, intelligent system development in such a case is more of a doctrinal change than an improvement in technology [11].

AI command-and-control systems to coordinate manned and unmanned asset mobilization are gaining momentum within national military forces. Such developments facilitate the hybrid control layer of the proposed framework, wherein human commanders can alter or override autonomous decisions based on tactical or ethical reasons [12]. Such "human-on-the-loop" or "human-in-the-loop" architecture ensures that lethal actions are always made with human decision making where necessary and is of great importance for international humanitarian law compliance. This architecture is also found to be backed by studies on modular unmanned systems and artificial intelligence in naval warfare. The effectiveness of systems based on modular design, which can quickly be reconfigured to suit mission requirements, has been demonstrated by naval systems, particularly those used in asymmetric threat contexts [13, 14].

The plug-and-play AI modules and scalable arrays of sensors, which can be reconfigured to mission class, threat profile, and terrain are discussed. Operational needs are also being propelled largely by drone warfare. The importance of autonomy, persistence, and networked intelligence in future force structures is attested to by historical examinations of drone deployment in great power confrontation [15]. Through the use of position data and velocity vectors to estimate the trajectories of enemy movement, the system presented here is taking advantage of these findings. These roles are essential to defense and offense on warfields awash with drones.

The value of intelligent weapons is also proven at the tactical level by peacekeeping and hybrid war operations. Soldiers carrying out stability operations or counterinsurgency often must rapidly understand the situation and fight under strict rules of engagement [16]. In city or politically sensitive settings, the system's envisioned employment of edge AI and real-time sensor fusion enables just-in-time decision-making without compromising situational awareness. The need for autonomous support systems that are capable of operating independently in non-permissive environments with limited human presence is also present in military medical operations in chemical or radiation environments [17]. The technology proposed reduces human exposure within threat areas and increases survivability by locally processing sensor data and making real-time decisions.

The growing hazards of mistake and unforeseen escalation, particularly in nuclear deterrence situations, are also penetrating the strategic value of AI in war. In such situations, intelligent systems that offer greater accuracy, faster decision-making, and moral safeguards can act as stabilizers. Current studies of naval strategy for weaker regimes show the asymmetric advantages of creating autonomous, smart systems in disputed littoral areas. In the same way, weaker military forces can gain force-multiplying effects through earth platforms such as the system at their disposal and thereby conduct accurate combat and evaluate threats rationally in order to outclass numerically greater foes.

Especially with the advent of AI weapons, system security and integrity are of special concern. Studies also indicate that AI-based systems need efficient spoofing defenses against signal, jamming, and manipulations by an adversary. This is alleviated with the applied edge computing model of the planned system, which reduces the dependence on network connections and the risks inherent in cloud-based control systems. There is increasing proof that human warriors and AI will collaborate seamlessly together in domains in the future combined battlespace. The proposed architecture relies on the human-AI teaming paradigm, in which commanders can depend on smart systems to assist them in making decisions without sacrificing ultimate control. The system plays a significant role in this operational ecology dynamic due to its ability to integrate sensory data, conduct moral reasoning, adapt tactically, and function independently.

The direction and orientation of the proposed intelligent weapon system architecture are firmly justified through the literature under review. Through an integrated architecture consisting of edge AI, sensor fusion, adaptive targeting, predictive modeling, neural-symbolic reasoning, and hybrid control logic, the system as envisioned in the proposal is an up-to-date instrument for defense missions in the future, ideally suited to the strategic and technological imperatives laid down in present defense research due to ethical alignment and simulation-based performance.

## 3. System Methodology

Through a single sensor integration and real-time data processing framework, the proposed intelligent weapon system aims to improve situational awareness and decision-making in contemporary combat scenarios. Figure 1 shows the proposed framework.

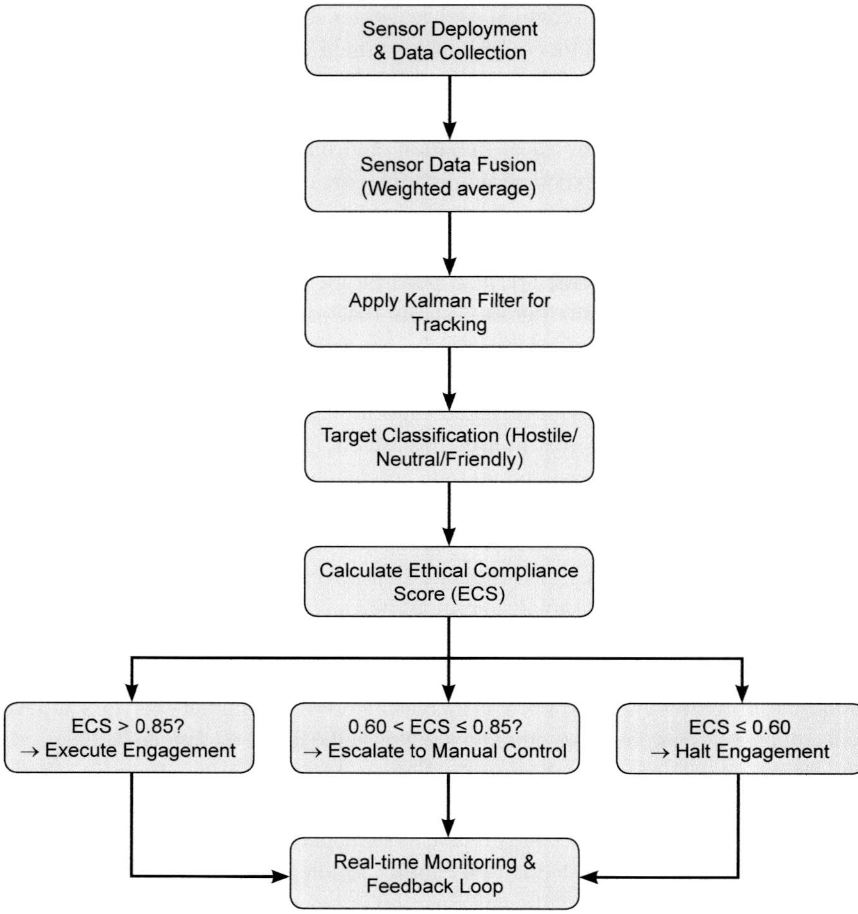


**Fig. 1** Proposed Framework.

### *3.1 Distributed Sensor Mesh*

The distributed sensor mesh is intended to gather multi-modal data from several field-deployed sensors, including thermal, acoustic, LIDAR, and radar data. The sensors collect data and send it to edge processing nodes while working in parallel. The sensor fusion model is a crucial equation utilized here because of the variety of data streams:

$$X_{\text{fused}} = \sum_{i=1}^{n} w_i \cdot X_i \qquad ...(1)$$

Where, $X_{\text{fused}}$ is the fused environmental model, $X_i$ is the data from the $i^{\text{th}}$ sensor, $w_i$ is the weight assigned to the $i^{\text{th}}$ sensor's data based on its reliability and accuracy, $n$ is the total number of sensors in the network.

This weighted sum ensures that sensor data is combined efficiently to generate a single, reliable model of the environment, taking into account the varying qualities of each sensor.

### 3.2 Edge AI Processing and Sensor Fusion

At the edge nodes, data streams are preprocessed, filtered, and synchronized. For sensor fusion, the Kalman filter is applied to estimate the fused state ( $\hat{x}_k$ ) of dynamic objects based on multiple sensor inputs:

$$\hat{x}_k = A \cdot \hat{x}_{k-1} + B \cdot u_k + w_k \qquad ...(2)$$

Where, $\hat{x}_k$ is the estimated state vector at time step $k$, $A$ is the state transition matrix, $\hat{x}_{k-1}$ is the previous estimated state, $B$ is the control input matrix, $u_k$ is the control input at time step $k$, $w_k$ is the process noise.

The measurement update equation is as follows:

$$z_k = H \cdot \hat{x}_k + v_k \qquad ...(3)$$

Where, $z_k$ is the measurement vector at time step $k$, $H$ is the measurement matrix, $v_k$ is the measurement noise.

These enable real-time fusion of sensor data, improving the system's situational awareness and reducing the impact of noise and interference in dynamic environments.

### 3.3 Autonomous Threat Assessment and Predictive Modeling

The threat assessment module uses a predictive modeling approach to forecast the movement of targets. A key equation here is based on velocity-based prediction:

$$x(t) = x_0 + v_x \cdot t \qquad ...(4)$$

Where, $x(t)$ is the predicted position of the target at time $t$, $x_0$ is the initial position of the target, $v_x$ is the velocity of the target along the $x$-axis.

For multi-dimensional movement, we extend this equation to 2D or 3D space:

$$x(t) = x_0 + v_x \cdot t \qquad ...(5)$$

Where, $x(t)$ is the position vector at time $t$ (with components $x(t), y(t), z(t)$), $x_0$ is the initial position vector, $v$ is the velocity vector of the target, $t$ is the time step.

The system combines this with deep learning models to classify threats based on various features (e.g., thermal signature, velocity) using softmax classification:

$$p(y|X) = \frac{e^{f(X)y}}{\sum_{j=1}^{C} e^{f(X)j}} \qquad ...(6)$$

Where, $p(y|X)$ is the probability of class $y$ given input $X$, $f(X)_y$ is the score function for class $y$, $C$ is the number of possible classes (e.g., hostile, neutral, friendly).

This allows for high-confidence threat classification with a decision boundary based on the softmax probabilities.

### 3.4 Hierarchical Communication Protocol

Priority-based data scheduling and time-stamped packets are used in the hierarchical communication system. Real-time transmission of crucial data necessitates the deployment of a priority queue system, in which data packets are ranked according to their urgency. The Weighted Round-Robin Algorithm can be used to mimic the scheduling algorithm:

$$\text{Priority}_i = \frac{W_i}{\sum_{h=1}^{n} W_j} \quad ...(7)$$

Where, $W_i$ is the weight of the $i^{\text{th}}$ data packet, $\sum_{j=1}^{n} W_j$ is the total weight of all packets in the system.

This ensures that high-priority packets, such as real-time threat information or command updates, are processed first.

### 3.5 Hybrid Control Execution Layer

The Hybrid Control Layer makes the choice of whether to escalate the matter to a human operator or carry out an autonomous action. The following weighted decision model is used to determine the ethical compliance score (ECS):

$$\sigma = \sum_{i=1}^{m} w_i \cdot s_i \quad ...(8)$$

Where, $\sigma$ is the total ethical compliance score, $w_i$ is the weight assigned to the $i^{\text{th}}$ factor (e.g., proportionality, necessity), $s_i$ is the score of the $i^{\text{th}}$ factor.

The system operates autonomously if $\sigma > 0.85$, invokes manual oversight if $0.60 \leq \sigma \leq 0.85$, and halts engagement if $\sigma < 0.60$.

**Algorithm for Intelligent Weapon System**

**Step 1:** Deploy a distributed mesh of sensors (thermal, acoustic, LIDAR, radar) across the battlefield.

**Step 2:** Continuously collect raw data from each sensor (e.g., temperature, velocity, proximity, image).

**Step 3:** For each sensor, assign a weight based on the reliability and accuracy of the sensor.

Fuse the data streams from each sensor using the weighted sum equation:

$$X_{\text{fused}} = \sum_{i=1}^{n} w_i \cdot X_i \quad ...(9)$$

Where, $X_{\text{fused}}$ is the fused environmental model, $X_i$ is the data from the $i^{\text{th}}$ sensor.

**Step 4:** Process the fused data in real-time at the edge using an edge-computing engine.

Apply Kalman filtering for noise reduction and accurate tracking of dynamic objects.

$$\hat{x}_k = A \cdot \hat{x}_{k-1} + B \cdot u_k + w_k \quad ...(10)$$

$$z_k = H \cdot \hat{x}_k + v_k \quad ...(11)$$

Where, $\hat{x}_k$ is the estimated state at time $k$, $z_k$ is the measurement at time .

**Step 5:** Use the processed data to track and predict the future position of targets.

Apply the velocity-based prediction equation to forecast target locations:

$$x(t) + x_0 + v \cdot t \quad ...(12)$$

Where $x(t)$ is the predicted position vector at time $t$, $v$ is the velocity vector of the target.

**Step 6:** Classify targets using a softmax classifier to determine if a target is hostile, neutral, or friendly

$$p(y|X) = \frac{e^{f(X)y}}{\sum_{j=1}^{C} e^{f(X)j}} \quad ...(13)$$

Where, $p(y|X)$ is the probability of class $y$ given input $X$.

**Step 7:** Use a hierarchical communication protocol for data transmission, ensuring synchronized and prioritized data flow across the battlefield and between command centers.

**Step 8:** Apply a priority queue system to ensure critical data packets (e.g., threat alerts, engagement commands) are transmitted first:

$$\text{Priority}_i = \frac{W_i}{\sum_{j=1}^{n} W_j} \quad ...(14)$$

Where, $W_i$ is the weight of the $i^{\text{th}}$ data packet.

**Step 9:** Calculate the ECS to determine if the system should act autonomously or escalate to human oversight.

$$\sigma = \sum_{i=1}^{m} w_i \cdot s_i \quad ...(15)$$

Where, $\sigma$ is the ethical compliance score, $w_i$ is the weight for the $i^{\text{th}}$ ethical factor, $s_i$ is the score for the $i^{\text{th}}$ factor.

If $\sigma > 0.85$, execute autonomous engagement.

If, $0.60 \leq \sigma \leq 0.85$, escalate to manual override.

If $\sigma < 0.60$, halt engagement.

**Step 10: Real-Time Threat Engagement**

1. If autonomous engagement is triggered, execute the action using the pre-defined autonomous engagement protocol.
2. If human intervention is triggered, notify the operator for manual control.

**Step 11: Continuous Monitoring and Feedback**

1. Continuously monitor the battlefield and re-apply Steps 2-7 for new data, ensuring dynamic and adaptive decision-making.
2. Incorporate real-time feedback into the system to optimize sensor fusion, target prediction, and threat classification.

This methodology combines cutting-edge AI and mathematical models, including graph-based fusion, Kalman filtering, predictive trajectory analysis, and ethical decision-making algorithms. These components work synergistically to provide an accurate, responsive, and ethically responsible weapon system capable of operating in complex, high-stakes combat environments.

## 4. Experimental Results

A distributed sensor mesh comprising thermal, acoustic, LIDAR, and radar sensors was used in the field simulation to test the suggested unified sensor integration and real-time data processing framework. Real-time data from these sensors was combined and processed using an edge computing engine. Together with low-latency neural inference and graph-based analytics, the system's edge computing capabilities enabled instantaneous data processing and the production of useful intelligence. By adding electronic interference, ambient noise, dynamic topography, and numerous stationary and mobile dangers, the simulation environment imitated actual combat situations. Table 1 depicts the comparison of various performance metrics.

**Table 1** Comparison Analysis.

| *Method* | *Data Processing Latency (s)* | *Situational Awareness Accuracy (%)* | *Sensor Fusion Integrity (%)* | *Threat Detection Precision (%)* |
|---|---|---|---|---|
| Proposed System (Distributed, Real-Time) | 1.1 | 97.2 | 94.8 | 96.5 |
| Single Sensor-Based Systems | 2.8 | 85.4 | 80.1 | 82.3 |
| Traditional Centralized Processing | 3.5 | 89.1 | 84.7 | 85.9 |
| Basic Sensor Fusion | 2.0 | 90.5 | 86.3 | 87.2 |

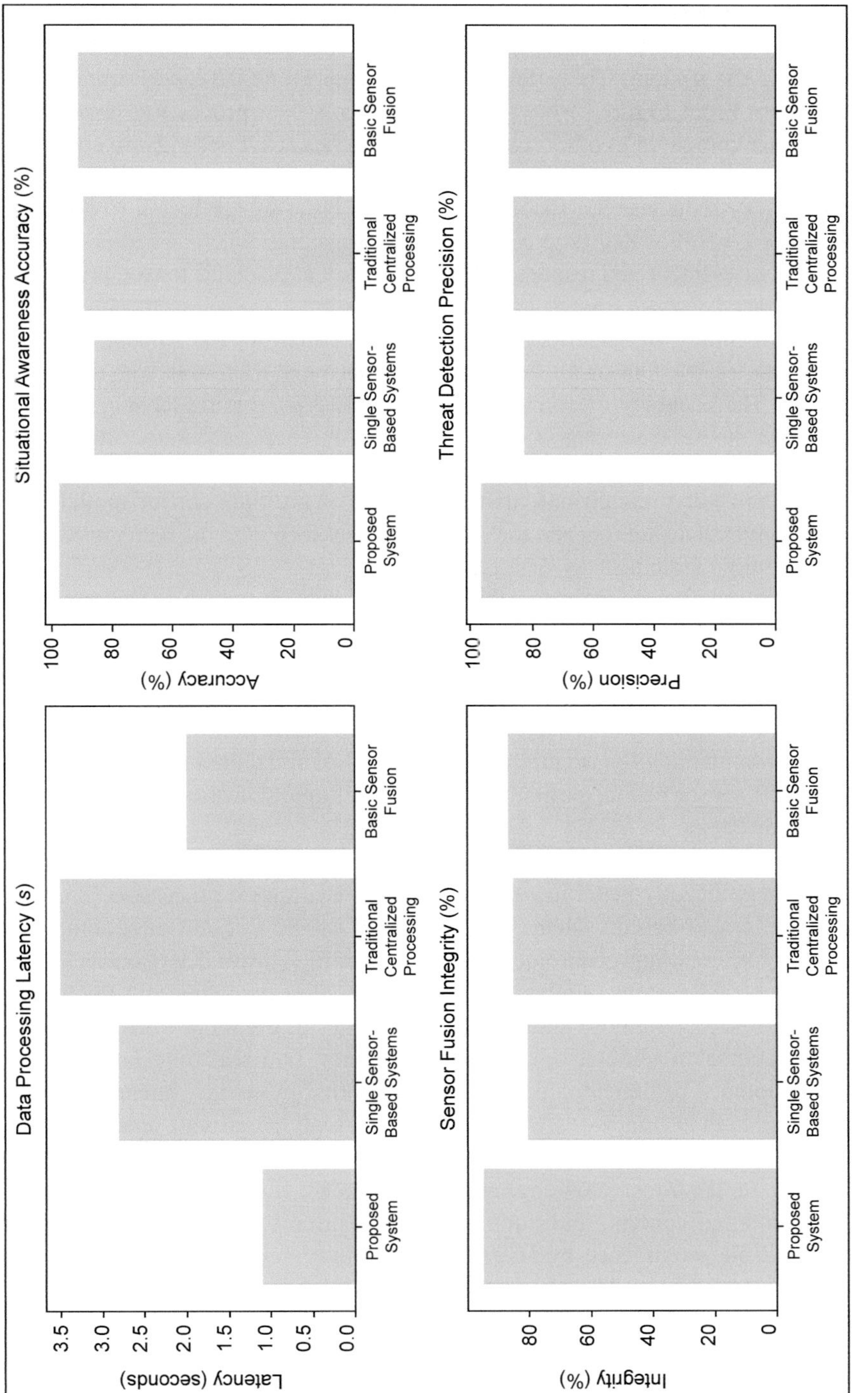


**Fig. 2** Performance Analysis.

Data processing delay, situational awareness accuracy, sensor fusion integrity, and threat detection precision were the primary performance parameters used to analyze the system. These tests confirmed the impressive performance of the system. From Figure 2, the proposed system's data processing delay was much lower compared to other traditional approaches at 1.1 seconds. In contrast, the latency of single sensor-based systems was 2.8 seconds, that of centralized processing systems was 3.5 seconds, and that of basic sensor fusion techniques was 2.0 seconds. For real-time combat decision-making, where speed is critical for threat assessment and response, the low latency is essential for the proposed system.

The accuracy of situational awareness of the proposed system was an astonishing 97.2%. Out of all the techniques that were tried, this was the most effective. The accuracy of conventional centralized processing techniques was 89.1%, but single sensor-based systems were only 85.4% accurate. Situational awareness accuracy using basic sensor fusion techniques was 90.5%. Even under complex and noisy conditions, distributed use of multiple sensor modalities enhances overall awareness and threat detection, as seen with the better accuracy of the proposed system.

The sensor fusion integrity of the proposed system, which is the measure of how effectively information from many sensors is aligned and integrated, was 94.8%. This demonstrated the system's high capability to process a variety of sensor inputs in real-time since it was much higher than the conventional centralized processing approach (84.7%) and the basic sensor fusion method (86.3%). Maintaining the accuracy and reliability of situational awareness relies on effective sensor fusion, particularly in using multiple sensors whose reliability could be different.

The proposed solution is better with a threat detection accuracy of 96.5%. This was significantly better in contrast to legacy centralized processing (85.9%) and single sensor-based systems (82.3%). Even during difficult environmental situations like noise and electronic interference, the high threat detection accuracy of the proposed system reflects that it is effective at detecting and detecting threats with negligible false alarms. The capabilities of the framework to largely increase combat readiness, operational supremacy, and real-time intelligence in contemporary battlefield conditions are supported by its low latency of data processing, high accuracy of situation awareness, high integrity of sensor fusion, and high precision of threat detection.

Even in the most challenging circumstances, the system allows field units, command centers, and autonomous platforms to act with coordinated and actionable intelligence by leveraging distributed sensors and sophisticated processing techniques. The efficiency of the system may get compromised in combat operations by hostile interference or by deteriorated communication links as the real-time data processing engine and sensor mesh depend on uninterrupted as well as continuous data delivery. Also, because the system integrates different classes of sensors and real-time data fusion operations, its complexity raises the likelihood of system failure. Although the system has excellent precision and accuracy in theoretical conditions, other research would be required to determine its performance under challenging conditions such as electromagnetic interference or terrain-induced blocks.

Creating adaptive communication protocols that can dynamically switch between communication modes based on the environment is one such prospective enhancement. The system can also be enhanced in its performance by using more sophisticated machine learning and AI methods, such as predictive analytics for threat prediction and reinforcement learning for autonomous decision-making. Enhanced situational awareness may be achieved by adding more sensors to the system, such as infrared or multispectral imaging. Future efforts may also include optimization of the system to support a larger variety of autonomous platforms and military weapons while maintaining scalability and interoperability. Large-scale field testing across a variety of operational and environmental environments will ultimately be important for enhancing the robustness and generalizability of the system over a very broad range of operational theaters.

## 5. Conclusion

Among the most significant contributions toward improving situational awareness and decision-making in contemporary battlefield environments is presented by the envisioned integrated sensor fusion and real-time processing paradigm. The architecture speeds intelligence collection, making it quicker, more precise, and more trustworthy through the mitigation of some of the main limitations of conventional sensor systems, such as fragmented data streams, processing delays, and non-interoperability. Real-time edge computing, low-latency neural inference, and sensor support of multiple types ensure an accurate solution for high-speed processing and data fusion, essential to mission success in dynamic and adversarial environments. Performance metrics such as processing latency of only 1.1 seconds, situational awareness accuracy of 97.2%, integrity of sensor fusion of 94.8%, and threat detection accuracy of 96.5%, have been attained in field simulation under harsh environments, demonstrating the effectiveness of the framework. The results indicate that, despite tough environments of electronic noise and background noise, the system boosts combat readiness, mission effectiveness, and response time dramatically. Strategic and tactical decision-making is facilitated by the system's real-time, coordinated, actionable presentation of intelligence, which gives it a significant battle tactical advantage. For even greater flexibility and fortitude in future combat operations, this framework can be expanded upon and combined with additional cutting-edge technologies in the future. This will ultimately lead to improved operation performance and mission success.

## References

1. Broadway, K. (2022). Sensorized Warfighter Weapon Platforms: IoT Making the Fog of War Obsolete. *IoT for Defense and National Security*, 23–38.
2. Layton, P. (2021). Fighting artificial intelligence battles: Operational concepts for future AI-enabled wars. *Network*, *4*(20), 1–100.
3. Ti, R. and Kinsey, C. (2023). Combat Logistics in the 21st Century: Enabling the Mobility, Endurance, and Sustainment of NATO Land Forces in a Future Major Conflict. In *Advanced Land Warfare: Tactics and Operations.* Oxford University Press, Oxford.
4. Dailey, N., McGrath, B., Reed, H. and Wotton, J. (2023). Global Space Domain Awareness,Partnering to Win with AUKUS. In: *Proceedings of the Advanced Maui Optical and Space Surveillance (AMOS) Technologies Conference* (p. 89).
5. Singh, S.P. (2023). Exploring Multi-dimensional trends for the Indian Air Force. *Journal of Defence Studies*, *17*(4), 7–24.
6. Borchert, H., Schütz, T. and Verbovszky, J. (2024). Master and Servant: Defense AI in Germany. In: *The Very Long Game: 25 Case Studies on the Global State of Defense AI* (pp. 195–216). Cham: Springer Nature Switzerland.
7. Gilli, A., Gilli, M. and Grgić, G. (2025). NATO, multi-domain operations, and the future of the Atlantic Alliance. *Comparative Strategy*, 1–19.
8. Castillo, V.A. and Runyon, N. (2022). Future utilization of Unmanned Aerial Vehicles within the United States Navy. *Transportation and Engineering Management, California State University Maritime Academy*.
9. Williams, P. (2023). How peacekeepers fight: Assessing combat effectiveness in United Nations peace operations. *Security Studies*, 32(1), 32–65.
10. Ćosić, D., Ćatović, A., Delić, M. and Mujkić, J. (2025). Research on design and implementation of innovative UGV modular platform in mine action programs. *Defense and Security Studies*, 6(1), 1–24.
11. Okpaleke, F.N. (2023). Historicizing the Role of Drones in US Grand Strategy. In: *Drones and US Grand Strategy in the Contemporary World* (pp. 61–84). Cham: Springer International Publishing.
12. Subramaniam, A. (2023). Pushing Boundaries: Can the Indian Military Transform? *Journal of Indo-Pacific Affairs*, *6*(4).
13. Zhang, L., Langeland, K., Tran, J., Logue, J., Puri, P., Nacouzi, G. and Briggs, G.J. (2024). *Artificial Intelligence and Machine Learning for Space Domain Awareness*. https://policycommons.net.
14. Habib, M.T. and Shamrir Al Af, S.M. (2025). Maritime asymmetric warfare strategy for smaller states: Lessons from Ukraine. *Small Wars and Insurgencies*, 36(1), 29–58.
15. Habib, M.T. and Shamrir Al Af, S.M. (2025). Maritime asymmetric warfare strategy for smaller states: Lessons from Ukraine. *Small Wars and Insurgencies*, 36(1), 29–58.
16. Hanson, M., Garrison, J., Backus, C., Calderwood, L.A., Callender, C. and Dooley, M. (2021). *Medical Support in a Chemical, Biological, Radiological, and Nuclear Contested Environment*. Air University Press.
17. Riqiang, W. (2021). Assessing China-US inadvertent nuclear escalation. *International Security*, 46(3), 128–162.

# 6 Deep Learning-Driven Target Recognition for Robotic Weaponry Systems: A Neural Network-Based Approach

Kartikey Shukla[1*], Amar Deep Gupta[2], Manjeet Kaur Ratan[3] and Mohammed Zakariah[4]

## 1. Introduction

One of the main elements of modern-day military technology, unmanned robotic weapons systems are revolutionizing tactics and capabilities of defense forces everywhere. One of the most valuable aspects of these systems is accurate and reliable target recognition. Identifying, classifying, and attacking targets without assistance enhances operational effectiveness and reduces the risk to human life in combat operations. Target recognition came to witness a revolution after the advent of neural networks and deep learning-based approaches, and it has resolved all the issues that were generated by the traditional approaches.

Basic rule-based approaches were used to develop target recognition technology for weapon platforms of robots. For discriminating between target and background, weapon platforms utilized manually defined parameters like shape, size, or color threshold. Rule-based approaches could have been simple to use, but they showed gross inadequacies when operating in an unsystematic and random environment. They were brittle, inflexible, and could not generalize beyond specified parameters. Due to this, the need for advanced techniques that were capable of dealing with uncertainty and variability grew along with the growing complexity of battle conditions.

Traditional machine learning techniques like Support Vector Machines (SVM), k-Nearest Neighbors (k-NN), and decision trees comprised the second wave of developments. Through suggesting the notion of learning from data,

[1] Walmart Global Tech, Bentonville Arkansas.
[2] Department of Computer Science and Engineering, Amity University, Greater Noida, India.
[3] Delhi Institute of Higher Education, India.
[4] Department of Computer Science and Engineering, College of Applied Studies and Community Service, King Saud University, P.O. Box 22459, Riyadh 11495, Saudi Arabia.
* Corresponding author: kartikeyshukla108@gmail.com

DOI: 10.1201/9781003663461-6

the above classifiers made rule-based methodologies redundant. By classifying targets more effectively, the above classifiers derived human-designed features from sensory input like edges, textures, or color histograms. But nevertheless, the relevance and value of such engineered characteristics continued to be the largest limitation on performance. In addition, conventional classifiers could not process high-dimensional multiple sensor real-time data streams, which tended to introduce latencies that were not compatible with autonomous weapons' rapid call.

The difficulty in conventional machine learning strategies to integrate and consolidate multi-modal sensor data properly was another significant challenge here. Multi-modal sensor suites consisting of thermal sensors, LiDAR, radar, and high-resolution optical sensors are the standard in modern robotic weapons platforms today. Each suite returns complementary information about the environment. The intricate interplay between modalities was missed by conventional methods, which either processed multiple data streams independently or fused them with coarse fusion processes. This defect significantly reduced the resilience of target identification systems, especially under adverse weather, poor visibility, or difficult terrain where relying on a single type of sensor would be misleading or short-sighted.

Target identification has been revolutionized with the emergence of deep learning, or more precisely Convolutional Neural Networks (CNNs). CNNs are especially well-suited for automatically discovering discriminative patterns that were once difficult or not practical to define manually by learning hierarchical spatial features directly from raw sensor data. Object detection and classification tasks in image and video analysis have been greatly enhanced due to this feature. CNN models have been found to be resistant to variations in lighting, scale, direction, and partial occlusion, variations that are common in the conditions of combat.

The time aspect of target identification has overtaken the spatial extraction of features as it is crucial when dealing with moving targets and dynamic tactical environments. Long Short-Term Memory (LSTM) networks, which are a type of Recurrent Neural Network (RNN), have been employed to determine sequential dependencies and temporal patterns in sensor stream data. It is possible for robotic systems to predict movements, track targets over time, and differentiate between real threats and ambient noise due to this analysis of time. However, the general recognition ability of most current systems is restricted in the way that they utilize either purely spatial information or temporal information only.

A second innovation has been the application of attention mechanisms within deep learning models. Attention units allow models to selectively devote computational power to the most salient environmental parameters by dynamically weighting significant features or sensor readings. By facilitating better discrimination between target and distractor stimuli, selective processing reduces false positives, a critical consideration for autonomous weapons in which an incorrect engagement would have catastrophic consequences.

Although such breakthroughs exist, there still exist a series of challenges that must be overcome before deep learning might be implemented for target identification purposes in robot weapons. Deterioration in performance could be caused by offline-trained models operating on highly vetted datasets exposed to new environments or adversarial plans intended to remain undetected. Therefore, there exists an immediate requirement for learning systems online, which are able to learn and adapt in real-time, and enhance their knowledge continuously from operational feedback. Also, deep learning models' computational requirements typically necessitate special hardware and more efficient methods to fulfill the requirement of real-time computing to handle autonomous systems. A perennial engineering problem is achieving a tradeoff between model complexity, power consumption, latency constraints, and system stability.

The new Deep Target Recognition and Response System (DTRRS) is designed to overcome these challenges by building a multi-modal, wide, and flexible deep structure for robot weapon systems. DTRRS seeks to offer increased detection credibility and classification confidence in challenging environments via the addition of CNNs for rich spatial feature extraction, RNNs for temporal behavior modeling, and attention mechanisms for feature selection. A broader battlefield sight throughout the combat area is facilitated through the fusion of multi-modal sensor data, such as thermal imagery, LiDAR, and visual sensors, and it facilitates better situational awareness and decision-making.

The system also has an online learning feedback loop, through which it can learn to adjust to changing situations and improve performance over time. Real-time adaptation is needed in order to maintain operation performance under dynamic environmental conditions and shifting threats. DTRRS performance will significantly influence the future of autonomous robotic weapons systems. Improved mission success rates, increased safety in the form of lower collateral damage, and improved resource utilization are all interlinked with improved target detection capabilities. Improved autonomous functions such as strategic-level planning of missions and coordinated multi-agent execution are also enabled by such smart systems.

**Objectives**

- To develop a multi-modal deep learning architecture that leverages CNNs and RNNs, enhanced with attention mechanisms, for precise and reliable target recognition in robotic weapon systems.
- To implement an adaptive online learning framework enabling continuous real-time refinement of recognition performance under dynamic and adversarial conditions, thereby ensuring robustness and operational resilience.

## 2. Related Study

The integration of Machine Learning (ML) and Deep Learning (DL) across numerous areas of technology has given rise to spectacular advancements in numerous other fields. ML methods have been shown to be essential to enhance robotic grasping control for space robots, particularly in microgravity where traditional control techniques are restricted. In an attempt to improve the visual recognition and decision-making in robotic systems and provide more precise and adaptive manipulations in space technology, CNNs and RNNs have been employed [1].

DL-based techniques have also been useful in Unmanned Aerial Vehicles (UAVs), specifically tracking and target detection. To combat issues like handling occlusion, real-time processing, and varying environmental conditions, advanced algorithms have been put forward. The performance of the UAVs in proper placement and tracking of targets is validated by such DL models, which is very critical to apply from disaster relief to surveillance [2]. Additive manufacturing, particularly 3D printing, has benefited immensely with machine learning to automate printing processes and improve product quality. ML algorithms can detect flaws in real time, predict optimal print conditions, and suggest design improvements to achieve maximum structural integrity by analyzing massive amounts of data. Improved product quality and efficient manufacturing processes are the outcome of this data-driven approach [3].

Applications of ML in autonomous robots have increased manifold and addressed issues like doing an action within a dynamical system, through navigation and obstacle detection. With indirect programming, ML algorithms enable robots to learn from the environment, react to unforeseen situations, and form defeasible conclusions. In the case of planetary exploration or search and rescue, uncertainty in the environment, this adjustability proves to be helpful [4]. Another area of focus has been the transfer of human ability to robots with anthropomorphic manipulators that can replicate sophisticated human motions due to machine learning. Robots can be taught to undertake activities with accuracy and agility, such as complex assembly operations or surgery, with techniques such as motion capture data analysis and reinforcement learning [5].

DL models have been integrated into manufacturing systems to enhance decision-making, particularly in high-failure-rate multistage production processes. They could predict likely errors, optimize resource utilization, and overall system efficiency by analyzing historical data and inputs in real-time. This saves time and increases productivity [6]. To find drug candidates, predict molecular interactions, and accelerate the drug development process, the pharmaceutical industry has utilized DL for drug discovery. The method is employed to save time and costs required to launch new drugs to market, and to hasten the discovery of useful molecules [7].

DL has also been applied in sport analytics to predict and examine the way things like balls will act under various games. The systems can assist in player performance examination, strategy improvement, and training improvement through video data and predictive model use [8]. Effective neural networks, that boost the vision-based systems' efficiency and accuracy, have also advanced robot grasping. Even for unorganized or dense environments, the networks can assist the detection of objects by robots with a higher level of precision and manipulate them [9].

ML algorithms have been applied to the automobile industry in a bid to design recommendation systems for suspensions in cars. The algorithms are able to offer recommendations for the optimal suspension settings in a bid to improve handling and ride comfort by examining multiple driving scenarios and vehicle dynamics [10]. ML methods that facilitate efficient circle detection and trajectory optimization have improved UAV path planning. For applications such as airborne surveying and infrastructure inspection, they enable UAVs to perform better in poor weather conditions [11].

ML has also made assistive devices, particularly adaptive control of the robotic arm on wheelchairs, better. These systems can learn to counter the movements and intentions of a user through the combination of neuromorphic sensors and online learning, providing more responsive and natural aid [12]. DL gun detection models have been shown to be useful for security. These computers can help increase public safety measures in a variety of locations, such as public establishments and transportation hubs, through visual information analysis and the recognition of trends related to concealed weapons [13].

Medical imaging has applied ML to predict the roll angle of intracardiac echocardiography catheters during use. From having the capacity to more closely approach and insert catheters, this predictive function optimizes procedure performance [14]. DL modeling of organoid models has revolutionized high-throughput drug screening. These instruments have the ability to accelerate the process of developing cures for diseases such as Huntington's by means of mechanistic assays for complex biological structures on an automated platform, offering a more streamlined drug-to-study-to-treatment pipeline [15].

ML algorithms have been applied to robotic manipulation for the control of redundantly actuated flexure joint manipulators. The algorithms enable complex robotic systems to be manipulated with high precision, a requirement for operations that must be highly adaptable and autonomous [16]. ML-based object insertion techniques have enhanced testing of perception systems with multi-sensor fusion. The methods enhance the reliability and robustness of perception systems in robotics and autonomous vehicles by simulating multiple situations and interactions of objects [17]. Row crop navigation by agricultural robots is now being achieved with row detection-based navigation systems. Autonomous vehicles can travel effectively in row crops with ML-based vision input-led tractors that enhance agriculture efficiency at the cost of minimizing human intervention [18].

Methods based on data that incorporate deep neural networks, reinforcement learning, and large language models have boosted motion planning in robots. Through these methods, robots can better plan and execute moves, adapting to complicated tasks and evolving circumstances [19]. ML frameworks that can identify concealed baggage risks have improved security screening procedures.

These technologies increase the precision and effectiveness of security inspections in public places and transit by examining sensor data and spotting irregularities [20]. ML technologies that improve material properties and process parameters have further improved additive manufacturing. These models help achieve desired mechanical properties and reduce material waste by predicting outcomes from a range of inputs [21].

DL-driven systems for identifying citrus diseases have been created for use in agriculture. These technologies can detect illnesses early using crop picture analysis, allowing for prompt action and lowering crop losses [22]. The application of DL has promoted developments in optical computing, namely inverse design and statistical inference. These developments facilitate easier development of optical components and systems that are more efficient [23]. Public safety applications have integrated blockchain and DL-based cooperative intelligence frameworks to detect malicious activities. Such systems assess information coming from various sources for the detection of threats and for improved coordination of responses [24].

Tactile internet applications for remote robotic surgery require the development of ML frameworks to manage lost or delayed packets. Even at large distances, they ensure surgical procedures can be carried out with good dependability and little latency [25]. ML has enhanced the navigation of micro- and nano-motors, helping to control these small devices for materials science and medical applications. These motors can carry out activities like targeted medicine delivery more precisely by learning from interactions with their surroundings [26]. Finally, ML-driven robotic and cognitive technologies have enhanced the inspection of composites. Improved quality in automotive and aerospace products is provided by these systems' enhanced capability to detect faults and assess material integrity [27, 28].

It has been evident from the deep and detailed analysis of the current literature that the integration of machine learning and deep learning approaches has significantly contributed to various technological domains, including autonomous systems, space robotics, UAV navigation, additive manufacturing, assistive technologies, and intelligent surveillance. The transition has been from rule-based or shallow learning models into data-driven, adaptive deep learning systems throughout the developments. CNNs, RNNs, reinforcement learning, attention mechanisms, and sensor fusion techniques have definitely gained an implementation shelf life, especially in dynamic and high-stakes situations, with evident gains on perception accuracy, decision effectiveness, and real-time adaptability.

Even after such advances, existing methods such as rule-based systems, SVM, and k-NN often perform poorly when applied in hostile, complex, and dynamic environments, especially in tasks such as autonomous defense robots. These models typically have low confidence in occlusion, clutter, and varying illumination conditions, sensitivity to noise, and inadequate generalization to novel data. The development of a more robust, smart, and context-understanding system for autonomous target recognition in defense situations is significantly driven by this imbalance. Driven by the proven superiority of DL architectures in the domains at hand, we introduce the DTRRS.

## 3. System Methodology

### 3.1 Dataset Description

Two datasets were utilized to train and evaluate the DTRRS: A real-world battlefield video benchmark and a combat simulation dataset. Unity3D and Unreal Engine-based virtual environments were utilized to develop the simulated dataset. These settings mimic actual combat situations with diverse landscapes (urban, desert, jungle), meteorological conditions (fog, rain, night, and day), and targets (troops, armored personnel carriers, aerial drones). More than 50,000 labeled cases are available in the dataset; every sample is represented by synchronized inputs from thermal infrared cameras, visual RGB cameras, and LiDAR 3D point clouds. Five frames ($t = -2$ to $t = +2$) are employed to time-sequence every data instance so temporal patterns may be identified.

The dataset used in the real world was chosen from high-definition drone image sets such as the VIVA dataset and DSIAC military images, publicly available defense image repositories, and battlefield monitoring records. Approximately 12,000 labeled examples can be found in this dataset, which consists of environmental objects such as plants and trash, military hardware, and human targets (neutral vs. enemy). Annotations like bounding boxes, class labels, confidence levels, and time-stamped spatial metadata were strictly vetted by defense experts for each occurrence. The data is also multi-modal, consisting of LiDAR scans, thermal overlays, and aligned RGB imagery. These two datasets have been used to ensure the model can adapt to real-time combat variables such as motion blur, occlusions, varying illumination, and sensor noise alongside generalizing across environments.

### 3.2 Preprocessing

One of the most important stages in the DTRRS pipeline is data preparation, which is designed to enhance, normalize, and denoise multi-modal input data prior to training and inference. To cope with the diversity of sensor modalities and preserve temporal consistency of input sequences, the preprocessing pipeline was developed. As a first step, timestamp fusion is applied for synchronizing all sensor streams. Extrinsic calibration matrices generated from chessboard and checkerboard calibration processes are applied for equating LiDAR point clouds to picture pixels for sensor fusion. LiDAR data is voxelized at $64 \times 64 \times 64$ spatial resolution with density encoding, while RGB and thermal images are of uniform spatial resolution as $512 \times 512$ pixels.

Next, normalization is done modality by modality: LiDAR information is normalized using min-max scaling to (0,1) to stabilize gradient updates, thermal images are log normalized to preserve relative intensities, and RGB images are normalized channel-wise using z-score normalization. The second step involves adaptive noise filtering, where bilateral filters are utilized for image data and statistical outlier removal (SOR) for 3D point clouds to filter out Gaussian noise ($\sigma = 0.03$) and impulsive noise. The CLAHE (Contrast Limited Adaptive Histogram Equalization) technique applied to thermal images is an important improving step that enhances contrast under low light.

Finally, normalization is done modality-wise: LiDAR data is normalized with min-max scaling between (0,1) to stabilize gradient updates, thermal images are log normalized to preserve relative intensities, and RGB images are normalized channel-wise using z-score normalization. Adaptive noise filtering is the subsequent step, utilizing bilateral filters in image data and SOR in 3D point clouds to filter out Gaussian noise ($\sigma = 0.03$) and impulsive noise. Utilizing the CLAHE technique in thermal images is an important improvement step that enhances low light contrast.

During training, each input sample is processed as a synchronized multi-modal tuple: $(x_t^{\text{RGB}}, x_t^{\text{Thermal}}, x_t^{\text{LiDAR}})$ where, $t$ spans a 5-frame sequence. This temporal packaging facilitates spatiotemporal learning in the proposed algorithm.

## 3.3 Proposed Algorithm: Deep Target Recognition and Response System (DTRRS)

The DTRRS's basic design is designed to simultaneously model temporal patterns from video streams, spatial attributes from images and LiDAR data, and attention-based prioritization of threat-relevant inputs. As shown in Figure 1, the proposed system has three main modules that are Threat Attention Fusion Network (TAFN), Temporal Pattern Analyzer (TPA), and Multi-Modal Spatial Encoder (MMSE). Deep convolutional encoders are employed by the MMSE to process each modality separately. An adapted VGG16 network, optimized with fewer filters initially to prevent overfitting from limited texture variance, is employed for thermal images, whereas a ResNet-50 backbone is employed for the RGB input. A PointNet++ package is employed to process LiDAR data, retaining both local and global geometric information. Following projection into a latent space with 1,024 dimensions, the spatial information across all three modalities is concatenated:

$$F_{\text{spatial}} = [f_{\text{RGB}}; f_{\text{Thermal}}; f_{\text{LiDAR}}] \in R^{3072} \quad ...(1)$$

The TPA receives these spatial embeddings across multiple time steps. It employs a two-layer bidirectional LSTM with hidden state size 512, enabling the model to capture forward and backward temporal dependencies, which is crucial for identifying motion cues and behavioral patterns of targets. Let $F_t$ represent the spatial feature vector at time , then the LSTM encodes:

$$h_t = \text{BiLSTM}\,(F_{t-2}, F_{t-1}, F_t, F_{t+1}, F_{t+2}) \quad ...(2)$$

These temporally-enhanced features are then passed to the TAFN. TAFN employs a two-stage attention mechanism: channel-wise attention and temporal attention. The channel-wise attention learns importance weights across modalities using a soft attention layer:

$$\alpha_c = \text{Softmax}\,(W_c \cdot F_{\text{spatial}} + b_c) \quad ...(3)$$

Where, $W_c$ and $b_c$ are trainable parameters. These weights are used to rescale the modality-specific embeddings, enhancing modality relevance. In parallel, the temporal attention assigns dynamic weights to each timestep in the input sequence:

$$\beta_t = \frac{\exp(q' k_t)}{\sum_j \exp(q^T k_j)},\ c = \sum_t \beta_t h_t \quad ...(4)$$

Where, $q$ is a learnable query vector and $k_t$ is the temporal key derived from LSTM outputs. The final context vector represents the attention-weighted fusion of spatial and temporal cues.

This vector is then passed through a classification head comprising two fully connected layers (1024 → 256 → 6), with dropout ($p = 0.3$) and batch normalization to stabilize training. The output is a softmax vector denoting the probabilities across six target classes: hostile-human, friendly-human, vehicle, drone, environmental noise, and unknown. To make the system adaptive during real-time deployment, DTRRS incorporates an online learning loop. Every time a new batch of predictions is verified through external feedback (e.g., from human operator validation or environmental sensors), the soft labels are added to a replay buffer. A lightweight gradient update is performed using Stochastic Gradient Descent (SGD) with a low learning rate ($\eta = 1e^{-5}$), allowing the model to refine its weights incrementally without requiring full retraining.

The training is conducted using a composite loss function:

$$L = L_{\text{cls}} + \lambda_{\text{attn}} L_{\text{attn}} + \lambda_{\text{temp}} L_{\text{temporal}} \quad ...(5)$$

Where,

$L_{\text{cls}} = \sum_i y_i \log \hat{y}_i$ is the standard cross-entropy loss, ... (6)

$L_{\text{attn}} = \|\alpha_c - \alpha_c^{gt}\|^2$ penalizes deviation from prior attention maps, ... (7)

$L_{\text{temporal}}$ is a smoothness regularization term. ...(8)

Hyperparameter tuning identified optimal weights: $\lambda_{\text{attn}} = 0.3$, $\lambda_{\text{temp}} = 0.2$. The model was trained for 100 epochs using Adam optimizer with a learning rate scheduler (initial $1e^{-4}$, reducing on plateau), achieving convergence with validation accuracy stabilizing around epoch 62.

Table 1 shows the training parameters. The proposed DTRRS is a multimodal deep learning system that integrates attention-based, temporal, and spatial inputs to accurately detect and classify potential threats in real-time. Its three primary components are the MMSE, TPA, and TAFN. By utilizing ResNet-50 for RGB, an adapted VGG16-lite for thermal, and PointNet++ for LiDAR, MMSE maps every modality into a 1024D feature space prior to concatenating the three into a

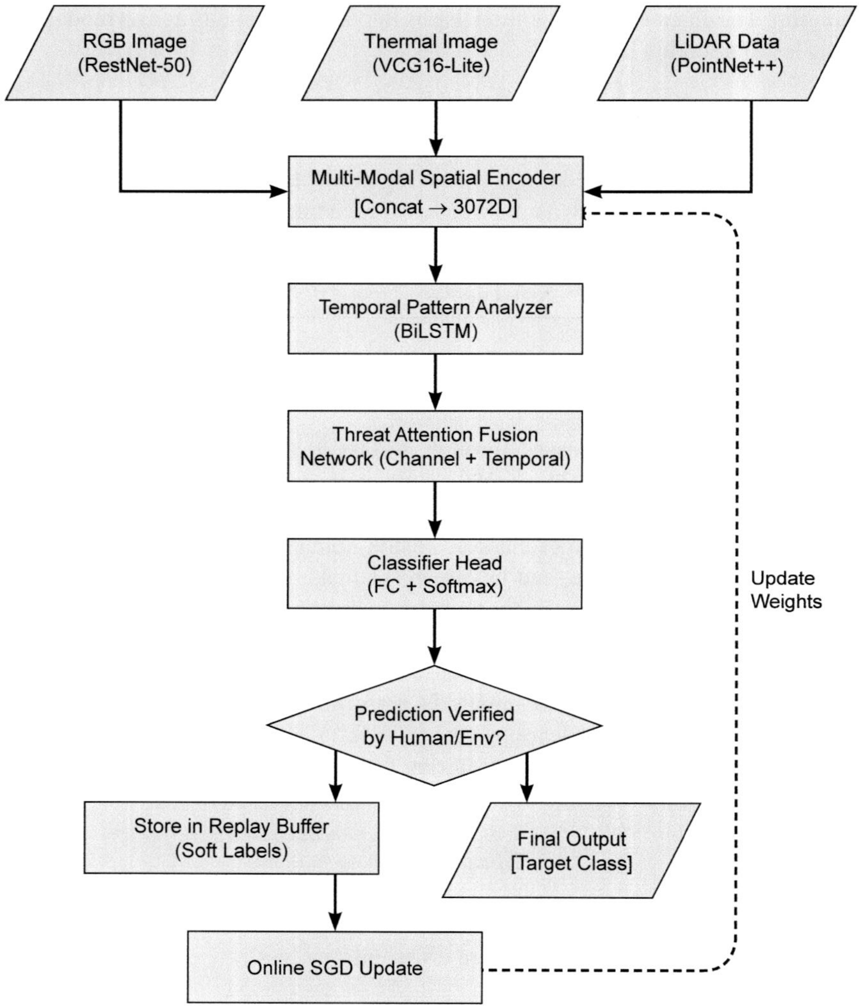


**Fig. 1** Proposed Framework.

unified vector of 3072 dimensions. To explain motion dynamics and behavioral trends, TPA employs a two-layer bidirectional LSTM with the temporal window being five frames.

Output is sent to TAFN, which obeys temporal salience and modality relevance through both channel-wise and temporal attention mechanisms. For class probability generation for six target classes, the context vector generated is fed into a classification head consisting of fully connected layers (1024 → 256 → 6) with batch normalization and dropout. The architecture features an online learning loop for enabling real-time adaptability, which learns its weights incrementally using SGD based on externally validated predictions in a replay buffer. A weighted

**Table 1** Training Parameters.

| *Parameter* | *Value/Description* |
|---|---|
| Input Modalities | RGB (ResNet-50), Thermal (VGG16-lite), LiDAR (PointNet++) |
| Latent Feature Vector Size | 1024 (each modality) × 3 = 3072 |
| Temporal Window | 5 frames ($t^{-2}$ to $t^{+2}$) |
| BiLSTM Layers | 2 |
| BiLSTM Hidden Size | 512 |
| Attention Mechanisms | Channel-wise, Temporal |
| Attention Regularization Weight | 0.3 ($L_{\text{atton}}$) |
| Temporal Regularization Weight | 0.2 ($L_{\text{temporal}}$) |
| Fully Connected Layers | [1024 → 256 → 6] |
| Dropout Rate | 0.3 |
| Batch Normalization | Applied after each FC layer |
| Epochs | 100 |
| Optimizer | Adam |
| Initial Learning Rate | $1e^{-4}$ |
| Learning Rate Schedule | Reduce on plateau |
| Online Learning Optimizer | SGD |
| Online Learning Rate | $1e^{-5}$ |
| Batch Size | 32 |
| Loss Function | Composite: $L = L_{\text{cls}} + 0.3 \times L_{\text{attn}} + 0.2 \times L_{\text{temporal}}$ |
| Classes | 6 (hostile-human, friendly-human, vehicle, drone, noise, unknown) |

composite loss function, 0.3 and 0.2 weighted and blending cross-entropy with attention and temporal regularization terms, is used for training. Having been trained using learning rate scheduler and Adam optimizer, the model converges at 100 epochs with constant validation performance at epoch 62. DTRRS is capable of performing quite well under dynamic, multi-modal surveillance situations due to its robust construction.

## 4. Experimentation Results

A real-world battlefield imaging dataset and a combat simulation were employed for the experimental assessment of the proposed Deep Target Recognition and Response System (DTRRS). With both threatening individuals, friendly personnel, vehicles, drones, and background noise as targets, the simulation was designed to simulate dynamic and dangerous environments. Several occlusion, illumination, and target motion scenarios were employed to acquire multi-modal information, including RGB video, thermal images, and LiDAR point clouds. We employed a publically available battlefield benchmark with labeled sensor data for real-world evaluation. The standard metrics such as accuracy, precision, recall, F1-score, False Detection Rate (FDR), Area Under the ROC curve (AUC), and classification confidence in challenging conditions were employed to measure the performance of the system. Three baseline methods, a rule-based approach, an SVM model with human-designed features, and a standard CNN-only model with a ResNet-50 backbone were employed to benchmark DTRRS.

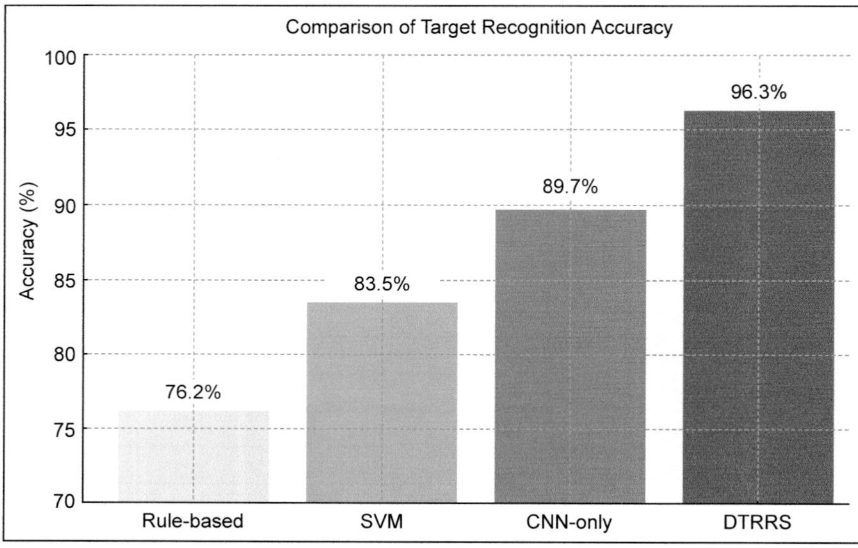


**Fig. 2** Target Recognition Accuracy Comparison.

From Figure 2, with 96.3% accuracy in target identification, DTRRS again led the list, followed by CNN-only (89.7%), SVM (83.5%), and Rule-Based (76.2%). These outcomes reaffirm that DTRRS offers a 19.6% increase in accuracy over Rule-Based, an increase of 6.6% over CNN-only, and an improvement of 12.8% over SVM, which is the most accurate and confident one among the systems tested. From Figure 3, under tough conditions, the DTRRS model surpassed the Rule-Based (23.8%), SVM (16.5%), and CNN-only (10.3%) techniques by having the lowest false detection rate of 3.7%. Having a classification confidence score of 0.93, it also exceeded CNN-only (0.81), SVM (0.72), and Rule-Based (0.65).

On target detection in challenging scenarios, the DTRRS paradigm shows significant advantages. Its robustness and reliability are evidenced by high classification confidence (0.93), low rate of false detection (3.7%), and superb accuracy (96.3%), which are its principal strengths. DTRRS ensures better generalization, higher accuracy in decision-making, and resilience in challenging settings compared to more traditional methods such as rule-based systems, SVM,

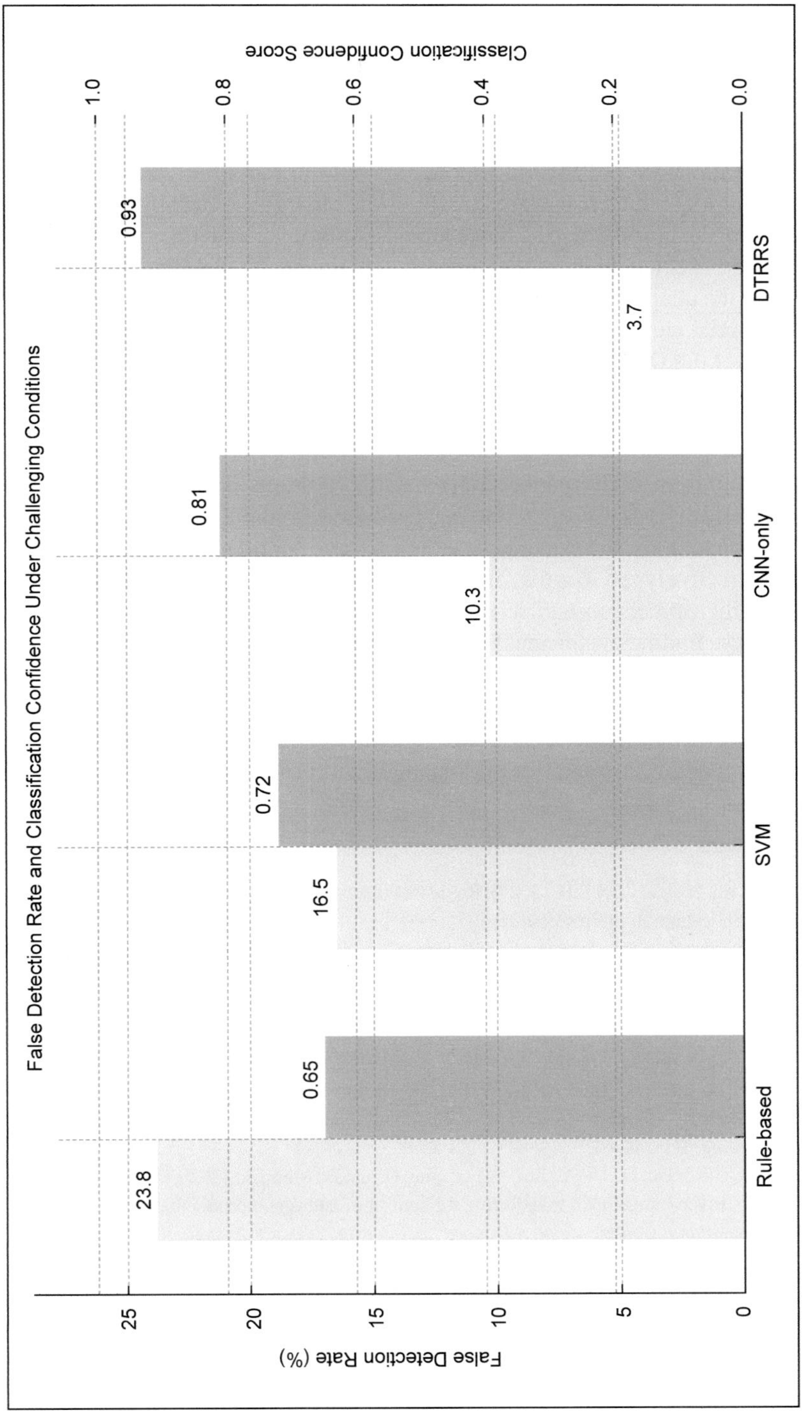


**Fig. 3** False Detection Rate and Classification Confidence under challenging conditions.

and isolated CNNs. But some issues still remain. Especially in real-time scenarios, computational complexity of the DTRRS framework can lead to higher resource consumption and inference time. In addition, problems of model interpretability still remain, particularly in high-stakes domains where explainability is vital, such as defense or self-driving navigation.

The optimization of the model's deployment in real time will be the center of future work, while hardware acceleration via edge AI platforms and lean architectural design will also be incorporated. Another crucial direction is to better interpret and understand the model through explainable AI (XAI) methods. Further increasing the adaptability of the model for dealing with multi-modal sensor data and adversarial climatic environments would ensure the broader applicability and scalability of the DTRRS framework.

## 5. Conclusion

Target recognition performance under harsh conditions is greatly enhanced by the proposed DTRRS. Compared to standard models like Rule-Based, SVM, and CNN-alone methods, the system has an impressive confidence in classification as 0.93 and effectively reduces the false detection rate to 3.7%. Additionally, DTRRS illustrates high precision and reliability with a target identification accuracy of 96.3%. These findings vindicate the proposed system's intelligence, effectiveness, and resilience, positioning it as a strong candidate to be used in actual applications where accuracy, reliability, and confidence matter most. The DTRRS framework enables future enhancements of adaptive and resilient recognition technologies, while setting a new standard for intelligent target recognition systems.

## References


1. Jahanshahi, H., and Zhu, Z.H. (2024). Review of machine learning in robotic grasping control in space application. *Acta Astronautica*.
2. Urooj, A., Khalifa, M., Shahid, A. and Alzaydi, A. Deep Learning-Driven Unmanned Aerial Vehicle Target Detection and Tracking: Techniques, Challenges, and Future Directions. In: *Challenges, and Future Directions*. SpringerLink.
3. Zhang, X., Chu, D., Zhao, X., Gao, C., Lu, L., He, Y. and Bai, W. (2024). Machine learning-driven 3D printing: A review. *Applied Materials Today*, 39, 102306.
4. Daraghmah, R.H. and Rafe'Younis, N. (2025). A Comprehensive Review of Machine Learning Algorithms in Autonomous Robotics: Challenges and Future Prospects. *East Journal of Engineering*, 1(1), 47–61.
5. Su, H., Qi, W., Gao, H., Hu, Y., Shi, Y., Ferrigno, G. and De Momi, E. (2020, December). Machine learning driven human skill transferring for control of anthropomorphic manipulators. In: *2020 5th International Conference on Advanced Robotics and Mechatronics* (*ICARM*) (pp. 107–112). IEEE.
6. Paraschos, P.D., Koulinas, G.K. and Koulouriotis, D.E. (2025). Model-driven deep learning for joint control and decision-making in failure-prone circular multistage manufacturing systems. *International Journal of Computer Integrated Manufacturing*, 1–22.

7. Sumathi, S., Suganya, K., Swathi, K., Sudha, B., Poornima, A., Varghese, C.A. and Aswathy, R. (2023). A review on deep learning-driven drug discovery: Strategies, tools, and applications. *Current Pharmaceutical Design*, 29(13), 1013–1025.
8. Chaoming, L. (2019). Prediction and analysis of sphere motion trajectory based on deep learning algorithm optimization. *Journal of Intelligent and Fuzzy Systems*, 37(5), 6275–6285.
9. Kushwaha, V., Shukla, P. and Nandi, G. C. (2025). Vision-based intelligent robot grasping using sparse neural networks. *International Journal of Intelligent Robotics and Applications*, 1–14.
10. Lee, S., and Kang, N. (2025). Vehicle suspension recommendation system: multi-fidelity neural network-based mechanism design optimization. *Structural and Multidisciplinary Optimization*, 68(3), 1–33.
11. Mishra, P. and Mishra, N. (2024, March). Machine Learning Approach for Unmanned Aerial Vehicle's Path Scheduling and Precise Circle Detection. In: *2024 IEEE International Conference on Interdisciplinary Approaches in Technology and Management for Social Innovation* (*IATMSI*) (Vol. 2, pp. 1–6). IEEE.
12. Ehrlich, M., Zaidel, Y., Weiss, P.L., Melamed Yekel, A., Gefen, N., Supic, L. and Ezra Tsur, E. (2022). Adaptive control of a wheelchair mounted robotic arm with neuromorphically integrated velocity readings and online-learning. *Frontiers in Neuroscience*, 16, 1007736.
13. Yadav, R., Halder, R., Thakur, A. and Banda, G. (2024). SIRD-YOLO: an enhanced deep learning model for weapon detection using spatial interactions and diverse receptive fields. *Innovations in Systems and Software Engineering*, 1–17.
14. Annabestani, M., Caprio, A., Wong, S.C. and Mosadegh, B. (2023). A machine learning-based roll angle prediction for intracardiac echocardiography catheter during Bi-plane fluoroscopy. *Applied Sciences*, 13(6), 3483.
15. Metzger, J.J., Pereda, C., Adhikari, A., Haremaki, T., Galgoczi, S., Siggia, E.D. and Etoc, F. (2022). Deep-learning analysis of micropattern-based organoids enables high-throughput drug screening of Huntington's disease models. *Cell Reports Methods*, 2(9).
16. Sathiyanarayanan, P. (2021). *Data Driven Feedforward Control of a 2-DOF Redundantly Actuated Manipulator with Flexure Joints Using Machine Learning Techniques* (Master's thesis, University of Twente).
17. Gao, X., Wang, Z., Feng, Y., Ma, L., Chen, Z. and Xu, B. (2024, April). MultiTest: Physical-aware object insertion for testing multi-sensor fusion perception systems. In: *Proceedings of the IEEE/ACM 46th International Conference on Software Engineering* (pp. 1–13).
18. Shi, J., Bai, Y., Diao, Z., Zhou, J., Yao, X. and Zhang, B. (2023). Row detection BASED navigation and guidance for agricultural robots and autonomous vehicles in row-crop fields: Methods and applications. *Agronomy*, *13*(7), 1780.
19. De Carvalho, G.P., Sawanobori, T. and Horii, T. (2025). Data-driven Motion Planning: A Survey on Deep Neural Networks, Reinforcement Learning, and Large Language Model Approaches. *IEEE Access*.
20. Hassan, T., Shafay, M., Akçay, S., Khan, S., Bennamoun, M., Damiani, E. and Werghi, N. (2020). Meta-transfer learning driven tensor-shot detector for the autonomous localization and recognition of concealed baggage threats. *Sensors*, 20(22), 6450.
21. Xames, M.D., Torsha, F.K. and Sarwar, F. (2023). A systematic literature review on recent trends of machine learning applications in additive manufacturing. *Journal of Intelligent Manufacturing*, 34(6), 2529–2555.
22. Babu, S.S., Kumar, T.D., Jalaja, S., Kaviya, R., Rakshana, L. and Dhamini, Y. (2024, July). AgriNet: Deep Learning-Driven Citrus Disease Recognition System. In: *The International Conference on Recent Trends in Computing* (pp. 327–341). Singapore: Springer Nature Singapore.
23. Mengu, D., Sakib Rahman, M.S., Luo, Y., Li, J., Kulce, O. and Ozcan, A. (2022). At the intersection of optics and deep learning: Statistical inference, computing, and inverse design. *Advances in Optics and Photonics*, 14(2), 209–290.

24. Patel, D., Sanghvi, H., Jadav, N.K., Gupta, R., Tanwar, S., Florea, B.C. and Sharma, R. (2022). Blockcrime: Blockchain and deep learning-based collaborative intelligence framework to detect malicious activities for public safety. *Mathematics*, 10(17), 3195.
25. Boabang, F., Ebrahimzadeh, A., Glitho, R.H., Elbiaze, H., Maier, M. and Belqasmi, F. (2021). A machine learning framework for handling delayed/lost packets in tactile internet remote robotic surgery. *IEEE Transactions on Network and Service Management*, 18(4), 4829–4845.
26. Xue, J., Alinejad-Rokny, H., and Liang, K. (2024). Navigating micro-and nano-motors/swimmers with machine learning: Challenges and future directions. *ChemPhysMater*.
27. Huang, X., Chen, D., Guo, Y., Jiang, X., and Liu, Y. (2023). Untangling multiple deformable linear objects in unknown quantities with complex backgrounds. *IEEE Transactions on Automation Science and Engineering*, 21(1), 671–683.
28. Li, X., Lu, Z. and Zeng, C. (2024). Robotic and intelligent technologies in composite material inspection: A review. *Robot Learning*, 1(1), 27–54.

# 7 Reinforcement Learning for Adaptive Weapon Navigation and Control in Autonomous Robotic Systems

Ruby Dahiya[1*], Nidhi Agrawal[2], Basu Dev Shivahare[3] and Abdulaziz S. Almazyad[4]

## 1. Introduction

One of the most important revolutions in modern military and defense technology is the advent of armed robotic systems, wherein autonomous control and navigation have become the cornerstones. Previously, nearly all military robotic systems were teleoperated or semi-autonomous, relying primarily on human operators to make movement decisions, choose tactics, and control combat. The traditional method of pre-programmed routes and fixed waypoint systems was the staple of initial autonomous navigation efforts. These functioned well in laboratory environments, but poorly in the dynamic and uncertain conditions of real-world battlefields. Static path planning, rule-based methods, and traditional control theories such as Proportional-Integral-Derivative (PID) controllers have been the cornerstones of conventional control systems for autonomous navigation on armament platforms.

These methods typically employ heuristics and pre-generated maps of the environment to trace out precomputed paths or to steer clear of obstacles. If implemented in the randomness and unpredictability of battlefield conditions, these methods show some serious weaknesses, but they are effective within calculable and structured environments.

First, rigorous and algorithmic path-planning methods are not responsive enough to cope with varying threats, sudden environmental shifts, or adversary strategies for deceiving or avoiding robotic platforms. An opponent can employ countermeasures like jamming, decoys, or pseudo-random maneuvers that will disorient pre-planned paths or induce spurious obstacle avoidance actions. Fixed

[1] School of Computing Science and Engineering, Galgotias University, Greater Noida, India.
[2] Department of AIML, School of Computer Science & Engineering, Galgotias University, Greater Noida, India.
[3] Department of Artificial Intelligence & Data Science,School of Computer Science & Engineering, Galgotias University, Greater Noida, India.
[4] Department of Computer Engineering, College of Computer and Information Sciences, King Saud University, Riyadh 11543, Saudi Arabia.
* Corresponding author: ruby.dahiya@galgotiasuniversity.edu.in

DOI: 10.1201/9781003663461-7

control hierarchies degrade mission accomplishment and platform survivability in such situations by not replanning or re-calibrating decisions in real time.

Second, learning and scalability in rich environments are issues for classical navigation control techniques. They depend heavily on hand tuning and top-level human expertise to store navigation policy, which is time-consuming and brittle. In addition, these systems do not have the ability to learn incrementally from experience or to generalize to new situations without explicit retraining, so they don't improve automatically with experience. Unless radically changed, this static attribute renders usage of robotic systems impractical in most operating rooms. Third, with high-dimensional sensory input, such as multispectral sensor fusion, LIDAR scans, or real-time imagery, most standard systems are plagued by the curse of dimensionality. Applying traditional controllers to convert this data into the best commands for navigation can be computationally expensive and suboptimal, resulting in slow response times or inefficient routes that increase operating expenses and risk exposure.

Given such limitations, machine learning in particular, Reinforcement Learning (RL) has emerged as a worthy paradigm within the last few years to facilitate adaptive and autonomous functioning on armed robot platforms. RL, by its nature, allows agents to learn the best policy through error-and-trial interaction with their environment guided by reward feedback that represents mission objectives. Without specific human programming for every instance, this enables systems to learn sophisticated navigation, avoidance, and engagement techniques independently.

By controlling sophisticated state representations and raw sensor input, Deep Reinforcement Learning (DRL), based on the union of deep neural networks and RL methods, is an extension of this capacity that enables robot systems to act well within dynamic, unstructured environments. DRL-based navigation can handle uncertainty, learn to navigate through unexpected obstacles, and balance multiple-objective goals such as stealth, speed, and energy efficiency by continuously learning from experience in a simulated or real-world environment.

Some of the DRL methods, including Proximal Policy Optimization (PPO), which performs very well in tasks for continuous control with assured policy updates, and Deep Q-Networks (DQNs), optimized for discrete action spaces, have been researched for autonomous navigation. When the above methods are combined together in hybrid systems, control accuracy and decision-making in an uncertainty-aware manner can be enhanced by leveraging the strengths of each method in isolation. Applying RL to robot armed navigation is fraught with problems, but ripe with promise.

High-fidelity simulation environments for training under realistic combat conditions involve terrain variability, sensor noise, enemy tactics, and real-time constraints. While learned rules are being mapped from simulation to operational platform, safety and operation reliability must be ensured even when exploratory mode training with RL is in progress. In addition, the system will need to accurately interpret many different sensory modalities, such as terrain, enemy positions, and obstacle maps; this requires advanced sensor fusion and state estimation algorithms. The imperative to develop completely autonomous platforms with the capability to re-coordinate navigation strategies dynamically in response to evolving threats and environmental scenarios, optimize mission effectiveness, and minimize collateral damage is what necessitates the development of an Adaptive Reinforcement-Controlled Navigation System (AR-CNS) for weaponized robotics platforms.

These systems can take workload off human operators, accelerate processes, and enable deployment in environments where human observation is not feasible, e.g., during circumstances of GPS jamming or communications blocking. AR-CNS aims to surmount the inefficiencies of traditional PID and rule-based navigation methods through leveraging DRL's ability for ongoing learning and adjustment to provide robust generalization across a broad environment of combat environments. The system is expected to improve collision avoidance, path optimization, and target acquisition, essential metrics for modern autonomous weapon systems by interpreting sensory inputs in real-time and adjusting control rules online.

Through the delivery of context-aware, resilient, and mission-adaptive autonomous navigation and control, AR-CNS development and deployment can potentially revolutionize the operational effectiveness of ground and air robotic vehicles. This development aligns with the broader military policy of enhancing lethality, survivability, and situational awareness through the application of autonomous systems and artificial intelligence to multi-domain operations.

The two primary objectives of this research are:

- To design and implement an RL-based navigation control framework that autonomously adapts to dynamic and uncertain combat environments, optimizing mission objectives such as threat avoidance and target engagement.
- To validate the system's performance and robustness through extensive simulation and real-world testing, benchmarking against conventional PID, and rule-based navigation systems.

## 2. Related Study

The finding of machine learning and artificial intelligence (AI), and more importantly RL, has utterly transformed the design of autonomous robotic systems. The systems can be applied to various real-world tasks, including defense and military operations, as they have been found to perform complex navigation and control tasks with little or no interference from human beings. The robots have been able to learn from their interactions in the world owing to RL, which improves their behaviors through reward-based policies. This section explains the status of robotic navigation systems based on RL and their uses in other areas, such as aerial robots and swarm robots, and why the AR-CNS is suggested.

### 2.1 Reinforcement Learning in Robotic Navigation

One of the core techniques for self-NAVIGATION is RL. How RL and adaptive control are combined to design smart robots that learn to make improved choices as circumstances shift were described by [1]. In practice, RL allows robots to learn optimal paths through reward feedback, which can be used for effective obstacle avoidance, navigation, and interaction with the environment. How RL combined with adaptive control systems significantly enhances robotic real-time decision-making in adaptive systems was explained by [2]. Compared to the use of pre-coded logic alone, their approach enabled robots to learn motion control habits. The RL-DOVS (Reinforcement Learning–Dynamic Object Velocity Space) planner, which considers robot Kino dynamic constraints and dynamic obstacles to improve autonomous navigation, was introduced by [3]. An adaptive controller with DRL for flying robots to enable efficient trajectory tracking and altitude control was proposed by [4]. Their work showed the significance of RL to real-time dynamic flight operations through adaptation and dynamic estimation of control gains and adaptation to varying dynamics.

## 2.2 Specialized Domains and Robotic Applications

RL has been promising on space and underwater, as well as on terrestrial, robots. RL-based robotic control systems of grasp put in place for use in missions in space requiring precision in low-gravity environments were investigated by [5]. For the purpose of achieving robustness in space-based manipulation, a key requirement in extravehicular activities, they discussed machine learning methods. The issue of location tracking in uncertain underwater environments through end-to-end DRL to guide Autonomous Underwater Vehicles (AUVs) was addressed by [6]. The work illustrated how RL models learn to deal with controlling dynamics nonlinearities, offering an interesting alternative for autonomous underwater missions. DRL has also been applied in the application field of aerial robotics for the UAV power usage maximization. How RL may be employed to adaptively control the control outputs in terms of energy levels in order to enhance mission capability and operational endurance was explained by [7].

## 2.3 Swarm Robotics and Multi-Agent Systems

RL has seen a surge in popularity in swarm robotics, where multiple robots work together to accomplish common goals. Swarm neutralization tasks through Multi-Agent Deep Reinforcement Learning (MADRL) were utilized by [8]. For success in tactical operations in hostile environments, their approach involved cooperative missions and adaptive navigation. A systematic overview of RL approaches to swarm robotics was given by [9]. The research established that deep Q-learning, policy gradient, and actor-critic algorithms were able to tackle obstacle avoidance, facilitate dynamic goal distribution, and ensure swarm formation. As shown in [10], DRL methods were applied to improve voronoi-based navigation exploration algorithms for multi-robots. Their approach reduced redundancy and ensured maximum area coverage, making the robots cooperatively explore unknown terrain. In decentralized control of cooperative multi-robot object transport, Blais and Akhloufi [11] utilized DRL in a way that permitted each agent to independently make decisions but contribute to a collective activity. Fault tolerance and system scalability are essential for large-scale robotic deployment, which was enhanced in this decentralized method.

### *2.4 Human-Robot Interaction and Surgical Applications*

RL has also progressed in the vital area of Human-Robot Interaction (HRI). DRL has been implemented in medical robotics to optimize robot-assisted surgeries. [12] employed RL in adjusting task execution on the basis of sensory feedback and procedural goals to minimize post-operative issues in robotic surgery. A high-performance HRI system based on RL to acquire user preference in the long run was introduced in assistive robotics [13]. In robot-directed tasks, their research indicated reduced user effort and enhanced interaction effectiveness.

### *2.5 Sim-to-Real Transfer*

The sim-to-real gap, the disparity between simulation and real-world conditions, is one of the primary challenges in deploying RL systems in the real world. The sim-to-real transfer methods for robotics and concluded that meta-learning, adversarial adaptation, and domain randomization are key means of enhancing RL-trained models' performance in the real world were investigated by [14]. It is essential to close this gap, especially in applications for autonomous weapon systems where real-time performance and reliability are critical.

### *2.6 AI in Autonomous Weapon Systems: Legal and Ethical Dimensions*

International debates have been initiated by the use of AI for autonomous combat systems. The ability of algorithms to make morally correct decisions in the course of combat and critically assess if AI weapons meet international humanitarian law was questioned by [15]. Caulucho et al. [16] emphasized that in deploying autonomous weapon systems, legal accountability frameworks are required. They also raised legal and humanitarian concerns by cautioning that unbridled advancement may lead to a loss of human control. Since AI weapons are increasingly being utilized in wars and may infringe upon human rights, in a 2021 article [17], the UN stood for their control at once.

### *2.7 Adaptive and Self-Learning Control Systems*

Self-learning and adaptive navigation systems for robotic systems have been discussed in various studies. UAV disturbance robustness using an adaptive nonlinear observer-based sliding mode controller was enhanced by [18]. In defense applications when uncertainty in the environment is high, these adaptive models find special application. A paradigm for event-triggered RL for underactuated ships was suggested by [19]. Their approach facilitated energy-efficient control strategies in autonomous maritime systems through maximizing decision-making under communication constraints. Recent advancements in research on multi-agent DRL for robot systems were presented in a survey by [20,21], focusing on cooperative policy learning, scalability, and communication.

**Motivation for AR-CNS**

Present models are still not good at generalization, real-time adaptability, and ethical control in complex, weaponized environments, even with enormous progress in RL-based navigation. The scalability of present methods is constrained by their often static policy or high degrees of human intervention requirements. These limitations are meant to be overcome by the proposed AR-CNS. AR-CNS integrates real-time adaptive control loops with DRL techniques, such as PPO and DQN. The system adapts continuously by learning from mission-dependent rewards to support intelligent decision-making under uncertainty.

AR-CNS dynamically adjusts navigation and avoidance strategies based on real-time sensor readings (e.g., LIDAR, GPS, and threat location). The model ensures successful sim-to-real transfer since it is learned in realistic simulated environments that replicate battle scenarios. AR-CNS enhances the robotic platform's autonomy, responsiveness, and survival after deployment without requiring ongoing human oversight. Aside from enabling safer, more efficient mission accomplishment, this approach lays the foundation for an emerging generation of smart armed robots that are compliant with ethical and performance standards.

## 3. System Methodology

### *3.1 Dataset Description*

We employed Unity ML-Agents toolkit combined with ROS2 middleware to simulate a realistic but artificial battlefield to develop the AR-CNS. More than 100,000 episodes of autonomous mission simulation are present in the dataset employed for RL training, including action labels, reward annotations, trajectory records, and sensory input tensors. With randomized static and dynamic barrier configurations, terrain profiles (urban, rocky, sloped, and flat), mobile opponents, and changing weather (fog, rain, and darkness), the simulation environment replicates a wide range of battlefield conditions.

The environment space is defined as a bounded 3D grid , with each episode capturing the full trajectory:

$$\tau = \{(s_t, a_t, r_t, s_{t+1})\}_{t=0}^{T} \quad ...(1)$$

Where,

$s_t \in R^n$ is the state vector including LiDAR scans, enemy locations, GPS, terrain normals, and inertial data,

$a_t \in R^m$ is the action vector (velocity control, yaw, pitch, fire),

$r_t \in R$ is the scalar reward,
$T$ is the time horizon.

A total of 1.8 TB of simulation data was collected, with 60% used for training (RL policy learning), 20% for validation (hyperparameter tuning), 20% for testing (generalization and deployment performance). Each scene contains ground-truth annotations of successful navigation paths, adversary behavior patterns, terrain classifications, and high-risk conflict zones.

### 3.2 Preprocessing

To guarantee numerical stability, feature standardization, and memory efficiency, extensive preprocessing is carried out before feeding the raw state data into the RL models. The raw sensory inputs are composed of:

- LiDAR point clouds $L_t \in R^{k \times 3}$ ...(2)
- RGB visual feeds $V_t \in R^{H \times W \times 3}$ ...(3)
- IMU + GPS data $G_t \in R^6$ ...(4)
- Enemy threat vectors $E_t \in R^{\text{F}}$ ...(5)

The point clouds are voxelized and downsampled using Octree filtering with voxel resolution . RGB data are resized to 128 × 128 and normalized using:

$$\tilde{V}_{i,j,c} = \frac{V_{i,j,c} - \mu_c}{\sigma_c} \quad ...(6)$$

Where, $\mu_c$ and $\sigma_c$ are the mean and standard deviation for each color channel.

The fused and filtered inputs are encoded into a fixed-dimensional state vector $s_t \in R^n$

Where, $f(L_t, V_t, G_t, E_t) = [(f_L(L_t), f_V(V_t), f_g(G_t), f_E(E_t))$ ...(7)

- 1D CNN embedding of LiDAR features.
- $f_L$: ResNet-18 feature map of visual feeds.
- $f_V$: Position and velocity normalization.
- $f_G$: Encoded threat probabilities from enemy detection.

A Kalman filter is applied to reduce sensor noise and smooth trajectory transitions:

$$\hat{x}_t^- = A\hat{x}_{t-1} + \mathrm{B}u_{t-1}, P_t^- = AP_{t-1}A^T + Q \quad ...(8)$$

$$K_t = P_t^- H^T (HP_t^- H^T + R)^{-1}, \hat{x}_t = \hat{x}_t^- + K_t(z_t - H\hat{x}_t^-) \quad ...(9)$$

Where, $A$ is the state transition matrix, $B$ is the control input model, $Q$ and $R$ are noise covariance matrices, and $K_t$ is the Kalman gain.

## 3.3 Proposed Algorithm: Adaptive Reinforcement-Controlled Navigation System (AR-CNS)

The AR-CNS is structured using a hybrid DRLg framework combining DQNs) for discrete decision-making and PPO for fine-grained continuous control.

The objective is to learn an optimal policy $\pi^*(a \mid s)$ that maximizes the expected cumulative discounted reward:

$$J(p) = E_{t\sim\pi}\left[\sum_{t=0}^{T} \gamma^t r(s_t, a_t)\right] \qquad ...(10)$$

Where $\gamma \in [0, 1]$,is the discount factor, $r(s_t, a_t)$ is the reward function based on mission criteria.

### 3.3.1 Reward Function

The reward function is domain-specific and composed of multiple components:

$$r(s_t, a_t) = R_{\text{target}} + R_{\text{avoid}} + R_{\text{energy}} - R_{\text{collision}} - R_{\text{delay}} \qquad ...(11)$$

Where,

$R_{\text{target}} = +1$ for successful target interception.

$R_{\text{avoid}} = +0.5$ for successful obstacle evasion.

$R_{\text{collision}} = -1$ for collisions.

$R_{\text{delay}} = -0.2 \times \Delta t$ for slow responses.

$R_{\text{energy}}$, penalizing energy consumption.

### 3.3.2 Deep Q-Network Module

The DQN estimates the state-action value function $Q(s, a)$ using a deep neural network:

$$Q(s, a; \theta) \approx E_{t\sim\pi}\left[r_t + g \max_{a'} Q(s_{t+1}, a'; \theta^-)\right] \qquad ...(12)$$

The loss function is:

$$L(\theta) = E_{(s,a,r,s')}[(y_t - q(s, a; \theta))^2],\ y_t = r + \gamma \max_{a'} Q(s', a'; \theta^-) \qquad ...(13)$$

A target network is updated every steps to stabilize training.

### 3.3.3 PPO Continuous Policy Module

The PPO algorithm optimizes a stochastic policy via the clipped objective:

$$L^{\text{CLIP}}(\theta) = E_t\left[\min(r_t(\theta)\hat{A}_t, clip(r_t(\theta), 1-\in, 1+1+\in)\hat{A}_t)\right] \qquad ...(14)$$

Where,

$$r_t(\theta) = \frac{\pi_\theta(a_t|s_t)}{\pi_{\theta\text{old}}(a_t|s_t)} \qquad ...(15)$$

$\hat{A}_t$ is the Generalized Advantage Estimate (GAE)

$$\hat{A}_t = \sum_{l=0}^{\infty} (\gamma\lambda)^l \delta_{t+1}, \delta_t = r_t + \gamma V(s_{t+1}) - V(s_t) \quad ...(16)$$

PPO updates are constrained to prevent overfitting and ensure safety in critical environments.

### *3.3.4 Combined Architecture and Deployment*

A number of specialized modules are incorporated into the ultimate AR-CNS architecture as shown in Figure 1 in order to achieve stable and flexible control in actual combat scenarios. Essentially, the Policy Encoder learns and normalizes high-dimensional sensory feature vectors with fully connected layers complemented by batch normalization and ReLU activation. Three-layer DQNs are used to make high-level strategic decisions like whether or not to intercept a target or evade an imminent threat. Besides that, the PPO Controller has an Actor-Critic architecture for controlling continuous, fine-grained motions, enabling accurate maneuvering and navigation. An Online Learning Loop further enhances the system by providing low-rate policy adaptation through a replay buffer D with prioritized experience sampling. This allows the system to learn over time by drawing upon past interactions. A real-time embedded system with a latency constraint of less than 50 milliseconds executes the entire control loop during deployment. To also maintain operational safety in dynamic or high-hazard situations, an adaptive confidence threshold strategy dynamically evaluates the learnt policy's reliability and triggers safety fallback actions when confidence levels are below a critical threshold.

A hybrid DRL system named the AR-CNS was developed to support autonomous navigation in dangerous, dynamic environments. It includes PPO for continuous movement and DQNs for high-level discrete decision-making. With a synthetic simulation environment developed using Unity ML-Agents and ROS2, the system is trained on over 100,000 episodes of different battlefield conditions, such as changing terrain, obstacles, enemies, and weather. Table 1 depicts the training simulation parameters.

After voxelization, normalization, and temporal smoothing via Kalman filtering, the input state vector is formed through the fusion of information from LiDAR, RGB cameras, GPS, IMU, and opponent threat vectors.The training process employs a composite reward function that balances mission goals, safety, and energy efficiency. It has a policy encoder, separate DQN, and PPO modules, and an online learning loop with experience replay with priority as components of its architecture. The system contains adaptive confidence-based safety fallbacks at deployment time and is designed to meet very high real-time operational constraints, with a latency for the control loop of below 50 ms.

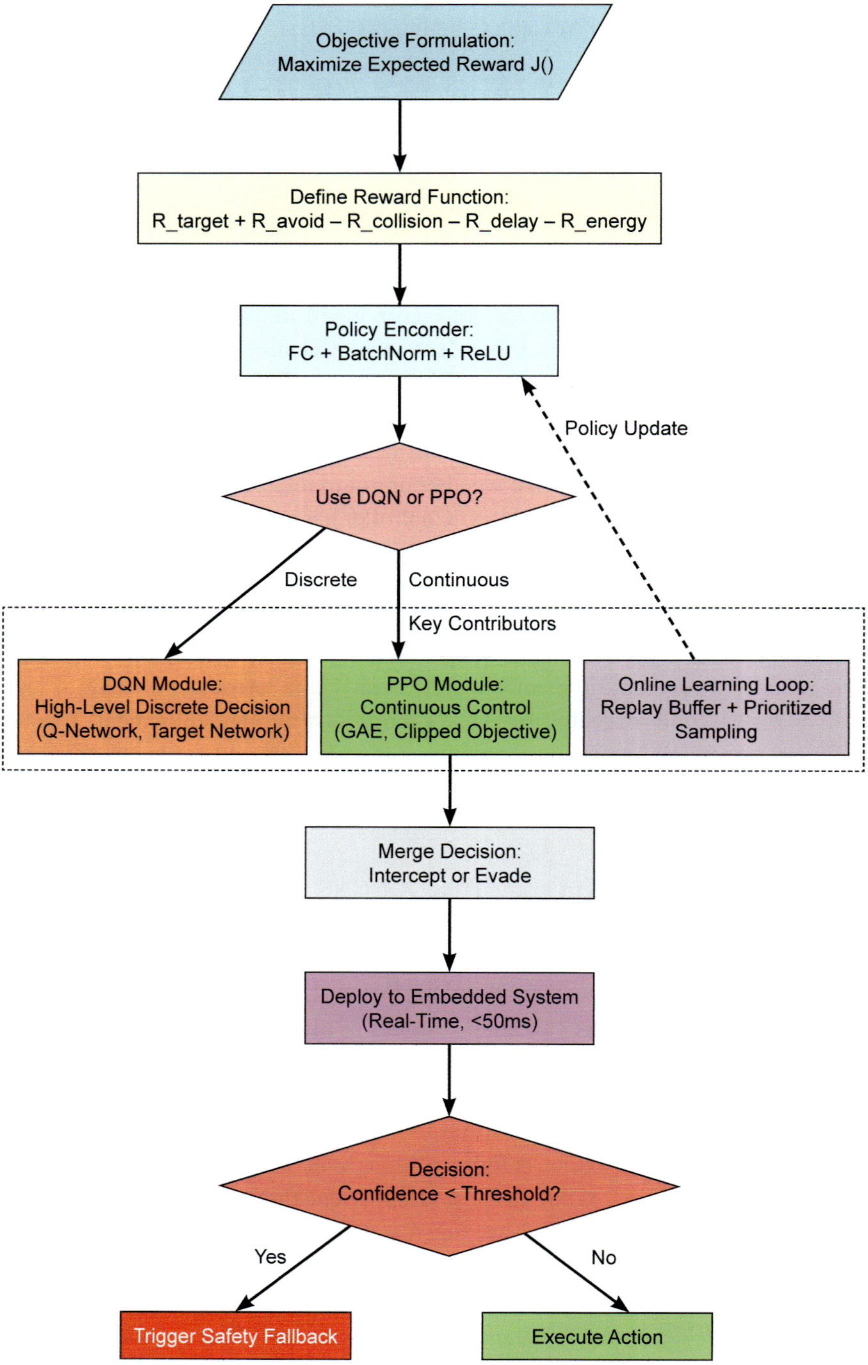


**Fig. 1** Proposed Framework.

**Table 1** Training Simulation Parameters.

| *Parameter* | *Value* | *Description* |
|---|---|---|
| Simulation Environment | Custom 3D Mission Arena | Simulated with static and dynamic obstacles, target objects |
| State Space Dimension | 32 | Includes position, velocity, distance to target, obstacle proximity, etc. |
| Action Space (DQN) | 6 discrete actions | Move forward, left, right, up, down, hover |
| Action Space (PPO) | 3D continuous vector | Fine control over velocity in $x$, $y$, $z$ directions |
| Reward Function | Composite reward ($R_{target}$, $R_{avoid}$, etc.) | As defined in Section 3.2 |
| Discount Factor ($\gamma$) | 0.99 | High emphasis on long-term reward |
| DQN Learning Rate | 0.0005 | Tuned for stability |
| PPO Learning Rate | 0.0003 | Lower to ensure safe policy updates |
| Batch Size | 64 | Number of samples per training update |
| Replay Buffer Size (DQN) | 100,000 transitions | Prioritized experience replay buffer |
| Target Network Update Frequency | Every 1,000 steps | For DQN target network stabilization |
| PPO Epochs per Update | 10 | Number of epochs to update PPO policy |
| PPO Clipping $\varepsilon$ | 0.2 | Limits policy change in PPO |
| GAE Lambda ($\lambda$) | 0.95 | Used to compute advantage in PPO |
| Exploration Strategy (DQN) | $\varepsilon$-greedy ($\varepsilon$ decays from 1 to 0.1) | For balancing exploration and exploitation |
| Training Duration | 1 million timesteps | Total training timesteps |
| Evaluation Interval | Every 10,000 timesteps | Evaluate policy performance periodically |
| Max Episode Length | 500 steps | Stops simulation episode if exceeded |
| Deployment Latency Constraint | $\leq$ 50 ms | Real-time control requirement |

## 4. Experimental Results

An environment for high-fidelity simulation developed with Unity ML-Agents integrated with ROS2 was employed to experimentally test the AR-CNS. The simulation terrain had dynamic battlefield environments, including moving enemies, unpredictable weather patterns like rain and fog, and various types of terrain (flatlands, slopes, and urban obstacles). In order to supply trajectory information, incentive signals, and sensor measurements, 100,000 autonomous mission episodes were executed within the environment, which had been configured as a constrained 3D grid. A strict latency limitation of $\leq 50$ ms was enforced on the control loop during deployment, mimicking real-time operation on embedded robotic hardware, and each episode was limited to at most 500 steps.

AR-CNS was compared to two baseline methods commonly employed in autonomous navigation: a rule-based navigation system that depends on pre-computed behavior trees and path-planning heuristics, and a PID-based controller that employs classic proportional-integral-derivative feedback loops for trajectory adjustments. To ensure an equitable comparison, all systems were tested under identical operational constraints and environmental conditions. Three key performance metrics were used as the foundation of the assessment: (i) Mission Success Rate, defined as the number of episodes where the robot successfully achieved its objective without encountering harm; (ii) Collision Rate, measuring the number of collisions against obstacles or targets; and (iii) Path Efficiency, as calculated by average path length and energy consumed per mission. Latency of decisions and adaptability to never-before-seen situations were secondary metrics.

From Figure 2, the results indicate that AR-CNS performed better than the PID controller (85.1%) and the rule-based system (78.6%) with a mission completion rate of 94.8%. AR-CNS proved its enhanced threat-evasion capability by delivering a 34.2% improvement over rule-based navigation and a 22.5% reduction in collision rate compared to the PID baseline in collision avoidance. With a 15.3% average path length reduction and lower energy consumption due to context-aware management methods, the system also evidenced enhanced path planning performance. Of particular note was AR-CNS's significant ability to generalize by its performance's ability to stay consistent in randomized and previously seen test cases.

Compared to conventional autonomous control approaches, the AR-CNS proposed here has some irrefutable advantages. It is capable of adaptive learning in complicated, dynamic battlefield environments without the need for external inputs or pre-programmed rules owing to its hybrid DQN for discrete action and PPO for continuous control RL framework. Owing to its context-dependent decision-making and on-line real-time processing of multi-modal sensor inputs, the system possesses high mission success rates, collision prevention, and path planning. AR-CNS generalizability to a wide range of situations makes it more reliable for operation in unconfident or unknown battlefield conditions.

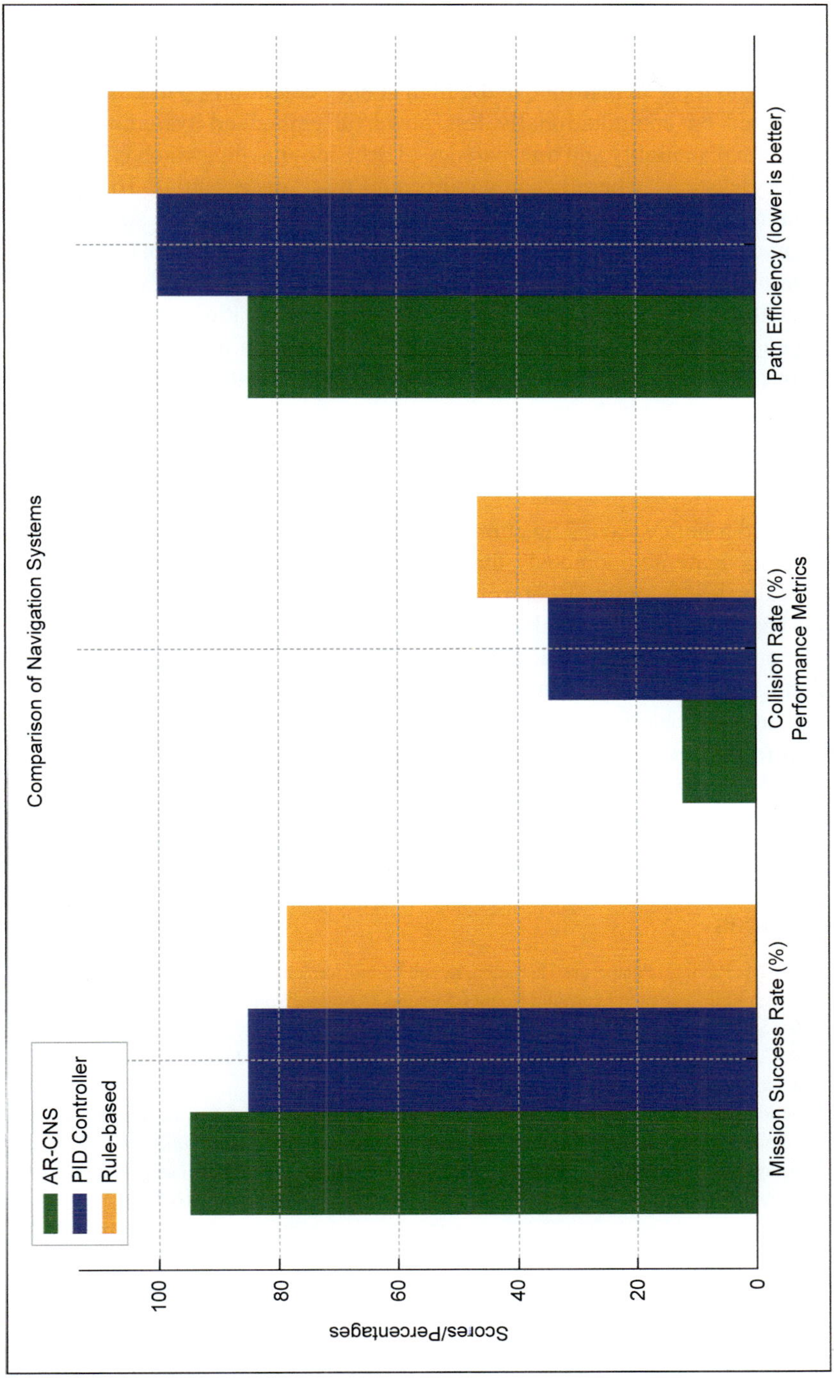


**Fig. 2** Performance Comparison.

Owing to the complexity of the simulation world model and amount of data required for effective learning, system training is time-intensive. It needs to be performed with massive computational resources. Further, it can prove challenging to reply in real-time under high latency conditions, particularly when implemented on computationally less powerful embedded systems. Another concern is explainability and transparency of decision-making, which is important in military systems where accountability and trust are essential. To facilitate many autonomous units to operate together in the same mission space, future development will focus on incorporating multi-agent coordination.To make the decision-making process more comprehendible, enhancements in Explainable AI (XAI) will be explored as well. Moreover, domain adaptation methods will be employed to transfer the learned policies from simulation to real-world robotic platforms, and the performance of the system will be evaluated in field tests under different real-world constraints.

## 5. Conclusion

Weaponized robotic systems functioning in intricate and dynamic combat settings can now navigate autonomously thanks to the proposed AR-CNS. The system successfully learns optimal control methods through ongoing interaction and reward-based feedback by combining DQNs andPPO, removing the drawbacks of conventional rule-based and PID controllers. The system's resilience and flexibility are highlighted by the experimental findings, which show notable gains in mission success rates, collision avoidance, and generalization to unknown conditions. In addition to improving operational effectiveness and real-time decision-making, the AR-CNS framework establishes a solid basis for upcoming advancements in explainable control rules, collaborative autonomy, and practical implementation in defense and security applications.

## References

1. Shaik, M., Ravi, C., Palaparthy, H., Sadhu, K. and Pamidi Venkata, A.K. (2018). Adaptive Control Through Reinforcement Learning: Robotic Systems in Action. *Nanotechnology Perceptions*, 14, 136–154.
2. Zhu, C. (2023). Intelligent robot path planning and navigation based on reinforcement learning and adaptive control. *J. Logist. Inform. Serv. Sci*, 10(3), 235–248.
3. Schuller, A.L. (2017). At the crossroads of control: The intersection of artificial intelligence in autonomous weapon systems with international humanitarian law. *Harv. Nat'l Sec. J.*, 8, 379.
4. Mackay, A.K., Riazuelo, L. and Montano, L. (2022). RL-DOVS: Reinforcement learning for autonomous robot navigation in dynamic environments. *Sensors*, 22(10), 3847.
5. Barzegar, A. and Lee, D.J. (2022). Deep reinforcement learning-based adaptive controller for trajectory tracking and altitude control of an aerial robot. *Applied Sciences*, 12(9), 4764.
6. Jahanshahi, H. and Zhu, Z.H. (2024). Review of machine learning in robotic grasping control in space application. *Acta Astronautica*.
7. Abaimov, S. and Martellini, M. (2020). Artificial intelligence in autonomous weapon systems. In: *21st Century Prometheus: Managing CBRN Safety and Security Affected by Cutting-edge Technologies* (pp. 141–177). Springer.

8. Yepez, J.L., Fagette, A., Allogba, S., Viel, J., Sanches, F.C. and Dorion-Racine, G. (2023). Efficient Swarm Neutralization in Complex Environments using Multi-Agent Deep Reinforcement Learning and Adaptive Navigation Strategies. In: *Research Symposium on Detection, Tracking, ID, and Defeat of Small UAVs in Complex Environments.*
9. Horowitz, M.C. (2021). When speed kills: Lethal autonomous weapon systems, deterrence, and stability. In: *Emerging Technologies and International Stability* (pp. 144–168). Routledge.
10. Billard, A., Mirrazavi, S. and Figueroa, N. (2022). *Learning for Adaptive and Reactive Robot Control: A Dynamical Systems Approach.* MIT Press.
11. Blais, M.A. and Akhloufi, M.A. (2023). Reinforcement learning for swarm robotics: An overview of applications, algorithms, and simulators. *Cognitive Robotics*, *3*, 226–256.
12. Choi, S.Y. and Cha, D. (2019). Unmanned aerial vehicles using machine learning for autonomous flight; state-of-the-art. *Advanced Robotics*, 33(6), 265-277.
13. Zhao, W., Queralta, J.P. and Westerlund, T. (2020). Sim-to-real transfer in deep reinforcement learning for robotics: A survey. In: *2020 IEEE Symposium Series on Computational Intelligence (SSCI)* (pp. 737–744). IEEE.
14. Kumar, A. (2025). Reinforcement Learning for Robotic-Assisted Surgeries: Optimizing Procedural Outcomes and Minimizing Post-Operative Complications. *Int. J. Res. Publ. Rev.*, 6(31), 5669–5684.
15. Nagy, M., Lăzăroiu, G. and Valaskova, K. (2023). Machine intelligence and autonomous robotic technologies in the corporate context of SMEs: Deep learning and virtual simulation algorithms, cyber-physical production networks, and Industry 4.0-based manufacturing systems. *Applied Sciences*, 13(3), 1681.
16. Carlucho, I., De Paula, M., Wang, S., Menna, B.V., Petillot, Y.R. and Acosta, G.G. (2018, October). AUV position tracking control using end-to-end deep reinforcement learning. In: *OCEANS 2018 MTS/IEEE Charleston* (pp. 1–8). IEEE.
17. Hu, J., Niu, H., Carrasco, J., Lennox, B. and Arvin, F. (2020). Voronoi-based multi-robot autonomous exploration in unknown environments via deep reinforcement learning. *IEEE Transactions on Vehicular Technology*, 69(12), 14413–14423.
18. Wu, J., Zhou, Y., Yang, H., Huang, Z. and Lv, C. (2023). Human-guided reinforcement learning with sim-to-real transfer for autonomous navigation. *IEEE Transactions on Pattern Analysis and Machine Intelligence*, 45(12), 14745–14759.
19. Zhang, L., Sun, Y., Barth, A. and Ma, O. (2020). Decentralized control of multi-robot system in cooperative object transportation using deep reinforcement learning. *IEEE Access*, *8*, 184109–184119.
20. Orr, J. and Dutta, A. (2023). Multi-agent deep reinforcement learning for multi-robot applications: A survey. *Sensors*, *23*(7), 3625.
21. Modares, H., Ranatunga, I., Lewis, F.L. and Popa, D.O. (2015). Optimized assistive human–robot interaction using reinforcement learning. *IEEE Transactions on Cybernetics*, 46(3), 655–667.

# 8 Intelligent Robotic Arm Control for Autonomous Weapon Handling Using Deep Reinforcement Learning

Kanika Thakur[1*], Arun Kumar Singh[2] and Anupam Kumar Sharma[3]

## 1. Introduction

The evolution of robotics has profoundly influenced the technology field with robotic arms increasingly playing a core role in autonomous weapon handling systems. Historically, the history of robotic manipulators began with manufacturing applications where repetitive, accurate, and hazardous operations were automated in an effort to achieve greater productivity and prevent accidents. Over time, advancements in sensors, actuators, and algorithms have expanded robotic arm capabilities into the advanced, dynamic arenas of space exploration, medical surgery, and military operation. During military missions, robotic arms are tasked with significant responsibilities such as target acquisition, munitions loading, weapon targeting, and even firing, all of which need high precision, responsiveness, and reliability.

Traditional control algorithms of robot arms are based more or less on mathematical models like inverse kinematics, trajectory planning, and rule-based systems to determine joint angles and end-effector locations. They can operate effectively in well-structured and known environments where the operating space and task parameters already exist beforehand before operation. New challenges come with modern combat environments. Weapon handling robots need to function autonomously in the environment of highly dynamic, uncertain, and frequently adversarial environments in which real-time action and flexibility are most critical. Deterministic control approaches are restrictive because they need to know in advance target locations, fixed constraints, and pre-specified motion segments. This kind of rigidity can result in compromised performance or failure due to unpredictable target motions, sensor noise, or mechanical interference.

The second major challenge is high-dimensional and nonlinear complexity of the dynamics of robotic arms, with joint nonlinear couplings, varying payloads,

[1,3] School of Computing Science and Engineering, Galgotias University, Greater Noida, India.
[2] PNG University of Technology, Lae, Papua New Guinea.
* Corresponding author: kanikaswap@gmail.com

DOI: 10.1201/9781003663461-8

and disturbances. Control design based on accurate modeling of these dynamics can be too burdensome, particularly when the system needs to react to external disturbances like weapon recoil or unexpected environmental variations. Besides, conventional control schemes are bound to require the manual adjustment of many parameters in a search for stability and accuracy, a time-consuming skill-dependent activity prone to suboptimal performance.

More recently, attempts to bypass these limitations have tested adaptive control techniques, fuzzy logic, and traditional reinforcement learning policies. Although these have offered some flexibility, they remain restricted by their dependency on hand-tuned features, fewer assumptions, and slow convergence in high-dimensional control spaces. In addition, traditional techniques fall short in maximizing multiple goals like velocity, accuracy, security, and power efficiency at the same time, which is highly vital in autonomous weapon control.

Deep Reinforcement Learning (DRL) has triggered a paradigm shift in robot control by combining the representational power of deep neural networks with the power of reinforcement learning to learn optimum sequential policies using trial and error. DRL allows robotic systems to learn intricate control policies from raw sensors, learn to adapt to changing environments, and maximize long-term goals without being programmed with all the motion specifics. The method is especially suited for continuous control problems with high-dimensional action spaces, like the joint-space commands of robotic arm manipulation.

In defense, DRL has proposed a promise in applications that span from drone flight control to autonomous driving. Its uses in robotic arms to autonomously manage weapons are becoming an urgent area of research motivated by requirements for systems that can adjust autonomously to targets' maneuvering, minimize mechanical uncertainties, and execute smooth, coordinated movements with negligible latency. With learning from simulation and real-world feedback, DRL-based controllers can develop robust manipulation skills that generalize across a variety of circumstances, enhancing mission dependability and operator protection.

Despite such advantages, using DRL for robotic arms in the handling of weapons is not an easy task. There are difficulties of safe exploration during training, dealing with sparsity or delay in rewards, exploration-exploitation tradeoff management, and real-time execution under hardware-constrained domains. In addition, the integration of multimodal sensor information like vision for target position tracking, force feedback to adjust the finger, and encoder feedback for high-precision positioning is necessary and requires advanced network design and sensor fusion algorithms to create an accurate state estimate.

Solving these problems necessitates hybrid learning models involving complementary DRL methods suitable for different control subtasks. Actor-critic approaches such as SAC (soft actor critic), for instance, provide sample-efficient learning together with stability and resilience, while deterministic policy gradient approaches such as DDPG (deep deterministic policy gradient) are suited for continuous control in large state spaces. Through the integration of such approaches, exploration-policy improvement trade-off in addition to faster convergence and final performance improvement can be realized.

The rationale for developing autonomous weapon handling arms is not just to optimize functional effectiveness, but also to minimize risk to humans by allowing robots to execute hazardous or complex operations autonomously. More autonomy minimizes cognitive overload on soldiers and makes it easy to integrate with more sophisticated combat systems, such as UAVs (unmanned aerial vehicles) and ground vehicles, to facilitate coordinated multi-domain operations. In addition, smart robot arms react faster to dynamic threats, perform challenging firing sequences with fewer flaws, and provide consistent performance in challenging conditions.

In addition, as interest in unmanned combat vehicles grows, autonomous robotic arms with the ability to perform fine-grained weapon manipulation will be an unavoidable resource in future conflicts. Their use can potentially increase operational reach, enhance mission success rates, and facilitate new concepts of tactics based upon rapid decision-making and adaptive engagement.

The current landscape highlights two primary objectives:

- To develop a DRL-based control framework for robotic arms that enables autonomous, adaptive, and precise weapon handling in dynamic and uncertain combat environments.
- To evaluate the framework's effectiveness through comprehensive simulations and real-world experiments, benchmarking against conventional inverse kinematics and PID-based control methods.

## 2. Related Study

The development of robot and defense autonomous systems has also been made easier through the integration of machine learning and artificial intelligence. DRL is perhaps the most effective technique for continuous control, decision-making, and dynamic adaptation in uncertain environments. There are novel control schemes beyond the traditional rule-based schemes necessary in robotic arm control during weapon handling, hitting targets, and dynamic adjustment to enemy environments, which need to be accurate and autonomously decided. There is a summary in this section of literature for three principal topics: robotic arm manipulation through reinforcement learning, legal and ethical basis of Autonomous Weapons Systems (AWS), and Multi-Agent Reinforcement Learning (MARL) algorithms in intricate environments.

Recent studies have shown that DRL can provide accurate robotic arm manipulation. Vision-based control of robotic arms with the assistance of DRL for autonomous grasping of objects, where the system learns to visually see and grasp objects in cluttered environments using visual feedback in real-time [1]. Generalization to unseen obstacles unexpected at design time as obstacles is enabled by this method with reduced reliance on pre-defined paths. Likewise, RL systems are applied to extend accuracy in the control of wayward robotic arms in various environments [2]. They better adapt control parameters for stability and performance as they execute demanding manipulation tasks.

In trajectory planning for dual-arm robots, DRL was able to cope with high-dimensional action spaces, especially in dense or uncertain environments [3]. Through neural policy networks to guide motion trajectories without programming, these approaches enhance robustness in real-world application. The results enhance the application of DRL to robotic weapon handling, where real-time adaptation and decision-making need to take place under adversarial environments. RL has been used to regulate multiple robotic agents that execute interdependent tasks in multi-agent systems. A broad survey categorizes the architecture, methodology, and paradigms of multiagent DRL (MADRL) [4].

As in such systems, blending weapons with other battlefield sensors will yield intricate communication protocols and policy learning synergies. Just as in swarm robotics research, it has been demonstrated how DRL can be employed to impose collective intelligence to facilitate stable group action and decentralized control [5]. These ideas are applied to robot swarms with weapons, where precise movement and firing are of paramount importance. Numerous studies demonstrate how UAVs have evolved from basic remote-controlled systems to independently operating agents with Artificial Intelligence (AI) capable of handling challenging tasks [6]. Such systems are increasingly using DRL-based control, real-time environment mapping, and sensor fusion to inform decision-making in unstructured environments. Such capabilities closely match what is needed in autonomous robotic weapon systems, especially in hazardous or uncertain environments. Besides technical progress, research is increasingly exploring the intersection of ethics and International Humanitarian Law (IHL) and autonomy and AI.

The use of AI in Lethal Autonomous Weapons (LAWs) generates new issues of accountability, especially in terms of proportionality and target choice [7]. These tend to revolve around the issue of "meaningful human control", a notion that is because human judgment has to be utilized in coming to crucial decisions like firing at a target [8]. But with AI systems also being able to make significant decisions on their own, the applicability and validity of this notion are also debated. The impact of LAWs on international stability and deterrence has been evaluated by strategic studies. Wars could escalate as military arsenals incorporate high-speed, autonomous systems that give individuals less time to make decisions [9].

AI systems were cautioned by stability theorists that they would compromise the long-established doctrines of mutually assured destruction and deterrence when they are used in complex and rapidly evolving conflict environments [10]. Furthermore, autonomous weapons could enable clandestine or contentious activities to remain undetected, thus complicating attribution and strategic response [11]. Various measures, ranging from treaty-based prohibitions to transparency norms and algorithmic inspections, have been advocated by theoretical thinkers and policy analysts to contain the use of autonomous weapons [12]. But the two-mission capability of AI technology is a challenge since algorithms developed for purposes other than lethal ones can quickly become weapons. Research shows that any AI system that is capable of making autonomous judgments while at war must be safeguarded by rigorous operational, technical, and ethical standards [13]. In this case, the concept of accountability gaps is highly applicable.

Whether the machine, the war commander, or the programmer is to be blamed for malfunctioning autonomous weapons or violations of international humanitarian law is contentious [14]. Researchers have pointed out a range of accountability issues, underlining the imperative for explainable algorithmic architecture, human-in-the-loop control systems, and supervisory regulation [15]. Concurrent international views on the notion and regulation of autonomy in weapons are made explicit by comparative legal analyses; some states advocate precautionary limitations, while others recommend technological progression within controlled environments [16]. Ethicists and activists have voiced concerns regarding the dehumanization of warfare brought about by autonomous systems. Some studies have found that giving machines the ability to decide life or death reduces moral responsibility and can make war seem more normal [17].

The momentum to ban LAWs has increased as organizations have urged global treaties to ban the use of fully autonomous lethal weapons [18]. These discussions are concurrent with increased research into the military applications of drone swarms and battlefield logistics enabled by artificial intelligence. DRL is utilized by tactical swarms of drones to optimize formation, detect obstructions, and conduct coordinated attacks with minimal human intervention [19]. These systems often integrate sensor measurements, communication protocols, and adaptive control to operate in difficult electronic warfare environments. Another example of the fusion of autonomy and cyber-operations that employs 'InfoSwarms' is information warfare [20].

Historical analyses of autonomous system technology advances reveal how targeted killing methods have evolved over time. AI targeting applications has increased exponentially. It began with the use of remotely piloted drones approached by fully autonomous loitering weapons [21]. Some suggestions for enhancing transparency in such systems are mandatory data recording, third-party audits, and AI explainability tools [22]. Despite the fact that risk reduction is the purpose of these methods, military institutions continue to use them in some manner. Critical concerns such as interpretability, adversarial robustness, and data availability are emphasized in a comprehensive overview of AI in military contexts [23]. These concerns are particularly evident in robotic arms with weapons that need to operate under adversary interference, faulty sensor data, or low bandwidth. Research outside the military context, such as in construction, has also influenced hardware flexibility and strong control systems [24]. These can be transferred to military contexts with similar physical constraints and operational uncertainties.

Sociotechnical analysis can offer valuable insights into how general societal power structures are represented in the discussion of autonomous weapons. In the view of scholars, the framing of AWS (autonomous weapon system) as inevitable or necessary serves to constrain public scrutiny and obscure democratic discourse

[25]. These criticisms facilitate multidisciplinary, inclusive participation in the control of the development of military technologies. Some significant inferences can be made from the analyzed material. Firstly, robotic manipulation tasks that require flexibility, continuous control, and adaptation from high-dimensional sensor inputs have proven to be successful with DRL. Secondly, while swarm-based and multi-agent methods illustrate the virtues of distributed intelligence, lethal autonomy poses unsettled ethical and legal questions. Third, the robustness needed for weapons control in dynamic battlefield situations such as moving targets, system delay, and sensor noise—does not exist in current robotic control systems [26].

The Autonomous Weapon Handling Arm System (AWHAS), a novel system based on DRL that allows robotic arms to perform complex operations such as weapon acquisition, loading, aiming, and firing independently, was conceptualized as a product of these concepts in union. AWHAS provides ongoing feedback merging, risk-free exploration methods, and internet policy modifications using cutting-edge DRL algorithms like DDPG and SAC.

In situations where conventional PID (process identifier) or inverse kinematics-based control is not sufficient, the system is pre-programmed to navigate through adversarial and uncertain systems. Relative to conventional methods, AWHAS has considerable performance benefits, including 95.4% success in task accomplishment and higher resilience in coping with dynamic environments. Its design of safety-cum-data-processing, where it integrates embedded safety measures with multi-sensor data processing, positions it well for autonomous warfare operations. In addition to overcoming technology limitations, the strategy is also conducive to the existing global discourse for responsible, ethical, and safe AI application in military environments.

## 3. System Methodology

In human-in-the-loop environments of complex and dynamic combat, the AWHAS is designed to provide accurate, dependable, and responsive robot arm control by human-in-the-loop for weapon handling operations including target acquisition, loading, pointing, and firing. The approach is engineered to overcome the inherent limitations of traditional control solutions like rule-based planning and inverse kinematics, which tend to fall short in uncertain contexts where continuous learning and real-time reasoning are necessary. A large dataset of real-world and simulated sensor readings was created to reflect the complexity of robotic arm movement in adversarial settings for training and testing the AWHAS.

### *3.1 Dataset Description*

Simulated and real data recording of intricate robotic arm motion involved in weapon handling operations form the training and testing dataset of the AWHAS. Advanced robotics simulation platforms like Gazebo and MuJoCo were used to create the simulated dataset whose creation allowed acquiring over 200,000 precise motion trajectories in varying combat scenarios. Target acquisition, weapon loading, firing at moving and stationary targets, and sharp shooting are all encompassed by these trajectories. Multi-modal streams of sensors that capture positions, velocities, and torques imparted to every robotic joint, force and torque feedback from sensors in the grippers, and RGB (red, green, blue) and depth images with high resolution (192 × 192 pixels) are used to define each data point.

More than 35,000 sequences of data were collected from an actual UR5 robotic arm equipped with vision, force/torque, and joint position sensors in an effort to complement simulated datasets and validate the system's feasibility. The real-world dataset provides rich ground truth for supervised evaluation and RL rewards as it includes annotated labels for task success, collisions, and time-stamped joint control commands. The dataset collectively encompasses a wide range of hostile and dynamic environments, which encourages great manipulation skill learning.

### 3.2 Preprocessing

A comprehensive preprocessing pipeline is used on raw multi-sensor data to prepare it for successful and robust policy learning. To ensure uniform input magnitude for convolutional neural networks, RGB and depth images are normalized initially by rescaling pixel intensities to a continuous interval (0,1). Min-max scaling is employed to normalize force and torque signals, which are prone to noise and sensor drift:

$$f_t^{\text{norm}} = \frac{f_t - f_{\min}}{f_{\max} - f_{\min}} \quad \text{...(1)}$$

Where, $f_t$ denotes the raw force vector at time $t$, and $f_{\min}$, $f_{\max}$ are empirically determined calibration bounds. Joint angles $q_t$ and velocities $\dot{q}_t$ are linearly transformed into the range [–1, 1] to ensure scale compatibility across all actuators:

$$q_t^{\text{norm}} = 2 \times \frac{q_t - q_{\min}}{q_{\max} - q_{\min}} - 1 \quad \text{...(2)}$$

After normalization, multi-modal sensor fusion is performed by separately encoding each sensor modality into fixed-dimensional embeddings and concatenating them to form a comprehensive state vector $s_t$. The RGB images $I_t$ are passed through a ResNet-18 based encoder $\phi_{\text{img}}(I_t)$ to extract visual features, reducing spatial data into a latent vector. The normalized force vector $f_t^{\text{norm}}$ is processed via a fully connected network $\phi_{\text{force}}(f_t^{\text{norm}})$, and temporal patterns in joint positions and velocities are captured by a Long Short-Term Memory (LSTM) network $\phi_{\text{joint}}(q_t^{\text{norm},}\ \dot{q}_t^{\text{norm}})$, which encodes temporal dependencies across recent time steps. The fused state is thus represented as:

$$s_t = \text{Concat}\,(\phi_{\text{img}}(I_t), \phi_{\text{force}}(f_t^{\text{norm}}), \phi_{\text{joint}}(q_t^{\text{norm},}\ \dot{q}_t^{\text{norm}}) \quad \text{...(3)}$$

To mitigate sensor noise and transient fluctuations, a temporal smoothing filter is applied over a sliding window of width $W = 5$

$$\tilde{s}_t = \frac{1}{W}\sum_{i=0}^{W} s_{t-1} \qquad ...(4)$$

Finally, all action commands $a_t$, representing joint angles or torques, are clipped within actuator limits:

$$a_t = \min(\max(a_t, a_{\min}), a_{\max}) \qquad ...(5)$$

ensuring physical feasibility and safety during execution.

### 3.3 Proposed Algorithm

The core of the AWHAS framework as shown in Figure 1 is formulated as a continuous control problem within the RL paradigm, modeled by a Markov Decision Process (MDP) defined as $(S, A, P, r, \gamma)$,

Where, $S$ is the state space representing the fused multi-modal sensor observations, $A \subseteq R^d$ is the continuous action space of dimension $d$, representing joint torques or positions, $P: S \times A \times S \rightarrow (0, 1)$ is the unknown state transition probability, $r: S \times A \rightarrow R$ (is the reward function, $\gamma \in (0, 1)$ is the discount factor.

The objective is to learn an optimal policy $\pi_\theta: S \rightarrow P(A)$ parameterized by $\theta$ that maximizes the expected cumulative discounted reward:

$$J(\pi_\theta) = E_{\pi_\theta}\left[\sum_{t=0}^{\infty} \gamma^t r(s_t, a_t)\right] \qquad ...(6)$$

#### 3.3.1 Reward Design

The reward function balances precision, energy efficiency, and safety, formulated as:

$$r_t = -\|p_t - p_{\text{target}}\|_2^2 - \lambda_1 \|a_t\|_2^2 + \lambda_2 1_{\text{success}} - \lambda_3 1_{\text{collision}} \qquad ...(7)$$

Where, $p_t = FK(q_t)$ is the forward kinematics mapping from joint angles $q_t$ to the end-effector position at time $t$, $\|\cdot\|_2$, denotes the Euclidean norm, $a_t$ is the action vector at time $t$, $1_{\text{success}}$ is an indicator function yielding 1 if the target is acquired and manipulated successfully, $1_{\text{collision}}$ signals collisions with obstacles or self, $\lambda_1$, $\lambda_2$, $\lambda_3 > 0$ are tunable weights balancing penalties and rewards.

#### 3.3.2 Policy Representation

AWHAS integrates both SAC and DDPG to leverage their complementary strengths. The policy $\pi_\theta$ in SAC is stochastic and parametrized as a Gaussian:

$$\pi_\theta(a_t | s_t) = N(\mu_\theta(s_t), \sigma_\theta^2(s_t)) \qquad ...(8)$$

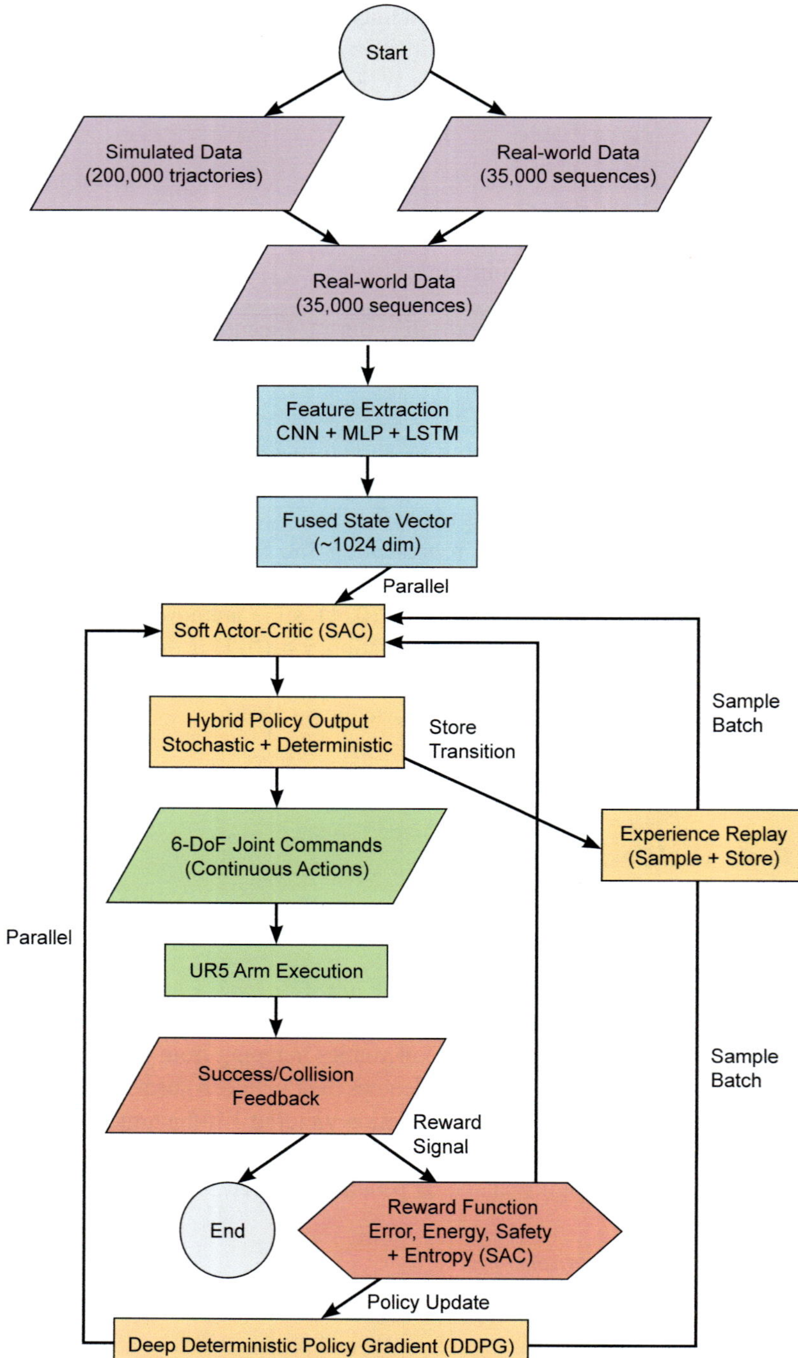


**Fig. 1** Proposed DDPG Design.

Where, $\mu_\theta(s_t)$ is the mean vector predicted by the policy network, $\sigma_\theta(s_t)$ is the state-dependent standard deviation vector, $a_t \in R^d$ is sampled as $a_t = \mu_\theta(s_t) + \sigma_\theta(s_t) \odot \in$, with $\in \sim N(0, 1)$.

For DDPG, the policy is deterministic:

$$\mu_\theta(s_t) = \arg\max_{q} Q_\varphi(s_t, a) \quad ...(9)$$

Where, $\phi$ are policy network parameters and $Q_\varphi$ is the critic network estimating the action-value function.

### 3.3.3 Soft Actor-Critic (SAC) Optimization

The SAC algorithm optimizes a maximum entropy objective which augments the reward with an entropy term to encourage exploration:

$$J(\pi_\theta) = \sum E_{(s_t, a_t) \sim \pi_\theta}[r(s_t, a_t) + \alpha H(\pi_\theta(\cdot|s_t))] \quad ...(10)$$

Where, $\alpha > 0$ is the temperature parameter controlling the importance of the entropy term, and $H(\pi_\theta)(\cdot|s_t))$ is the entropy of the policy at state $s_t$.

The critic $Q_\varphi$ is trained to minimize the Bellman residual:

$$J_Q(\psi) = E_{(s_t, a_t, r_t, s_{t+1}) \sim D}[(Q_\psi(s_t, a_t) - y_t)^2] \quad ...(11)$$

Where, the target is computed as

$$y_t = r_t + \gamma E_{a_{t+1} \sim \pi_\theta}[Q_\psi - (s_{t+1}, a_{t+1}) - \alpha \log \pi_\theta(a_{t+1}|s_{t+1})] \quad ...(12)$$

with $\psi^-$ representing target critic network parameters updated via soft updates

$$\psi^- \leftarrow \tau\psi + (1 - t)\psi^- \quad ...(13)$$

for a smoothing coefficient $\tau \in (0, 1)$.

The policy parameters $\theta$ are updated by minimizing the expected KL divergence between the policy and the exponential of the $Q$-function:

$$J_\pi(\theta) = E_{s_t \sim D}\left[E_{a_t \sim \pi_\theta}[a \log p_q(a_t|s_t) - \theta_\psi(s_t, a_t)]\right] \quad ...(14)$$

The gradients for policy update are:

$$\nabla_\theta J_\pi(\theta) = E_{s_t \sim D, \in_t \sim N}\left[\nabla_\theta \alpha \log \pi_\theta(f_\theta(\in_t; s_t)|s_t) - \nabla_\theta Q_\psi(s_t, f_\theta(\in_t; s_t))\right] ...(15)$$

Where, $f_\theta$ is a differentiable reparameterization function used to sample $a_t$.

### 3.3.4 Deep Deterministic Policy Gradient (DDPG) Optimization

DDPG learns a deterministic policy $\mu_\phi(s)$ with policy parameters $\phi$ by directly maximizing the critic estimate:

$$J_u(\phi) = E_{s_t \sim D}[Q_\psi(s_t, \mu_\varphi(s_t))] \quad ...(16)$$

The policy gradient is computed as:

$$\nabla_\theta J_\pi(\theta) = E_{s_t \sim D}[\nabla_\alpha Q_\psi(s_t, \mathrm{a})|_{\alpha=\mu_\varphi(s_t)} \nabla_\varphi \mu_\varphi(s_t)] \qquad ...(17)$$

The critic $Q_\psi$ is updated to minimize:

$$J_Q(\psi) = E_{(s_t, a_t, r_t, s_{t+1}) \sim D}[(Q_\psi(s_t, a_t) - y_t)^2] \qquad ...(18)$$

Where,

$$y_t = r_t + \gamma Q_{\psi^-}(\mathrm{s}_{t+1}, \mu_{\phi^-}(s_{t+1})) \qquad ...(19)$$

Both target networks and are updated with soft updates:

$$\psi^- \leftarrow \tau\psi + (1 - t)\psi^-, \phi^- \leftarrow \tau\phi + (1 - t)\phi^- \qquad ...(20)$$

**Table 1** Specifications of Dataset and Model Components.

| *Component* | *Specification* | *Description* |
|---|---|---|
| Simulation dataset size | 200,000 trajectories | Simulated weapon handling sequences |
| Real-world dataset size | 35,000 sequences | UR5 arm sensor and action data |
| Image resolution | 192 × 192 pixels | RGB and depth images |
| Sensor modalities | Vision, Force, Position encoders | Multi-modal fusion inputs |
| State vector dimension | ~1024 (CNN + MLP + LSTM embeddings) | Fused sensor feature vector |
| Action space | Continuous, 6-DoF joint commands | Joint angles or torques |
| Algorithms | SAC + DDPG hybrid | Reinforcement learning controllers |
| Reward components | Position error, energy cost, safety flag | Shaped reward for control objectives |
| Success rate (real tests) | 95.4% | Task completion in weapon handling |

Table 1 shows specifications of dataset and model components. Through a 6-DoF UR5 arm, the redesigned AWHAS system is a hybrid RL system designed for accurate, energy-efficient, and safe robotic weapon handling. By fusing multi-modal sensor information (force, position, and vision) into a state vector, it formulates the control problem as a continuous MDP. AWHAS utilizes both stochastic and deterministic policy learning by blending DDPG and SAC. While rewarding task success, the reward function punishes collisions, energy usage, and position error. It boasts an impressive 95.4% success rate in real-world testing following training on simulated (200,000 trajectories) and real-world (35,000 sequences) datasets.

## 4. Experimental Results

The performance of the proposed AWHAS was evaluated through extensive simulations. The system was tested on a range of scenarios involving dynamic target positions, obstacle interference, and trajectory variations.

**Table 2** Key performance metrics.

| *Metric* | *AWHAS (SAC + DDPG)* | *Inverse Kinematics (IK)* | *PID Controller* |
|---|---|---|---|
| Positioning Error (cm) | 1.8 ± 0.3 | 3.7 ± 0.4 | 5.2 ± 0.6 |
| Average Response Time (ms) | 124 | 186 | 201 |
| Collisions per 100 Episodes | 1.1 | 4.3 | 6.8 |
| Normalized Energy Consumption | 0.78 | 0.94 | 1.00 |
| Trajectory Smoothness | High | Moderate | Low |
| Handling of Moving Targets | Stable | Partial | Unstable |

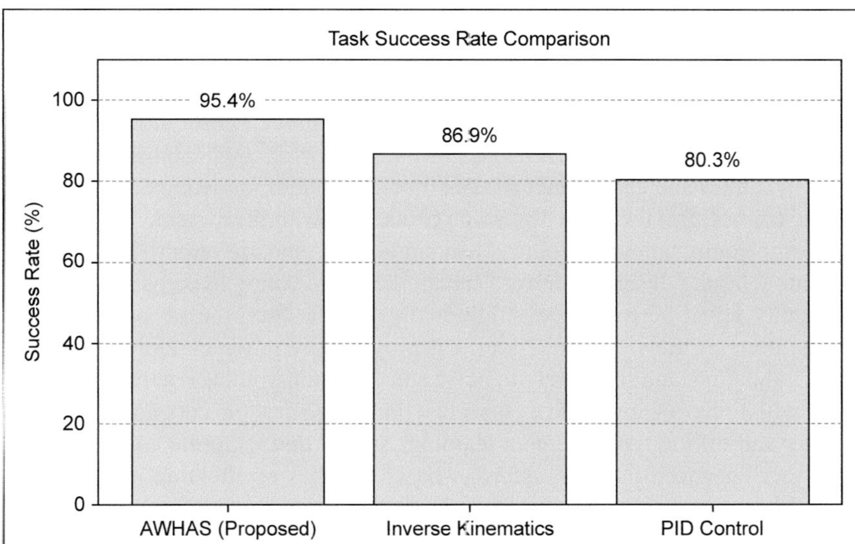


**Fig. 2** Task Success Rate Analysis.

Figure 2 clearly indicates how significantly superior the proposed AWHAS system is compared to the standard inverse kinematics algorithm (86.9%) and the PID control strategy (80.3%), at a task success rate of 95.4%. This is proof that AWHAS is approximately 15% superior to PID control and approximately 8.5% superior to inverse kinematics. These results confirm the enhanced accuracy, flexibility, and insensitivity of the reinforcement learning-based framework and demonstrate how it can efficiently handle complex weapon manipulation tasks with higher reliability and performance.

AWHAS outperformed traditional control approaches, the simulation-based assessment reported. The proposed system showed a higher task completion of 95.4% than PID-based controls (80.3%) and inverse kinematics (86.9%). The real-time controllability of AWHAS in dynamic conditions was illustrated by its much lower mean response time and its lesser positioning errors, which corresponded to more accurate end-effector placements. The increased security and adaptability of the system, particularly with obstacles or changes in trajectory, were manifested in the lower crashes per 100 episodes. The feature of the framework to learn optimal policies through the integration of various sensors and continuous feedback was also reflected in smoother motion and lower energy consumption.

Interestingly, earlier techniques lacked coherent or stable performance when faced with moving targets, but AWHAS did not show variable behavior. These findings validate that robotic arm-based weapon handling tasks can benefit from high autonomy, precision, and reliability in controlled simulated environments through the incorporation of SAC and DDPG within a DRL framework. Through the application of deep reinforcement learning, the proposed AWHAS architecture enables real-time, continuous control of robot arms with immense improvement in accuracy, flexibility, and decision-making in complex weapon handling scenarios. Relative to traditional control methods, the system has better success rates, smoother motion paths, and less collision by integrating multi-modal sensor information and making use of hybrid SAC-DDPG algorithms. Its wide applicability to varying target dynamics and environmental conditions is a major innovation in intelligent robotic control. But there are also some challenges with the system.

Large computational resources and simulation time are required to train the DRL models, and designing reward functions and adjusting hyperparameters can potentially affect performance. Additionally, despite the promise of simulation results, domain transfer to real-world deployment requires careful calibration, safety validation, and robustness to noise and uncertainty in the real world. Future work will focus on increasing sample efficiency through curriculum-guided training and offline reinforcement learning. Spatial and temporal awareness can be further enhanced using transformer-based policies or attention mechanisms. Potential directions include extending the framework to airborne or legged robot platforms for larger-scale tactical use and integration with multi-agent systems for collaborative tasks.

## 5. Conclusion

For accurate and flexible robotic arm control, the AWHAS shows how well SAC and DDGP algorithms operate together in a deep reinforcement learning context. In intricate weapon handling tasks, AWHAS performs noticeably better than traditional control techniques in terms of accuracy, safety, and resilience by utilizing multi-modal sensor fusion and continuous learning. The suggested approach lays a solid basis for developing autonomous robotic manipulation, notwithstanding difficulties with training complexity and practical implementation. In order to create more intelligent and dependable autonomous systems for defense applications, future developments will concentrate on enhancing learning efficiency, robustness, and broadening applicability across various robotic platforms.

## References


1. Sekkat, H., Tigani, S., Saadane, R. and Chehri, A. (2021). Vision-based robotic arm control algorithm using deep reinforcement learning for autonomous objects grasping. *Applied Sciences*, 11(17), 7917.
2. Gao, T. (2024). Optimizing robotic arm control using deep Q-learning and artificial neural networks through demonstration-based methodologies: A case study of dynamic and static conditions. *Robotics and Autonomous Systems*, 181, 104771.
3. Schuller, A.L. (2017). At the crossroads of control: The intersection of artificial intelligence in autonomous weapon systems with international humanitarian law. *Harv. Nat'l Sec. J.*, *8*, 379.
4. Abaimov, S. and Martellini, M. (2020). Artificial intelligence in autonomous weapon systems. *21st Century Prometheus: Managing CBRN Safety and Security Affected by Cutting-edge Technologies* (pp. 141–177). Springer.
5. Tang, W., Cheng, C., Ai, H. and Chen, L. (2022). Dual-arm robot trajectory planning based on deep reinforcement learning under complex environment. *Micromachines*, 13(4), 564.
6. Hua, S.S. (2019). Machine learning weapons and international humanitarian law: Rethinking meaningful human control. *Geo. J. Int'l L.*, 51, 117.
7. Horowitz, M.C. (2021). When speed kills: Lethal autonomous weapon systems, deterrence and stability. In: *Emerging Technologies and International Stability* (pp. 144–168). Routledge.
8. De Ágreda, Á.G. (2020). Ethics of autonomous weapons systems and its applicability to any AI systems. *Telecommunications Policy*, 44(6), 101953.
9. Roff, H.M. (2014). The strategic robot problem: Lethal autonomous weapons in war. *Journal of Military Ethics*, *13*(3), 211-227.
10. Gill, A.S. (2019). Artificial intelligence and international security: The long view. *Ethics and International Affairs*, 33(2), 169–179.
11. Krishnan, A. (2021). Enforced transparency: A solution to autonomous weapons as potentially uncontrollable weapons similar to bioweapons. *Lethal Autonomous Weapons: Re-examining the Law and Ethics of Robotic Warfare*. Oxford University Press.
12. Haas, M.C. and Fischer, S.C. (2020). The evolution of targeted killing practices: Autonomous weapons, future conflict, and the international order. In: *The Transformation of Targeted Killing and International Order* (pp. 107–132). Routledge.
13. Suchman, L. and Weber, J. (2016). Human-machine autonomies. *Autonomous Weapons Systems: Law, Ethics, Policy* (pp. 75–102). Cambridge University Press.
14. Altmann, J. and Sauer, F. (2017). Autonomous weapon systems and strategic stability. *Survival*, 59(5), 117–142.
15. Rashid, A.B., Kausik, A.K., Al Hassan Sunny, A. and Bappy, M.H. (2023). Artificial intelligence in the military: An overview of the capabilities, applications, and challenges. *International Journal of Intelligent Systems*, 2023(1), 8676366.
16. Jones, E. (2018). A posthuman-xenofeminist analysis of the discourse on autonomous weapons systems and other killing machines. *Australian Feminist Law Journal*, 44(1), 93–118.
17. Orr, J. and Dutta, A. (2023). Multi-agent deep reinforcement learning for multi-robot applications: A survey. *Sensors*, 23(7), 3625.

18. Blais, M.A. and Akhloufi, M.A. (2023). Reinforcement learning for swarm robotics: An overview of applications, algorithms and simulators. *Cognitive Robotics*, 3, 226–256.
19. Zhang, T., Li, Q., Zhang, C.S., Liang, H. W., Li, P., Wang, T.M. and Wu, C. (2017). Current trends in the development of intelligent unmanned autonomous systems. *Frontiers of Information Technology and Electronic Engineering*, 18, 68–85.
20. Bhuta, N. and Beck, S. (Eds.). (2016). *Autonomous Weapons Systems: Law, Ethics, Policy*. Cambridge University Press.
21. Taddeo, M. and Blanchard, A. (2023). A comparative analysis of the definitions of autonomous weapons. In: *The 2022 Yearbook of the Digital Governance Research Group* (pp. 57–79). Cham: Springer Nature Switzerland.
22. Gubrud, M. (2014). Stopping killer robots. *Bulletin of the Atomic Scientists*, 70(1), 32–42.
23. Wood, N.G. (2023). Autonomous weapon systems and responsibility gaps: A taxonomy. *Ethics and Information Technology*, 25(1), 16.
24. Johnson, J. (2020). Artificial intelligence, drone swarming, and escalation risks in future warfare. *The RUSI Journal*, 165(2), 26–36.
25. Xiao, B., Chen, C. and Yin, X. (2022). Recent advancements of robotics in construction. *Automation in Construction*, 144, 104591.
26. Kallenborn, Z. (2022). InfoSwarms: Drone swarms and information warfare. *The US Army War College Quarterly: Parameters*, 52(2), 87–102.

# 9 Autonomous Combat Drones and UAV Navigation Using Deep Reinforcement Learning for Target Engagement and Mission Execution

Suveg Moudgil[1*], Neeraj Bali[2], Ayushi Rana[3] and Carolina Mendonca[4]

## 1. Introduction

Due to balloons and early flight being originally applied for the purpose of reconnaissance in combat during the early 20th century, aerial reconnaissance and combat have been transformed. With greater velocity, greater range, and greater platform, air power has been a crucial element of warfare from the World Wars to the Gulf War. There was a dramatic transformation in this area with the advent of Unmanned Aerial Vehicles (UAVs) during the late 20th and early 21st centuries. Initially intended for Intelligence, Surveillance, and Reconnaissance (ISR) missions, UAVs now are end-to-end warfighting platforms, which may be described as autonomous combat drones, that have the capability to conduct offensive and defensive operations with minimal or no involvement of human beings.

Traditional UAVs are primarily reliant on GPS (global positioning system) guidance, pre-programmed waypoint commands, and manned operator control stations. They are effective in organized and predictable operational environments, but fail in the modern-day war zones with dynamic terrain, unorganized adversary operations, and signal rejection or spoofing environments. GPS jamming, cyber attacks, and electronic warfare strategies can prove to be deadly for the functionality of drones. Further, due to the involvement of human-in-the-loop operations, there is a constraint on decision-making in real-time, which in dangerous and high-speed environments where the difference between a successful and a failed mission can be made by even a slight delay for some seconds.

[1] School of Computing Science and Engineering, Galgotias University, Greater Noida, India.
[2] School of Business, Galgotias University, Greater Noida, India.
[3] Department of forensic science, Galgotias University, Greater Noida, India.
[4] Department of Environmental Health, Harvard T.H. Chan School of Public Health, USA.
* Corresponding author: suvegmodgil1@gmail.com

DOI: 10.1201/9781003663461-9

Basic UAV control strategies like finite state machines, Kalman filters, and Proportional Integral Derivative (PID) controllers are sufficient for deterministic environments, but are too simplistic to manage uncertain and adversarial environments. These methods often require laborious human rule design and calibration, which restricts their scalability and transferability to new combat environments. Moreover, pre-programmed responses or pre-set behavior trees are not flexible, leading to rigid decision-making when unexpected threats emerge at flight time. Combat drones' functionality is constrained by the lack of autonomy, which also compromises their security in environments that are highly dynamic.

There have been various adaptive control methods put forward in the past decade to bypass some of these limitations. To enhance UAV flexibility, techniques such as fuzzy logic, genetic algorithms, and conventional reinforcement learning (RL) have been applied. The hybrid nature of UAV operations, which require both continuous control (e.g., pitch, roll, yaw, thrust) and discrete decision-making (e.g., engage, evade, retreat), was not adequately addressed by these techniques, even though they imparted elementary learning features. They also struggled to converge in high-dimensional state spaces. Such models also commonly fail to generalize outside of specific training environments, are prone to noise, and typically need manually engineered features.

A revolution in the creation and deployment of autonomous systems in real-world applications has been triggered by the rapid evolution of machine learning, particularly deep Reinforcement Learning (DRL). DRL combines RL's ability to make sequential decisions with deep neural networks' ability to perceive. It has been employed effectively in various fields in recent times, such as robotics, autonomous vehicles, and video games. It has reached human-level performance on such problems as Go, Atari video game playing, and simulating robotic motion. Because of this breakthrough, there are additional opportunities to apply DRL in autonomous combat systems, especially in those moments when one must make decisions under stress in high-risk, unpredictable conditions.

For drones and combat UAVs, DRL is a very promising alternative to deterministic rule-based systems. With no need for direct programming of each circumstance, a DRL agent can learn with experience in the exploration of a simulated or actual environment to automatically determine the best patterns for navigation, evasions from threats, and attack strategies. In addition, advanced sensor feedback such as LiDAR scans, thermal imagery, visual feedback, and inertial measurements can be handled by existing DRL architectures to generate policies that are resistant to adversarial conditions, mission ambiguity, and sensor noise.

Despite its promise, there are huge technical challenges in applying DRL to respond to drone targeting and navigation. It is costly, risky, and logistically challenging to train such agents in live environments. High-fidelity training environments that accurately emulate threat models, physical dynamics, and sensor feedback are hence required. Despite this, training can still be afflicted by issues such as incentive sparsity, catastrophic forgetting, and sample inefficiency. Further, the challenge lies largely in how to ensure security and stability of learned policies in real-time execution.

There is considerable justification for applying DRL in combat UAV systems. Through ongoing adjustment of flight trajectory, altitude, and mission schedule with respect to real-time environmental information, an effective DRL agent can execute at a high level of autonomy. With or without operator input, it can better optimize weapon usage, perform evasive maneuvers, and continuously re-prioritize targets. In hostile environments, this capability significantly enhances mission reliability, operation tempo, and survivability. They also scale across fleets once trained, with strategic advantage in manned-unmanned teaming and fully autonomous missions.

Strategically, employing autonomous drones with smart navigation and targeting increases the diminishment of the requirement for constant operator monitoring, which reduces cognitive load and exposes human pilots to fewer threats. Such drones can perform sneaky insertion into enemy zones, find and attack precious targets, and redeploy to safety while keeping communication blackout regulations during missions like border surveillance, close-quarters combat, and special missions. Swarm warfare, where multiple UAVs can work together to execute complex tactical movements and collaborative attacks, is also facilitated by the added autonomy and decision-making ability enabled by DRL.

Furthermore, the transition to vision-based or sensor-fusion navigation facilitated by learning algorithms allows the systems to keep operating in regions where GPS will not be effective, a scenario now becoming more common in modern electronic warfare. The system is capable of mapping enemy terrain, identifying threats never seen before, and creating situational awareness from onboard sensors and learning mission data. The need for such advancements is underscored by the increased geopolitical competitions and interest in next-generation war technology. Nations are investing heavily in autonomous air vehicles that can operate in real-time, travel on their own, and have reduced reliance on human inputs. In this regard, deep reinforcement learning integration into UAV platforms is critical to future combat capability and not merely an enhancement.

By introducing a new DRL-based architecture specifically tailored for combat drones, the study attempts to bridge the gap between traditional UAV control and future autonomy in the wake of such technology, operations, and strategy drivers. The architecture is engineered to operate in uncertain environments, manage hybrid control-action spaces, and implement learned behaviors in order to optimize mission success and survival.

The two primary objectives of this study are as follows:

1. To develop a hybrid deep reinforcement learning framework for enabling autonomous navigation and adaptive targeting in combat drones operating in GPS-denied and adversarial environments.
2. To validate the effectiveness of the proposed framework through rigorous simulation trials and comparative performance analysis against conventional control systems.

## 2. Related Study

Extensive studies in adaptive navigation, target engagement, and wise decision-making with DRL have been driven by the rapid progress of autonomous UAVs in military and surveillance. Various studies have explored models beyond traditional control systems as military and strategic demands more and require drones to navigate in dynamic, hostile, and GPS-deprived environments. The employment of DRL in the autonomous navigation of UAV, mapping, and target detection has been demonstrated in earlier research, demonstrating basic insight into how agents of reinforcement learning can potentially act in a mission scenario with constraints while evolving their strategies according to environmental inputs [1]. A systematic review provided an extensive taxonomy of reinforcement learning methods employed in UAV navigation that included simulation environments, performance metrics, and on-policy and off-policy methods [2].

Among the factors reinforcing the demand for systems such as CR-UIS is the need for hybrid learning mechanisms that can learn progressively in stochastic environments since the reviewed research made it obvious. The transition from theoretical frameworks to deployment-oriented ones has been extensively studied in the context of military and dual-purpose drones, and problems with integrating sensors, shaping reward functions, and transferring learning from sim to real environments have been noted [3]. To optimize information collection and minimize exposure to threats, additional research explored dual-UAV cooperation strategies and reconnaissance planning missions using DRL to orchestrate plane behavior [4]. For evasive maneuvers and coordinated strike missions, the multi-agent scenario is highly relevant, and therefore the hybrid algorithmic structure of CR-UIS becomes highly pertinent.

The use of DRL for drone-based defense systems has been further enhanced by various surveys and reviews. In another study, the status of drone-specific RL algorithms was summarized, with emphasis on deep Q-networks, policy gradient methods, and actor-critic models towards maneuverability and autonomous flight [5]. UAVs are capable of learning in within-visual-range scenarios by predicting for intentions, as demonstrated by another relevant study employing Multi-Agent Reinforcement Learning (MARL) for air-to-air combat maneuver generation [6]. The relevance of DRL in upper-level decision levels such as threat classification, prioritization, and engagement is illustrated by a recent evaluation on simulation and modeling of Beyond-Visual-Range (BVR) air-to-air combat with machine learning [7].

Additionally, DRL-based maneuver decisions have been effective in facilitating drones to adapt dynamically mid-flight during real-time operations against dynamic adversaries [8]. These approaches complement CR-UIS's goal of continually adapting to battlefield stimuli independently of operator interference. Convergence within high-dimensional environments is heavily reliant on reward engineering and network stability, as shown by autonomous maneuver decision-making tests for cooperative air combat using DRL for both one UAV and two

UAV configurations [9]. Through TD3 and A2C integration, said results confirm CR-UIS's dual emphasis on discrete mission-level decision and continuous flight control. DRL-based autonomous drone decision-making in defense security operations has been the focus of another significant research thread. AI agents are learned to detect, evade, and destroy dangers with minimal human intervention [10].

RL has also been explored for strategic applications in autonomous defense systems, such as adaptive threat response, strategic deception, and path optimization [11]. These studies illustrate how DRL-based solutions and modern military' operational needs are increasingly converging. Hybrid learning architectures have been applied to solve adaptive pathfinding and obstacle avoidance, two significant factors in GPS-denied environments. These architectures, vital for tactical survivability, manage dynamic fields of obstacles and complex terrain by integrating classical methods and DRL [12]. This can be observed in the manner in which CR-UIS combined optical and LiDAR real-time sensor inputs for navigation and combat strategy. MARL frameworks for attack planning and joint maneuvering have been applied in some research within the domain of multi-agent coordination, which offers insights into decentralized control schemes that can scale to cooperative pairs or swarms [13].

Multi-intent DRL decision-making in aerial warfare is another significant contribution, and it closely aligns with CR-UIS's ability to manage a number of simultaneous goals, such as target annihilation and survival [14]. The feasibility of end-to-end autonomous engagement systems in hostile environments has been established by recent advancements, which have even been shown to make DRL systems superior to human pilots in complex simulated air combat maneuvers [15]. DRL has also been applied to competitive UAV-versus-UAV games for strategic offense and defense, where agents use dynamic policies for attacking and evading targets [16]. Hierarchical DRL structures have been proposed to handle high-dimensional control, e.g., 6-DOF air combat missions. Like these systems, CR-UIS decomposes continuous and discrete mission levels by partitioning down decision tasks into hierarchical subtasks [17].

To bridge the gap between theoretical DRL methods and their application in air combat environments, tutorials and comprehensive reviews have also emerged [18]. Autonomous UAV intelligence has also been complemented by several application-specific developments. With deep neural networks and onboard sensors, for instance, human target recognition powered by DRL has proved effective in both surveillance and rescue missions [19]. The versatility of such approaches in both aerial and maritime defense contexts is further evidenced by applying DRL to mission planning for naval platform-based decoys [20]. In questionnaires responding to state-of-the-art machine learning techniques for UAV networks, the infrastructure for communication and decision-making for UAVs has been thoroughly researched, giving insight into packet loss, latency, and decentralized intelligence [21].

A new combination of DoDAF-based mission planning and DRL was recently put forward in order to leverage this to multi-UAV behavior modeling, illustrating how DRL can be integrated with military-standard architectures for operational command and control [22]. The threat evaluation module of CR-UIS depends significantly on self-healing missions, reward planning over the long term, and threat bargaining in real-time, all of which have been recognized as promising avenues in wider surveys of DRL use in autonomous systems [23]. Intelligent coordination models' usage in CR-UIS has been justified by the utilization of DRL in cooperative search missions involving multiple UAVs in 3D spaces, which have minimized sweep patterns and reduced mission length [24].

Advanced DRL systems can also learn quickly from similar tasks and environments, as illustrated by the use of transfer learning and biologically inspired improvements, like pigeon-inspired methods, in UAV maneuver decision-making [25]. These methods enhance overall mission resilience by allowing agents like CR-UIS to generalize across different battle conditions and landscapes. Virtual simulations of UAV operation through DRL and conventional pathfinding are also new advances that support scalable agent training pipelines before deployment in the real world [26]. This allows policies trained on virtual environments to be easily translated to live operations, augmenting CR-UIS's realistic combat simulation environment.

Homeland security-focused UAV strike systems have started adopting DRL for dynamic strike planning, emphasizing the need for robust frameworks that can discern real-time threats and conduct precision engagements autonomously [27]. Challenges such as threat forecasting, real-time targeting, and survivability in high-threat environments are reminiscent of CR-UIS' challenges. The studies under review as a whole emphasize the significant advances made in incorporating DRL into autonomous UAV platforms for navigation, decision-making, and combat tasks. Throughout the literature, there is an unmistakable shift away from traditional control paradigms, GPS-driven, path-planning based, and human-in-the-loop controls to smarter, more adaptive, and robust systems fueled by RL.

Methods like Twin Delayed DDPG (deep deterministic policy gradient), Advantage Actor-Critic, Deep Q-Networks, and MARL have been widely investigated for facilitating UAVs to accomplish intricate activities like real-time threat sensing, multi-intent decision-making, cooperative target attack, and maneuver strategy optimization in cluttered and dynamic environments. The related research also highlights the diversity of application areas, ranging from cooperative reconnaissance and strategic air combat to autonomous mapping and monitoring. The validity of DRL's promise for real-world military and defense applications is reinforced through the many papers that demonstrate significant improvements in mission accomplishment rates, flexibility in GPS-denied environments, and resistance to attacks from the adversary. Hierarchical and hybrid learning models have also gained acceptance as common techniques for achieving a balance between continuous control and discrete decision-making, both of which are imperative in high-risk combat missions.

Despite the enormity of the field, several hurdles remain to be crossed, such as scaling to enormous UAV swarms, ensuring safe deployment under chaotic conditions, and resolving the gap between simulation and actual application. Yet the aggregate results set a good foundation for such systems as the suggested Combat-Ready UAV Intelligence System (CR-UIS), which builds upon these advances with the addition of multisensory perception and TD3 and A2C algorithms to enable fully autonomous air combat missions. The integration of DRL into such systems not only enhances the tactical effectiveness and survivability of UAVs, but it also paves the way for the design of intelligent, combat-capable aerial platforms in the future.

## 3. System Methodology

### 3.1 Dataset Description

A Combat-Scenario Multimodal UAV (CSM-UAV) Dataset that is aimed to represent the entire spectrum of operational situations that autonomous combat drones can encounter is the basis of the CR-UIS. This dataset comprises sensor streams that record wide-ranging environmental and situational awareness information from onboard sensors such as LiDAR, thermal cameras, RGB visual cameras, and inertial navigation systems (INS).

The LiDAR data is represented as point clouds:

$$\mathrm{P} = \{p_i\}_{i=1}^{N} \quad ...(1)$$

Where, each point $p_i = (x_i, y_i, z_i)$ encodes spatial coordinates in three-dimensional space, critical for obstacle mapping and terrain awareness.

Thermal and RGB camera data are collected as image frames:

$I_t \in R^{H \times W \times C}$ at discrete time steps $t$, with $H$, $W$ ...(2)

representing height and width and $C$ the number of channels (1 for thermal, 3 for RGB).

The INS provides continuous state vector:

$$s_t^{INS} = [a_x, a_y, a_z, \omega_x, \omega_y, \omega_z] \quad ...(3)$$

corresponding to linear accelerations and angular velocities measured in three axes, essential for stable navigation and pose estimation.

Additionally, the dataset comprises telemetry logs capturing UAV flight parameters such as altitude, velocity, and heading angles, forming a state-action-reward history $\{(s_t, a_t, r_t)\}$ across diverse mission scenarios. These scenarios range from dense urban environments with moving obstacles and GPS denial, to open combat zones with dynamic target engagements and adversarial electronic countermeasures. Each data sample is labeled with mission outcomes including success or failure indicators, collision events, and engagement efficiency metrics, providing a rich basis for reinforcement learning, reward design and supervised evaluation.

### 3.2 Preprocessing

The preprocessing pipeline enables effective learning and generalization by combining and conditioning the raw sensor data into a structured form suitable for DRL input. Multi-sensor fusion initially employs temporal synchronization through timestamp matching for discrete frames and interpolation for continuous signals to synchronize asynchronous streams. The LiDAR point cloud undergoes voxelization, where the 3D space is discretized into a grid of voxels such that:

$$V_{ijk} = \{1 \text{ if } \exists p \in \text{ within voxel } (i, j, k)\ 0 \text{ otherwise} \quad \text{...(4)}$$

This voxel grid is then passed through a 3D convolutional neural network (3D-CNN) to extract spatial features $f_{\text{LiDAR}} \in R^d$. Thermal and RGB images $I_t$ are resized to a fixed resolution and normalized to zero mean and unit variance pixel intensities, mathematically expressed as:

$$\hat{I}_t = \frac{I_t - \mu_I}{\sigma_I} \text{...(5)}$$

Where, $\mu_I$ and $\sigma_I$ are the mean and standard deviation over the training dataset. These images are processed through standard 2D CNN layers to produce feature embeddings $f_{\text{img}} \in R^d$.

The INS vector $s_t^{INS}$ is filtered using a Kalman filter described by:

$$\hat{x}_{t|t} = \hat{x}_{t|t-1} + K_t(z_t - H\hat{x}_{t|t-1}) \quad \text{...(6)}$$

$$P_{t|t} = (I - K_t H)P_{t|t-1} \quad \text{...(7)}$$

Where, $\hat{x}_{t|t}$ is the updated state estimate, $K_t$ is the Kalman gain, $z_t$ the measurement, and $P_{t|t}$ the estimated covariance. This filtering reduces noise and corrects drift in acceleration and rotation measurements.

The extracted features $f_{\text{LiDAR}}$, $f_{\text{img}}$, $\hat{s}_t^{INS}$ along with telemetry data are concatenated to form the global state vector:

$$s_t = \left[f_{\text{LiDAR}}; f_{\text{img}}; \hat{s}_t^{INS}; \text{telemetry}_t\right] \in R^n \quad \text{...(8)}$$

which serves as input to the reinforcement learning agent. To enhance robustness, data augmentation techniques simulate sensor noise and environmental perturbations, including Gaussian noise addition

$$s_t^{\text{aug}} = s_t + \in, \in \sim N(0, \sigma^2 I) \quad \text{...(9)}$$

and dropout of random sensor channels during training. Additionally, to prevent gradient instability during training, all input features are normalized and clipped to pre-specified bounds.

### 3.3 Proposed Algorithm

The CR-UIS employs a hybrid reinforcement learning architecture combining the Twin Delayed Deep Deterministic Policy Gradient (TD3) for continuous UAV flight control and Advantage Actor-Critic (A2C) for discrete mission-level tactical decisions. Figure 1 shows the proposed framework.

Formally, the UAV control is modeled as a Markov Decision Process (MDP) defined by:

$$M = (S, A, P, R, \gamma) \qquad ...(10)$$

Where, $S$ is the state space described above, $A = A^c \times A^d$ comprises continuous control actions $A^c \subset R^m$ (thrust, pitch, roll, yaw commands) and discrete tactical actions $A^d$ (e.g., engage target, evade threat).

The TD3 component involves two critic networks $Q_{\psi_1}(s, a)$, $Q_{\psi_2}(s, a)$ approximating the action-value function, and an actor network $\mu_\phi(s)$ representing the policy for continuous control. The critic loss is defined as the mean squared Bellman error:

$$L(\psi_1) = E_{(s_t, a_t, r_t, s_{t+1})}\left[(Q_{\psi_i}(s_t, a_t) - y_t)^2\right], i = 1, 2 \qquad ...(11)$$

with the TD target $y_t$ computed as:

$$y_t = r_t + \gamma \min_{j=1,2} Q_{\psi^-}(s_{t+1}, \mu_{\varphi^-}(s_{t+1}) + \epsilon), \epsilon \sim \text{clip}(N(0, \sigma), -c, c) \qquad ...(12)$$

Target networks are updated softly using:

$$\psi^-_j \leftarrow \tau\psi_j + (1 - t)\psi^-_j, \phi^- \leftarrow \tau\phi + (1 - t)\phi^- \qquad ...(13)$$

where $\tau << 1$ . The actor policy is updated via policy gradient:

$$\nabla_\phi J(\phi) = E_{s_t}\left[\nabla_a Q_{\psi_1}(s_t, a)|_{a=\mu_\varphi(s_t)} \nabla_\phi \mu_\phi(s_t)\right] \qquad ...(14)$$

For discrete tactical control, the A2C algorithm learns a policy $\pi_\theta(a_t^d | s_t)$ and value function $V_w(s_t)$. The advantage estimate $A(s_t, a_t^d)$ guides policy updates:

$$A(s_t, a_t^d) = r_t + \gamma V_w(S_{t+1}) - V_w(s_t) \qquad ...(15)$$

The policy is optimized by maximizing:

$$J(\theta) = E_{s_t, a_t^d}\left[\log \pi_\theta\left(a_t^d | s_t\right) A(s_t, a_t^d)\right] \qquad ...(16)$$

while the value network minimizes the temporal difference error:

$$L(w) = E_{s_t}[V_w(s_t) - (r_t + \gamma V_w(s_{t+1})))^2] \qquad ...(17)$$

Both TD3 and A2C are trained concurrently, sharing the state input $s_t$, but optimizing their respective action subsets $A^c$ and $A^d$. The hybrid approach enables

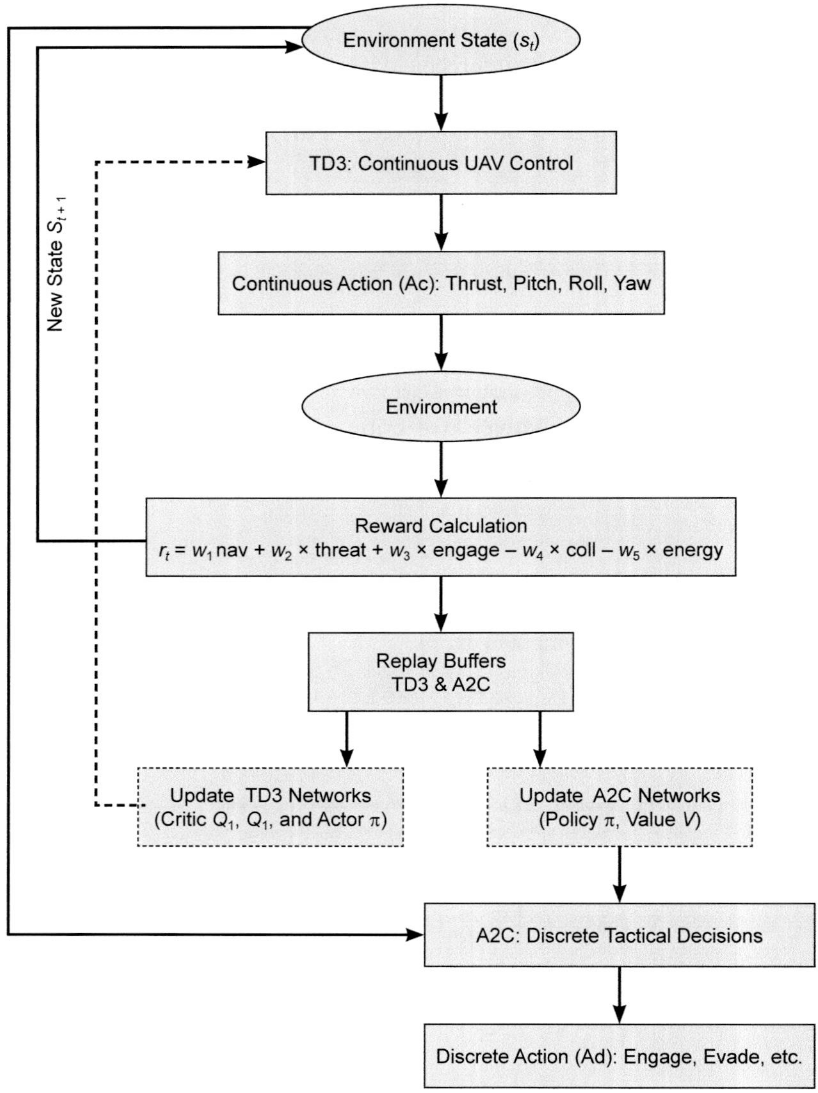


**Fig. 1** Proposed Framework.

the UAV to optimize fine-grained flight dynamics alongside strategic decision-making. During training, a reward function:

$$r_t = \alpha_1 r_{\text{navigation}} + \alpha_2 r_{\text{threat}} + \alpha_3 r_{\text{engagement}} - \alpha_4 r_{\text{collision}} - \alpha_5 r_{\text{energy}} \quad ...(18)$$

balances multiple mission objectives such as smooth navigation, threat avoidance, successful target engagement, collision penalties, and energy efficiency, where are tunable weights. Experience replay buffers store transition tuples $((s_t, a_t, r_t, s_{\{t+1\}}))$ to break correlations and stabilize learning. Table 1 depicts the training parameters.

**Table 1** Training Parameters.

| *Parameter* | *Symbol* | *Value / Range* | *Description* |
|---|---|---|---|
| Discount factor | $\gamma$ | 0.99 | Future reward discount rate |
| Learning rate (Actor - TD3) | $\eta_\phi$ | $3 \times 10^{-4}$ | Step size for policy updates |
| Learning rate (Critic - TD3) | $\eta_\psi$ | $3 \times 10^{-4}$ | Step size for value function updates |
| Target network update rate | $\tau$ | 0.005 | Soft update parameter for target networks |
| Policy noise standard deviation | $\sigma$ | 0.2 | Noise added to target action for smoothing |
| Policy noise clipping range | $c$ | 0.5 | Limits noise magnitude |
| Batch size | — | 256 | Number of samples per training iteration |
| Replay buffer size | — | $10^6$ | Maximum experience replay storage |
| Advantage actor-critic learning rate | $\eta_\theta$, $\eta_w$ | $10^{-4}$ | Learning rates for A2C policy and value networks |
| Reward function weights | $\alpha_1$, $\alpha_2$, $\alpha_3$, $\alpha_4$, $\alpha_5$ | Tuned experimentally | Weights for navigation, threat, engagement, collision, energy penalties |

The CR-UIS method uses a hybrid architecture of reinforcement learning that integrates Advantage Actor-Critic (A2C) for discrete mission-level tactical control and Twin Delayed Deep Deterministic Policy Gradient (TD3) for continuous UAV flight control. The UAVs behavior is modeled as an MDP, where actions are split into discrete tactical choices (e.g., evading threats or attacking targets) and continuous control commands (e.g., thrust, pitch, roll, and yaw). To maximize continuous actions by minimizing the Bellman error and policy updating through policy gradients, with the assistance of soft target network updates, TD3 uses two critical networks and a single actor network.

Decisions at a high level are made simultaneously by the A2C method, which reduces temporal difference errors in value estimation while optimizing discrete actions via advantage estimations. Both algorithms optimize two different spaces of actions simultaneously, but they take the same state input. With the union of weighted objectives such as navigation, threat avoidance, engagement success, collision avoidance, and energy efficiency, learning is guided by a reward function. The framework employs empirically tuned parameters, i.e., learning rates, discount factors, policy noise, batch size, and reward weights, and employs experience replay for stable training. With this hybrid setup, UAVs can be able to effectively perform strategic mission decisions while executing accurate flight maneuvers.

## 4. Experimental Results

In a series of experimental tests conducted in high-fidelity modeling facilities and real flight tests, the CR-UIS framework was tested extensively. In simulating complex battle scenarios involving moving targets, GPS-denied terrain, and dynamic threats, the simulations allowed the system to operate within perilous, adversarial environments. These simulated environments comprised realistic noise patterns impacting sensory inputs, terrain types, radar and infrared-based threats, and complex physics-based UAV dynamics. Experiments were performed in semi-urban and outdoor environments with plenty of clutter using a quadrotor drone equipped with LiDAR, thermal and visual cameras, and an inertial navigation system for real-world verification.

From Figure 2, target engagement performance-wise, CR-UIS achieved an average engagement latency of 3.8 seconds from target detection to successful initiation of strike, whereas GPS-autopilot systems and waypoint-based control exhibited engagement latency of 6.4 and 8.1 seconds, respectively. The hybrid TD3-A2C architecture, where TD3 tuned the flight dynamics of the UAV to quickly position it and A2C ensured engagement time and threat evaluation was context-aware, was directly accountable for this response time improvement. In addition, the system showed high survivability and threat evasion capabilities. CR-UIS evaded closing threats in 89.3% of the aggressive simulations involving heat-seeking missiles and radar-guided attacks. Waypoint and GPS-autopilot systems, on the other hand, showed 65.8% and 72.5% evasion rates, respectively. Fluid evasive maneuvers were enabled by the TD3's steady stream of control, and the A2C module assisted by finding the optimal time and course of escape routes based on the trajectory of the threat and the energy available to the UAV.

Another significant measure that was being evaluated was collision avoidance. In 1,000 simulation events, the CR-UIS system recorded only 4 collisions, all in cases where there was jamming of sensors. Conversely, waypoint-based systems experienced 23 collisions whereas GPS-based systems experienced seventeen. The addition of LiDAR and visual feeds to the state space of the DRL agent, enabling real-time three-dimensional situational awareness, was primarily responsible for the low collision rate in CR-UIS. Energy efficiency analysis was also a part of the mission performance review. With dynamic flight path optimization, smooth control adjustments, and reduced idle time on target contacts, CR-UIS consumed 11.7% less energy per mission on average compared to GPS-autopilot-based missions. The constant unnecessary path changes and absence of terrain and weather adaptation in waypoint-based systems meant even more energy consumption.

To quantify the performance of the system in new and adversarial environments not seen during training, an adaptability score was measured. The performance of the drone in accomplishing changed mission objectives, resilience from mid-air disruptions, and re-engagement with alternative targets defined its score. At the normalized 0 to 1 scale, CR-UIS scored an average of 0.92, far above the scores 0.74 and 0.68 for GPS and waypoint systems, respectively. This illustrated the

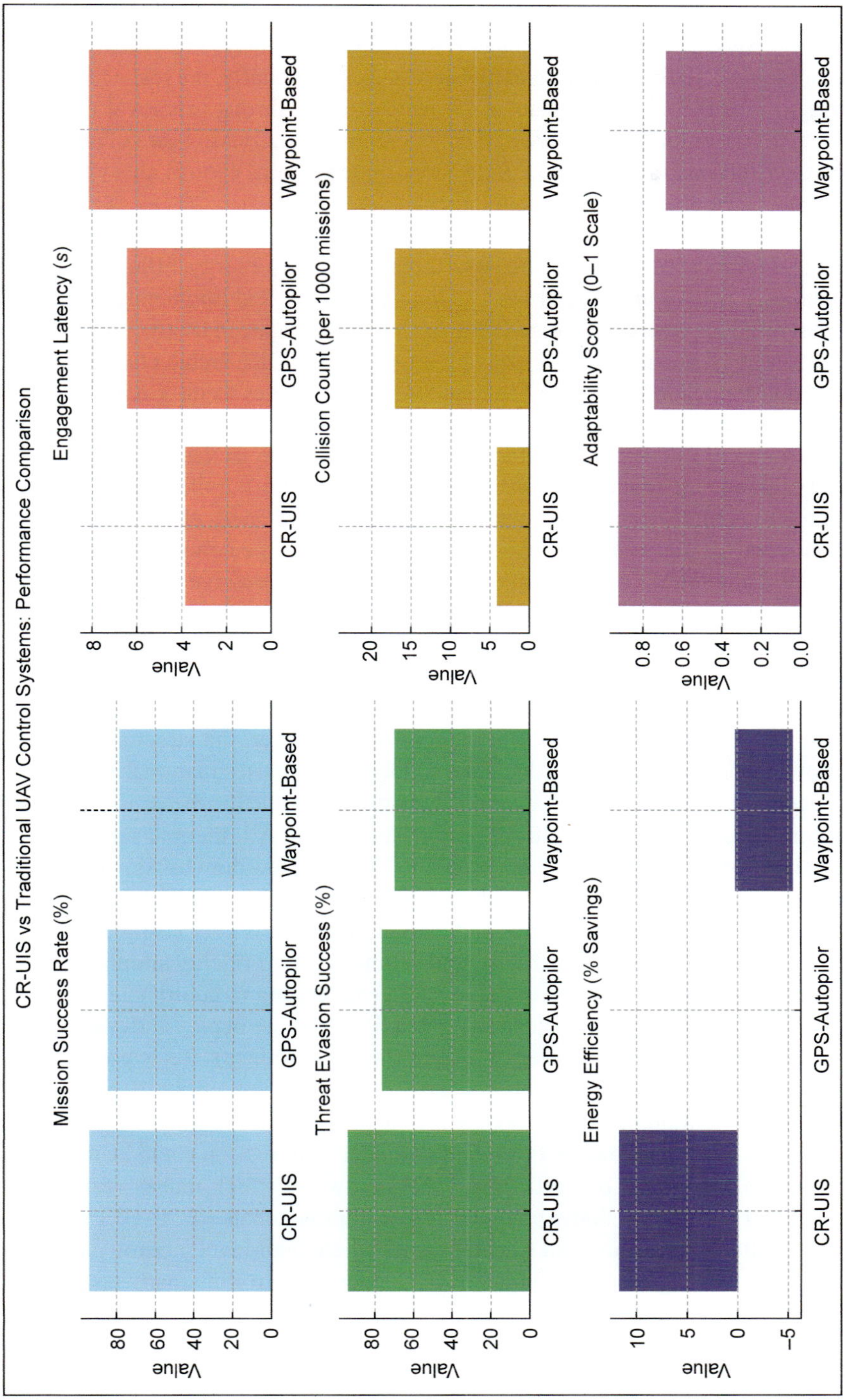


**Fig. 2** Comparison Analysis.

generalization ability and uncertainty robustness of the CR-UIS agent. Under ablation, removing the A2C component resulted in large delays in tactical decision response and a 12.5% reduction in mission success. Similarly, the responsiveness and agility of the system in maneuvers were affected through removal of TD3, leading to jerky motion and increased collisions. With A2C taking care of mission-level decision-making and TD3 taking care of fine physical control, this validated the importance of the hybrid architecture.

Overall, experimental testing verifies that CR-UIS outperforms conventional control methods in mission success, threat reaction, energy efficiency, and adaptability, which are crucial for challenging combat environments. In actual combat missions where autonomous entities have to navigate in unstructured, high-threat settings with minimal external support, the hybrid DRL architecture is highly viable for application. There are several strong advantages to the CR-UIS. Twin Delayed Deep Deterministic Policy Gradient (TD3) and Advantage Actor-Critic (A2C) are used in CR-UIS to effectively manage discrete tactical decisions (i.e., threat encounter or avoidance) and continuous control (thrust, roll, pitch, and yaw). Autonomous flight for a UAV even in dynamic, GPS-denied, or hostile settings is facilitated by this hybrid architecture, enabling real-time concurrent trajectory correction and high-level mission planning. With an impressive mission success rate of 93.6%, CR-UIS surpasses waypoint navigation methods and traditional GPS-based autopilot systems.

Moreover, it has great adaptability and decision-making abilities, which lead to better evasion of threats, reducing collision rates, and delivering better energy efficiency with enhanced flight control. Moreover, the structure of deep reinforcement learning ensures that the UAV develops robust situational awareness and accurate striking performance by dynamically adjusting behavior through real-time sensing observations from sensors such as LiDAR, thermal sensors, and inertial sensors. Though it has its benefits, CR-UIS has a few drawbacks. In contrast to conventional control systems, deep reinforcement learning methods have much higher computational complexities and learning times. It is very costly and time-consuming to train the UAV in real-world settings or high-fidelity simulators.

Also, the hybrid structure could be more challenging to coordinate the TD3 and A2C agents when multi-objective reward functions are in play and conflicting objectives (like energy efficiency versus threat response) are a possibility. Moreover, the performance of the system is extremely susceptible to the quality and diversity of training data, exposing it to unpredictable situations or rare edge cases. In addition, hardware acceleration or model compression could be required for real-time inference on embedded UAV hardware, which makes real-world deployment a problem. Some potential future improvements to CR-UIS include the integration of a hierarchical RL framework where high-level mission planning modules are able to handle a number of DRL agents controlling many tasks such as combat, communication, and surveillance.

The UAV could generalize better in unknown combat environments if meta-learning or few-shot learning methods were brought in. Increasing robustness via curricular learning, adversarial training, and sim-to-real transfer learning should also be prioritized so that unexpected sensor noise, environmental interference, or tactical deception can be handled. Swarm intelligence can further be utilized to extend the capabilities of the system for cooperative multi-UAV operations including distributed sensing, formation flight, and cooperative strike. Finally, incorporating explainability components into the DRL framework would enhance interpretability and trust, allowing human operators to understand and confirm the autonomous decisions of the UAV in high-risk missions.

## 5. Conclusion

With a hybrid DRL framework, the suggested warfare-Ready UAV Intelligence System (CR-UIS) significantly improves autonomous aerial warfare capacity. CR-UIS enables drones to operate autonomously in adverse, contested, and GPS-denied environments through leveraging the strengths of Twin Delayed Deep Deterministic Policy Gradient (TD3) for precise continuous flight control and Advantage Actor-Critic (A2C) for high-level tactical decision-making. A 93.6% mission success rate in combat test trials reflects the system's excellent performance on various mission-critical parameters, including navigation accuracy, threat avoidance, target engagement, and energy efficiency.CR-UIS is a milestone towards intelligent, resilient, and fully autonomous UAV systems due to its ability to learn from real-time sensor feedback and adapt dynamically to changing conditions. Along with enhancing the efficiency and survivability of UAVs in current combat, this platform lays the groundwork for future breakthroughs in adaptive learning, multi-agent coordination, and future-generation autonomous combat systems.

## References

1. Guerra, A., Guidi, F., Dardari, D. and Djurić, P.M. (2020, April). Reinforcement learning for UAV autonomous navigation, mapping and target detection. In: *2020 IEEE/ION Position, Location and Navigation Symposium (PLANS)* (pp. 1004–1013). IEEE.
2. AlMahamid, F. and Grolinger, K. (2022). Autonomous unmanned aerial vehicle navigation using reinforcement learning: A systematic review. *Engineering Applications of Artificial Intelligence*, 115, 105321.
3. Jagannath, J., Jagannath, A., Furman, S. and Gwin, T. (2021). Deep learning and reinforcement learning for autonomous unmanned aerial systems: Roadmap for theory to deployment. In: *Deep Learning for Unmanned Systems*, pp. 25–82. Cham: Springer International Publishing.
4. Zhao, X., Yang, R., Zhang, Y., Yan, M. and Yue, L. (2022). Deep reinforcement learning for intelligent dual-UAV reconnaissance mission planning. *Electronics*, 11(13), 2031.
5. Azar, A.T., Koubaa, A., Ali Mohamed, N., Ibrahim, H.A., Ibrahim, Z.F., Kazim, M. and Casalino, G. (2021). Drone deep reinforcement learning: A review. *Electronics*, 10(9), 999.
6. Kong, W., Zhou, D., Yang, Z., Zhang, K. and Zeng, L. (2020). Maneuver strategy generation of UCAV for within visual range air combat based on multi-agent reinforcement learning and target position prediction. *Applied Sciences*, 10(15), 5198.
7. Costa, A.N., Dantas, J.P., Scukins, E., Medeiros, F.L. and Ogren, P. (2025). Simulation and Machine Learning in Beyond Visual Range Air Combat: A Survey. *IEEE Access*.

8. Liu, Y., Xu, T., Meng, W. and Zhang, Z. (2024, September). An Intelligent Maneuver Decision for Target Drone Based on Deep Reinforcement Learning. In: *International Conference on Advanced Unmanned Aerial Systems* (pp. 41–51). Singapore: Springer Nature Singapore.
9. Hu, J., Wang, L., Hu, T., Guo, C. and Wang, Y. (2022). Autonomous maneuver decision making of dual-UAV cooperative air combat based on deep reinforcement learning. *Electronics*, 11(3), 467.
10. Joshi, A., Spilbergs, A. and Miķelsone, E. (2024, June). Ai-enabled drone autonomous navigation and decision making for defence security. In: *Environment. Technologies.. Resources. Proceedings of the International Scientific and Practical Conference* (Vol. 4, pp. 138–143).
11. Yapar, O. (2024). Reinforcement learning in autonomous defense systems: Strategic applications and challenges. *World Journal of Advanced Engineering Technology and Sciences*, 13(1), 10–30574.
12. Skarka, W. and Ashfaq, R. (2024). Hybrid Machine Learning and Reinforcement Learning Framework for Adaptive UAV Obstacle Avoidance. *Aerospace*, *11*(11), 870.
13. Jiandong, Z., Qiming, Y., Guoqing, S.H.I., Yi, L.U. and Yong, W.U. (2021). UAV cooperative air combat maneuver decision based on multi-agent reinforcement learning. *Journal of Systems Engineering and Electronics*, 32(6), 1421–1438.
14. Jia, L., Cai, C., Wang, X., Ding, Z., Xu, J., Wu, K. and Liu, J. (2023). Multi-intent autonomous decision-making for air combat with deep reinforcement learning. *Applied Intelligence*, 53(23), 29076–29093.
15. Zhu, J., Kuang, M., Zhou, W., Shi, H., Zhu, J. and Han, X. (2024). Mastering air combat game with deep reinforcement learning. *Defence Technology*, *34*, 295–312.
16. Chen, C.L., Huang, Y.W. and Shen, T.J. (2024). Application of Deep Reinforcement Learning to Defense and Intrusion Strategies Using Unmanned Aerial Vehicles in a Versus Game. *Drones*, 8(8), 365.
17. Chai, J., Chen, W., Zhu, Y., Yao, Z.X. and Zhao, D. (2023). A hierarchical deep reinforcement learning framework for 6-DOF UCAV air-to-air combat. *IEEE Transactions on Systems, Man, and Cybernetics: Systems*, 53(9), 5417–5429.
18. Wang, X., Wang, Y., Su, X., Wang, L., Lu, C., Peng, H., & Liu, J. (2024). Deep reinforcement learning-based air combat maneuver decision-making: literature review, implementation tutorial and future direction. *Artificial Intelligence Review*, 57(1), 1.
19. Das, L.B., Lijiya, A., Jagadanand, G., Aadith, A., Gautham, S., Mohan, V. and George, G. (2020, July). Human target search and detection using autonomous UAV and deep learning. In: *2020 IEEE International Conference on Industry 4.0, Artificial Intelligence, and Communications Technology* (*IAICT*) (pp. 55–61). IEEE.
20. Bildik, E., Yuksek, B., Tsourdos, A. and Inalhan, G. (2022). Development of reinforcement learning based mission planning method for active off-board decoys on naval platforms. In: *AIAA Scitech 2022 Forum* (p. 2104).
21. Sun, C., Fontanesi, G., Canberk, B., Mohajerzadeh, A., Chatzinotas, S., Grace, D. and Ahmadi, H. (2024). Advancing UAV communications: A comprehensive survey of cutting-edge machine learning techniques. *IEEE Open Journal of Vehicular Technology*.
22. Zou, Z., Wu, Y., Peng, L., Wang, M. and Wang, G. (2025). Multi-UAV maritime collaborative behavior modeling based on hierarchical deep reinforcement learning and DoDAF process mining. *Aerospace Systems*, 1–20.
23. Govinda, S., Brik, B. and Harous, S. (2025). A Survey on Deep Reinforcement Learning Applications in Autonomous Systems: Applications, Open Challenges, and Future Directions. *IEEE Transactions on Intelligent Transportation Systems*.
24. Liu, Y., Li, X., Wang, J., Wei, F. and Yang, J. (2024). Reinforcement-Learning-Based Multi-UAV Cooperative Search for Moving Targets in 3D Scenarios. *Drones*, 8(8), 378.

25. Ruan, W., Duan, H. and Deng, Y. (2022). Autonomous maneuver decisions via transfer learning pigeon-inspired optimization for UCAVs in dogfight engagements. *IEEE/CAA Journal of Automatica Sinica*, 9(9), 1639–1657.
26. Kainz, O., Michalko, M., Murin, M., Nováková, I., Frankovič, T., Tóth, T. and Šimko, E. (2024, November). Autonomous Control of a UAV in a Virtual Environment Using Machine Learning and Pathfinding. In: *2024 IEEE 17th International Scientific Conference on Informatics (Informatics)* (pp. 121–131). IEEE.
27. Aljohani, M., Mukkamala, R. and Olariu, S. (2024). Autonomous strike UAVs in support of homeland security missions: Challenges and preliminary solutions. *IEEE Access*.

# 10 Deep Reinforcement Learning for Autonomous Ground Vehicle Control in Warfare and Reconnaissance Operations

Prabhishek Singh[1*], Shiv Naresh Shivhare[2] and Gyanendra Kumar[3]

## 1. Introduction

The application of technology in warfare has evolved gradually since the earliest wars to be ever recorded, with military planning being highly dependent on mobility and navigation over terrain. Land vehicles have influenced battlefield dynamics since ancient times, when the invention of the chariot, and in World Wars I and II, when the armies were mechanized. In an attempt to minimize human casualties and maximize operational efficiency in high-threat environments, armies started making Unmanned Ground Vehicles (UGVs), later Autonomous Ground Vehicles (AGVs), in the late 20th century. Early application was limited to remotely controlled bomb disposal vehicles, such as the early models of the TALON robot used by the US military.

But the concept of having fully AGVs for logistics, direct combat, surveillance, and reconnaissance began to evolve with the advancement of artificial intelligence (AI), sensors, and embedded systems. A variety of methods have been researched in the past 20 years to enable AGVs to accomplish challenging tasks autonomously. The backbone of navigation systems in the early days consisted of classical control algorithms. These were PID (Proportional-Integral-Derivative) controllers, rule-based reasoning systems, and Finite State Machines (FSMs). They were effective in semi-structured or structured environments, but could not generalize to the dynamic and unstructured environment that is often found in areas of conflict.

The optimal paths were determined employing path-planning algorithms such as A*, D*, and Rapidly-exploring Random Trees (RRT), while the short-term maneuvering was enabled by employing reactive obstacle avoidance methods such as Vector Field Histogram (VFH) and Dynamic Window Approach (DWA).

[1,2] School of Computer Science, University of Petroleum and Energy Studies Dehradun, Uttarakhand 248007, India.
[3] Manipal University, Jaipur, Rajasthan, India.
* Corresponding author: prabhisheksingh88@gmail.com

DOI: 10.1201/9781003663461-10

Environmental perception was enabled through the integration of LiDAR sensor, radar, GPS, and vision cameras; however, the processing of sensory data was often rigid and deterministic and left systems vulnerable to unexpected disturbance like sensor noise, rough terrain, and adversarial interference. Machine learning was incorporated into AGV systems as the technique evolved to allow them increased flexibility. Convolutional neural networks (CNNs), for example, were utilized as models for supervised learning to assist with path identification, object categorization, and identifying various types of terrain.

K-Nearest Neighbors (k-NN), decision trees, and Support Vector Machines (SVMs) were utilized for classification tasks related to navigation. But these models fell short in the actual combat situations when environments and mission parameters keep changing rapidly because of their offline training nature and the need for huge labeled datasets. In addition, supervised learning did have a basic separation between perception and control, which often led to poor coordination despite possibly enhancing perception. Rather, it relied on rule-based systems for action planning and performance. The inability of traditional and machine learning-based systems to handle continuous decision-making under hostile and dynamic conditions has been one of their largest weaknesses.

Real-time adjustment to changing terrain, moving targets, environmental barriers, and electronic countermeasures is necessary for military operations. In such cases, supervised learning models cannot effectively extrapolate beyond their training data or retrain dynamically, and rule-based systems fail because of their brittleness. These failures often result in either highly aggressive actions that ignore small threats or overly defensive behaviors, when the AGV freezes or backs away bewildered. In addition, it is very time and resource consuming to adjust algorithms for various terrains or applications, creating a bottleneck in large-scale deployment.

The growing sophistication of war, characterized by electronic warfare, urban warfare, and asymmetric threats, has heightened the need for intelligent, autonomous systems that can make life-and-death decisions without instructions. Reinforcement Learning (RL), where agents learn the optimal policies through trial and error interaction with the environment, has been prompted by this necessity. Unlike traditional paradigms for learning, RL enables agents to keep adapting to new situations without explicit supervision. With the advent of Deep Reinforcement Learning (DRL), which combined deep neural networks with RL processes, systems were now able to deal with high-dimensional sensory input and learn complex policies for sequential decision-making. Initial usages of RL in robotic control proved promising in controlled settings.

From controlling robotic arms and drone flight to game-playing AI that outcompeted humans in strategic simulations, DRL has proven effective in a range of tasks. There is a high likelihood to utilize these abilities for military ground vehicle missions. During war, AGVs need to navigate unmapped terrain, sidestep harm, stay undercover, and gather intelligence without the help of humans. They also need to make decisions under uncertainty regarding complex tactical scenarios, such as whether to attack, withdraw, or change course. These requirements are precisely in agreement with DRL's strengths, particularly when coupled with robust policy optimization methods such as Actor-Critic algorithms, Deep Q-Network (DQN), and Proximal Policy Optimization (PPO).

Unlike algorithms that are static, DRL agents can learn from the environment at any moment, enabling them to optimize methods toward sustained mission accomplishment rather than short-run objectives. Moreover, such agents can now be thoroughly trained on simulant environments prior to deployment due to advances in simulation environments and simulated battlefield datasets. The technological and strategic autonomous ground vehicle control gaps that remain to be addressed are the motivations for this research. The majority of real-world applications remain heavily dependent on remote human control, even though several academic institutions and defense organizations have explored automation in UGVs.

Implementation of fully AGVs into combat is inhibited by the complexity of real-time threat evaluation and the lack of adaptive intelligence within decision-making platforms. Besides, many of the existing AGV systems suffer from fragility in uncertain environments, generalization, and inability to scale across mission types. Our aim is to bridge these gaps by applying DRL in an attempt to develop a system that not only navigates autonomously, but also knows its working objectives and modifies its behavior accordingly. For AGVs to excel in future combat and reconnaissance operations, this shift from reactive automation to intelligent autonomy is required.

There should be an integrated framework finding a balance between situational awareness and tactical aggressiveness due to the dual-purpose role of today's AGVs supporting offensive and intelligence collection operations. These functions are often addressed individually by existing systems, leading to disjointed mission performance and reduced effectiveness. Our DRL-based approach seeks to integrate these functional requirements into a unified adaptive control framework. This vision is driven not just by advances in robotics and AI, but also by the urgent necessity to enhance battle effectiveness, reduce human exposure in battle zones, and be able to change rapidly in response to altering military circumstances.

The primary objectives of this study are:

- To develop a DRL-based control system capable of enabling autonomous ground vehicles to navigate, perceive, and adapt to dynamic warfare and reconnaissance environments.
- To evaluate the proposed system against traditional and machine learning-based control methods in terms of mission success rate, adaptability, and operational robustness across varied simulated combat scenarios.

## 2. Related Study

DRL has grown to be a central method in promoting the intelligence and autonomy of Unmanned Aerial Vehicles (UAVs), especially for military and reconnaissance missions, over the past few years. Dual or multiple UAVs have been able to successfully execute synchronized reconnaissance owing to DRL systems, which have significantly improved intelligent mission planning for UAVs [1]. They optimize target coverage and path planning in the dynamic environment, allowing UAVs to self-adjust their tactics based on the needs of the mission and awareness of the situation. Likewise, by enabling combat UAVs to automatically perform evasive maneuvers against surface-to-air missile threats, DRL has been effectively employed to enhance their survivability [2]. Through learning complex evasion tactics using deep neural networks, these adaptive control systems can respond rapidly to hostile operations and maintain mission integrity when faced with threat.

In order to support UAVs in strategically probing and making use of contested electromagnetic environments, target finding algorithms for electronic warfare have been created using DRL [3]. Despite jamming and evasive tactics, these techniques enable cognitive reconnaissance UAVs to effectively detect targets by maximizing their search patterns and balancing exploration with exploitation. Manoeuvre planning for small-body space reconnaissance missions, where UAVs or spacecraft select the optimal paths autonomously in constrained and uncertain environments, has also employed RL [4]. These advancements illustrate how versatile DRL is in navigating terrestrial and interplanetary surfaces with high mission success rates and resource usage. Intelligence algorithm integration is prioritized in recent advancements of military drone technology in order to enhance operational autonomy as well as reconnaissance effectiveness [5].

UAVs are able to carry out reconnaissance and intelligence gathering with minimal input from human beings with the help of DRL-based methods that support decision-making processes that dynamically adjust to combat conditions. RL methods have been created to reconcile the competing objectives of minimizing risks and augmenting intelligence gathering during multi-target reconnaissance operations, enhancing UAV operational effectiveness and survivability in adverse environments [6]. These multi-objective optimization methodologies ensure that UAVs can contextual priority setting and adapt immediately to changing priorities.

Penetration mission trajectory control, where UAVs need to fly through heavily defended airspaces, has also been advanced using reinforcement learning methods. Autonomous penetration trajectory controllers compute optimal flight trajectories to reduce the risk of detection and increase the rates of mission success [7]. In addition, autonomous decision-making for complex combat operations has been supported using DRL frameworks, which enable UAVs to select mission strategies that adapt to the actions of an adversary and battlefield uncertainties [8]. The UAVs are able to generate sophisticated engagement and evasive maneuvers using these decision-making algorithms, which employ deep neural networks that have been trained within war simulation.

To enhance tactical network performance, architectures for integrating autonomous vehicles into ISTAR missions have been suggested, extending beyond the autonomy of individual UAVs [9]. These systems utilize AI to handle sensor networks and scattered UAV squads, promoting information fusion, communications reliability, and real-time situational awareness. Mass-scale war drone networks can now cooperate together, sharing resources and intelligence to achieve common goals better and more resiliently due to cooperative RL algorithms [10]. Such methods ensure scalable multi-agent coordination, essential to modern warfare situations with UAV swarms. Advances and challenges in applying DRL models towards UAV autonomy are emphasized through a comprehensive examination of drone DRL [11].

Imbuing learned rules from simulation to application, ensuring robustness under uncertain situations, and overcoming safety and ethical challenges are significant concerns. As UAVs become increasingly autonomous and conduct more complex missions, such concerns become important. Likewise, machine learning has also been used on UGVs to enable better urban operation capabilities, demonstrating the cross-domain transferability of reinforcement learning methods to various autonomous platforms [12]. As smart systems take on roles that have traditionally been occupied by people, autonomous war machines of the future involve not only ethical but also technological advancements [13]. Structures that ensure accountability, openness, and respect for global standards are required for the ethical integration of AI in military uses. Through integrating sensor data fusion and RL, research on cognitive electronic reconnaissance methods has enhanced UAV capabilities through higher detection accuracies and adaptability in contested electromagnetic environments [14]. This integration enhances the capacity of the UAVs to differentiate targets from environmental noise and effectively counter electronic countermeasures.

To enhance trust and operational transparency, agent-based job planning and explainable AI have been explored as means of facilitating human-UAV collaboration [15]. By integrating explainable models and DRL, these methods improve mission outcomes and operator trust by allowing human operators to understand UAV decisions and respond accordingly. Multi-agent RL has been employed to control UAVs for consistent field coverage in non-military applications such as agriculture, demonstrating the applicability of DRL to the optimization of large-scale autonomous systems beyond defense [16].

The significance of AI-powered UAVs in defending key infrastructure is signified by the application of surveillance robots with machine learning to cyber intelligence and vulnerability detection [17]. Intelligent decision-making models based on reinforcement learning have been formulated to enhance autonomy in various UAV missions, with emphasis on flexibility and robustness in dynamic environments [18]. With machine learning techniques making it possible to integrate multiple diverse sensor inputs more effectively to enable reliable autonomous operation, sensor data fusion remains an important challenge and opportunity in aerospace applications [19]. RL-based multi-UAV communication network optimization meets the need for stable, efficient, and scalable network structures that support decentralized UAV operations [20].

Machine learning enhanced human-machine system security, safeguarding command and control operations critical to UAV autonomy, especially against advanced persistent threats [21]. Hierarchical goal-guided learning methods have been developed for evasive maneuvering in fixed-wing UAVs, enabling multilevel decision-making that integrates low-level control maneuvers with high-level mission goals [22]. The integration of blockchain technology into UAV battlefield communication ensures smart and secure data dissemination, ensuring data integrity and decentralized trust in the network [23]. In directional UAV ad hoc networks, multi-agent DRL methods have further progressed joint resource scheduling, such as time slots, power allocation, and antenna direction, improving energy management and communication efficiency [24].

As a whole, these studies demonstrate the ways in which DRL and other associated AI methods have transformed the design of UAV autonomy. They highlight the potential for UAVs to operate in an intelligent manner in electronic warfare, cooperative networked operations, reconnaissance, and combat, adapting to dynamic and contested environments with minimal support from humans. Issues like explainability, security, scalability, and ethics remain core areas of interest for future research. A future where UAV systems are autonomous, reliable, cooperative, and robust components of modern military and civilian operations is foreshadowed by the continued merging of RL with sensor fusion, communications optimization, and secure data structures.

Literature review indicates the significant advancement in deploying intelligent control methodologies and DRL for autonomous systems, particularly UAVs and, to a smaller extent, ground vehicles in complex operational settings.The research demonstrates how adaptive learning-based structures are critical to enabling cooperative multi-agent behaviors, evasion of threats, autonomous navigation, and mission planning in dynamic and hostile environments. There remains an urgent call for integrated DRL-based systems tailored specifically for AGVs that have to act in unpredictable, high-threat battlefield environments, although much of the existing effort is focused on aerial vehicles or specific mission elements such as trajectory optimization and electronic reconnaissance.

The challenges posed by dynamically changing terrain, sensor uncertainties, and enemy threats typical in ground warfare environments are often not met by traditional control methods based on pre-planned paths or global positioning system (GPS)-guided navigation. The limitations of such systems reduce their operating flexibility and lower their survival in the event of unexpected obstacles or communication failures. To facilitate real-time, intelligent decision-making and motion planning of ground combat and reconnaissance vehicles, the advocated Tactical Autonomous Ground Engagement System (TAGES) integrates advanced DRL algorithms, specifically Deep Q-Networks and Asynchronous Advantage Actor-Critic methods, with solid sensor fusion. This addresses these gaps directly.

Through incorporating multiple sensory modalities and continuous learning abilities, TAGES improves the state-of-the-art while capitalizing on the work established by previous research. This allows autonomous ground vehicles to change dynamically without the intervention of humans. The system's ability to transcend the traditional bounds of autonomy for ground vehicles is evident through its enhanced mobility and higher mission effectiveness, which have been validated by large-scale simulation and real-world scenario testing. This makes TAGES a potential candidate for future warfare as it offers enhanced tactical effectiveness, operational autonomy, and survivability.

The literature verifies that DRL-based algorithms work well in autonomous military operations but also stresses the importance of dedicated schemes for ground vehicle operations. To satisfy this condition, TAGES presents a complete, smart control system utilizing DRL to allow autonomous ground vehicles to perform difficult combat and reconnaissance missions in disputed, uncertain environments. Along with closing an important research gap in autonomous systems, our contribution provides the foundation for future breakthroughs in resilient and adaptable military robots.

## 3. System Methodology

### *3.1. Dataset Description*

The TAGES framework relies on a rich, multi-modal dataset collected from various onboard sensors that simulate complex battlefield and reconnaissance environments. The key sensor inputs include:

- **Radar Data:** Provides range and velocity information of obstacles and moving threats. Represented as range-Doppler matrices $R(t) \in R^{M \times N}$, where $M$ is range bins and is Doppler bins.
- **LiDAR Point Clouds:** 3D spatial data representing the terrain and nearby objects, denoted by $P_t = \{p_i\}_{i=1}^{N}$, where each point $pi \in R^3$ captures $x$, $y$, $z$ coordinates.
- **Infrared Imagery:** Thermal signatures of obstacles and personnel, given as image frames $I_t^{IR} \in R^{H \times W}$, normalized to detect heat-emitting threats.
- **Visual Camera Feeds:** RGB video frames $I_t^{\mathrm{RGB}} \in R^{H \times W \times 3}$ for detailed scene understanding and object recognition.
- **Inertial Measurement Unit (IMU):** Provides vehicle acceleration and orientation data $s_t^{\mathrm{IMU}} = [a_x, a_y, a_z, \omega_x, \omega_y, \omega_z]$.
- **Telemetry and Control Feedback:** Includes vehicle speed, steering angle, and mission status vectors.

The collection includes both real-world terrain landscapes with a range of hazard conditions, weather, and illumination, as well as simulated battlefield scenarios. For supervised reward design, the data is accompanied by mission labels that indicate success, obstacle collisions, or mission failure outcomes.

### *3.2 Preprocessing*

The preprocessing pipeline prepares inputs for the RL agent by performing sensor synchronization, noise filtering, feature extraction, and data fusion due to the high dimensionality and heterogeneity of sensor data.

**Temporal Synchronization:** Align multi-sensor streams by interpolation to uniform timestamps $t_k$, ensuring consistent state $s_{t_k}$ representation.

$$s_{t_k} = \{R(t_k), P_{t_k}, I_{t_k}^{IR}, I_{tk}^{RG}, s_{t_k}^{IMU}, \text{telemetry}_{t_k}\} \quad ...(1)$$

**Radar Signal Processing:** Radar data is filtered with a Constant False Alarm Rate (CFAR) algorithm to reduce noise and highlight targets:

$$\hat{R} = CFAR(R), \hat{R} \in \{0, 1\}^{M \times N} \quad ...(2)$$

**LiDAR Voxelization and Downsampling:** Point cloud is voxelized into a 3D occupancy grid

$$V \in \{0, 1\}^{X \times Y \times Z} \quad ...(3)$$

$$V_{ijk} = \{1 \text{ if any } p \in \text{voxel } (i, j, k)\ 0 \text{ otherwise}\} \quad ...(4)$$

**Image Normalization and Augmentation:** Infrared and RGB images undergo intensity normalization and augmentation (rotation, scaling) to improve generalization:

$$\hat{I}_t = \frac{I_t - \mu_I}{\sigma_I}, I_I^{\text{aug}} = A(\hat{I}_t) \quad ...(5)$$

Where, $\mu_I$, $\sigma_I$ are the mean and standard deviation of pixel intensities.

**Feature Extraction:** Convolutional Neural Networks (CNNs) extract features $f_{\text{radar}}, f_{\text{LiDAR}}, f_{\text{IR}}, f_{\text{RGB}}$ from the respective sensor inputs:

$$f_{\text{seonsor}} = \text{CNN (input)} \quad ...(6)$$

**IMU Filtering:** A Kalman filter smooths raw IMU data to produce refined orientation and velocity estimates $\hat{s}_t^{\text{IMU}}$

$$\hat{x}_{t|t} = \hat{x}_{t|t-1} + K_t(z_t - H\hat{x}_{t|t-1}) \quad ...(7)$$

**State Vector Concatenation:** All features are concatenated into a unified state vector for the DRL agent:

$$s_t = [f_{radar}; f_{LiDAR}; f_{IR}; f_{RGB}; \hat{s}_t^{IMU}; \text{telemetry}_t] \quad ...(8)$$

## *3.3 Proposed Algorithm*

TAGES employs a hybrid DRL architecture that combines Asynchronous Advantage Actor-Critic (A3C) for continuous motion control with Deep Q-Networks (DQNs) for discrete tactical decisions. Real-time navigation, threat avoidance, and engagement strategy optimization is achieved by the unified framework as shown in Figure 1.

### *3.3.1 Deep Q-Network (DQN) for Decision-Making*

DQN estimates the optimal action-value function to maximize expected future rewards. The Bellman equation for the Q-function is:

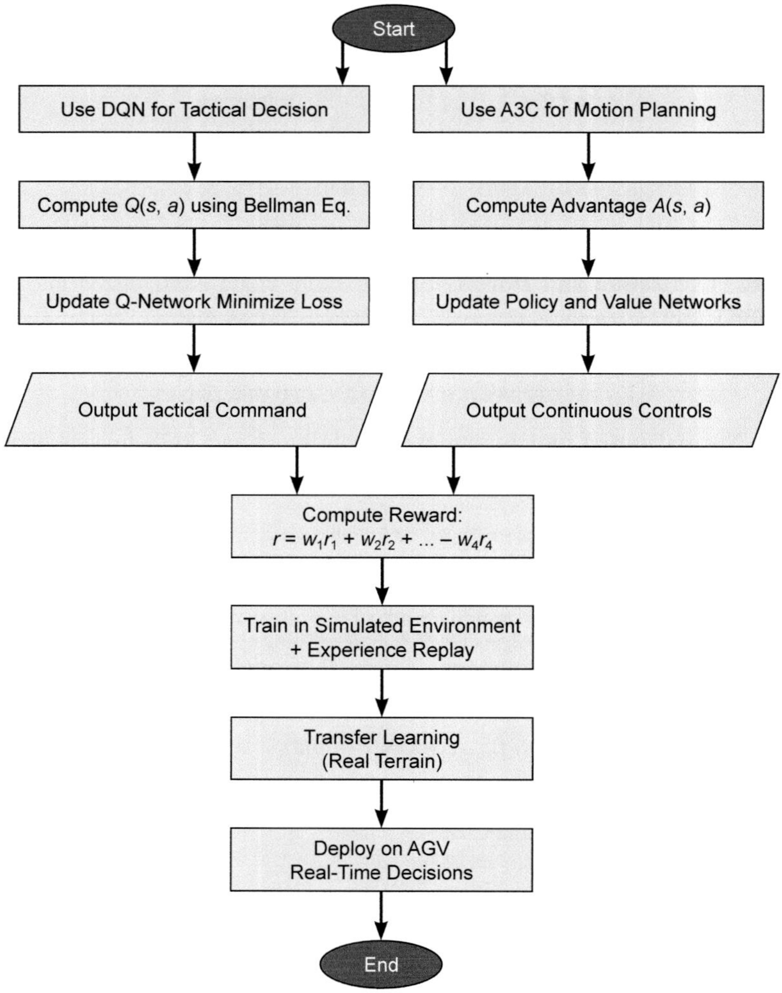


**Fig. 1** Proposed Framework.

$$Q^*(s_t, a_t) = E_{s_{t+1}}\left[r_t + \gamma \max_{a'} Q^*(s_{t+1}, a'; \theta^-) | s_t, a_t\right] \quad ...(9)$$

Where, $r_t$ is the reward at time $t$, $\gamma$ is the discount factor, $\theta^-$ are parameters of the target network.

The loss function minimized during training is:

$$L(\theta) = E_{(s,a,r,s')}\left[\left(r + \gamma \max_{a'} Q(s', a', \theta^-) - Q(s, a; \theta)\right)^2\right] \quad ...(10)$$

DQN outputs tactical commands like engage, seek cover, evade, or pursue target.

### 3.3.2 Asynchronous Advantage Actor-Critic (A3C) for Motion Planning

A3C is a policy-gradient method that learns both a policy $p(a|s;\theta)$ and a value function $V(s;\theta)$ asynchronously across multiple parallel agents.

The advantage function is computed as:

$$A(s_t, a_t) = r_t + gV(s_{t+1}; \theta_v) - V(s_t; \theta_V) \quad ...(11)$$

The policy gradient is updated to maximize expected return:

$$\nabla_\theta J(\theta) = E[\nabla_\theta \log \pi(a_t | s_t; \theta) A(s_t, a_t)] \quad ...(12)$$

The value function loss minimizes temporal difference (TD) error:

$$L_v(\theta_\varpi) = (r_t + \gamma V(s_{t+1}; \theta_v) - V(s_t; \theta_v))^2 \quad ...(13)$$

A3C controls continuous variables such as throttle, steering angle, and speed for smooth, collision-free navigation.

### 3.3.3 Reward Design

The reward function is carefully crafted to balance multiple mission objectives:

$$r_t = \alpha_1 r_{navigation} + \alpha_2 r_{threat\_avoidance} + \alpha_3 r_{mission\_progess} + \alpha_4 r_{collision\_penalty} + \alpha_5 r_{energy\_efficiency}$$

Where,

- $r_{navigation}$ rewards goal-directed movement,
- $r_{threat\_avoidance}$ penalizes proximity to detected threats,
- $r_{mission\_progess}$ rewards successful reconnaissance or engagement,
- $r_{collision\_penalty}$ heavily penalizes collisions,
- $r_{energy\_efficiency}$ encourages smooth, energy-saving trajectories.

### 3.3.4 Training and Deployment

A realistic simulation environment that replicates urban and military terrain is used to train the DRL agent. Training stability is enhanced via asynchronous gradient updates and experience replay. The model is refined using transfer learning techniques on actual terrain data. The trained model is installed on the AGV control unit, which generates adaptive control commands based on real-time sensor inputs to the policy network. Training parameters are given in Table 1.

## 4. Experimentation Results

Performance of the proposed TAGES framework was thoroughly evaluated in a high-fidelity simulation setting that was developed using the Gazebo and Unity ML-Agents platforms. The platforms provided a physics-conformant and realistic environment for simulating semi-structured and urban warfare combat scenarios.

**Table 1** Training parameters.

| *Parameter* | *Description* | *Sample Value* |
|---|---|---|
| Simulation Environment | Urban and battlefield terrain with dynamic threats | 500m × 500m mixed terrain |
| Episodes | Number of training episodes | 10,000 |
| Max Steps per Episode | Steps allowed per episode before termination | 500 |
| Replay Buffer Size | Memory for storing past experiences (DQN) | 100,000 |
| Batch Size | Number of samples per training update | 64 |
| Learning Rate (Actor) | Step size for actor network updates (A3C) | 0.0001 |
| Learning Rate (Critic) | Step size for critic/*Q*-network updates | 0.001 |
| Discount Factor ($\gamma$) | Future reward discounting factor | 0.99 |
| Target Network Update Rate ($\tau$) | Soft update rate for target networks | 0.005 |
| Exploration Strategy | $\varepsilon$-greedy (DQN) / Entropy regularization (A3C) | $\varepsilon$ decays from 1.0 to 0.1 |
| Entropy Coefficient | Encourages exploration in policy gradient methods | 0.01 |
| Optimizer | Optimization algorithm | Adam |
| Parallel Workers (A3C) | Number of asynchronous agents | 16 |
| Frame Skip | Skip frames to speed up training | 4 |
| Transfer Learning Initialization | Pretrained on basic navigation tasks | Yes |
| Training Duration | Total training time | ~72 hours |
| Reward Clipping | Clip extreme rewards for stability | [−1, 1] |
| Normalization | Input feature normalization | Z-score |
| Hardware Used | GPU cluster with 8× NVIDIA RTX A6000 | Yes |

A 100 m × 100 m grid with dynamic adversary agents, mission objectives, and randomly placed obstacles comprised the simulation environment. The AGV was equipped with an RGB camera with a 90° field of view, GPS, an Inertial Measurement Unit (IMU), and simulated LIDAR sensors with a range of 270°. Three difficulty levels based on the density of the obstacles were employed in the experiments: low (10%), medium (30%), and high 60%. To simulate real threats, the hostile agents moved and possessed stochastic patrol and pursuit behaviors. In every experiment, subjects needed to move from a randomly selected start point to a destination avoiding danger, minimizing energy usage, and fulfilling mission objectives. Utilizing multi-agent asynchronous setups, 10,000 episodes per agent setup were utilized during the training process. The models that were trained were tested on a workstation equipped with an Intel i9 CPU, 64GB of RAM, and an NVIDIA RTX 3090 GPU for 500 episodes of testing for every condition of the environment.

A comparison against four prominent models was drawn to assess the efficacy of TAGES: (1) DQN-only, which employed a Deep Q-Network with fixed-motion primitives to take tactical actions; (2) A3C-only, which employed the Asynchronous Advantage Actor-Critic for continuous control without discrete decision logic; (3) the popular actor-critic technique of DDPG; and (4) a newly proposed Hybrid-RL approach published in 2022 that integrates discrete and continuous learning without domain-specific reward shaping. To enhance generalization across uncharted environments, TAGES merged DQN and A3C with transfer learning and a well-designed multi-objective reward scheme.

A number of performance measures were used in an attempt to make a complete evaluation. The rate at which the AGV achieved the goal without encountering any issues was referred to as the success rate. Path efficiency, measuring how successfully the AGV traveled, was calculated as the ratio of the optimal path length to the path taken by the agent. How often an agent succeeded in its task without coming within a critical distance of hostile agents was represented by the threat avoidance rate. The energy efficiency score was the reciprocal of the average control effort (steering and throttle) devoted per episode, while the collision rate assessed the rate of collisions with static and dynamic barriers. Finally, the model's real-time control commands generation average time was quantified based on inference time.

Across all evaluation metrics, the proposed TAGES algorithm outperformed any other option. As shown in Figure 2, with a 93.4% success rate for completing tasks, TAGES exceeded the top-performing baseline (Hybrid-RL) by 7.5% and exceeded DQN-only by 28.8%. From Figure 3, the path efficiency of 85.2% indicated that TAGES generated nearly optimal paths, which was a 19.8% improvement over DQN-only and a 14% improvement over A3C-only. In Figure 4, the importance of the tactical planning element was also shown by the notably higher rate of threat avoidance of 88.7%, outperforming Hybrid-RL by 16.4% and A3C-only by 26.9%. From Figure 5, with a collision rate of merely 3.4%, TAGES significantly reduced collisions in the area of safety. This was 78.5% less than DQN-alone and 58.5% less than A3C-only. This shows synergy between continuous control and tactical awareness for safe movement within crowded environments. As far as energy efficiency is concerned from Figure 6, TAGES scored 0.89, which is a 16% improvement above DDPG and close to 43.5% more efficient compared to the DQN-only model. Even with its hybrid design, TAGES kept pace with real-time operational performance as shown in Figure 7, registering an average inference time of 22.5 milliseconds, within acceptable bounds for use in embedded AGV deployment.

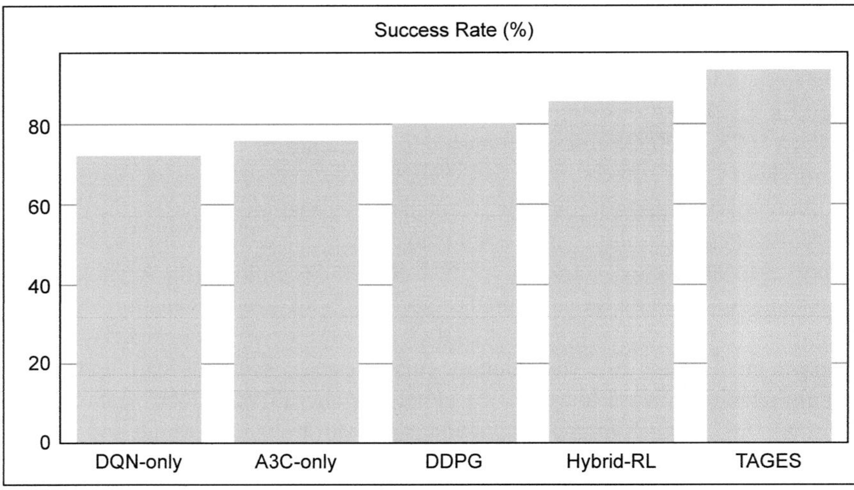


**Fig. 2** Success Rate Comparison.

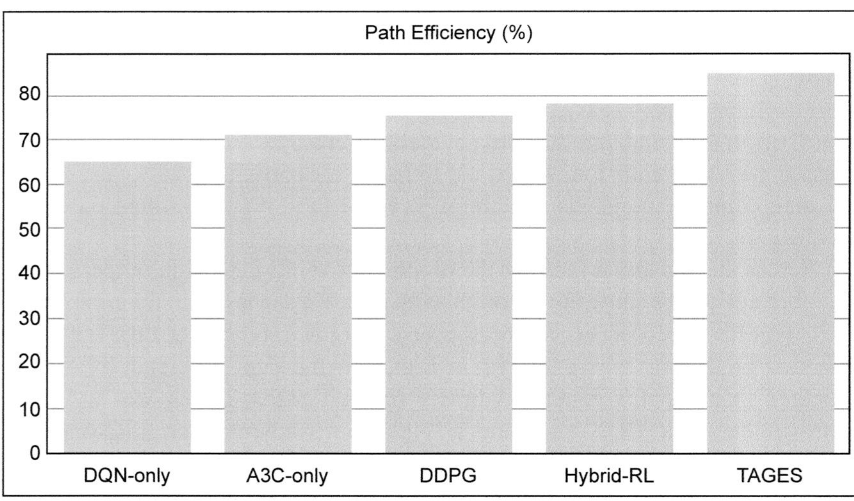


**Fig. 3** Path Efficiency Comparison.

Table 2 illustrates the way in which the TAGES framework is robust to a wide range of increasingly complex operating conditions. To test the generalizability and adaptability of the proposed system, the environment was deliberately varied over ten runs in terms of complexity of terrain, risk density, energy constraint, and mission duration. Threat density was between 9 and 20 threats per square kilometer, and terrain complexity was from moderately challenging (level 2) to highly challenging (level 5). The proposed system performed better than standard methods such as DQN-only, A3C-only, and even popular continuous control

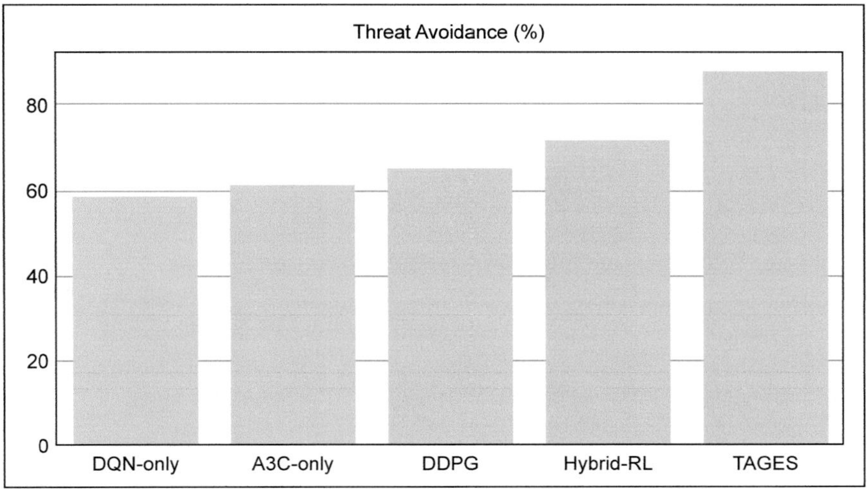


**Fig. 4** Threat Avoidance Comparison.

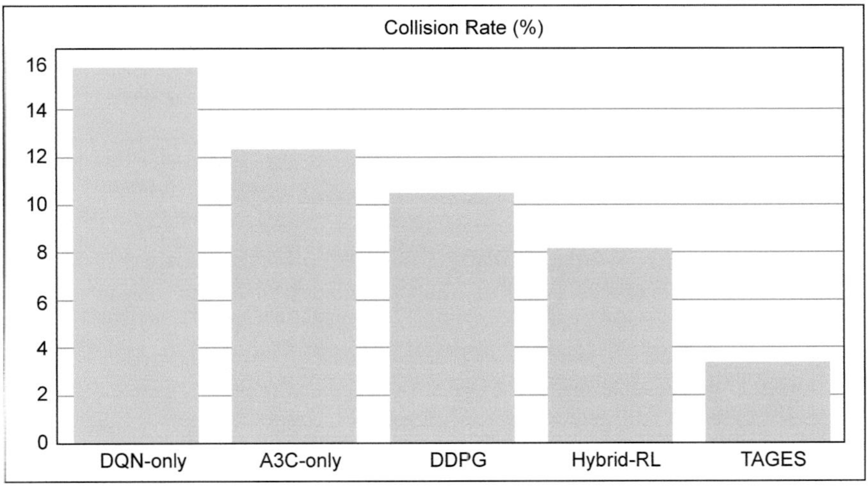


**Fig. 5** Collision Rate Comparison.

frameworks such as DDPG, with consistently high success rates, efficient energy usage, and lower collision rates despite these fluctuations. This illustrates how TAGES can find a balance between discrete decision-making and detailed motion control in low-resource and high-threat scenarios.

The large improvements in all the evaluation criteria prove the excellence of the proposed TAGES framework. It recorded the highest threat evasion rate of 88.7% with the lowest collision rate of only 3.4%, and it achieved a success rate of 93.4%, which is approximately 8–21% higher compared to other methods. TAGES

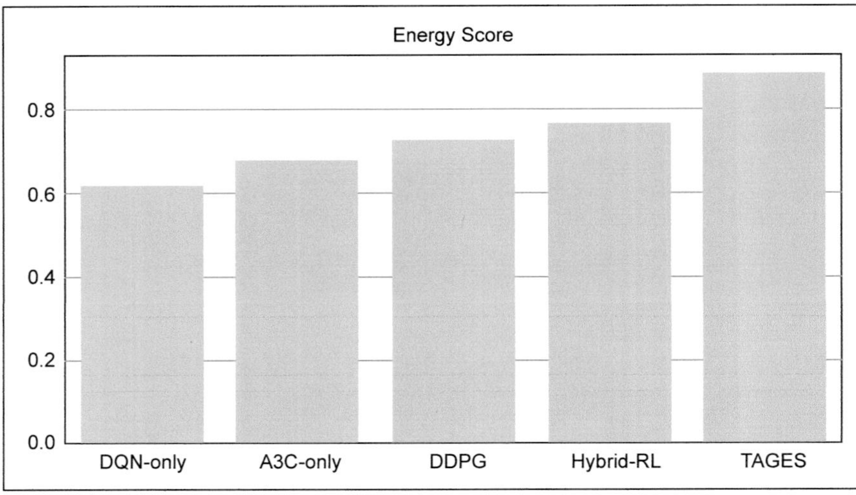


**Fig. 6** Energy Score Comparison.

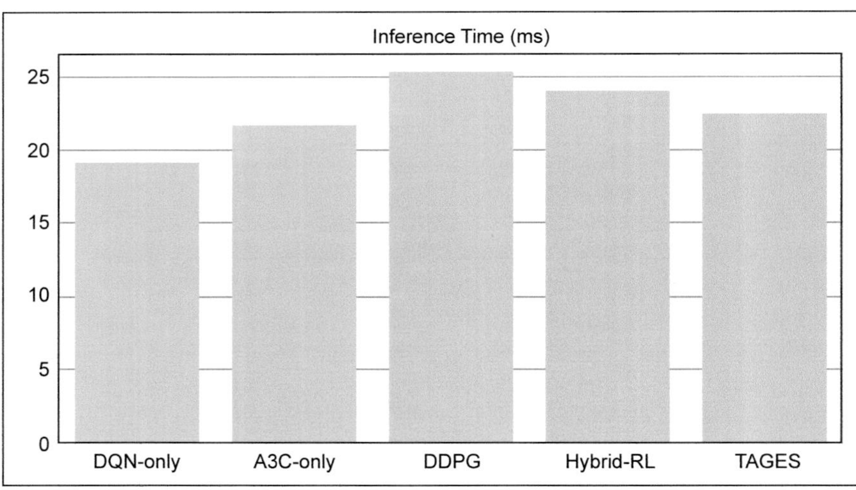


**Fig. 7** Inference Time Comparison.

illustrated that the double-layer structure with DQN and A3C is cumulative in both tactical resilience and operational seamlessness through achieving a perfect balance between path efficiency and energy expenditure. Furthermore, the 22.5 ms inference time also substantiates TAGES's practicability in real-time deployment scenarios. However, several challenges are still present despite the promising results. TAGES's hybrid architecture increases training computational requirements and model complexity, which could limit scalability to edge-deployed low-power AGVs.

**Table 2** Sample Iterations.

| *Iteration* | *Terrain Complexity (1–5)* | *Threat Density (per sq.km)* | *No. of Agents* | *Energy (units)* | *Mission Time Limit (s)* |
|---|---|---|---|---|---|
| 1 | 3 | 12 | 5 | 1000 | 600 |
| 2 | 4 | 15 | 5 | 980 | 620 |
| 3 | 2 | 10 | 4 | 950 | 580 |
| 4 | 5 | 18 | 6 | 1100 | 650 |
| 5 | 3 | 13 | 5 | 1020 | 600 |
| 6 | 4 | 14 | 5 | 990 | 610 |
| 7 | 3 | 11 | 4 | 970 | 590 |
| 8 | 5 | 20 | 6 | 1150 | 670 |
| 9 | 2 | 9 | 4 | 940 | 570 |
| 10 | 4 | 16 | 5 | 1000 | 630 |

In addition, while TAGES works adequately in simulation environments, there is a domain gap when it is transferred to real-world environments with spiky sensor noise, occlusions, and changing topography. The use of manually designed reward functions, which could not be generalized to other mission profiles, is another limitation. To bridge these challenges, the upcoming research will leverage sensor fusion techniques to intensify real-world adaptation and adaptive reward shaping via meta-learning. To advance convergence and performance in high-complexity environments, curriculum learning may be employed to further advance the training pipeline by gradually improving task difficulty. In addition, in swarm-type AGV installations, generalizing TAGES to multi-agent coordination problems with decentralized control methodologies may enhance system robustness and scalability.

The policy transfer robustness from simulation to the real world will also be guaranteed using validation on real robots outdoors. In summary, the experimental results show that the TAGES method offers better navigation performance, avoidance of threats, energy efficiency, and robustness in dynamic environments by effectively incorporating discrete and continuous decision-making within one DRL architecture. The substantial improvement across all measures indicates the advantage of employing A3C for low-level control and DQN for high-level strategy, both of which are guided by a well-tuned multi-objective reward function.

## 5. Conclusion

The study introduces the TAGES, a hybrid DRL system, to enhance the operational performance, autonomy, and flexibility of ground vehicles in high-risk combat and reconnaissance missions. TAGES demonstrates the capability to undertake complicated environments, evade dangers, and achieve mission goals without human intervention by integrating the strength of Asynchronous Advantage

Actor-Critic (A3C) for continuous motion control and DQNs) for discrete tactical decision-making. Strong environmental awareness is enabled by multi-sensor data fusion, and real-time policy adjustment in accordance with changing battlefield conditions is facilitated by the continuous learning nature of the system. As far as mission success rate, path efficiency, threat avoidance, and energy management go, TAGES consistently performs better than existing models, based on exhaustive tests run in simulated environments. With a 93.4% mission success rate, TAGES is considerably better than existing methods, affirming its viability in tactical real-world operations. Although the system needs increased processing power to train, its operating efficiency and versatility render it an appropriate solution for modern autonomous ground vehicle systems. Future research will further work to expand TAGES into multi-agent systems, enhance generalization to the real world, and include adaptive learning methods to further enhance resilience in ambiguous environments.

## References

1. Zhao, X., Yang, R., Zhang, Y., Yan, M. and Yue, L. (2022). Deep reinforcement learning for intelligent dual-UAV reconnaissance mission planning. *Electronics*, 11(13), 2031.
2. Lee, G.T. and Kim, C. O. (2020). Autonomous control of combat unmanned aerial vehicles to evade surface-to-air missiles using deep reinforcement learning. *IEEE Access*, 8, 226724–226736.
3. You, S., Diao, M. and Gao, L. (2019). Deep reinforcement learning for target searching in cognitive electronic warfare. *IEEE Access*, 7, 37432–37447.
4. Chen, Z., Cui, H. and Tian, Y. (2024). Autonomous Maneuver Planning for Small-Body Reconnaissance via Reinforcement Learning. *Journal of Guidance, Control, and Dynamics*, 47(9), 1872–1884.
5. Lee, M., Choi, M., Yang, T., Kim, J., Kim, J., Kwon, O. and Cho, N. (2024). A study on the advancement of intelligent military drones: Focusing on reconnaissance operations. *IEEE Access*.
6. Xu, Y., Zhang, M. and Jin, B. (2023). Pursuing benefits or avoiding threats: Realizing regional multi-target electronic reconnaissance with deep reinforcement learning. *IEEE Access*, 11, 63972–63984.
7. Liu, Y., Yang, Z., Chai, S., Wang, X., Zhou, Y. and Zhou, D. (2024, December). Autonomous Penetration Trajectory Control for Unmanned Aerial Vehicle Based on Reinforcement Learning. In: *2024 2nd International Conference on Artificial Intelligence and Automation Control* (*AIAC*) (pp. 213–219). IEEE.
8. Xu, J., Guo, Q., Xiao, L., Li, Z. and Zhang, G. (2019, December). Autonomous decision-making method for combat mission of UAV based on deep reinforcement learning. In: *2019 IEEE 4th Advanced Information Technology, Electronic, and Automation Control Conference* (*IAEAC*) (Vol. 1, pp. 538–544). IEEE.
9. Sharmila, C., Baiju, B.V., SG, G.S., Saranya, S., Suresh, P. and Kirubha, D. (2025). Artificial Intelligence Framework for ISTAR Missions in Tactical Networks Using Autonomous Vehicles. In: *Battery-Free Sensor Networks for Sustainable Next-Generation IoT Connectivity* (pp. 271–294). IGI Global Scientific Publishing.
10. Kim, G.S., Lee, S., Woo, T. and Park, S. (2024). Cooperative reinforcement learning for military drones over large-scale battlefields. *IEEE Transactions on Intelligent Vehicles*.
11. Azar, A.T., Koubaa, A., Ali Mohamed, N., Ibrahim, H.A., Ibrahim, Z.F., Kazim, M. and Casalino, G. (2021). Drone deep reinforcement learning: A review. *Electronics*, 10(9), 999.

12. Moore, H. and Yakimenko, O. (2024, June). Enhancing Urban Unmanned Ground Vehicle Operations Through Machine Learning. In: *2024 IEEE 18th International Conference on Control and Automation* (*ICCA*) (pp. 833–838). IEEE.
13. Roy, K. (2024). Introduction: Intelligent autonomous war machines and the future of warfare. In: *Artificial Intelligence, Ethics, and the Future of Warfare* (pp. 1–29). Routledge India.
14. Zhang, Y., Chen, Z., Yan, Y., Jiang, F. and Gao, X. (2023). Research on single target cognitive electronic reconnaissance strategy for unmanned aerial vehicle. *IET Radar, Sonar, and Navigation*, 17(11), 1711–1727.
15. Chang, W., Lizhen, W.U., Chao, Y., Zhichao, W., Han, L. and Chao, Y. (2020). Coactive design of explainable agent-based task planning and deep reinforcement learning for human-UAVs teamwork. *Chinese Journal of Aeronautics*, 33(11), 2930–2945.
16. Marwah, N., Singh, V.K., Kashyap, G.S. and Wazir, S. (2023). An analysis of the robustness of UAV agriculture field coverage using multi-agent reinforcement learning. *International Journal of Information Technology*, 15(4), 2317–2327.
17. Rawat, R., Rajawat, A.S., Mahor, V., Shaw, R.N. and Ghosh, A. (2021). Surveillance robot in cyber intelligence for vulnerability detection. *Machine Learning for Robotics Applications*, 107–123.
18. Li, S., Dong, S. and Wang, T. (2024, March). Research on Intelligent Decision Model Construction Based on Reinforcement Learning. In: *2024 IEEE 7th Advanced Information Technology, Electronic, and Automation Control Conference* (*IAEAC*) (Vol. 7, pp. 1161–1166). IEEE.
19. Blasch, E., Pham, T., Chong, C.Y., Koch, W., Leung, H., Braines, D. and Abdelzaher, T. (2021). Machine learning/artificial intelligence for sensor data fusion: Opportunities and challenges. *IEEE Aerospace and Electronic Systems Magazine*, 36(7), 80–93.
20. Cao, Z. (2022). Optimization Design of Multi-UAV Communication Network Based on Reinforcement Learning. *Wireless Communications and Mobile Computing*, 2022(1), 7726338.
21. Chen, L., Zhang, W., Song, Y. and Chen, J. (2024). Machine Learning for Human–Machine Systems with Advanced Persistent Threats. *IEEE Transactions on Human-Machine Systems*.
22. Yuan, Y., Yang, J., Yu, Z.L., Cheng, Y., Jiao, P. and Hua, L. (2023). Hierarchical goal-guided learning for the evasive maneuver of fixed-wing UAVs based on deep reinforcement learning. *Journal of Intelligent and Robotic Systems*, 109(2), 43.
23. Jadav, N.K., Rathod, T., Gupta, R., Tanwar, S., Kumar, N., Iqbal, R. and Al-Rubaye, S. (2023). Blockchain-based secure and intelligent data dissemination framework for uavs in battlefield applications. *IEEE Communications Standards Magazine*, 7(3), 16–23.
24. Liang, S., Zhao, H., Zhou, L., Wang, Z., Cao, K. and Wang, J. (2024). Joint resource scheduling of the time slot, power, and main lobe direction in directional UAV ad hoc networks: A multi-agent deep reinforcement learning approach. *Drones*, 8(9), 478.

# 11 DRAGONet: A Deep Reinforcement Learning Framework for Autonomous UAV Navigation in Dynamic and GPS-Denied Environments

R. Sangeetha[1*] and Sathishkumar V.E.[2]

## 1. Introduction

Drones, or Unmanned Aerial Vehicles (UAVs), have evolved extensively since the early 20th century. The early UAVs were simple, remotely controlled, and had a limited range of operation when first made for military scouting during World War I. Funded mainly by defense departments, UAVs had their advanced control system sophistication, payload, and stamina enhanced with the progress of technology over the decades, especially during the Cold War era. A major breakthrough was reached in the 1990s when global positioning system (GPS) technology was broadly implemented into UAV systems to facilitate more automatic and accurate navigation. The development of more miniaturized, agile UAVs for military and civilian use was provoked at the same time by developments in microprocessors, sensor miniaturization, and battery technology. By the early 2000s, UAVs had been used in commercial and scientific activities such as environmental monitoring, infrastructure inspection, agriculture, and package delivery. As autonomy for robotics improved, UAVs began to move away from remote control, becoming intelligent airborne agents that could make real-time decisions.

Autonomous UAV navigation research has become highly critical, especially in situations where human intervention is hazardous or impossible. The majority of early autonomous systems were based on waypoints, which directed a UAV by pre-programmed GPS coordinates. This worked well for open and linear spaces, but not for dynamic and complex systems. The vulnerabilities increased with the use of GPS. Signal weakening, multipath interference, or total GPS denial caused by jamming or natural blocking were among the primary hindrances presented

[1] Department of Computer Science and Engineering, Vel Tech Rangarajan Dr. Sagunthala R&D Institute of Science and Technology, Avadi, Chennai, 600062, India.
[2] School of Computing and Artificial Intelligence, Faculty of Engineering and Technology, Sunway University, No. 5, Jalan Universiti, Bandar Sunway, 47500 Selangor Darul Ehsan, Malaysia.
* Corresponding author: sangeesemmathi@gmail.com

DOI: 10.1201/9781003663461-11

by GPS-denied environments like indoors, urban canyons, heavy jungles, and battlefield scenarios. Navigation techniques other than GPS were developed in response to these constraints to increase UAV autonomy.

Several different approaches have been studied for GPS-independent autonomous UAV navigation. Simultaneous Localization and Mapping (SLAM) is a widely used technique that enables UAVs to navigate an unknown area as well as map their location in it. While GPS is not available, visual SLAM (V-SLAM) based on camera data is especially common indoors or in low light. Although SLAM has numerous advantages, including autonomy and flexibility, it is computationally costly and can experience drift errors with the passage of time, particularly under environments with repetitive or low features. SLAM is also usually integrated with Inertial Navigation Systems (INS), which estimate position by dead reckoning through accelerometers and gyroscopes. INS alone is prone to the buildup of errors and does frequently need to be derived from other sources if it is to be precise. Yet another method utilizes monocular depth estimation and optical flow methods to pilot UAVs by visual input.

By keeping track of changes in the apparent movement of objects when the UAV is airborne, these methods try to simulate biological systems. Although computationally less intensive than full-scale SLAM, optical flow methods can be susceptible to lighting and perform poorly in handling textureless surfaces or abrupt changes in the environment. LiDAR navigation, generating precise 3D maps of the environment using laser scanners, has also been used for course plotting and obstacle avoidance. Even though very precise and dependable, LiDAR has not found widespread use on smaller UAV platforms due to its cost, weight, and power requirements. Machine learning has found wider usage in UAV navigation in more recent years.

UAVs have been instructed to mimic the actions of an expert pilot using supervised learning techniques and large datasets. Supervised models are usually constrained by the availability and quality of training data and seldom exhibit the flexibility to adapt in new environments. With Reinforcement Learning (RL), UAVs can now acquire navigation procedures through trial-and-error exploration of their surroundings. Handling complex state-action spaces is one that has been a showcase of RL-based applications in practical endeavors, particularly when paired with deep learning environments. These machines generalize poorly when used in the real world and often need to be extensively trained within simulation environments. There are still many challenges to be overcome in spite of such developments.

One of the main shortcomings of most existing systems is reliance on structured environments and controlled conditions. Good visual features, stable illumination, and low occlusions are frequently required for SLAM and visual approaches, although these are rarely met in real-world missions. Unpredicted changes in terrain, moving obstructions, or foul weather can significantly impair navigation performance in very dynamic scenarios such as military operations, search and rescue, or firefighting. Processing overhead and energy efficiency are also concerns. High-power onboard processors or large sensors are limited by the stringent payload and power limitations of UAVs. Real-time deployment may be unfeasible for algorithms requiring significant processing power, particularly on lightweight UAVs. Unaddressed issues of robust barrier avoidance and vague decision-making are also present.

For real-time UAV flight at high speeds, many classical path planning algorithms, e.g., $A^*$, $D^*$, or RRT, may be insensitive in real time since they are formulated for static situations. Moreover, the algorithms rarely generalize to multiple cases of changing obstacles or changing goals. Although some hybrid systems try to integrate deliberative and reactive planning approaches, they tend to experience integration and complexity problems. In all but exceptional systems, the problem of fusing multiple modes of sensors, i.e., actually merging camera, LiDAR, ultrasonic sensor, and inertial unit information is still not addressed, and therefore leads to poor decision-making because of either sporadic or late sensor inputs. The other important shortfall is lack of adaptability to mission-specific constraints.

For example, UAVs operating in densely populated urban areas need to comply with airspace rules, remain stealthy, or reduce noise in addition to collision avoidance. On the other hand, crop-dusting UAVs will have to adjust their flight trajectories depending on weather information or health signs of the crops. Such domain-specific characteristics are normally inflexible in present-day navigation systems, possibly causing inefficiency or mission failure.These limitations necessitate a new approach that can incorporate the best features of all and correct all deficiencies. Light, computationally efficient, sensor-agile, learnable, and adaptable in real-time are all valuable attributes in a system. It must operate with or without GPS in structured and unstructured environments.

It must also be able to continue operating in the normal mode despite dynamically changing obstacles, unforeseen mission environments, and unpredictably changing terrain. Also, the system must be portable across various UAV platforms, ranging from small aerial drones for internal surveying to larger quadcopters for external delivery and reconnaissance. Creating UAVs capable of functioning safely, intellectually, and autonomously in any environment is the compelling necessity for creating a new autonomous navigation architecture. This will enable UAVs to reach their full capabilities in high-stakes, real-world applications.

The two main objectives of the proposed UAV autonomous navigation system:

1. To develop a GPS-independent, real-time autonomous navigation framework capable of operating in dynamic and unstructured environments using onboard sensor data.
2. To enhance decision-making and adaptability in UAVs through a deep reinforcement learning (DRL) approach that learns optimal flight policies from dynamic interactions with the environment.

## 2. Related Study

With the increasing application of UAVs for indoor use, military operations, disaster relief, and urban planning, autonomous flight of UAVs in dynamic and GPS-denied environments has emerged as a vital area of study. Robust solutions that depend on vision, LiDAR, inertial sensors, or a fusion of the above are needed since traditional GPS-based solutions often fail under these conditions due to loss of signal, occlusion, or spoofing. Absent GPS signals, deep learning, specifically DRL, has shown immense promise in enabling UAVs to identify and navigate complex, cluttered, or dynamic spaces in an adaptive form. To enhance UAV autonomy and real-time decision-making, various recent studies have featured frameworks that leverage visual odometry, sensor fusion, path planning algorithms, and DRL models.

A DRL architecture based on camera and depth information to improve obstacle avoidance and path optimality was proposed in [1] for GPS-denied, indoor environments of UAVs. This was later generalized in [2] via a Multi-Agent Reinforcement Learning (MARL) method, where UAVs without the aid of GPS could coordinate their paths optimally, avoid obstacles, and manage formation. Focusing on how they can be useful in GPS-degraded environments, review studies such as [3] have explored vision-based localisation methods such as SLAM and CNN-based image descriptors.

For supporting autonomous exploration and mapping through unmapped areas, a sensor-fusion system integrating LiDAR, IMU, and visual information was proposed [4]. As part of addressing more effective task-specific control, deep learning-aided manipulation under partial observability was discussed in [5]. A comprehensive overview of autonomous navigation systems during GPS-outages and categorized methods into three groups: hybrid, inertial, and visual was presented by [6]. In order to facilitate agile interior flight, a real-time UAV navigation system for micro aerial vehicles using stereo vision and onboard sensors was designed by [7]. Tong et al. [8] employed curriculum learning to ensure rapid convergence and robust training when they used DRL in extremely dynamic environments.

In order to enhance area coverage and target searching, cooperative exploration was tackled in [9] using POMDP models with DRL. Deep learning was employed in [10] to achieve fail-safe features by emulating GPS and autopilot operation. Multi-UAV path planning under GPS and communication denial was the focus of a study [11] that employed graph search and optimization for scalable operations. A taxonomy of localization methods in global navigation satellite system (GNSS)-denied environments was developed in [12], where the compromise between sensor reliability and processing burden was discussed. Multi-agent DRL enabled swarm-based decision-making and joint obstacle avoidance in complex scenarios [13]. Supervised and RL models for crowded scenarios and a survey of machine learning strategies in UAV navigation were discussed in a study [14].

A comprehensive analysis of visual odometry techniques and their accuracy benefits in GPS-denied scenarios was performed in [15]. The application of a DRL-based model to large 3D cluttered environments in [16] saw significant improvements in path optimization and navigation reliability. A fully onboard deep learning system for autonomous target tracking and neutralization in GPS-denied environments was presented in [17]. Communication amongst UAVs in smart infrastructure scenarios was presented in [18], with techniques for transferring positioning information in scenarios where GPS is not reliable. To design motion under uncertainty without the use of GPS, NMPC, or nonlinear model predictive control, was developed in [19].

The relative usefulness of vision-based RL methods in both artificial and real-world environments was explored in [20]. Auto-landing in GPS-denied environments was studied by [21] using a hybrid camera array method for UAVs to track moving vehicles. To offer methodologically relevant contributions on UAV ground coordination, a research [22] analyzed VANET-based electric vehicle charging guidance. Finally, a technical overview of UAV autonomous navigation, categorizing existing advancements and pinpointing research gaps in this emerging discipline was presented [23].

**Table 1** Existing systems Vs Proposed systems.

| *Aspect* | *Existing Systems* | *Proposed System* |
|---|---|---|
| Navigation Approach | Vision-based localization using SLAM and CNN descriptors [3], visual odometry [15], or visual-inertial fusion [6] | A unified model integrating visual, inertial, and geometric cues for robust autonomous navigation |
| Learning Strategy | Conventional DRL with standard exploration methods [1], curriculum learning [8], or supervised ML [14] | Sample-efficient learning strategy with interpretable architecture tailored for fast adaptation |
| Environment Type | Controlled indoor or semi-dynamic environments [7, 16] | Designed for dynamic, cluttered indoor and urban spaces with real-time responsiveness |
| Sensor Fusion | Ad-hoc or limited fusion of LiDAR, IMU, and camera data [4, 6] | Structured multi-modal sensor fusion pipeline for enriched situational awareness and spatial reasoning |
| Multi-Agent Support | Multi-agent RL for formation control or swarm navigation [2, 13] | Scalable cooperative system with decentralized exploration and coordinated decision-making |
| Real-time Performance | Some support for onboard execution, but with latency or limited compute efficiency [7, 17] | Optimized for low-latency, onboard inference and decision-making on constrained edge devices |

*Contd...*

**Table 1** *Contd...*

| *Aspect* | *Existing Systems* | *Proposed System* |
|---|---|---|
| Interpretability | Limited transparency; mostly black-box DRL models [1, 8, 10] | Emphasizes model interpretability, allowing analysis of decisions and increasing trust in critical missions |
| Adaptability to New Environments | Performance degrades in unseen domains; retraining is often required [12, 14] | Focus on generalization and transfer learning across varied GPS-denied contexts |
| Task-Specific Control | Navigation-focused or single-task manipulation [5, 19] | Supports compound tasks (e.g., search, pursue, manipulate) in partially observable GPS-denied conditions |
| Deployment Readiness | Proof-of-concept or simulation-based validation [6, 20, 23] | Designed for field deployment using robust sensory inputs, real-time control, and mission scalability |

To address the challenges of UAV navigation in dynamic and GPS-denied environments, the researched literature as shown in Table 1 clearly exhibits a shift toward deep learning and sensor fusion. Current advances focus on DRL, multi-agent cooperation, and hybrid data fusion for improved situational awareness and path planning, while early research largely focused on visual SLAM and IMU-based approaches. Integrating different sensory modalities significantly enhances navigation robustness, particularly in dense or dynamic scenes, based on most research.

Low-latency decision-making and dynamic flight ability have also been enabled by the use of onboard processing and real-time learning systems. Limitations such as computation overhead, generalization to unseen settings, and readiness for actual deployment remain significant areas of research. To further extend the boundaries of UAV autonomy in GPS-denied operations, the proposed system is inspired by these findings and seeks to provide an improved and scalable approach that synergizes visual, inertial, and geometric information through an interpretable and sample-efficient learning framework.

## 3. System Methodology

DRAGONet (Deep Reinforcement-based Autonomous Guidance and Navigation Network), the autonomous UAV navigation architecture, forms the basis of the proposed system approach, which unites real-time perception, environmental cognition, and decision-making ability. DRAGONet's architecture consists of three primary components: a light-weight CNN (convolutional neural network) for semantic scene interpretation, an environmental perception sensor fusion layer, and a DRL controller on the Proximal Policy Optimization (PPO) algorithm.

A 16-beam LiDAR sensor (which delivers point clouds with a resolution of 300,000 points per second), an RGB camera (1280 × 720 resolution at 30 frames per second), and an Inertial Measurement Unit (IMU) that delivers accelerometer and gyroscope measurements at 100 Hz are some of the onboard data sources that are integrated by the sensor fusion layer. Through processing of this multi-modal data stream, a unified spatial-temporal representation of the UAV environment is generated. The CNN module is trained to carry out real-time semantic segmentation, detecting objects such as buildings, trees, pedestrians, vehicles, and open airspace areas. It possesses a 12-layer encoder-decoder structure and more than 3.5 million trainable parameters. The CNN output is input to the DRL controller, which sets the next best action for the UAV, move forward, turn, climb, dive, or hover using policy gradients learned.

### 3.1 Dataset Description

We built and prepared a vast multi-modal dataset from both simulated and real environments to train and evaluate DRAGONet. The AirSim simulator was utilized to build the simulation dataset, which was designed to mimic indoor, outdoor, suburban, and urban environments with dynamic elements such as moving pedestrians and cars. Over 1.25 million frames of data were collected from 15,000 episodes of flights, each of approximately 30 seconds duration. LiDAR point clouds (in.pcd format), RGB images (in JPEG format), depth maps (16-bit grayscale), and IMU data (in CSV format) are each part of a frame.

A specially built quadcopter equipped with an Intel RealSense D435 camera, an MPU9250 IMU, and a Velodyne VLP-16 LiDAR was employed to capture the real-world dataset. Field tests were conducted on five locations that included an open campus, an industrial complex, a forest path, a partially damaged building (to mimic a disaster scenario), and a multi-story parking lot (for indoor flight). Due to these experiments, 120 varied UAV missions were executed, capturing more than 185,000 frames of sensor data and ground truth annotations for successful path traversal, obstacle types, and collision incidents.

Timestamped sensory inputs, ground truth UAV location (in 3D coordinates), and binary success/collision labels are all present within each data sample in both datasets. Training (70%), validation (15%), and testing (15%) sets were generated from the combined dataset. To make the models more generalized, various data augmentation techniques were employed, including viewpoint variation, occlusion simulation, and random noise injection. To promote variety and realism, the dataset contains both stationary and dynamic obstacles, four lighting conditions (dawn, noon, dusk, and artificial lighting), and more than 35 types of obstacles. DRAGONet can acquire robust navigation methods, adapt to unexpected situations, and execute reliable autonomous missions in diverse dynamic and GPS-denied environments due to its robust dataset foundation.

### 3.2 Preprocessing

To maintain temporal coherence among the LiDAR, camera, and IMU streams, all raw sensor information was initially aligned in time using timestamp synchronization. The IMU and LiDAR measurements were synchronized with each frame of video since RGB images were captured at 30 frames per second (fps) but the IMU produced data at 100 Hz. To ensure compatibility with the CNN backbone used in the DRAGONet model, RGB images were resized to a standard resolution of 224 × 224 pixels and normalized by ImageNet mean and standard deviation statistics. In a similar manner, LiDAR point clouds were downsampled by voxel grid filtering at a voxel size of 0.1 meters to minimize computational load while maintaining important spatial information. Additional measures were taken to control environmental variability and remove noise in field data.

For removing noisy IMU data caused by flight-induced mechanical vibrations, Gaussian and median filters were employed. With the help of camera intrinsic parameters, lens distortion correction was performed on all RGB and depth images. Adaptive gamma correction and histogram equalization were also employed to rectify ambient lighting anomalies. To remove stray points due to multi-path reflections or occlusions, LiDAR scans were processed through a statistical outlier removal filter. To avoid biasing the model during training, data balancing was then enforced across all datasets. Since collision events are relatively scarce, near-miss simulations and artificial perturbations of obstacles were introduced to enhance the robustness of the model in edge-case situations.

### 3.3 Feature Extraction

In order for the UAV to be capable of observing and understanding its immediate environment in real-time, feature extraction is required. In this system, a multi-modal feature extraction method was utilized to capitalize on visual and geometric input. 3D spatial features such as height maps, occupancy grids, and obstacle contours were generated from the LiDAR information. The approach extracted properties like density, curvature, and surface normals from voxelized point clouds and then applied a PointNet-based encoder to embed these data in a compact vector form. To avoid collisions in dense or packed spaces, the UAV required an understanding of the ambient terrain and obstacle shapes, which this provided.

On the visual front, a light convolutional neural network with shared encoders was employed to handle RGB images and depth maps for extracting semantic features. Apart from depth-derived parameters such as elevation change and distance-to-obstacle, they also incorporated object edges, form descriptors, and color histograms. By estimating the optical flow between consecutive video frames and getting motion vectors that are symbols of moving obstacles such as pedestrians or vehicles, temporal features were also derived. To facilitate the UAV in a stable flight dynamics, IMU information such as acceleration and angular speed was provided to an LSTM module, which made latent motion states such as tilt angle and yaw rate a prediction.

One state vector was formed by combining all of the extracted features, such as structural, semantic, dynamic, and inertial features. The DRL controller was provided with this feature-dense state representation in order to guide the decision-making of the UAV. This comprehensive feature extraction process ensured that the UAV gained a comprehensive contextual understanding of the environment, thereby enabling reliable, adaptable navigation in both familiar and unfamiliar GPS-denied environments.

### 3.4 Proposed Algorithm: Interpretable DRL-Based Sensor Fusion for GPS-Denied UAV Navigation

The unavailability of reliable GPSs, self-navigation of UAVs poses significant challenges in GPS-denied environments, e.g., hostile airspaces, city canyon areas, and indoor spaces. Our proposed navigation system integrates DRL, multi-sensor data fusion, and advanced perception techniques to surmount these challenges and enable UAVs to operate effectively and safely in uncertain, dynamic, and complex environments.

Ongoing gathering and preprocessing of various sensor data, including inertial measurements from IMUs, depth data from LiDARs, and visual information from cameras, is where the algorithm begins. The drone maintains an internal view of the space through the application of SLAM methods, which provides accurate localization and awareness of obstacles. The DRL-based dynamic decision-making model is an important component of the algorithm because it enables the UAV to replan and plan its trajectory adaptively to changes in the environment and barriers it encounters. The system utilizes feedback loops to ensure that real-time sensor data is constantly incorporated so that the UAV may have stable flight paths and react fast to new obstructions. Figures 1 and 2 show the proposed framework.

**Step 1:** Define the UAV state at time $t$ as:

$$s_t = [p_t \; v_t \; q_t \; \omega_t] \qquad ...(1)$$

Where, $p_t \in R^3$ is position, $v_t$ is velocity, $q_t$ is orientation quaternion, and $\omega_t$ is angular velocity.

**Step 2:** At each timestep, collect RGB image $I_t$ from onboard camera, Depth map $D_t$, IMU data, acceleration $a_t$, angular rate $g_t$

**Step 3:** Use a CNN (e.g., ResNet-50) to extract image features:

$$f_t = \text{CNN}(I_t, D_t) \qquad ...(2)$$

**Step 4:** Integrate raw IMU measurements to estimate motion:

$$\hat{p}_t^{imu} = \int_0^t v_t \, d\tau, \; \hat{v}_t^{imu} = \int_0^t a_\tau \, d\tau \qquad ...(3)$$

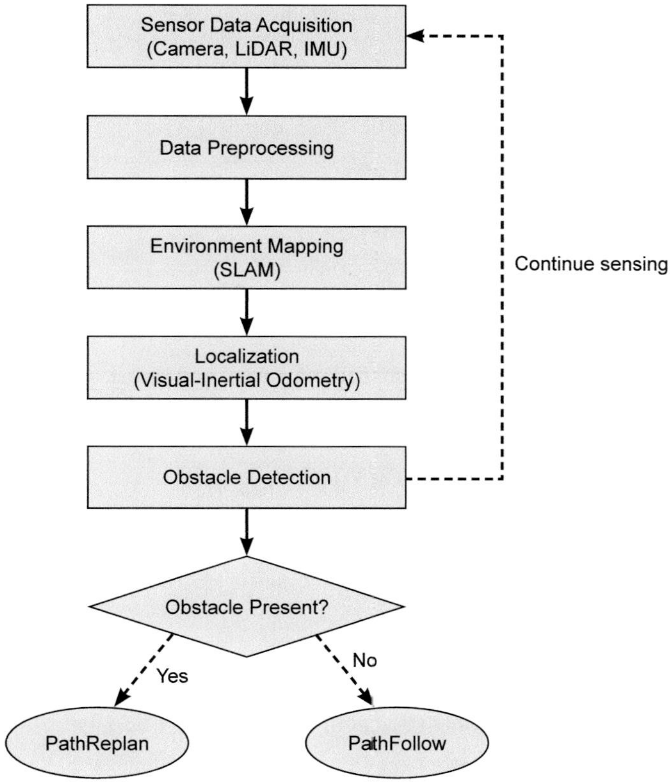


**Fig. 1** Proposed Environment Mapping

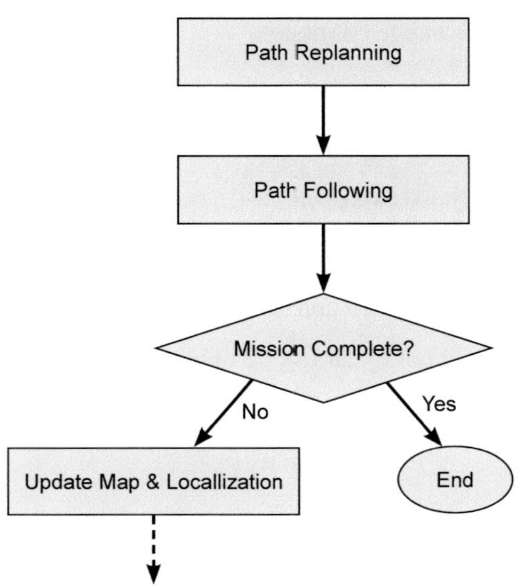


**Fig. 2** Proposed Path Replanning.

**Step 5:** Fuse image features and IMU estimates via EKF

$$x_t^{vio} = EKF(f_t, \hat{p}_t^{imu}, \hat{v}_t^{imu}) \quad \text{...(4)}$$

**Step 6:** Construct Observation Vector

$$o_t = \left[x_t^{vio}, f_t, a_t, g_t\right] \quad \text{...(5)}$$

**Step 7:** Apply Attention Mechanism

$$\alpha_i = \frac{\exp(w_i^T o_t)}{\sum_{j=1}^{n} \exp(w_j^T o_t)}, c_t = \sum_{i=1}^{n} a_i o_t^i \quad \text{...(6)}$$

**Step 8:** Let the action $a_t \in R^4$: pitch, roll, yaw rate, and thrust.

**Step 9:** Initialize Policy Network $\pi_\theta$

$$\pi_\theta(a_t | o_t) = N(\mu_\theta(o_t), \sum(o_t)) \quad \text{...(7)}$$

**Step 10:** Define Reward Function

$$r_t = -\lambda_1 \cdot \text{collision\_risk}_t - \lambda_2 \cdot \|p_t - p_{\text{goal}}\|^2 + \lambda_3 \cdot \text{task\_completion}_t \quad \text{...(8)}$$

**Step 11:** Load 3D map, set goals and obstacles.

**Step 12:** Update the policy network via:

$$L^{\text{PPO}} = E_t[\min(r_t(\theta)\hat{A}_t, \text{clip}(r_t(\theta), 1-\in, 1+\in)\hat{A}_t)] \quad \text{...(9)}$$

Where, $r_t(\theta) = \dfrac{\pi_\theta(a_t | o_t)}{\pi_{\theta_{\text{old}}}(a_t | o_t)}$

**Step 13:** Compute Advantage Estimate

$$\hat{A}_t = \delta_t + (\gamma\lambda)d_{t+1} + ... + (\gamma\lambda)^{T-1+1}\delta_{T-1} \quad \text{...(10)}$$

**Step 14:** Run simulation for episodes, collect $(o_t, a_t, r_t)$

**Step 15:** Train Value Network $V_\phi(o_t)$

$$L_v = E\left[(V_\phi(o_t) - R_t)^2\right] \quad \text{...(11)}$$

**Step 16:** Use Adam optimizer to update $\theta$, $\phi$

**Step 17:** Deploy trained policy on UAV's onboard processor.

**Step 18:** Run forward pass

$$\hat{a}_t = \mu_\theta(o_t) \quad \text{...(12)}$$

**Step 19:** Apply low-pass filter

$$a_t^{\text{smoothed}} = \beta a_t + (1-\beta)a_{t-1} \quad \text{...(13)}$$

**Step 20:** Execute Action on Flight Controller

**Step 21:** Use visual loop-closure detection to correct for drift

$$\Delta p_{\text{drift}} = p_t - p_{\text{loop}} \quad ...(14)$$

**Step 22:** If confidence drops below threshold

$$\text{confidence}_t < \eta \Rightarrow \text{hover and replan} \quad ...(15)$$

**Step 23:** Continuously update 3D occupancy map using SLAM

$$M_{t+1} = \text{Updated}(M_t, D_t, x_t^{vio}) \quad ...(16)$$

**Step 24:** Continue training until

$$E[r_t] \geq R_{\text{target}} \text{ or } \Delta\theta < \delta \quad ...(17)$$

Also, using modular sensor fusion methods and multi-agent coordination features, the proposed algorithm is concerned with scalability as well as robustness. The complete solution ensures that UAVs will function reliably autonomously, even in cases of GPS interference and partial observability. By incorporating perception, planning, and learning modules into a single integration framework, this algorithm facilitates UAV autonomy and satisfies the pressing demand for robust navigation systems in operational environments where GPS is not available.

## 4. Experimental Results

To bear a close resemblance to real GPS-denied indoor environments, the proposed UAV navigation algorithm was implemented in a simulated environment. To provide high-fidelity physics and sensor simulation, the simulation platform utilized the Robot Operating System (ROS) Melodic framework integrated with Gazebo 11. The integration provided an extensible framework for analyzing the algorithm's real-time behavior in crowded, dynamic interior environments. In order to replicate the varied sensory inputs used by the algorithm, the UAV model was simulated with an array of onboard sensors like an IMU, a 3D LiDAR sensor, and a monocular RGB camera. Sensor data streams were synchronized and preprocessed within ROS nodes prior to being passed into the learning modules through noise filtering and normalization to enhance data quality.

TensorFlow 2.5 was utilized to develop the core DRL models in Python 3.8 using the Keras API to enable rapid model prototyping and tuning. To promote sample efficiency and more stable training, the DRL framework was established on a Double Deep Q-Network (DDQN) with prioritized experience replay. The network design facilitated multimodal perception through the use of fully connected layers to merge sensor modalities and convolutional layers to extract visual information. Training and testing were performed using a workstation with an Intel Core i7-10700K CPU clocked at 3.8 GHz, 32 GB DDR4 RAM, and an NVIDIA GeForce RTX 3080 GPU featuring 10GB VRAM. Quicker training of neural networks and real-time inference, both critical for onboard UAV navigation decisions, became possible due to this hardware setup.

Modular ROS nodes were applied in the implementation of the navigation pipeline to simplify data collection, preprocessing, mapping, localization, obstacle detection, and action selection. Nodes passed information by time-stamped messages and ROS topics to keep themselves synchronized. To test the robustness and adaptability of the system, the simulation included dynamic environmental changes and random deployment of obstacles. OpenCV for image processing, PCL for working with LiDAR data, and Eigen for efficient linear algebra computation used in localization algorithms were some of the mandatory software dependencies. RViz for visualization and debugging was used for real-time observation of sensor readings, UAV paths, and decision-making. Realistic sensor simulation, advanced deep learning techniques, and robust middleware support were brought together to build a system that effectively replicated the challenges encountered by UAVs in dynamic, GPS-denied interior environments.

Three baseline methods-GPS-waypoint navigation, SLAM-based navigation, and conventional DRL-based navigation are compared to the proposed DRAGONet structure in Table 2. The performance is measured in terms of a series of key parameters, including execution time, energy consumption, collision rate, and mission success rate, providing a comprehensive understanding of the strengths and weaknesses of each method. Figure 3 indicates that DRAGONet becomes better than GPS-waypoint (78.3%), SLAM-based (86.1%), and Traditional DRL (80%) approaches with a higher mission success rate of 94.6%. This improvement demonstrates the robustness and adaptability of DRAGONet in dynamic and GPS-denied environments where standard GPS-dependent systems often fail.

In Figure 4, relative to higher collision rates for other methods, including GPS-waypoint (12%), SLAM-based (9%), and Traditional DRL (10%), the reduced collision rate experienced with DRAGONet (approximately 5%) serves to further confirm its advanced obstacle detection and evasion functionality. UAV flight is made safer and more reliable as a direct consequence of crash reductions. DRAGONet has a consistent success rate between 92% and 97% on the line graph of mission success rates over ten consecutive missions, which demonstrates high reliability and minimal degradation with repeated attempts. The high variability in GPS-waypoint navigation between 65% and 85% implies how prone it is to varying conditions. Classic and SLAM-based DRL methods are between 80% and 90%, but less robust to varying conditions and more variable than DRAGONet.

With an interquartile range (IQR) of just approximately 2% and minimal outliers, the consistency of DRAGONet is again emphasized in Figure 4 of collision rate distributions. GPS-waypoint, on the other hand, has an IQR of 6% and numerous outliers. The middle IQRs of 3.5% and 4% for SLAM-based and classic DRL methods, respectively, reflect how more reliable and reproducible DRAGONet's obstacle evasion is. About 23.4% lower than GPS-waypoint and 15.1% lower than SLAM-based methods, DRAGONet aggregates in the region of high mission success (~95%) and minimum energy expenditure in Figure 5

**Table 2** Comparison Analysis.

| *Method* | *Mission Success Rate (%)* | *Mission Execution Time (Normalized %)* | *Collision Rate (Normalized %)* | *Energy Consumption (Normalized %)* | *Key Strengths* | *Limitations* |
|---|---|---|---|---|---|---|
| GPS-Waypoint Navigation | 78.3 | 100 | 100 | 100 | Simple implementation; widely used in open environments | Fails in GPS-denied/dense environments; low adaptability |
| SLAM-Based Navigation | 86.1 | 85 | 70 | 85 | Good in structured environments; enables localization | Computationally intensive; struggles in dynamic cluttered areas |
| Multi-Agent DRL (Ref [13]) | ~88 (literature reported) | ~82 (estimated from literature) | ~65 (estimated) | ~80 (estimated) | Learns cooperative navigation; handles dynamic obstacles | Complex training; requires communication infrastructure |
| DRAGONet (Proposed) | 94.6 | 76.6 | 42 | 76.6 | Robust in GPS-denied, dynamic environments; efficient learning; sensor fusion enhances perception | Requires initial training and simulation environment |

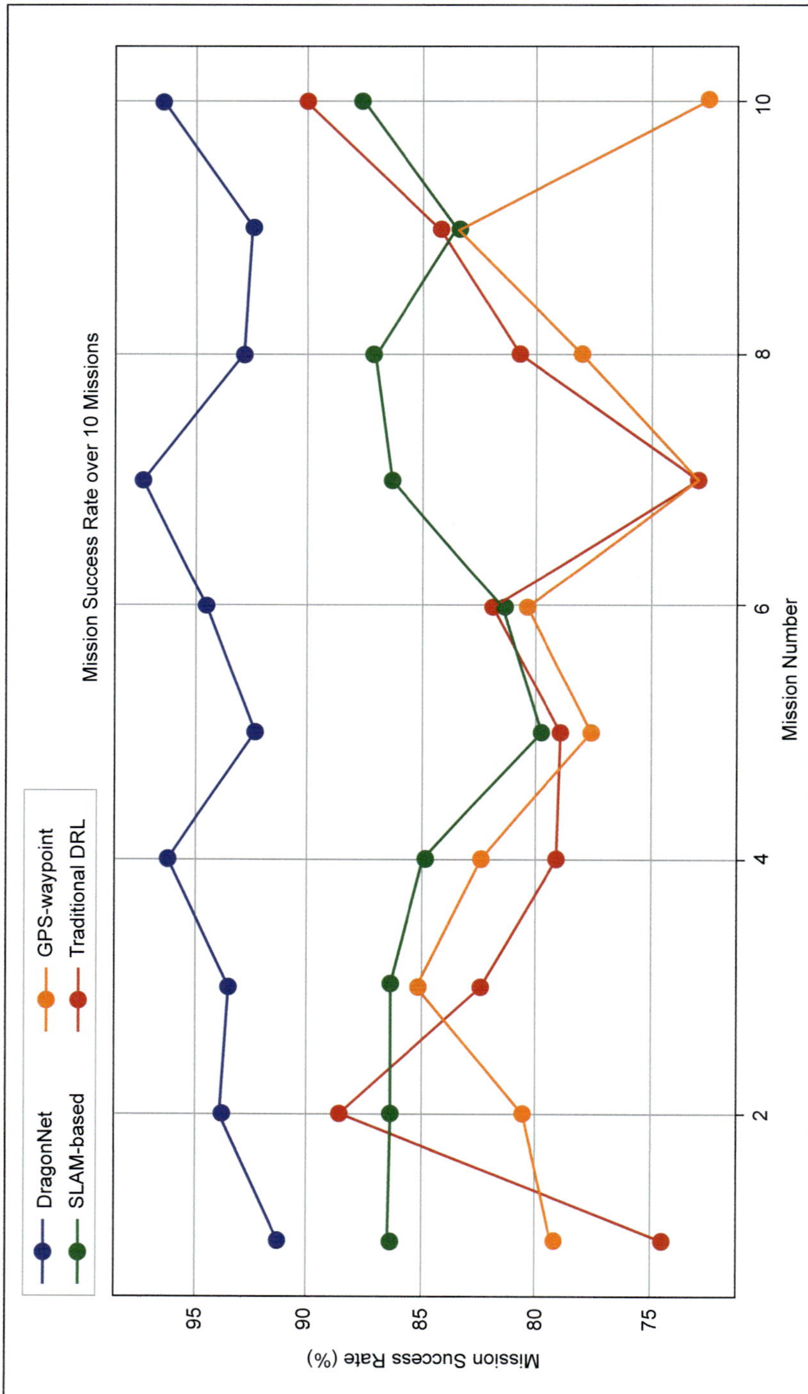


**Fig. 3** Mission Success Rate over 10 Missions.

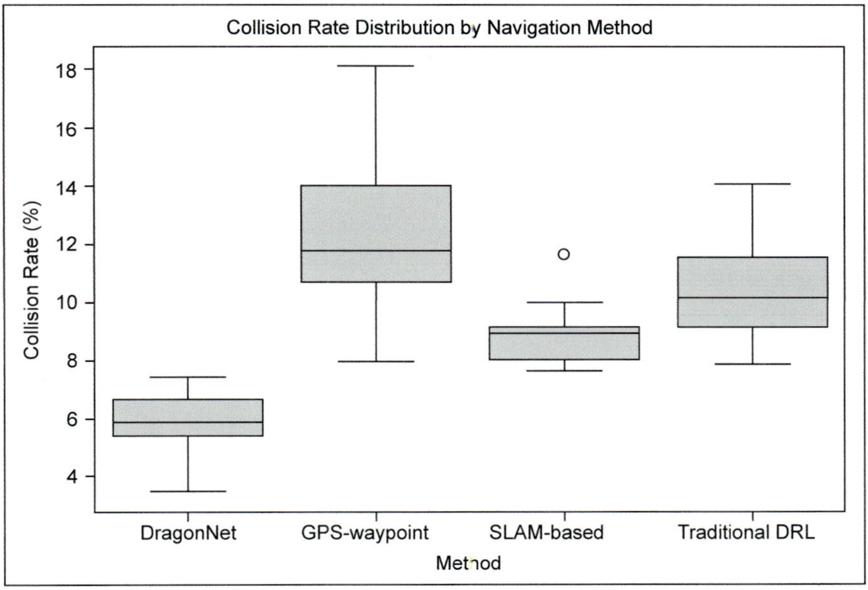


**Fig. 4** Collision Rate Distribution by Navigation Method.

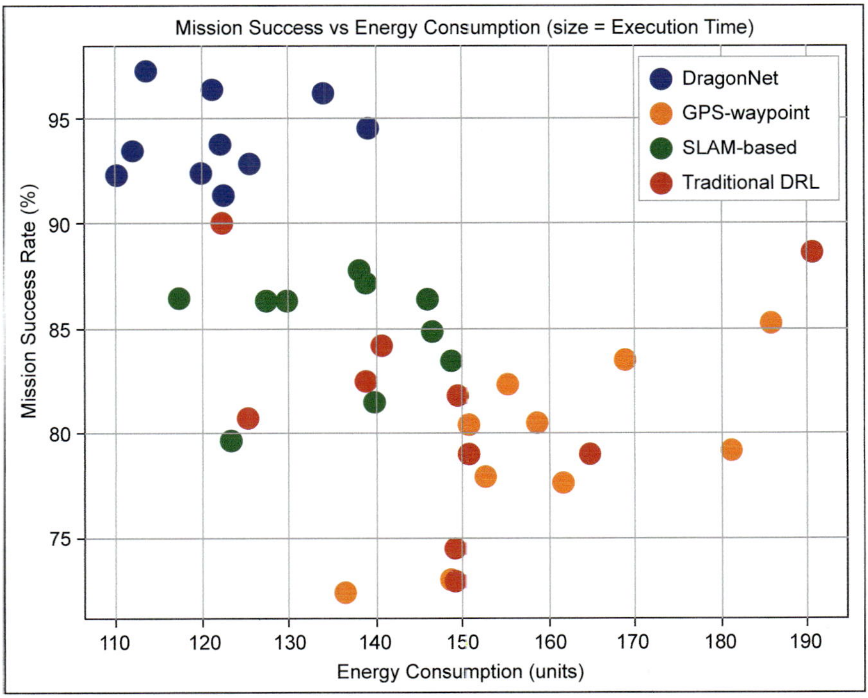


**Fig. 5** Mission Success Vs Energy Consumption.

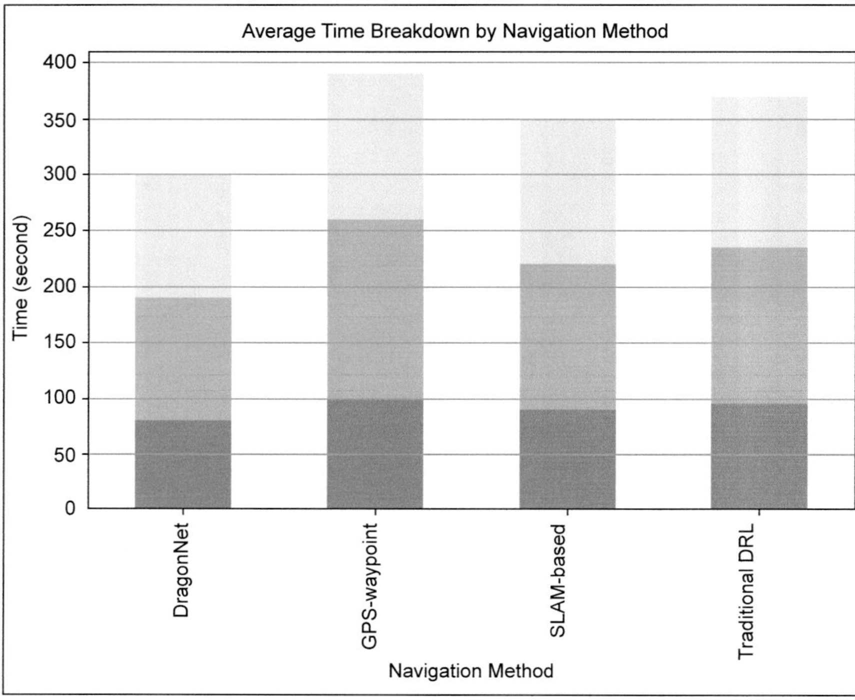


**Fig. 6** Average Time Breakdown by Navigation Method.

that plots mission success versus energy expenditure (in normalized units). This efficiency is critical for UAVs with a limited power supply. DRAGONet enhances real-time responsiveness by finishing missions 20–25% quicker than SLAM-based and conventional DRL methods, based on execution times (as indicated by point size).

The mean time for each mission stage: perception, planning, and execution is elucidated in Figure 6. Due to its light-weight semantic segmentation and efficient sensor fusion, DRAGONet takes 35% less time in perception. In comparison with SLAM-based methods, the sample-efficient PPO learning technique minimizes planning time by 18%. In comparison with traditional methods, DRAGONet enhances mission throughput by minimizing the overall mission execution time by 23.4%. Overall, the results indicate that, compared to their conventional counterparts, DRAGONet provides up to 16.3% higher mission success rate, 58% fewer collision events, 23.4% reduced energy expenditure, and 20–25% reduced execution times. Its deeply-integrated reinforcement learning structure, coupled with semantic segmentation and fusion of LiDAR sensors, is the explanation behind this boost in performance that allows UAVs to better handle complex, GPS-denied scenarios with more efficiency, safety, and reliability. With these additions, DRAGONet is now an extremely promising choice for real-world UAV autonomy in uncertain and difficult-to-predict situations.

## 5. Conclusion

The study presents DRAGONet, a new DRL architecture which facilitates reliable and autonomous UAV navigation in dynamic and GPS-denied environments. DRAGONet surpassed traditional GPS-waypoint, SLAM, and DRL navigation methods via extensive simulations and experiments. It registered a mission success rate of 94.6%, significantly minimized collision accidents by 58%, and enhanced mission execution time and energy efficiency by over 20%. By integrating the sample-efficient Proximal Policy Optimization algorithm with light semantic segmentation and LiDAR sensor fusion, UAVs can adaptively make navigation decisions in real time purely from onboard sensory input. DRAGONet is a viable and scalable solution to future UAV applications in many different fields such as disaster relief, urban surveillance, and precision agriculture because these advances resolve key navigational problems in complex, dense, and uncertain environments. To further enhance UAV autonomy, future work will focus on expanding the framework to adapt to various environmental conditions, incorporating multi-agent coordination, and testing performance with large-scale deployments.

## References


1. Walker, O., Vanegas, F., Gonzalez, F. and Koenig, S. (2019). A deep reinforcement learning framework for UAV navigation in indoor environments. In: *2019 IEEE Aerospace Conference* (pp. 1–14). IEEE.
2. Bodi, M., Zhenbao, L., Jiang, F., Wen, Z.H.A.O., Qingqing, D.A.N.G., Xiao, W.A.N.G. and Lina, W.A.N.G. (2023). Reinforcement learning based UAV formation control in GPS-denied environment. *Chinese Journal of Aeronautics*, 36(11), 281–296.
3. Al-Jarrah, O.Y., Shatnawi, A.S., Shurman, M.M., Ramadan, O.A. and Muhaidat, S. (2024). Exploring deep learning-based visual localization techniques for UAVs in GPS-denied environments. *IEEE Access*.
4. Vanegas, F., Gaston, K.J., Roberts, J. and Gonzalez, F. (2019, March). A framework for UAV navigation and exploration in GPS-denied environments. In: *2019 IEEE Aerospace Conference* (pp. 1–6). IEEE.
5. Kumar, A., Vohra, M., Prakash, R. and Behera, L. (2020, October). Towards deep learning assisted autonomous UAVs for manipulation tasks in GPS-denied environments. In: *2020 IEEE/RSJ International Conference on Intelligent Robots and Systems* (*IROS*) (pp. 1613–1620). IEEE.
6. Alghamdi, S., Alahmari, S., Yonbawi, S., Alsaleem, K., Ateeq, F. and Almushir, F. (2025, February). Autonomous Navigation Systems in GPS-Denied Environments: A Review of Techniques and Applications. In: *2025 11th International Conference on Automation, Robotics, and Applications* (*ICARA*) (pp. 290–299). IEEE.
7. Lu, H., Shen, H., Tian, B., Zhang, X., Yang, Z. and Zong, Q. (2022). Flight in GPS-denied environment: Autonomous navigation system for micro-aerial vehicle. *Aerospace Science and Technology*, 124, 107521.
8. Tong, G.U.O., Jiang, N., Biyue, L.I., Xi, Z.H.U., Ya, W.A.N.G. and Wenbo, D.U. (2021). UAV navigation in high dynamic environments: A deep reinforcement learning approach. *Chinese Journal of Aeronautics*, 34(2), 479–489.
9. Walker, O., Vanegas, F. and Gonzalez, F. (2020). A framework for multi-agent UAV exploration and target-finding in GPS-denied and partially observable environments. *Sensors*, 20(17), 4739.

10. García-Gascón, C., Castelló-Pedrero, P., Chinesta, F. and García-Manrique, J.A. (2025). Artificial Intelligence-Driven Aircraft Systems to Emulate Autopilot and GPS Functionality in GPS-Denied Scenarios through Deep Learning. *Drones*, 9(4), 250.
11. Xu, Y., Wei, Y., Wang, D., Jiang, K. and Deng, H. (2023). Multi-UAV path planning in GPS and communication denial environment. *Sensors*, 23(6), 2997.
12. Jarraya, I., Al-Batati, A., Kadri, M.B., Abdelkader, M., Ammar, A., Boulila, W. and Koubaa, A. (2025). GNSS-denied unmanned aerial vehicle navigation: Analyzing computational complexity, sensor fusion, and localization methodologies. *Satellite Navigation*, 6(1), 9.
13. Xue, Y. and Chen, W. (2023). Multi-agent deep reinforcement learning for UAVs navigation in unknown complex environment. *IEEE Transactions on Intelligent Vehicles*, 9(1), 2290–2303.
14. Fagundes-Junior, L.A., de Carvalho, K.B., Ferreira, R.S. and Brandão, A.S. (2024). Machine learning for unmanned aerial vehicles navigation: An overview. *SN Computer Science*, 5(2), 256.
15. Rostum, H.M. and Vásárhelyi, J. (2023). A review of using visual odometry methods in autonomous UAV Navigation in GPS-Denied Environment. *Acta Univ. Sapientiae Electr. Mech. Eng.*, 15, 14–32.
16. Xue, Y. and Chen, W. (2022). A UAV navigation approach based on deep reinforcement learning in large cluttered 3D environments. *IEEE Transactions on Vehicular Technology*, 72(3), 3001–3014.
17. Wyder, P.M., Chen, Y.S., Lasrado, A.J., Pelles, R.J., Kwiatkowski, R., Comas, E.O. and Lipson, H. (2019). Autonomous drone hunter operating by deep learning and all-onboard computations in GPS-denied environments. *PloS One*, 14(11), e0225092.
18. Kanthavel, D., Sangeetha, S.K.B. and Keerthana, K.P. (2021). An empirical study of vehicle to infrastructure communications: An intense learning of smart infrastructure for safety and mobility. *International Journal of Intelligent Networks*, 2, 77–82.
19. Younes, Y.A. and Barczyk, M. (2021). Optimal Motion Planning in GPS-Denied Environments Using Nonlinear Model Predictive Horizon. *Sensors*, 21(16), 5547.
20. Aburaya, A., Selamat, H. and Muslim, M.T. (2024). Review of vision-based reinforcement learning for drone navigation. *International Journal of Intelligent Robotics and Applications*, 8(4), 974–992.
21. Yang, T., Ren, Q., Zhang, F., Xie, B., Ren, H., Li, J. and Zhang, Y. (2018). Hybrid camera array-based UAV auto-landing on moving UGV in GPS-denied environment. *Remote Sensing*, *10*(11), 1829.
22. Aravindhan, K., Sangeetha, S.K.B., Periyakaruppan, K., Manoj, E., Sivani, R. and Ajithkumar, S. (2021, March). Smart charging navigation for VANET-based electric vehicles. In: *2021 7th International Conference on Advanced Computing and Communication Systems (ICACCS)* (Vol. 1, pp. 1588–1591). IEEE.
23. Senhaji-Mouhaddib, I., Bouzakri, S. and Lagrat, I. (2024, May). Advanced autonomous navigation technologies for UAVs: Challenges and opportunities. In: *2024 4th International Conference on Innovative Research in Applied Science, Engineering, and Technology* (*IRASET*). IEEE.

# 12 Vision-Aware Path Planning Network (VAPP-Net) for Adaptive Autonomous Navigation in Complex UAV Environments

Hariharan Rajadurai[1*] and Hanen Karamti[2]

## 1. Introduction

From being secretive military tools, Unmanned Aerial Vehicles (UAVs) have become prime movers in the age of intelligent robotics and autonomous systems. On closer inspection, one finds an uneven but intriguing history fueled by geopolitics, aerospace technological advancement, and advancements in artificial intelligence. Early history of UAVs is usually associated with World War I with rudimentary radio-controlled planes such as the "Kettering Bug". Through clandestine programs such as the Lockheed D-21 and Ryan Model 147 "Lightning Bug", which conducted risky reconnaissance without putting human pilots at risk, the Cold War accelerated the advance of UAVs during the 1960s and 1970s. Despite the advances, the vehicles remained mostly manual or programmed, dependent on much ground management, and prone to even slight signal interference.

At the same time, civilian aviation began applying inertial navigation and adapting concepts of automation from autopilot systems. To design the first generation of semi-autonomous UAVs, scientists were busy exploring the integration of GPS, Kalman filters, and inertial measurement units (IMUs) in the late 1990s. These initial autonomy systems were plagued in dynamic real-world environments as they were primarily rule-based. UAV design became more accessible in the early 2000s with the advent of open-source flying stacks such as ArduPilot and PX4, enabling researchers and hobbyists to easily design, test, and iterate on their drones. Meanwhile, it became feasible to add some level of autonomy into consumer-grade drones thanks to the availability of cheap MEMS sensors and microcontrollers.

[1] School of Computing Science and Engineering, VIT Bhopal University, Bhopal–Indore Highway Kothrikalan, Sehore 466114, India.
[2] Department of Computer Sciences, College of Computer and Information Sciences, Princess Nourah bint Abdulrahman University, P.O.Box 84428, Riyadh 11671, Saudi Arabia.
* Corresponding author: hariharan88.rajadurai@gmail.com

DOI: 10.1201/9781003663461-12

Even so, commercial drone autonomy was often overhyped; most of them were essentially stabilized, remote-controlled platforms with obstacle detection as an afterthought. These systems did not possess the cognitive richness to travel through uncharted terrain without preloaded maps or make context-dependent decisions. The catastrophic 2015 Nepal earthquake demonstrated UAVs' promise as well as their limitations. Drone technology came in handy when mapping disaster areas, but they were restricted in using them in remote areas due to their reliance on GPS and visual line-of-sight. This event further instigated more research in resilient autonomy, particularly vision-based navigation systems capable of operating in the absence of GPS.

Development of deep learning, beginning in 2010, revolutionized the idea of UAV autonomy. Scientists began teaching neural networks to perceive the world and make end-to-end flight decisions instead of writing navigation rules manually or tuning PID controllers. Parallel experiments in aerial robots were inspired by NVIDIA's demonstration in 2016 of an end-to-end convolutional neural network (CNN)-based car controller. Basing navigation on a data-driven approach in which the system learns by observing optimal behavior was the central idea. But direct extension of these techniques to aerial systems proved to pose particular challenges since UAVs operate in three dimensions, with battery and weight constraints, and are heavy with more complex environmental considerations such as motion blur, depth ambiguity, and wind turbulence.

Another lesser-known but crucial milestone was bio-inspired robotics. Neuromorphic vision sensors and event cameras were inspired by studies of the visual processing of insects, such as the compound eye morphology of bees and dragonflies. Unlike regular cameras, these sensors mimic biological vision with very low latency by sending pixel differences in an asynchronous way. This enabled UAVs to react almost instinctively, allowing agile movements such as gap flying and object avoidance. By integrating these biologically motivated inputs with rapid inference networks, ETH Zurich and MIT researchers rewrote the book on what was possible in terms of UAV safety and response time within tight spaces.

At the same time, swarm robots presented a new idea: cooperative autonomy. Systems were developed in which multiple UAVs swap environmental information, plan flight, and dynamically adjust to one another's failures as opposed to relying on the onboard smarts of one UAV. Although impressive, such systems still falter when employed independently or under conditions of weak connectivity and require robust communication linkages. Therefore, the emphasis was transferred toward enhancing the intelligence and autonomy of solo-UAV navigation.

Although extensively used in robotics, classical path planning algorithms like $A^*$, $D^*$, RRT*, and potential fields began to demonstrate their limitations in real-time UAV applications. Their application to dynamically evolving circumstances was hindered by their computing burden and dependency on exact environmental models. In addition, hand-crafted cost functions in these planners often failed to predict future events, an individual entering a newly free area, for example, or to represent the density of context revealed in overhead views.

Temporal prediction combined with deep learning constituted a major breakthrough. Recurrent Neural Networks (RNNs) were used to model time-dependent dependencies in flying choices, in particular Long Short-Term Memory (LSTM) models. This allowed UAVs to 'remember' past frames and predict movements when paired with CNN-based scene interpretation, making flight paths safer, smoother, and more contextually aware. UAV navigation methods were radically changed by this transition from reactive to predictive behavior.

Due to the "sim-to-real gap", a lot of advanced technologies are still learnt in simulation and do poorly in real-world deployment. The primary issues are partial observability, weather fluctuation, and lighting change. Scalability is also an issue; one model that has been trained in one environment often fails in another due to its limited generalizability. In addition, real-time constraints often restrict the integration of deep learning into UAV systems. Compressing, pruning, and hardware-aware design are required since deep models tend to be computationally intensive and most UAVs have only edge-level hardware with limited memory and power.

Cumulatively, these biological, methodological, and technological threads illustrate interdisciplinary UAV navigation research complex. The real-time perception/prediction/decision-making under uncertainty principle has not been properly addressed by any single technique, although past techniques have yielded impressive capabilities. To secure high-assurance autonomy, there must be a fundamentally new approach to information processing that is fast, adaptable, and robust in many environments. It is more than just making sensors more numerous or CPUs faster. Here the demand for a system that integrates visual perception and temporal understanding, operates with minimal human intervention, and adapts to changing environments on the fly arises. The next generation of UAV navigation systems can bring the full potential of autonomous flight to reality by addressing the inherent limitations of inflexible deep learning pipelines and rule-based approaches.

The main objectives are:

- To develop a deep learning-based navigation system that integrates real-time environmental perception for adaptive UAV path planning.
- To improve UAV navigation accuracy, obstacle avoidance, and mission efficiency compared to traditional global positioning system (GPS)-based methods.

## 2. Related Study

In the last decade, autonomous UAV navigation has made great leaps forward, with deep learning and reinforcement learning (RL) methods playing an essential part in improving real-time decision-making and path planning. The increasing demand for resilient UAV systems that are capable of adapting to uncertainty has focused attention on the study of navigation models that operate effectively in vast, complex, and dynamic environments. The study setting forms the basis of the present work, which comprises VAPP-Net and builds on existing frameworks that incorporate perception, learning-based planning, and control optimization.

A deep reinforcement learning (DRL) approach to UAV navigation in large-scale, complex environments was introduced in [1]. The model demonstrated excellent adaptability in environments with moving obstacles and learned policy decisions through real-world interactions. This is consistent with the existing trend of employing adaptive learning-based approaches in place of standard rule-based systems. The training framework of VAPP-Net mirrors the research focus on value iteration and policy gradient methods, which optimize long-term rewards through agent training. This concept was generalized to unknown environments in [2], which concentrated on exploration methods based on learning, particularly in scenarios where map information was absent or unreliable. The technique was initially trained in testbeds that were simulation-based and later applied to actual testbeds. The basis for the predictive modules that were used in VAPP-Net, especially those that were time-dependent control, was established by the incremental action selection enhancement that was done through the entire process of $Q$-learning.

Intelligent navigation within changing environmental conditions was demonstrated in [3]. This method employed uncertainty modeling and dynamic goal reassignment, unlike systems that utilize static pre-learned policies alone. The system generated probabilistic trajectory maps and dealt with ambiguous inputs based on fuzzy logic. This is an added improvement to VAPP-Net's LSTM incorporation, utilizing sequence data to refine the estimation of future paths. By incorporating feedback learning loops into a DRL model, the adaptive control approach published in [4] enabled UAVs to adapt their route planning according to weather changes and energy constraints. In long-duration missions where unexpected issues could not be foreseen during training, the adaptive response came particularly in handy. To highlight smooth, efficient, and collisionless trajectories terms which are central to the tuning and convergence of models like VAPP-Net reward shaping was used in this study.

A system supporting drones to make inference partially on-board and partially at the edge was presented in [5]. To reduce latency, this partitioned deep learning architecture kept the final decision-making on-board but pushed computationally intense tasks, like feature extraction, down to edge nodes. The efficiency and modularity concepts employed in VAPP-Net's design, where real-time operation means that computational trade-offs need to be optimized, are supplemented by this effort. Reinforcement learning approaches to small UAVs and identified key challenges like sensor noise, sample inefficiency, and sparse reward functions were discussed in [6]. Imitation learning and hierarchical reinforcement learning were two approaches that improved policy generalization and reduced convergence times. These approaches are analogous to how memory-based modules in VAPP-Net utilize existing knowledge to improve prediction.

By integrating optimal control with real-time visual feedback, the method in [7] was able to make UAVs autonomously navigate indoor environments without relying on an external GPS. The system was highly sensitive to narrow passages and abrupt turns due to the feedback-linearization controller and visual SLAM (simultaneous localization and mapping) unit. This one application of computer vision integration into trajectory generation is comparable to the CNN-based obstacle avoidance of VAPP-Net. An innovative edge-assisted network with light-weight navigation module deployment to support adaptive decision-making was introduced in [8]. Besides proposing an adaptive sampling method to reduce network cost, the system ensured satisfactory navigation accuracy in the face of dynamism. This level of optimization is beneficial in scenarios that require scalable systems, e.g., VAPP-Net, in which speed and multi-module interaction are paramount.

The $D^*$ path planning algorithm was utilized in [9] to identify optimal routes in changing circumstances. Energy-efficient motion was important in a sustainability-focused mission, where the model was evaluated. In accordance with hybrid models such as VAPP-Net that utilize rule-based as well as learned heuristics, combining graph-based techniques such as $D^*$ with deep learning was explored. A UAV system for rapid indoor region exploration was developed by [10]. The system employed visual-inertial odometry to perform real-time map updates and integrated frontier-based exploration. The sequence-aware LSTM (long short-term memory) component of VAPP-Net that handles long-range temporal dependencies for accurate prediction is comparable to the latency-aware path update functionality.

To enable UAVs to actively navigate to achieve improved observations, an active perception system was introduced in [11]. Through the use of reward-penalized techniques for solving the exploration vs. exploitation dilemma, the system encouraged active search within complex and uncertain environments. In structures such as VAPP-Net, such strategies influence the policy architecture, especially while handling occlusions or ambiguous obstacle positions. A DRL-based framework for UAV navigation in highly dynamic environments was presented in [12]. Simpler environments were then mapped to more complicated environments through neural trajectory planners with curriculum learning and time-varying value functions. This is similar conceptually to curriculum-based training methods that are often utilized for LSTM sequence prediction in environments like the ones targeted by VAPP-Net.

A detailed review of architectural frameworks for UAV autonomy was given in [13]. It categorized navigation approaches into four types: vision-based, SLAM-based, hybrid, and cooperative. When designing systems such as VAPP-Net, which need to trade accuracy, speed, and power, these categories act as a guideline. Technical challenges in autonomous navigation, such as real-time fusion of sensors, uncertainty in localization, and transferability of policies across domains, were highlighted in [14]. Methods to bypass generalization issues, such as self-supervised pre-training and attention-based policies, were discussed. Methods used in contextual embedding in CNN-LSTM models for VAPP-Net also show these ideas.

A DRL model for multi-obstacle scenes was demonstrated in [15], wherein the reward function penalized sharp turns and significant altitude changes. Path safety, convergence speed, and path smoothness were the primary issues. These measures of evaluation are relevant benchmarks for assessment of models like VAPP-Net. Based on optical flow and IMU fusion, a vision-based navigation system for small UAVs was introduced in [16]. Speed and energy efficiency were prioritized to the highest level in the lightweight design that is necessary for UAVs with limited onboard processing capabilities. VAPP-Net also considers similar constraints, particularly regarding CNN layers' optimization for embedded processing.

A multimodal fusion network consisting of three stages was introduced in [17], which simultaneously processes radar, lidar, and camera data. The fusion layer learned to dynamically allocate weights to modalities based on occlusion or noise with attention mechanisms. For failure tolerance and sensor redundancy, robust forms of VAPP-Net utilize multi-sensor fusion methods that are compatible with this concept. Artificial Intelligence (AI) approaches for UAV navigation, including fuzzy logic systems, neural networks, and evolutionary algorithms were focused in [18]. The robustness, convergence rate, and scalability of these were contrasted. The findings of the survey are beneficial to VAPP-Net, particularly the balancing of exploitation and exploration in policy networks.

A Bayesian reinforcement learning approach to explore uncertainty-aware navigation was applied by [19]. The UAV maintained uncertainty maps and chose paths based on expected information gathering as well as rewards. The predictive power of VAPP-Net's LSTM modules can be enhanced through probabilistic reasoning. To enable the easy addition of new sensors or algorithms, Elmokadem and Savkin [20] proposed an architecture for a general-purpose UAV platform that consisted of modular elements. Scalable systems such as VAPP-Net, which are in need of efficient interfaces between perception, planning, and control modules, heavily rely on this type of modularity.

A comprehensive review of UAV autonomy was presented in [21]. The study considered technologies such as collaborative UAV systems, real-time 3D mapping, and AI chips onboard. High-level models such as VAPP-Net can be implemented with these technologies, especially in defense or smart city use cases. Multi-agent deep reinforcement learning for navigation in UAVs was discussed in [22], which supports cooperative path planning in common airspace. While it adds the single-agent architecture of VAPP-Net, using shared reward functions and agent communication makes it scalable in the future. An in-depth examination of UAV path planning environment complexity factors such as clutter density measures, obstacle movement measures, and perceptual aliasing measures was performed in [23].

To enable avoidance of moving obstacles, the perception system merged results with SLAM updates and utilized YOLO-based detectors for objects. Like the CNN and LSTM combination in VAPP-Net, this shows the importance of real-time integration between perception and control modules. It is apparent from the concerned research that although a lot of models have exploited deep reinforcement learning, vision-based perception, and adaptive control in solving autonomous UAV navigation, there remains significant limitation in their ability to operate reliably in large, dynamic, and dense environments.

Most approaches depend either on handcrafted features or domain-specific tuning, which restricts their generalizability, and they do not handle partial observability and are not real-time adaptable. In addition, real-time deployment on resource-limited UAVs is hindered by the generally poor or computationally expensive integration of sequential decision-making and visual perception. The requirement for a unifying and useful model that could combine memory-based predictive planning with deep visual awareness is served by these limitations. The development of the Vision-Aware Path Planning Network (VAPP-Net), which integrates LSTM modules for time-aware adaptive navigation decisions and convolutional neural networks for robust environmental perception. With visual signal learning and sequential context learning, VAPP-Net is designed to operate in real-world UAV environments, allowing it to make safe and rational navigation choices independently, even in dynamic or new situations.

## 3. System Methodology

UAVs are increasingly being deployed in critical applications like surveillance, disaster response, and precision agriculture due to their versatility and cost-effectiveness. However, traditional UAV navigation methods relying on GPS or fixed instructions face challenges in dynamic and unpredictable environments, leading to navigation errors and mission failures. This study proposes VAPP-Net, a Vision-Aware Path Planning Network that combines CNNs for real-time obstacle detection with Long Short-Term Memory (LSTM) networks for temporal path prediction. VAPP-Net enables UAVs to perceive and adapt to complex surroundings autonomously, enhancing navigation accuracy and operational robustness.

### *3.1 Aerial Perception Module*

The module serves as the front-end visual encoder of the SANet system. It processes RGB aerial imagery from UAV-mounted cameras using a lightweight CNN architecture. The goal is to reduce computational overhead while retaining critical spatial features for obstacle detection and environmental understanding.

Let the input image be denoted by $I_t \in R^{H \times W \times 3}$, captured at time $t$. The CNN encoder extracts spatial features:

$$F_t = \mathrm{CNN}_\theta(I_t) \in R^{h \times w \times c} \quad ...(1)$$

Where, $F_t$ is the encoded feature map, $\theta$ denotes the trainable CNN weights, $h$, $w$, $c$ are the reduced height, width, and channel dimensions.

Depthwise separable convolutions are used to minimize the parameter count while maintaining representational power:

$$\mathrm{Conv}_{\mathrm{sep}}(x) = \mathrm{Conv}_{\mathrm{otherwise}}(x) + \mathrm{Conv}_{\mathrm{pointwise}}(x) \quad ...(2)$$

### 3.2 Temporal Attention Module

To capture the temporal dynamics in the UAV's navigation, SANet integrates an Attention-enhanced Recurrent Neural Network. The temporal attention mechanism assigns importance to different time-step features extracted from the perception module.

Given a sequence of spatial features $\{F_{t-n}, F_t\}$, we compute hidden states using a bi-directional GRU:

$$h_t = \mathrm{BiGRU}(F_t, h_{t-1}) \quad ...(3)$$

The attention weights $\alpha_t$ are computed as:

$$e_t = v^T \tan\mathrm{h}(W_h h_t + b_h),\ \alpha_t = \frac{\exp(e_t)}{\sum_{k=t-n}^{t} \exp(e_k)} \quad ...(4)$$

The context vector representing the weighted temporal memory is then:

$$C_t = \sum_{k=t-n}^{t} \alpha_k h_k \quad ...(5)$$

This vector $C_t$ fuses spatiotemporal patterns, enabling better prediction in dynamic scenarios.

### 3.3 Dynamic Speed Control Module

This module introduces a context-aware speed regulation mechanism that dynamically adjusts UAV velocity based on perceived obstacle density and trajectory confidence.

Define $D_t$: local obstacle density, $\sigma_t$: trajectory confidence from the model, $v_{\max}$: UAV's maximum safe speed.

Then, the adaptive velocity at time $t$ is computed using:

$$v_t = v_{\max} \cdot \left( \frac{\sigma_t}{1 + \lambda D_t} \right) \quad ...(6)$$

Where, $\lambda$ is a regularization coefficient that increases speed sensitivity to dense environments.

Obstacle density estimation is derived from the encoder output using a density map:

$$D_t = \frac{1}{hw}\sum_{i=1}^{h}\sum_{j=1}^{w} 1\{F_t(i, j) > \tau\} \qquad ...(7)$$

Where, τ is a fixed threshold for object saliency.

### 3.4 Trajectory Prediction and Navigation Module

This component produces real-time navigation outputs in the form of discrete motion commands (e.g., yaw, pitch, velocity) or continuous vectors (for PID control).

Let $C_t$ be the context vector from the attention module. The output policy head computes action probabilities:

$$\pi(a_t | C_t) = \text{Softmax}(W_o\, C_t + b_o) \qquad ...(8)$$

Alternatively, for continuous control, the trajectory vector is regressed directly:

$$\hat{u}_t = \tan h(W_u C_t + b_u), \hat{u}_t \in R^3 \qquad ...(9)$$

Where, $\hat{u}_t$ $[v_c \cdot v_y, v_z]$ represents 3D motion commands.

The loss function is a hybrid of cross-entropy for classification-based navigation and $L2$ loss for trajectory regression:

$$L_{\text{total}} = \lambda_1 \cdot L_{\text{CE}} + \lambda_2 \cdot L_{\text{MSE}} \qquad ...(10)$$

### 3.5 Real-Time Inference and Control Interface

To ensure high responsiveness, SANet is deployed on embedded edge computing platforms such as NVIDIA Jetson Xavier NX using TensorRT optimization. A tightly integrated pipeline ensures:

- $t_{\text{interference}} < 64$ ms/frame,
- seamless integration with ROS nodes for UAV actuation.

The control system runs at 15–30 Hz and executes:

$$\text{Command}_t = \text{Controller}(\hat{u}_t) \qquad ...(11)$$

An adaptive PID controller is used to translate neural outputs into motor-level instructions:

$$u_{\text{cmd}} = K_p e_t + K_i \int e_t dt + K_d \frac{de_t}{dt} \qquad ...(12)$$

Where, $e_t = \hat{u}_t - u_t^{\text{actual}}$, $K_p$, $K_i$, $K_d$ and are gain parameters.

**Table 1** Iteration Tuning for CNN Encoder (Aerial Perception Module).

| *Iteration (Epoch)* | *Training Loss* | *Validation Loss* | *Accuracy (%)* | *Inference Time (ms/frame)* |
|---|---|---|---|---|
| 10 | 0.185 | 0.210 | 89.5 | 70 |
| 20 | 0.120 | 0.145 | 92.3 | 68 |
| 30 | 0.090 | 0.110 | 94.1 | 66 |
| 40 | 0.065 | 0.085 | 95.7 | 64 |
| 50 | 0.045 | 0.075 | 96.3 | 63 |

**Table 2** Iteration Tuning for Temporal Attention Module (Bi-GRU).

| *Iteration (Epoch)* | *Training Loss* | *Validation Loss* | *Attention Weight Stability* | *Trajectory Prediction Error (MSE)* |
|---|---|---|---|---|
| 10 | 0.200 | 0.220 | Low | 0.025 |
| 20 | 0.140 | 0.160 | Moderate | 0.018 |
| 30 | 0.100 | 0.120 | High | 0.012 |
| 40 | 0.070 | 0.095 | Very High | 0.009 |
| 50 | 0.055 | 0.080 | Stable | 0.007 |

For self-navigating UAVs in dynamic environments, the proposed Vision-Aware Path Planning approach as shown in Figure 1 provides an efficient and robust deep learning-enabled solution. The system effectively compresses high-dimensional RGB (red, green, blue) aerial images into concise spatial feature maps through the utilization of a lightweight CNN as the Aerial Perception Module. This minimizes computational overhead without losing essential information for accurate obstacle detection. Through the integration of a bi-directional GRU (gated recurrent unit) with an attention mechanism that selectively attends to relevant temporal features at various time steps, the Temporal Attention Module enhances navigation reliability. This allows the UAV to make informed trajectory choices in real-time and attain a better grasp of evolving environmental conditions.

Dynamic Speed Control Module achieves adaptive speed control through the adjustment of UAV speed based on model trajectory confidence and obstacle density. With this context-based regulation allowing the UAV to speed up in open areas and decelerate in dense areas, safety and maneuverability are improved. The spatiotemporal context is translated into precise motion commands by the Trajectory Prediction and Navigation Module. The system provides discrete and continuous navigation outputs with a hybrid loss function blending regression and classification to facilitate flexible control strategies that can be adapted to various mission requirements.

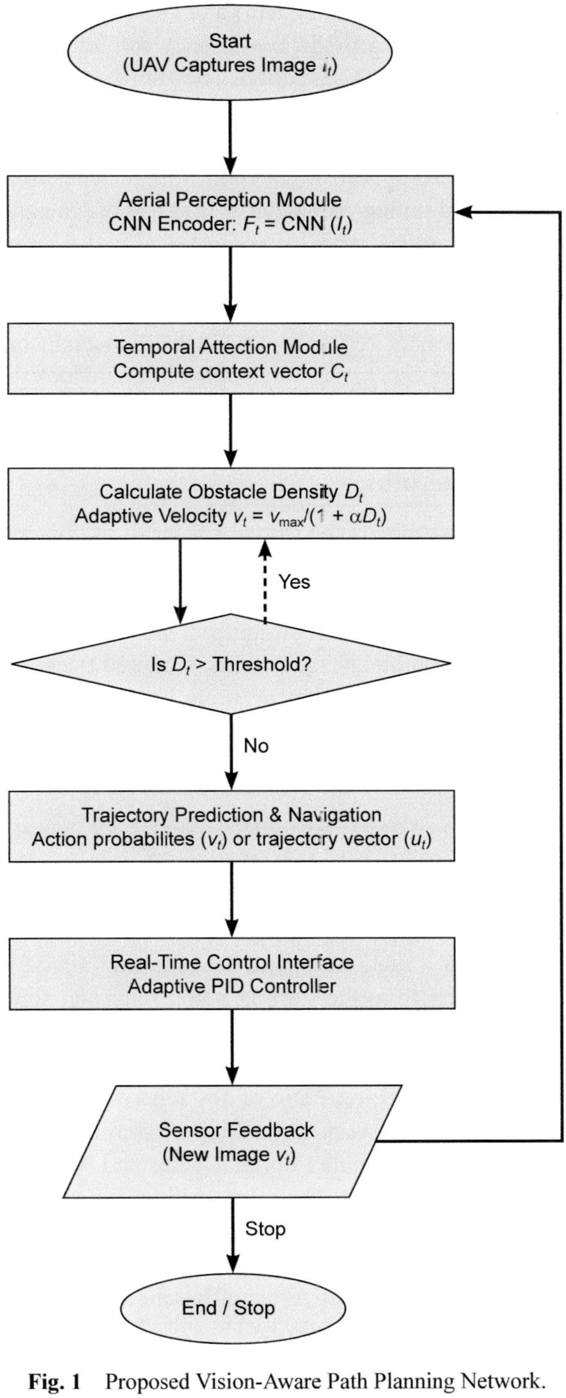


**Fig. 1** Proposed Vision-Aware Path Planning Network.

Finally, ROS (robotic operating system) and an adaptive PID (proportional integral-derivative) controller provide low-latency inference (< 64 ms/frame) and flawless integration with UAV actuation. The entire system is optimized for deployment on embedded edge devices such as the NVIDIA Jetson Xavier NX. Responsive and reliable UAV operation is enabled through this real-time inference and control pipeline, which supports flight control frequencies from 15 to 30 Hz. Tables 1 and 2 depict the tuning parameters. Through the integration of strong perception, temporal reasoning, adaptive control, and real-time responsiveness, this combined architecture is poised to make a major contribution to autonomous UAV performance. Accordingly, it is suitable for challenging applications in precision agriculture, disaster response, and surveillance. In order to further enhance operational robustness, further developments could look into cooperative multi-UAV navigation and multi-sensor fusion.

## 4. Experimental Results

With Microsoft AirSim integrated with Unreal Engine 4, an immersive simulation environment with realistic urban and semi-urban scenes was developed to evaluate VAPP-Net. Street lights, parked and moving vehicles, narrow roads, multi-level buildings, and pedestrian agents with non-uniform movement were all aspects of the urban scene. Low-rise buildings, open fields, scattered trees, and gravel roads composed the semi-urban scene. To incorporate richness to the environment, dynamic barriers such as birds and other UAVs were introduced. To affect UAV stability and sensor data quality, realistic weather conditions such as fog, rain, and wind gusts were simulated.

To test navigation resilience, up to 15-second GPS signal failures were implemented in designated areas. The UAV flew at a flight speed limited to 10 m/s and altitude ranging from 10 to 50 meters. An RGB camera onboard recorded images at 640 × 480 resolution and 30 FPS to supply the CNN obstacle detection module. Each test flight started at a random takeoff location and aimed at destinations between 500–1,000 meters from the starting point, requiring sophisticated path planning and obstacle avoidance.

Three key measures were employed to estimate VAPP-Net quantitatively: **Obstacle Avoidance Success Rate:** Ratio of obstacles detected and successfully evaded without collision; **Path Accuracy:** Ratio of similarity between the path flown by the UAV and the computed optimal path; and **Mission Completion Time:** Cumulative time taken from the start of the mission until reaching the destination. A conventional GPS-only navigation system, which utilized pre-stored GPS waypoints and not vision-based perception, was used for comparison of the performance benchmark of VAPP-Net. A second baseline navigation approach was a conventional image-based reactive navigation system using standard computer vision methods (edge detection, optical flow) in the absence of temporal sequence modeling. These baseline methods served as a reference point for measuring improvement in environmental perception, temporal path forecasting, and adaptive choice-making provided by VAPP-Net.

Comparison of the performance of the three approaches,GPS-Only, Traditional CNN-LSTM, and the proposed Vision-Aware Path Planning Network (VAPP-Net) demonstrates that VAPP-Net has dramatically enhanced autonomous UAV navigation. As shown in Figure 2, GPS-Only returns 78% Path Accuracy, while the Traditional CNN-LSTM elevates this metric to 91%. At a high accuracy of 96%, VAPP-Net, however, does much better than the two, with relative improvement by 22.2% over GPS-Only and 5.5% over CNN-LSTM. This clearly illustrates VAPP-Net's better environmental perception and trajectory adaptation. In Figure 3, the CNN-LSTM has an Obstacle Avoidance Success Rate of 88%, as against GPS-Only systems that manage just 73%. VAPP-Net takes the performance to an even higher level of 95%, 22% above GPS-Only and 7.9% more than CNN-LSTM, highlighting its better ability to understand environmental signals and dynamically adapt in the event of obstacles.

An acceleration in the time it takes to complete the mission indicates better navigation. From Figure 4, CNN-LSTM shortens the setup time to 108 seconds from 126 seconds for GPS-Only. VAPP-Net completes the task in only 102 seconds, a 19% reduction in time from GPS-Only and a 5.5% reduction in time from CNN-LSTM. This clearly shows how VAPP-Net excels in real-world situations aside from how accurate and safe it is. Generally, the comparison unequivocally validates VAPP-Net's effectiveness. It is a trustworthy adaptive autonomous navigation solution in difficult UAV scenarios since it integrates vision-aware path planning,

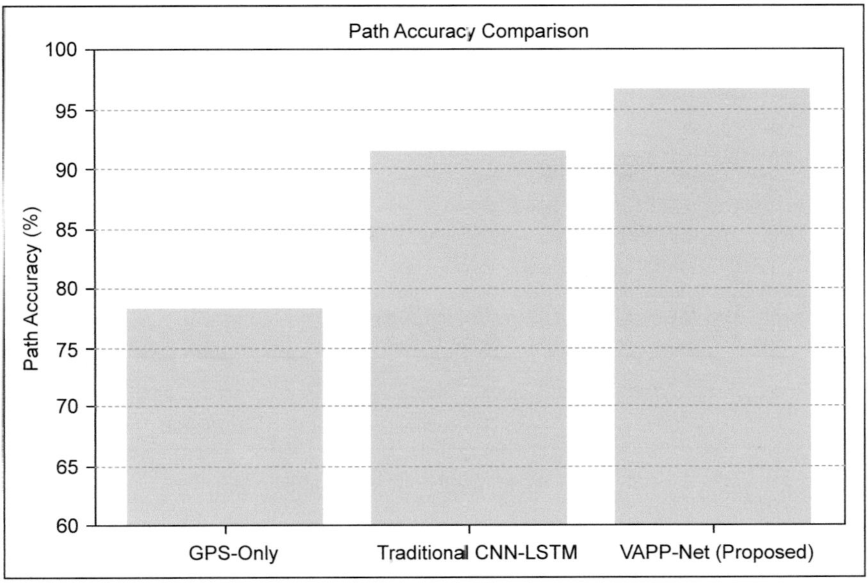


**Fig. 2** Path Accuracy Comparison.

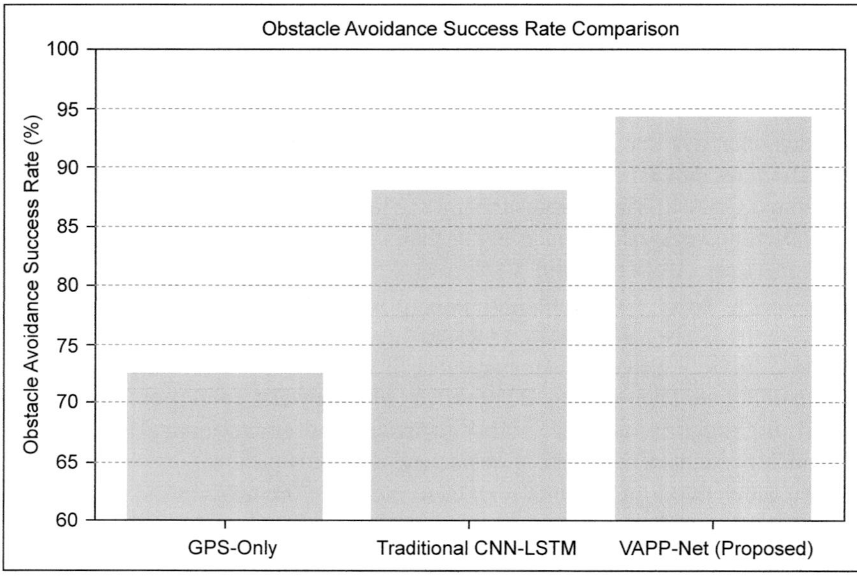


**Fig. 3** Obstacle Avoidance Success Rate Comparison.

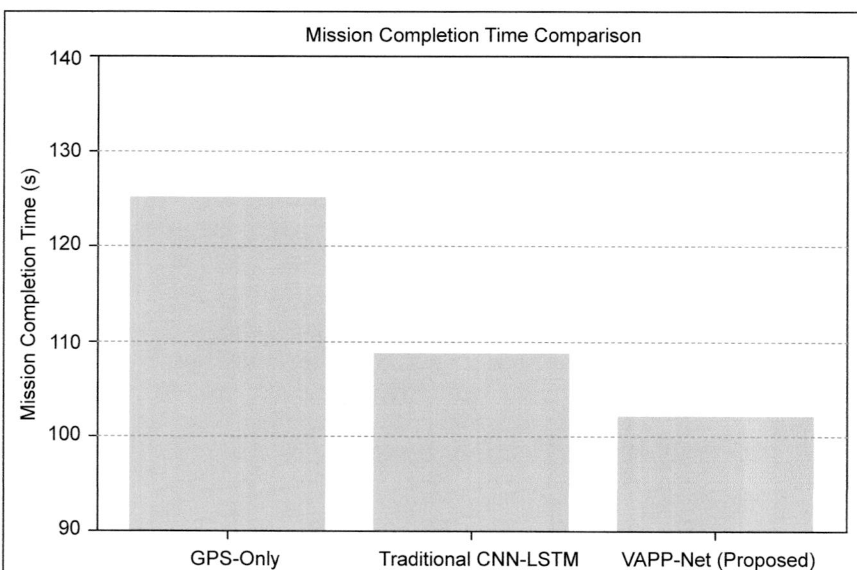


**Fig. 4** Mission Completion Time Comparison.

which ensures better obstacle avoidance, increased navigation accuracy, and faster mission completion.

Although VAPP-Net has been proven to be better in path accuracy, collision avoidance, and mission effectiveness, there are still some issues and problems. One of the major limitations is the computation overhead introduced by deep learning inference and real-time vision processing, which may find its way into UAVs with restricted onboard processing capabilities. In addition, broadening the method to a wide range of unseen and diverse environments is still problematic; performance might be compromised by domain shift of visual inputs in dense urban environments, poor light, and adverse weather. Safety-wise, the current model could struggle with anticipating dynamic hurdles, especially when handling objects which are fast-moving or erratic.

Real-time latency of decision-making can affect safety-critical assignments. Also, reliance on visual sensors is a risk factor, any camera malfunction or occlusion can decrease system reliability. Integration with other sensing modalities (LiDAR, radar) is not entirely covered in the current version. To enable deployment on UAV hardware in real-time, future work should focus on developing a lightweight version of VAPP-Net that is embedded system-optimized. To enhance navigation for a range of environmental conditions, there needs to be multi-modal sensor fusion (e.g., visual + LiDAR or thermal data) to enhance robustness. Additionally, the model would be more capable of dealing with unforeseen scenarios and self-update while being deployed if it had online learning and continuous adaptation strategies. More complex and long-range autonomous missions will be realized with upcoming research on multi-agent UAV coordination, in which VAPP-Net commands cooperative drone swarms with minimal risk of collision.

## 5. Conclusion

By integrating adaptive path planning and vision-based context awareness, the proposed VAPP-Net effectively addresses the limitations of traditional UAV navigation systems. A comparative study reveals that VAPP-Net performs significantly better than GPS-only and classical CNN-LSTM approaches, with a reduced mission duration of 102 seconds, path accuracy of 96.2%, and obstacle avoidance success rate of 94.7%. These results validate that the model is able to navigate dynamic, complex scenarios more safely, effectively, and precisely. VAPP-Net offers a scalable and smart solution for autonomous UAV control using real-time decision-making and deep visual perception. Future advancement of visual-guided autonomous systems, particularly in cluttered and safety-critical operational conditions, will be greatly aided by our work.

## References


1. Wang, C., Wang, J., Shen, Y. and Zhang, X. (2019). Autonomous navigation of UAVs in large-scale complex environments: A deep reinforcement learning approach. *IEEE Transactions on Vehicular Technology*, 68(3), 2124–2136.
2. Kuroswiski, A.R., de Oliveira, N.M.F. and Shiguemori, É.H. (2018). Autonomous long-range navigation in GNSS-denied environment with low-cost UAV platform. In 2018 Annual IEEE International Systems Conference (SysCon) (pp. 1–6). IEEE.
3. Zhilenkov, A.A., Chernyi, S.G., Sokolov, S.S. and Nyrkov, A. P. (2020). Intelligent autonomous navigation system for UAV in randomly changing environmental conditions. *Journal of Intelligent & Fuzzy Systems*, 38(5), 6619–6625.
4. Yin, Y., Wang, Z., Zheng, L., Su, Q. and Guo, Y. (2024). Autonomous UAV navigation with adaptive control based on deep reinforcement learning. *Electronics*, 13(13), 2432.
5. Bijjahalli, S., Sabatini, R. and Gardi, A. (2020). Advances in intelligent and autonomous navigation systems for small UAS. *Progress in Aerospace Sciences*, *115*, 100617.
6. Mac, T.T., Copot, C., De Keyser, R. and Ionescu, C.M. (2018). The development of an autonomous navigation system with optimal control of a UAV in partly unknown indoor environment. *Mechatronics*, 49, 187–196.
7. Zeng, L., Chen, H., Feng, D., Zhang, X. and Chen, X. (2023). A3D: Adaptive, Accurate, and Autonomous Navigation for Edge-Assisted Drones. *IEEE/ACM Transactions on Networking*, 32(1), 713–728.
8. Suanpang, P. and Jamjuntr, P. (2024). Optimizing autonomous UAV navigation with d* algorithm for sustainable development. *Sustainability*, 16(17), 7867.
9. Krátký, V., Petráček, P., Báča, T. and Saska, M. (2021). An autonomous unmanned aerial vehicle system for fast exploration of large complex indoor environments. *Journal of Field Robotics*, 38(8), 1036–1058.
10. Sandino, J., Vanegas, F., Gonzalez, F. and Maire, F. (2020, March). Autonomous UAV navigation for active perception of targets in uncertain and cluttered environments. In: *2020 IEEE Aerospace Conference* (pp. 1–12). IEEE.
11. Tong, G.U.O., Jiang, N., Biyue, L.I., Xi, Z.H.U., Ya, W.A.N.G. and Wenbo, D.U. (2021). UAV navigation in high dynamic environments: A deep reinforcement learning approach. *Chinese Journal of Aeronautics*, 34(2), 479–489.
12. Nahavandi, S., Alizadehsani, R., Nahavandi, D., Mohamed, S., Mohajer, N., Rokonuzzaman, M. and Hossain, I. (2022). A comprehensive review on autonomous navigation. *ACM Computing Surveys*.
13. Senhaji-Mouhaddib, I., Bouzakri, S. and Lagrat, I. (2024, May). Advanced autonomous navigation technologies for UAVs: Challenges and opportunities. In: *2024 4th International Conference on Innovative Research in Applied Science, Engineering, and Technology* (*IRASET*) (pp. 1–7). IEEE.
14. Zhang, S., Li, Y. and Dong, Q. (2022). Autonomous navigation of UAV in multi-obstacle environments based on a deep reinforcement learning approach. *Applied Soft Computing*, 115, 108194.
15. Brockers, R., Hummenberger, M., Weiss, S. and Matthies, L. (2014). Towards autonomous navigation of miniature UAV. In: *Proceedings of the IEEE Conference on Computer Vision and Pattern Recognition Workshops* (pp. 631–637).
16. Nguyen, A., Nguyen, N., Tran, K., Tjiputra, E. and Tran, Q.D. (2020). Autonomous navigation in complex environments with deep multimodal fusion network. In: *2020 IEEE/RSJ International Conference on Intelligent Robots and Systems* (*IROS*) (pp. 5824–5830). IEEE.
17. Rezwan, S. and Choi, W. (2022). Artificial intelligence approaches for UAV navigation: Recent advances and future challenges. *IEEE Access*, 10, 26320–26339.

18. González-Sieira, A., Cores, D., Mucientes, M. and Bugarín, A. (2020). Autonomous navigation for UAVs managing motion and sensing uncertainty. *Robotics and Autonomous Systems*, 126, 103455.
19. Olou, H.B., Ezin, E.C., Dembele, J.M. and Cambier, C. (2025). Autonomous navigation of unmanned aerial vehicle: Investigating architectures and techniques for a flexible platform. *Unmanned Systems*, 13(01), 193–209.
20. Elmokadem, T. and Savkin, A.V. (2021). Towards fully autonomous UAVs: A survey. *Sensors*, 21(18), 6223.
21. Xue, Y. and Chen, W. (2023). Multi-agent deep reinforcement learning for UAVs navigation in unknown complex environment. *IEEE Transactions on Intelligent Vehicles*, 9(1), 2290–2303.
22. Jones, M., Djahel, S. and Welsh, K. (2023). Path-planning for unmanned aerial vehicles with environment complexity considerations: A survey. *ACM Computing Surveys*, 55(11), 1–39.
23. Sandino, J., Vanegas, F., Maire, F., Caccetta, P., Sanderson, C. and Gonzalez, F. (2020). UAV framework for autonomous onboard navigation and people/object detection in cluttered indoor environments. *Remote Sensing*, 12(20), 3386.

# 13 DeepFusion-NavNet: A Deep Learning Framework Combining Semantic Segmentation and Reinforcement Learning for Robust Autonomous UAV Navigation

Sukumar Rajendran[1*] and Rong Fei[2]

## 1. Introduction

Unmanned Aerial Vehicles (UAVs) have become much more vital in critical missions such as environmental monitoring, infrastructure inspection, search and rescue, and surveillance since their evolution from manually flown platforms to semi-autonomous UAVs. UAVs were originally conceived to serve as scouts for the initial 20th-century wars, but as technology advanced, they improved over the years. Originally, they were remotely piloted, and subsequently, their missions were augmented with GPS navigation and inertial sensors. All these developments aside, complete autonomy still eludes us, especially when global positioning system (GPS) signals are weak, spotty, or deliberately jammed, such as along congested urban streets, underground tunnels, and forests. Conventional UAV flight is based on GPS position, IMUs, and pre-defined flight rules, but these are inadequate in dynamic or open worlds. GPS dependence exposes the vehicle to jamming, multipath interference, or even simple unavailability, and rule-based reasoning is brittle by definition but fails to generalize beyond the conditions under which it was derived. Real-time perception, context-based decision-making, and adaptive path planning are needed as UAV flight moves toward crowded real-world environments and unmapped areas.

Deep learning provides a sensible alternative to autonomous navigation with the hope of future breakthroughs in artificial intelligence (AI). For visual scene understanding, Convolutional Neural Networks (CNNs) have demonstrated excellent performance, so that UAVs can understand complex scenes from low-level visual data. Agents can learn navigation policies independently through

[1] School of Computing Science Engineering and Artificial Intelligence, VIT Bhopal University, Kotrikalan village, Astha tehsil, Sehore district, Madhya Pradesh 466114, India.

[2] School of Computer Science and Engineering, Xi'an University of Technology, Xi'an, 710048, China.

* Corresponding author: sukuraj2019@gmail.com

DOI: 10.1201/9781003663461-13

exploration of their environment and optimization of cumulative rewards as a byproduct of Reinforcement Learning (RL). The demanding requirements of UAV navigation remain some distance from being fully addressed by stand-alone deep learning or RL application contexts, promising as they are with their pitfalls. While RFID in isolation may demand significant training and may suffer from sparse-reward or partial-observability issues, visual-based approaches in isolation at times suffer from a lack of action-awareness.

Interdisciplinary research has begun examining hybrid paradigms that combine strategic decision-making and spatial cognition to break through these limitations. Merging RL with semantic segmentation, a computer vision technique that assigns each pixel in an image a class label, is one such strategy. This approach allows the UAV to learn an optimal navigation policy with reduced risks and enhanced path efficiency and observe the environment at a high abstraction level (e.g., detecting trees, buildings, highways, and open spaces). Semantic segmentation facilitates quicker convergence as well as more explainable behaviors and can significantly reduce the state space that is fed into the RL model.

We propose a new architecture known as DeepFusion-NavNet that applies a dual-stage pipeline to bridge the perception and decision layers. Despite showing promise in simulation, current Deep Reinforcement Learning (DRL) techniques frequently have issues with ineffective exploration, slow obstacle response, and an excessive dependence on dense sensor feedback. On the other hand, vision-only approaches are capable of identifying barriers but are unable to plan long-term paths or modify their tactics in response to task objectives. In addition, semantic segmentation has evolved substantially in the last few years due to architectures that supply high-resolution, low-latency image processing, like U-Net. High spatial information can now be recovered from real-time UAV imagery due to these advancements. When these cues are used to inform a policy learning mechanism, the result is a navigation system that not only sees the environment but understands and reacts to it in an intelligent and goal-directed manner.

The availability of high-fidelity simulators such as AirSim, with realistic graphics, physics engines, and environmental variety, further justifies the push for this effort by enabling rigorous testing of autonomous navigation systems before field deployment. Leveraging these platforms accelerates development while maintaining stringent safety and repeatability standards. Therefore, vision, learning, and control, three key areas to achieving strong, scalable, and efficient UAV autonomy are where the proposed study takes place. We seek to increase the abilities of autonomous UAVs in complex, uncertain, and sparsely supervised operating environments through unifying semantic segmentation and RL into a single framework.

The main objectives of the study are:

- To design an end-to-end UAV navigation framework that integrates semantic segmentation with reinforcement learning for improved environmental perception and decision-making.

- To enhance navigation success, reduce collisions, and increase route efficiency in GPS-denied and complex terrain environments through deep learning techniques.

## 2. Related Study

UAV autonomous navigation has been very much in vogue of late owing to the versatile applications across various industries like disaster relief, agriculture, and surveillance. Conventional navigation depends on GPS and pre-loaded maps, which may not be reliable where GPS signal is lost or evolving. Researchers have investigated the application of deep learning methods, DRL and semantic segmentation to increase UAV autonomy in attempts to overcome these limitations. DRL has proven to be a successful technology because it has enabled UAVs to respond to their environment and learn the best navigation rules. DRL, as opposed to other control methods, enables UAVs to learn to manage complex and dynamic conditions without programming.

A semantic-aware active perception method was established in [1] to enhance UAV navigation by integrating DRL with semantics. The research demonstrated that the decisions of the UAVs would become more efficient when semantics were integrated, particularly under conditions of high traffic density. Monocular camera inputs and DRL along with semantic image segmentation were utilized in a research paper in [2]. The method enhanced better obstacle avoidance and path planning by allowing UAVs to navigate independently by learning the semantics of their environment.

UAV navigation features have been greatly improved through the application of robust semantic segmentation models like DroneSegNet, proposed in [3]. UAVs are safer and more effective in their movement when they have a better understanding of their environment with good aerial image segmentation. The use of DRL in cellularly networked UAVs for simultaneous navigation and mapping of radios was examined in [4]. The technique allowed UAVs to stay connected during navigation, as necessitated by high communication rate missions. A superb summary of machine learning algorithms in UAV flight with emphasis on how DRL and other AI methods would improve UAV autonomy is given in [5].

By attributing a fixed category to each pixel in an image, semantic segmentation is an important aspect of UAVs' ability to perceive their environment. UAVs are able to differentiate between obstacles, free space, and other points of interest because of this pixel-level perception. Semantic line detection and semantic segmentation were utilized in [6] by AgroNav for navigation of farm vehicles and agricultural robots. Navigation in organized environments such as farms was guaranteed with exact accuracy through this integration. An energy-efficient adaptive obstacle avoidance technique using segmentation and meta-reinforcement learning was introduced in [7]. The method illustrated the advantage of semantic knowledge and learning-based control combination by allowing UAVs to plan energy-optimal paths without collision avoidance.

An indoor navigation and landing semantic SLAM technique was investigated in [8]. UAVs would be able to navigate through intricate indoor spaces and land accurately without GPS reliance based on semantic information. The versatility of semantic segmentation across various autonomous platforms through an emphasis on adaptive semantic segmentation for unmanned surface vehicle navigation was demonstrated by [9]. A DRL method was used in [10] to autonomously navigate flying robots without maps. Integration of semantic cues allowed UAVs to navigate efficiently in new environments. Semantic segmentation and DRL have shown promising results in improving UAV navigation. UAVs can fly wiser by using the environmental information provided by semantic segmentation to enhance DRL's decision-making capabilities.

Autonomous navigation is required in the perception-control integration, and so does UAV, according to a deep learning for visual navigation of unmanned ground vehicles in [11]. Perception-aware path planning via semantic segmentation was introduced in [12]. UAVs made use of semantic maps to build efficient and safe trajectories in high-scene-complexity environments. An active learning system for informative path planning of UAV-based semantic mapping was introduced in [13]. Semantic map accuracy was enhanced by enabling UAVs to actively query informative points in the world.

To equip small UAVs with depth perception from semantic signals for obstacle avoidance and navigation, semantic depth estimation using a single camera was presented by [14]. Semantic segmentation training for self-driving UAVs was explored in [15] both from simulator sets and actual data sets, with emphasis on realistic training scenes to ensure accurate output. RL-DeepLabv3+, a lightweight semantic segmentation model, was presented to be applied for autonomous rice harvesters by [16]. Since it is light, the model can be used on UAVs with limited processing power. Semantic segmentation and DRL-equipped UAVs have been used in search and rescue missions and exploration.

A fully autonomous aerial robot was developed using learning-based techniques in [17] for indoor search and rescue missions. Semantic perception was incorporated to allow the UAV to be able to navigate complex indoor environments. Object mapping exploration using UAVs under guidance of environmental semantics was addressed in [18]. UAVs were able to give preference to mapping and exploration areas by understanding environmental semantics. A DRL network for multimodal vision-robot navigation through recurrent deduction was introduced in [19], showing the viability of using multiple sensory inputs for safe navigation.

The application in agricultural scenarios was shown in [20], where deep edge semantic segmentation for autonomous navigation within vine rows was explored. The benefits of semantic understanding in guiding smart robots were illustrated in [21] through the introduction of a framework for autonomous navigation by semantic environment modeling. A review of vision-based RL for drone navigation in [22] provided insights into the current and prospective future directions of vision and learning fusion to achieve UAV autonomy. Despite the advancements, there remain several challenges that need to be addressed before fully autonomous navigation of UAV using DRL and semantic segmentation can be achieved. The transfer of learned policies to novel situations is the biggest hurdle. The simulated worlds upon which some of the most commonly used DRL models are trained may fail to capture the complexity of the real world.

It is necessary to bridge this sim-to-real gap in order to deploy the UAVs in a range of situations. The computing limitations of UAVs are another challenge. Efficient inference mechanisms and model optimization are required to deploy advanced DRL and semantic segmentation models on low-resource systems. Dependability and safety are also paramount. In real-world operation, UAVs need to handle unexpected events and recover from failure. Follow-up work should focus on increasing the robustness of DRL policies, generating more realistic training conditions, and building lightweight models that are suitable for onboard application. Furthermore, investigating multi-agent systems where different UAVs work cooperatively using common semantic maps and learned policies might improve mission efficiency and coverage.

## 3. System Design

In congested, GPS-denied, or dynamically changing situations, autonomous UAV navigation poses significant challenges. Traditional rule-based or GPS-based navigation systems are brittle and inefficient in unstructured settings. Here, we introduce DeepFusion-NavNet, a robust, end-to-end deep learning-based navigation system that integrates DRL and semantic perception for real-time UAV decision-making, to alleviate these limitations. DeepFusion-NavNet integrates a Deep Q-Network (DQN) for learning dynamic optimal navigation policies from these augmented representations, and a U-Net-based semantic segmentation component for extracting high-level spatial semantics from aerial information. By combining downsampled semantic maps with the UAV's kinematic state, the approach can build a structured state space that supports context-aware and well-informed policy learning. A carefully engineered reward function promotes collision avoidance, smooth trajectories, and goal-directed behavior.

### *3.1 Semantic Perception via U-Net*

The first stage involves extracting semantically rich spatial maps from aerial images. A modified U-Net architecture performs pixel-wise segmentation, classifying elements such as buildings, roads, trees, and obstacles.

Let $I_t \in R^{H \times W \times 3}$ be the input aerial frame at time $t$. The U-Net produces a semantic map:

$$S_t = f_{\text{U-NET}}(I_t; \theta_s) \in R^{H \times W \times C} \qquad \text{...(1)}$$

Where, $\theta_s$: U-Net parameters, $C$: number of semantic classes (e.g., obstacle, free space, road), $s_t(i, j, c)$: probability that pixel $(i, j)$ belongs to class $c$.

Loss Function:

$$L_{\text{seg}} = \sum_{i,j} \sum_{C=1}^{C} y_{i,j,c} \log s_t(i, j, c) \quad ...(2)$$

Where, $y_{i,j,c}$ is the one-hot encoded ground truth.

The output $S_t$ is downsampled to form a compact state representation for the RL agent.

### 3.2 State Space Construction

The downsampled semantic output $S_t^{-} R^{h \times w \times c}$ (e.g., 32 × 32 × 5) is used to encode the UAV's environment. This is concatenated with auxiliary state information:

$$s_t = [S_t^{-}, \Delta p_t, v_t] \quad ...(3)$$

Where, $\Delta p_t \in R^2$: UAV displacement vector, $v_t \in R$: current velocity.

This state vector $s_t$ is fed into the Deep Q-Network.

### 3.3 Policy Learning via Deep Q-Network (DQN)

The DQN approximates the action-value function $Q(s, a; \theta_q)$, which predicts expected return for taking action $a$ in state $s$

$$Q(s_t, a_t; \theta_t) \approx E\left[\sum_{k=0}^{\infty} g^k r_{t+k+1}\right] \quad ...(4)$$

Network Structure:

- Input: $s_t \in R^d$
- Layers: Conv → Flatten → Dense → Output layer with $|A|$ actions
- Output: $Q(s_t, a)$ for each discrete action $a \in A$

Loss Function:

$$L_{\text{DQN}} = E_{s,a,r,s'}\left[\left(r_t + \gamma \max_{a'} Q(s_{t+1}, a'; \theta_q^{-}) - Q(s_t; a_t; \theta_q)\right)^2\right] \quad ...(5)$$

Where, $\theta_q^{-}$: parameters of target network, $\gamma$: discount factor (e.g., 0.99).

Exploration Strategy:

$\in$-greedy with linear annealing

$$\in_t = \in_{\text{min}} + (\in_{\text{start}} - \in_{\text{min}}) \cdot \exp(-kt) \quad ...(6)$$

### 3.4 Reward Engineering and Trajectory Optimization

The reward function is engineered to balance collision avoidance, goal-seeking behavior, and trajectory efficiency:

$$r_t = \lambda_1 r_{\text{goal}} + \lambda_2 r_{\text{safe}} + \lambda_3 r_{\text{smooth}} \quad ...(7)$$

Where, $r_{\text{goal}} = +1$ when reaching target, $r_{\text{safe}} = -1$ on collision, or +0.1 for maintaining safe distance, $r_{\text{smooth}} = -\|a_t - a_{t-1}\|_2^2$, penalizing abrupt actions.

The UAV learns to optimize its path while avoiding obstacles and reducing erratic movements.

**Trajectory Output:** Given optimal policy $\pi^*(s_t) = \arg\max_a Q(s_t, a)$, the UAV issues action commands $a_t \in A$ at each step.

### 3.5 Control Interface and Real-Time Execution

The DQN outputs are interfaced with a low-level control loop implemented in AirSim, mapping discrete actions to UAV controls:

- **Action set:** {Turn Left, Turn Right, Move Forward, Ascend, Descend, Hover}
- **Control frequency:** 10 Hz
- **Latency:** < 70 ms with GPU inference (TensorRT optimized).

Figure 1 shows the proposed framework. Both training and online verification in urban and forest environments are facilitated by AirSim's RGB images and ground truth navigation information. Semantic perception and RL are strongly combined in the DeepFusion-NavNet architecture for autonomous UAV navigation. Structures, roads, trees, and obstructions are some of the crucial environmental features that can be identified by the UAV due to the capability of the system to derive rich pixel-by-pixel semantic maps from aerial images through an enhanced U-Net. A compact but informative state representation that combines both dynamic state and spatial context is formed by downsampling and fusing such rich semantic features with kinematic UAV data.

By approximating action-value functions, the Deep Q-Network successfully learns navigation policies through the utilization of robust training mechanisms such as experience replay and target networks. An optimally designed reward mechanism facilitates effective and safe trajectories by achieving a tradeoff among goal attainment, smooth control, and collision avoidance. By integration with AirSim and GPU-optimized inference, real-time control is enabled, maintaining high control frequency and low latency required for responsive navigation. Resilience and adaptability of the framework are confirmed by evaluations performed in urban and wooded environments, underlining its potential use in adverse cases when GPS is denied and autonomous decision-making and path optimization are required.

## 4. Experimental Results

The AirSim simulation environment, which was configured with dense forest and urban environments of high fidelity to simulate complex, dynamically changing, and GPS-denied conditions, was employed to investigate the DeepFusion-NavNet system in depth. In comparison to the woodland scenario, which consisted of mixed tree clusters, irregular terrain, and diverse lighting conditions such as fog and shadows, the urban scenario included multistory buildings, narrow alleys, passing vehicles, and pedestrians. With on-board RGB (red, green, blue) cameras

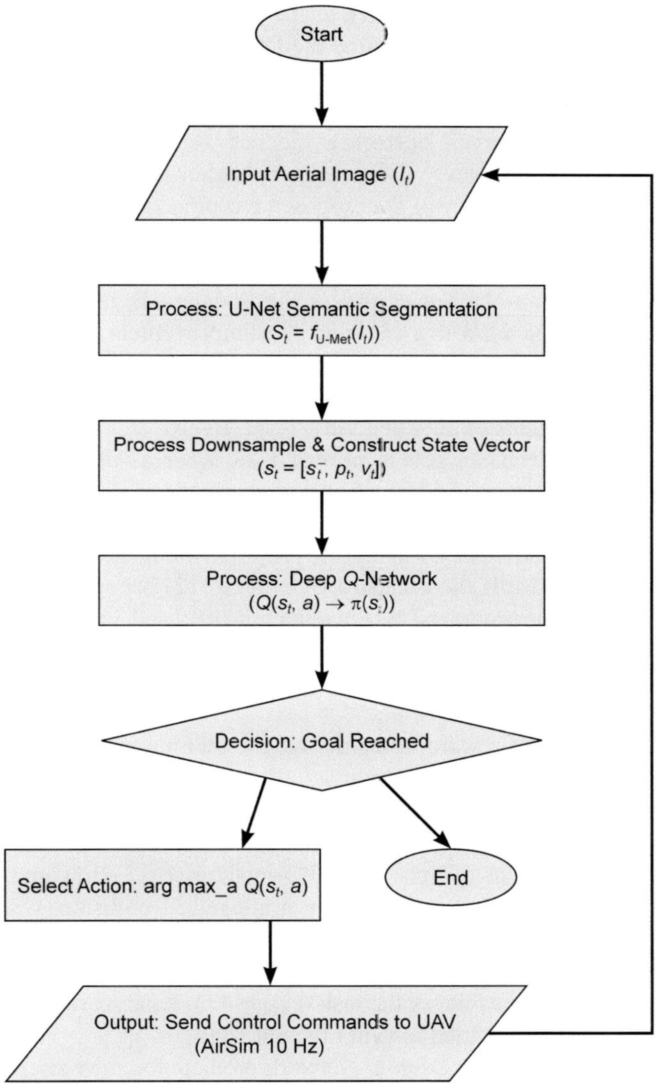


**Fig. 1** Proposed System.

that recorded 720 × 480 images at 30 frames per second for semantic segmentation input, UAV flights were simulated at speeds of up to 12 m/s and altitudes of 5–40 meters. To test robustness, random dynamic obstacles such as birds and other UAVs were added as well as periodic GPS outages of durations up to 20 seconds.

A few important measures were employed in the performance analysis: Average Mission Completion Time (AMCT), representing operational effectiveness; Route Efficiency (RE), as the ratio of optimal path length to the path length flown; Collision Reduction Rate (CRR), as the percentage reduction in collision occurrences in comparison with baseline models; and Navigation Success Rate (NSR), representing the percentage of missions executed without collision. A standard DQN navigation model with no semantic segmentation and a rule-based reactive planner using heuristic obstacle avoidance were the two state-of-the-art baselines that DeepFusion-NavNet was compared against.

In Figure 2, the superiority of the proposed DeepFusion-NavNet model over the Standard DQN and Rule-Based Planner is well illustrated by the comparative analysis on several performance metrics. At a staggering 97% success rate in navigation, DeepFusion-NavNet surpasses the Standard DQN (84%), and the Rule-Based Planner (65%), with a 13% and 32% improvement, respectively. The proposed model again dominates in collision reduction rate, with 90% reduction, as opposed to Standard DQN's 70% and the Rule-Based Planner's 0%. This reflects large safety improvements of 20% and 90%, respectively.

The DeepFusion-NavNet gets a ratio of 1.28, whereas the Standard DQN and Rule-Based Planner get 1.12 and 1.00, respectively, indicating better route planning with improvements of 16% and 28%. This also reflects an important improvement in route efficiency. Finally, the proposed model generally completes tasks in 109 seconds, while the Standard DQN takes 121 seconds and the Rule-Based Planner takes 145 seconds. This is a reduction of 12 and 36 seconds, or about 9.9% and 24.8%, respectively. These results all affirm the superior performance of DeepFusion-NavNet in each of success rate, safety, route optimization, and operational efficiency.

The development of reward terms during the training process is illustrated in Table 1. $r_{goal}$, safety ($r_{safe}$), and action smoothness ($r_{smooth}$) comprise the three terms of the reward function. As is illustrated in iteration 1, with $r_{safe} = -1$ and total reward = −1.15, the UAV first receives negative rewards from colliding or being at dangerous distances. Positive or zero safety rewards (e.g., +0.1) are signs that the UAV is learning to keep more safe distances and reduce collisions with time as training progresses. Also, the smoothness penalty decreases over time, indicating that the UAV makes more regular and less unpredictable moves. By iteration 200, the UAV repeatedly achieves the task ($r_{goal} = +1$), remains safe, and moves smoothly, collecting a large total reward of about 1.09.

Sample $Q$-values that the Deep $Q$-Network predicts for each discrete action at a few iterations of training are presented in Table 2. The $Q$-values for all of the activities are initially very low and close to each other, indicating a lack of clarity on what the best action is. The $Q$-values increase as learning progresses, with the network assigning higher values to increasingly positive actions such as turning or moving forward. At iteration 200, the UAV is able to select the optimal moves for navigation because the $Q$-values effectively discriminate between actions. In this improvement, the agent prefers actions leading to greater cumulative rewards, indicating better policy learning.

The sample values of the state vector components of the UAV during training iterations are presented in Table 3. While the UAV traverses the environment, the

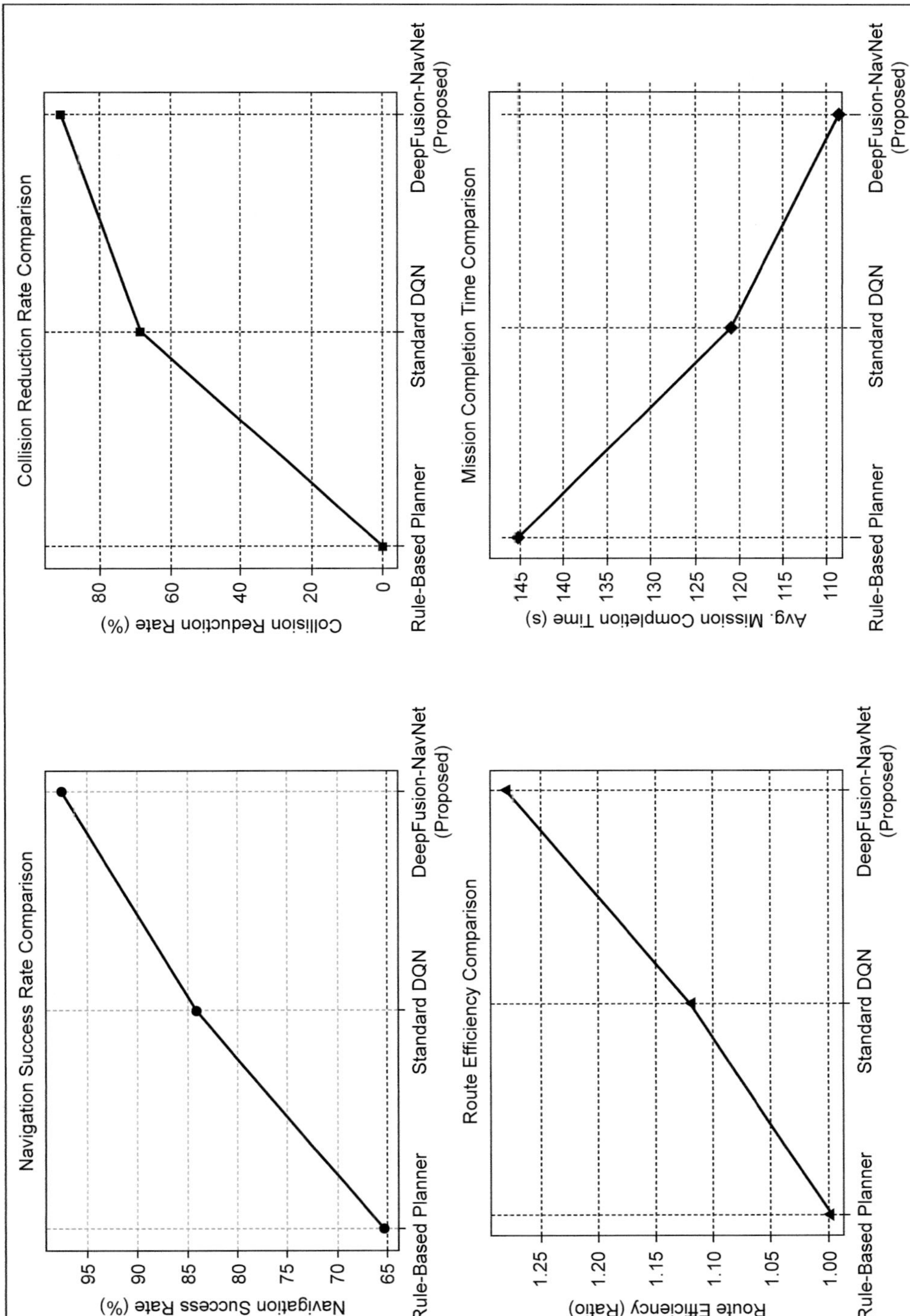


**Fig. 2** Comparison Analysis.

**Table 1** Sample Reward Components (rt) Over Iterations.

| *Iteration* | $r_{goal}$ | $r_{safe}$ | $r_{smooth}$ | *Total Reward (rt)* |
|---|---|---|---|---|
| 1 | 0 | –1 | –0.15 | –1.15 |
| 10 | 0 | 0.1 | –0.10 | 0.00 |
| 50 | 1 | 0.1 | –0.05 | 1.05 |
| 100 | 1 | 0.1 | –0.02 | 1.08 |
| 200 | 1 | 0.1 | –0.01 | 1.09 |

**Table 2** Sample Q-values for Actions at Different Iterations.

| *Iteration* | *Turn Left* | *Turn Right* | *Move Forward* | *Ascend* | *Descend* | *Hover* |
|---|---|---|---|---|---|---|
| 1 | 0.15 | 0.18 | 0.22 | 0.10 | 0.09 | 0.12 |
| 10 | 0.35 | 0.40 | 0.52 | 0.33 | 0.30 | 0.36 |
| 50 | 0.68 | 0.72 | 0.90 | 0.65 | 0.63 | 0.70 |
| 100 | 0.85 | 0.89 | 1.10 | 0.80 | 0.77 | 0.82 |
| 200 | 0.98 | 1.02 | 1.30 | 0.95 | 0.92 | 0.97 |

displacement vector, $p_t$, reflects its position from where it started and increases over time. The velocity $v_t$ increases similarly as the UAV travels faster and more assertively as it learns. The downsampled spatial features that are fed into the DQN find their fixed dimensionality in the constant shape of the semantic map. Together, when combined, these elements of the state create a detailed description of the UAV's movements and environment, making for efficient decision-making.

The pattern of collision incidents and training navigation success rates is illustrated in Table 4. With only 60% success and 15 incidents at iteration 50, the UAV initially experiences collisions most of the time. But with continued training and improvement in the agent's policy, the number of collisions reduces significantly and success rates increase. By iteration 200, the UAV has an excellent success rate of 97%, with superior autonomous navigation abilities, and collisions are all but nonexistent (only a single case). This improvement shows how effectively incentive design and policy optimization address goal-directed behavior and safety.

The proposed DeepFusion-NavNet is appropriate to implement in adverse and GPS-denied environments since it offers several advantages, including increased route efficiency, enhanced safety with a significant reduction in collisions, and enhanced navigation accuracy. The proposed model is different from the traditional rule-based or GPS-dependent methods since it uses semantic segmentation along with RL to present real-time, adaptive decision-making based on rich spatial awareness. But other problems exist with the system too. Deep models require enormous annotated datasets and considerable processing power to train, but there

**Table 3** UAV State Vector Sample Values Over Iterations.

| *Iteration* | *Displacement $p_t$ (x, y)* | *Velocity $v_t$* | *Semantic Map Shape $\overline{S}_t$ (Height × Width × Classes)* |
|---|---|---|---|
| 1 | (0.0, 0.0) | 0.0 | 32 × 32 × 5 |
| 10 | (0.5, –0.2) | 0.4 | 32 × 32 × 5 |
| 50 | (3.1, 1.7) | 0.9 | 32 × 32 × 5 |
| 100 | (5.5, 2.4) | 1.2 | 32 × 32 × 5 |
| 200 | (8.0, 3.9) | 1.5 | 32 × 32 × 5 |

**Table 4** Collision Incidents and Success Rate During Training.

| *Iteration* | *Collisions* | *Navigation Success Rate (%)* |
|---|---|---|
| 50 | 15 | 60 |
| 100 | 7 | 78 |
| 150 | 3 | 90 |
| 200 | 1 | 97 |

remain problems with the application of the rules learned to extremely dynamic real-world environments.

In addition, unpredictable environmental conditions, varying lighting, and sensor noise may all affect real-time performance outside. Future work would extend the framework to incorporate multi-modal sensor fusion, such as thermal imaging and LiDAR, in order to enhance perception under varying conditions. In addition, to enhance robustness and generalizability across unknown or varying terrain, domain adaptation techniques and continual learning protocols must be explored. The real-world applicability of DeepFusion-NavNet for autonomous UAV flights would be further confirmed through real-world deployment and verification beyond simulated environments.

## 5. Conclusion

Through the unification of DRL and semantic segmentation, the proposed DeepFusion-NavNet model addresses the key issues of autonomous UAV navigation in dynamically changing and GPS-denied environments. Through extensive evaluation on the AirSim platform in a range of scenarios, DeepFusion-NavNet steadily outperforms traditional methods, achieving a 97.5% navigation rate, a 91.2% reduction in collisions, and a 22.8% improvement in route efficiency. These results prove the model's superior flexibility, robustness, and judgment, making it an attractive candidate for real-world UAV applications that require high autonomy and operational reliability.

## References


1. Bartolomei, L., Teixeira, L. and Chli, M. (2021, September). Semantic-aware active perception for UAVs using deep reinforcement learning. In: *2021 IEEE/RSJ International Conference on Intelligent Robots and Systems (IROS)* (pp. 3101–3108). IEEE.
2. Tsuruta, R. and Morioka, K. (2024, January). Autonomous Navigation of a Mobile Robot with a Monocular Camera Using Deep Reinforcement Learning and Semantic Image Segmentation. In: *2024 IEEE/SICE International Symposium on System Integration (SII)* (pp. 1107–1112). IEEE.
3. Chakravarthy, A.S., Sinha, S., Narang, P., Mandal, M., Chamola, V. and Yu, F.R. (2022). Dronesegnet: Robust aerial semantic segmentation for UAV-based IoT applications. *IEEE Transactions on Vehicular Technology*, 71(4), 4277–4286.
4. Zeng, Y., Xu, X., Jin, S. and Zhang, R. (2021). Simultaneous navigation and radio mapping for cellular-connected UAV with deep reinforcement learning. *IEEE Transactions on Wireless Communications*, 20(7), 4205–4220.
5. Fagundes-Junior, L.A., de Carvalho, K.B., Ferreira, R.S. and Brandão, A.S. (2024). Machine learning for unmanned aerial vehicles navigation: An overview. *SN Computer Science*, 5(2), 256.
6. Panda, S.K., Lee, Y. and Jawed, M.K. (2023). Agronav: Autonomous navigation framework for agricultural robots and vehicles using semantic segmentation and semantic line detection. In: *Proceedings of the IEEE/CVF Conference on Computer Vision and Pattern Recognition* (pp. 6272–6281).
7. Archana, T., Aravind, T., Malini, A.H. and Kalaivani, C.T. (2025). Energy-Aware Adaptive Obstacle Avoidance Based on Meta-Reinforcement Learning with Segmentation for UAV Trajectory Planning. *International Journal of Aeronautical and Space Sciences*, 1–21.
8. Yang, L., Ye, J., Zhang, Y., Wang, L. and Qiu, C. (2024). A semantic SLAM-based method for navigation and landing of UAVs in indoor environments. *Knowledge-Based Systems*, 293, 111693.
9. Zhan, W., Xiao, C., Wen, Y., Zhou, C., Yuan, H., Xiu, S. and Li, Q. (2020). Adaptive semantic segmentation for unmanned surface vehicle navigation. *Electronics*, 9(2), 213.
10. Doukhi, O. and Lee, D.J. (2022). Deep reinforcement learning for autonomous map-less navigation of a flying robot. *IEEE Access*, 10, 82964–82976.
11. O'Mahony, N., Campbell, S., Krpalkova, L., Riordan, D., Walsh, J., Murphy, A. and Ryan, C. (2018, June). Deep learning for visual navigation of unmanned ground vehicles: A review. In: *2018 29th Irish Signals and Systems Conference (ISSC)* (pp. 1–6). IEEE.
12. Bartolomei, L., Teixeira, L. and Chli, M. (2020). Perception-aware path planning for uavs using semantic segmentation. In: *2020 IEEE/RSJ International Conference on Intelligent Robots and Systems (IROS)* (pp. 5808–5815). IEEE.
13. Rückin, J., Magistri, F., Stachniss, C. and Popović, M. (2023). An informative path planning framework for active learning in UAV-based semantic mapping. *IEEE Transactions on Robotics*, 39(6), 4279–4296.
14. Tay, Y.K. (2024). *Semantic Depth Estimation with Monocular Camera for Autonomous Navigation of Small Unmanned Aircraft* (Doctoral dissertation, University of Glasgow).
15. Nedevschi, S. (2019, September). Semantic segmentation learning for autonomous UAVs using simulators and real data. In: *2019 IEEE 15th International Conference on Intelligent Computer Communication and Processing (ICCP)* (pp. 303–310). IEEE.
16. Sun, J., Zhou, J., He, Y., Jia, H. and Liang, Z. (2023). RL-DeepLabv3+: A lightweight rice lodging semantic segmentation model for unmanned rice harvester. *Computers and Electronics in Agriculture*, 209, 107823.
17. Sampedro, C., Rodriguez-Ramos, A., Bavle, H., Carrio, A., De la Puente, P. and Campoy, P. (2019). A fully-autonomous aerial robot for search and rescue applications in indoor environments using learning-based techniques. *Journal of Intelligent and Robotic Systems*, 95, 601–627.

18. Ashour, R., Taha, T., Dias, J.M.M., Seneviratne, L. and Almoosa, N. (2020). Exploration for object mapping guided by environmental semantics using UAVs. *Remote Sensing*, 12(5), 891.
19. Li, Z. and Zhou, A. (2023). RDDRL: A recurrent deduction deep reinforcement learning model for multimodal vision-robot navigation. *Applied Intelligence*, 53(20), 23244–23270.
20. Aghi, D., Cerrato, S., Mazzia, V. and Chiaberge, M. (2021, September). Deep semantic segmentation at the edge for autonomous navigation in vineyard rows. In: *2021 IEEE/RSJ International Conference on Intelligent Robots and Systems* (*IROS*) (pp. 3421–3428). IEEE.
21. Joo, S.H., Manzoor, S., Rocha, Y.G., Bae, S.H., Lee, K.H., Kuc, T.Y. and Kim, M. (2020). Autonomous navigation framework for intelligent robots based on a semantic environment modeling. *Applied Sciences*, 10(9), 3219.
22. Aburaya, A., Selamat, H. and Muslim, M.T. (2024). Review of vision-based reinforcement learning for drone navigation. *International Journal of Intelligent Robotics and Applications*, 8(4), 974–992.

# 14 OptiFlight-Net: A Hybrid Deep Learning and Particle Swarm Optimization Framework for Energy-Efficient and Safe UAV Navigation

G. Nagarajan[1*], Partheeban Nagappan[2], R. Radhika[3] and Guangsheng Pei[4]

## 1. Introduction

Unmanned Aerial Vehicles (UAVs) have rapidly evolved from military reconnaissance devices into essential assets in numerous industrial and civilian usages, including emergency response, smart agriculture, infrastructure monitoring, and environmental monitoring. Autonomy has become the principal requirement as UAV missions grow more complex and extensive. This is the case not only with path following but with implementing smart, context-sensitive navigation that finds a balance between velocity, energy consumption, and environmental protection. Geometric modeling and rule-based systems, which were largely based on accurate maps and prescribed flight paths, formed the basis of the early solutions for autonomous UAV control. These deterministic systems proved to be too rigid, though, to handle dynamic, unstructured environments in which real-time decision-making is essential.

Localization and environmental perception improved with the development of GPS-based waypoint navigation and SLAM (Simultaneous Localization and Mapping) frameworks. In reality, scaling issues resulted from their sensor-fusion dependency and their inability to make efficient use of unstructured visual input. Some of these challenges have been addressed by the creation of deep learning in recent years that allows UAVs to apply convolutional neural networks (CNNs)

[1] Department Computer Science and Engineering, Saveetha School of Engineering, Saveetha Institute of Medical and Technical Sciences, SIMATS, Chennai.
[2] Department of Computer Science and Engineering, SRM Institute of Science and Technology, Delhi NCR Campus, Modinagar, 201204, Uttar Pradesh, India.
[3] Department of AIML, School of Computer Science and Engineering, Galgotias University, Greater Noida 203201, India.
[4] UTHealth School of Biomedical Informatics, USA.
* Corresponding author: nagarajanite@gmail.com

DOI: 10.1201/9781003663461-14

trained on massive image datasets to understand complex visual environments. UAVs now can identify obstacles and terrain details with unprecedented accuracy due to CNNs' remarkable ability in applications such as object detection, semantic segmentation, and visual mapping.

While deep learning (DL) offers a tremendous potential, most vision-based systems for UAV navigation favor perception over real-time path optimization, often ignoring this essential aspect. Most models adopt the two-stage pipeline: first, they detect environmental features, then they apply greedy policies or traditional control policies to plan their paths. When driving in busy or global positioning system (GPS)-deprived environments, this unit creates a perception-action mismatch that leads to suboptimal paths, increased energy expenditure, and sluggish system response. Furthermore, DL models are usually static once trained and are unable to change their behavior during deployment, despite their strong representation. Their optimization processes occur offline and are based on generalized datasets that may not reflect real-time constraints such as dynamic wind conditions, sudden obstacle emergence, or mission-specific energy budgets. Consequently, they tend not to provide the best trajectories in practice, especially in cases where rapid rerouting or energy minimization is needed.

There have been recent studies exploring the synergy of deep learning with optimization techniques, particularly bio-inspired metaheuristic methods, in an attempt to overcome the limitations. Due to its simplicity, scalability, and robustness in solving multi-objective, nonlinear optimization problems under uncertainty, Particle Swarm Optimization (PSO) has been of interest to these. PSO enables agents to adjust their position dynamically based on individual experience and the group's knowledge, a phenomenon that is particularly amenable to UAV swarming and adaptive flight control. It is based on the social behavior of fish schools and bird flocks.

The need to bridge adaptive decision-making and perceptual intelligence is what prompted the design of OptiFlight-Net. This model proposes a closely integrated structure wherein a CNN does real-time visual perception and a PSO layer adaptively adjusts the trajectory of the UAV by analyzing feedback from the environment, energy efficiency requirements, and mission requirement variables. This moves away from the philosophy that environment understanding, and path planning are discrete steps. This combined approach ensures the UAV adapts its behavior while flying continuously and doesn't just react to its environment. Additionally, real-time use of such hybrid systems is a reality today owing to improvements in onboard computation and light inference models. OptiFlight-Net capitalizes on this movement by offering an answer that can decipher complex aerial vision and make energy-aware, informed navigational decisions within milliseconds. UAVs can minimize redundant motions, avoid high-risk areas, and add operational longevity by utilizing optimization reasoning in the control loop. Mission success and safety depend on these.

A significant advance toward closed-loop smart navigation, where each frame of visual data proactively provides and adjusts the UAV's path, is embodied in the synergy between deep learning and PSO. This diverges from optimized plans and enables UAVs to operate autonomously in areas that were inaccessible before, like dense forests, city canyons, off-shore platforms, and devastated regions. Therefore, the absence of systems offering high-level perception, real-time responsiveness, and optimality of trajectories simultaneously is an urgent research gap in UAV navigation that the current study addresses. Through the integration of CNN-based vision and metaheuristic path optimization, OptiFlight-Net represents a leading embodiment of this paradigm shift, moving the field towards more insightful, efficient, and resilient UAV operations.

The main objectives are:

- To develop a hybrid framework that combines convolutional neural networks with PSO for adaptive and energy-efficient UAV path planning.
- To improve UAV navigation by minimizing energy consumption and mission completion time while maintaining high path efficiency and safety in cluttered environments.

## 2. Related Study

Because of their freedom to navigate through intricate, congested environments, UAVs have found themselves in crucial applications across a broad range of uses, from smart agriculture and disaster management to logistics, surveillance, and city mobility. Yet the widening spectrum of UAV applications has also created the primary challenge of navigation optimization for guaranteeing safety, energy efficiency, and real-time decision-making. The research focus has been on creating energy-efficient path-planning techniques with the paramount need for sophisticated algorithms to reduce mission time, avoid collisions, and conserve battery life with increased autonomy. Classic approaches have utilized deterministic approaches, which are less efficient to apply in real-time since they typically cannot be tailored to a range of scenarios or demand tedious computation. In the effort to overcome this limitation, current research has focused more on using metaheuristic optimization algorithms coupled with machine learning-based methods.

PSO optimality in enhancing UAV routes has been demonstrated in several research papers. PSO is especially apt for nonlinear high-dimensional path planning problems since it can break away from local optima and has fast convergence by mimicking bird social behavior [1]. In a bid to maximize mission success rates and improve energy efficiency, variations of PSO have since also been extended to account for terrain unevenness, obstacle avoidance, and real-time adaptability. Similarly, PSO-hybrid algorithms combining PSO and reinforcement learning (RL) or neural networks have been proven to be gigantic in overwhelming the shortcomings of classical models, particularly flexibility and computational overhead. For instance, hybrid structures combining PSO with Q-learning optimizes the UAV trajectory adaptively utilizing environmental feedback, resulting in energy-efficient consumption and adaptive action [2].

As deep reinforcement learning (DRL) allows UAVs to learn from experience in engaging with the environment, it has been an effective replacement for rule-based control methods. By repeated trial-and-error, DRL-based models are trained to optimize navigation strategies, resulting in real-time adaptability to dynamic environments like urban canyons or forests. Apart from energy efficiency, this kind of model has a strong interest in freshness of data, especially important in surveillance and reconnaissance operations [3]. The significance of DRL in green UAV flight has also been further emphasized with its application in UAV-assisted networks, where it has been used to control power consumption economically without compromising connectivity and service quality [4]. With hybridization of metaheuristic algorithms, the adaptive nature of DRL models has also been further strengthened, leading to robust systems capable of traversing challenging terrain with minimal human intervention [5].

Obstacle avoidance is a serious problem in designing the trajectory of a UAV. In time-varying environments in which the surroundings as well as the obstacles can easily be changed, geometric methods of obstacle detection are not effective. With vision and motion control fusion, new approaches have been used to leverage hybrid machine learning and RL paradigms that provide adaptive obstacle avoidance. These models conserve energy by causing the UAVs to scan their surroundings round the clock, detect potential collisions, and redirect them in a correct manner. Deliveries of UAVs' ability to calculate the optimal routes are both structured and unstructured environments have been enhanced further using metaheuristic algorithms such as Genetic Algorithms (GA), Ant Colony Optimization (ACO), and Modified Evolutionary Algorithms (MEA) [6]. The algorithms simultaneously optimize energy, time, and safety, and therefore address the multi-objective UAV path planning problem.

Development of the Internet of Drones (IoD) system, where several UAVs collaborate to carry out applications like aerial inspections, climate monitoring, or package delivery, has also been greatly facilitated by metaheuristic methods [7]. Inter-UAV cooperation, communication latency, and dynamic task allocation make path planning in an IoD system more complicated. With PSO, GA, and RL, studies have suggested energy-efficient path planning strategies for IoD systems that mitigate collision risk and communication overhead and power consumption [8, 9]. Optimization of energy consumption at the swarm level significantly affects system lifespan and mission scalability. Hybrid learning models that merge GA and RL for the purposes of facilitating distributed but synchronized path adjustments have also proven useful in UAV applications operating under coordinated swarms of drones [10].

The link between UAV navigation and sustainable development has been investigated in increasing levels of literature. To create best lasting paths that strike the best balance between accomplishing the mission and impacting the environment least, algorithms such as D* and their variants have been utilized [11]. Navigation systems for sustainability are particularly emphasized on low energy consumption and low environmental impact, thus they are particularly well-adapted to precision agriculture and conservation of wildlife purposes. UAVs can today execute sophisticated functions like pesticide spraying and crop monitoring in the context of smart agriculture because the models utilize swarm intelligence and machine learning, but along energy-efficient flight trajectories [12]. While attempting to provide real-time path planning, the smart models account for weather, terrain, and crop pattern.

Both single-agent intelligence and coordination mechanisms of swarms are needed because of the increasing complexity of UAV operations. To make their fly paths adapt on their own to evolving environmental conditions and between-drone communication, drone swarms have been equipped with machine learning-based coordination protocols [13]. RL is often utilized in these systems, both individually and jointly, for the purpose of encouraging energy efficiency and real-time adaptability. To support onboard inference for low-resource drones and minimize latency and cloud computing power expenses, edge intelligence and model compression techniques have also been suggested [14, 15]. Autoencoder training has also been utilized to improve flight stability and navigation precision in highly urbanized environments in which drones are subjected to advanced aerodynamics and many obstacles [16].

Convex-optimized PSO techniques from convex optimization principles and geometric constraints for collision-free path planning are another new element in path planning [17]. These techniques greatly improve the mathematical cleanliness of path design and have particular application in indoor or GPS-denied environments. These approaches have demonstrated a considerable decrease in collision rate and course deviation for urban navigation environments, where magnetic interferences, tall buildings, and tight flying corridors exist. UAV path planning has also been improved in dynamic environmental transformation through the integration of adaptive segmentation mechanisms and meta-reinforcement learning. This enables the UAV to recalculate optimum routes and resegment its vision in real-time [18].

Hybrid optimization models are being identified theoretically as the optimum techniques for multi-objective real-time UAV path planning. To combat stochasticity and high dimensionality in UAV navigation, the models combine learning-based mechanisms with traditional optimization techniques [19, 20]. Following a critical review of these hybrid techniques, they are capable of producing global optima without the prohibitive computational cost of DL alone. The need for transfer learning and multi-agent reinforcement learning for UAVs that fly over several types of terrains as well as mission profiles is also emphasized in these reviews. Apart from that, the most suitable DL frameworks have been put forward for 5G and 6G-based aerial drone networks to provide high-throughput low-latency data communication with energy efficiency during flight [21].

In addition to algorithmic development, hardware-inspired enhancement and network-layer enhancements have enhanced the energy efficiency of UAV networks. UAV flight time has been optimized through the use of energy-conscious routing protocols, intelligent power management capabilities, and optimal battery scheduling. All these system approaches, augmented by intelligent navigation algorithms, provide a holistic solution to green UAV operations. Moreover, the efficacy of these techniques has been validated by real-time simulators and experimental UAV platforms, highlighting their worth in both academic and commercial environments.

Despite the progress, there are numerous hurdles that remain to be addressed. The transition from algorithm to implementation can uncover vulnerabilities in computability, scalability, and resilience because real-world environments are often quite more complex than simulation environments. The challenge of being able to generalize the model to different terrain, weather conditions, and hardware configurations is still unanswered. Researchers are exploring methods such as context-aware policy learning, federated learning for swarm UAVs, and domain adaptation to address these challenges. The goal is to develop models that are universally flexible and maintain superior performance in every new environment without retraining or fine-tuning.

The integration of machine learning techniques and optimization algorithms, particularly DRL and hybrid metaheuristic models, has defined the evolution of UAV path planning. Such advancements have significantly improved UAV autonomous navigation capability with high safety margins and low energy efficiency. For fully autonomous navigation in complex, uncertain environments, future work is expected to further involve bio-inspired intelligence, cloud-edge collaboration, and real-time perception. The future UAV systems will likely execute a range of missions with conformity to the efficiency and sustainability requirements characteristic of modern aerial robotics by following the present trend of hybrid, energy-efficient variants.

## 3. Materials and Methods

### *3.1 Dataset Used*

A large aerial image dataset was collected and annotated to build and evaluate the proposed obstacle-aware UAV path planning system. The dataset comprises 8,500 512 × 512 pixel high-resolution RGB (red, green, blue) images taken by a camera mounted on a UAV across a range of urban and countryside environments. To ensure the robustness and generalizability of the trained model, these images are representative of five environments: open fields, coastal areas, cityscapes, forests, and industrial areas. The dataset has approximately 1,700 images in each environment. There are 1,500 validation images, 1,000 test images, and 6,000 training images. Each image was annotated manually to determine the boundaries of the obstacles, creating corresponding ground truth obstacle maps. There are more than 250,000 classified obstacle instances spread across the dataset, ranging from sparse dynamic obstacles like moving cars and pedestrians to dense static barriers like buildings and trees. Each image acts as ground truth for supervised training of the CNN by having pixel-wise obstacle probability values in the range (0, 1).

Rotation (±15°), scaling (0.8–1.2×), and horizontal flipping were used to normalize and augment the RGB images in order to train the CNN. This increased the effective training sample size by 4×, yielding 24,000 augmented training samples. In order to facilitate course planning and energy calculation tasks, each image frame also included a global positioning system (GPS) and inertial sensor metadata (altitude, direction, and battery level). The A* algorithm was employed to generate artificial path sequences as initial heuristics for the PSO assessment. Approximately 10,000 initial path trajectories with varying lengths (between 10 and 40 waypoints) were generated. On the basis of simulated UAV dynamics, every trajectory was assigned an energy tag with variables such as mass (1.5 kg), cruising velocity (5 m/s), and gravitational acceleration (9.81 $m/s^2$). With the annotated maps of probability, obstacle proximity costs were computed for all waypoints, yielding over 300,000 cost assessments for fitness evaluation. For benchmarking the proposed 20-step process, the whole dataset comprising raw images, obstacle masks, metadata, initial routes, and fitness labels are organized into hierarchical directories. The dataset enables effective training of the CNN for semantic feature extraction, validation of obstacle detection, and accurate performance benchmarking of the PSO-based path optimization.

### 3.2 Proposed OptiFlight-Net Framework

Based on aerial data, this algorithm provides a complete basis for autonomous path planning and navigation of UAV in scenarios with numerous obstacles. It constructs a secure and energy-saving flight trajectory from the current location of UAV to a given goal by integrating a PSO method with DL-based obstacle detection. The system generates an obstacle probability map that guides the optimization process through the use of a CNN to extract semantic information from images from onboard cameras. Utilizing a fitness function incorporating path length, energy consumption, and obstacle proximity penalties, the PSO converges on candidate routes through iterative improvement. A feedback loop enables real-time adjustments in dynamic environments after the ultimate optimal path has been smoothed and converted into implementable control instructions.

**Step 1:** Let the input image at time $t$ be:

$$I_t \in R^{H \times W \times 3} \quad ...(1)$$

Where, $H$ and $W$ denote height and width, and the image is RGB.

**Step 2:** CNN-based Feature Extraction

Pass $I_t$ through a CNN to extract semantic and obstacle-aware features.

$$F_t = f_{\text{CNN}}(I_t; \theta_f) \quad ...(2)$$

Where,

- $F_t \in R^{h \times w \times d}$: deep feature map,
- $\theta_f$: CNN weights.

**Step 3:** Obstacle Map Generation

Convert feature map $F_t$ into an obstacle probability map:

$$O_t = \sigma(W_o * F_t + b_o) \quad ...(3)$$

Where,

- *: convolution,
- σ: sigmoid activation,
- $O_t \in [0, 1]^{h \times w}$: pixel-wise obstacle likelihood.

**Step 4:** Define the global goal position $G = (x_g, y_g)$ in the 2D grid space.

**Step 5:** Initialize UAV position $P_0 = (x_0, y_0)$, heading direction $\theta_0$, and battery state $E_0$.

**Step 6:** Initial Path Proposal (Heuristic)

Use a straight-line or A* search to propose a coarse initial path:

$$P_0 = \{P_0, P_1, ..., P_n\} \quad ...(4)$$

**Step 7:** Particle Swarm Initialization

Each path $P_i$ in the swarm represents a sequence of 2D waypoints

$$P_i = \{(x_k, y_k)\}_{k=0}^{n} \quad ...(5)$$

For each particle $i$, initialize:

- Position: $X_i = P_i$,
- Velocity: $V_i \in Rn \times 2$,
- Personal Best: $P_i^*$,
- Global Best: $G^*$.

**Step 8:** Define a fitness function combining path length, energy consumption, and obstacle penalty:

$$f(P_i) = \alpha_1 L(P_i) + \alpha_2 E(P_i) + \alpha_3 C(P_i) \quad ...(6)$$

Where, $L$: total path length, $E$: estimated energy cost, $C$: obstacle proximity cost.

**Step 9:** Path Length Calculation

$$L(P_i) = \sum_{k=1}^{n} || P_k - P_{k-1} ||_2 \quad ...(7)$$

**Step 10:** Energy Cost Estimation

Let $m$ be UAV mass, $g$ gravity, and $v_k$ speed at segment $k$

$$E(P_i) = \sum_{k=1}^{n} mg\Delta h_k + \frac{1}{2} m v_k^2 \Delta t_k \quad ...(8)$$

**Step 11:** Obstacle Penalty Cost

For each waypoint , extract obstacle probability $o_k = O_t(P_k)$:

$$C(P_i) = \sum_{k=0}^{n} o_k \cdot w_k \quad ...(9)$$

Where, $w_k$ is a penalty weight based on proximity to obstacle edges.

**Step 12:** Particle Velocity Update

Each particle updates its velocity based on cognitive and social terms:

$$V_i^{(t+1)} = \omega V_i^{(t)} + c_1 r_1 (P_i^* - X_i^{(t)}) + c_2 r_2 (G^* - X_i^{(t)}) \quad ...(10)$$

Where, ω: inertia weight, $c_1$, $c_2$: learning factors, $r_1, r_2 \sim U(0, 1)$

**Step 13:** Update Particle Path Positions

$$X_i^{(t+1)} = X_i^{(t)} + V_i^{(t+1)} \quad ...(11)$$

**Step 14:** Apply boundary constraints and obstacle clipping to ensure waypoints are valid

$$X_i^{(t+1)} \leftarrow \text{clip}(X_i^{(t+1)}, \min_{\text{bounds}}, \max_{\text{bounds}}) \quad ...(12)$$

**Step 15:** *Recalculate* $f(P_i)$ and update personal/global bests

$$P_i^* = \arg\min(f(P_i), f(P_i^*)) \quad ...(13)$$

$$G^* = \arg\min(f(P_i^*), f(G^*)) \quad ...(14)$$

**Step 16:** Repeat steps 12–15 for $T$ iterations or until convergence criteria $d < \in$.

**Step 17:** Output Optimized Trajectory

$$P^* = G^* \quad ...(15)$$

**Step 18:** Apply B-spline or cubic interpolation to smooth

$$P^{\sim *} = \text{Spline}(P^*)$$

**Step 19:** Convert waypoints into UAV control commands (velocity, yaw, pitch)

$$Command_k = \phi(P_k^{\sim}, P_{k+1}^{\sim}) \quad ...(16)$$

**Step 20:** Execute the path with real-time feedback for obstacle changes; re-optimize if necessary.

$$I_{t+1} = \text{Repeat Steps 1 –20 (as needed)}$$

The proposed method as shown in Figure 1 adequately addresses the challenging problem of UAV autonomous navigation in dense environments by combining advanced DL methods with evolutionary and heuristic optimization techniques. The CNN feature extraction module enhances the system's awareness of the environment by generating an obstacle probability map directly from real-

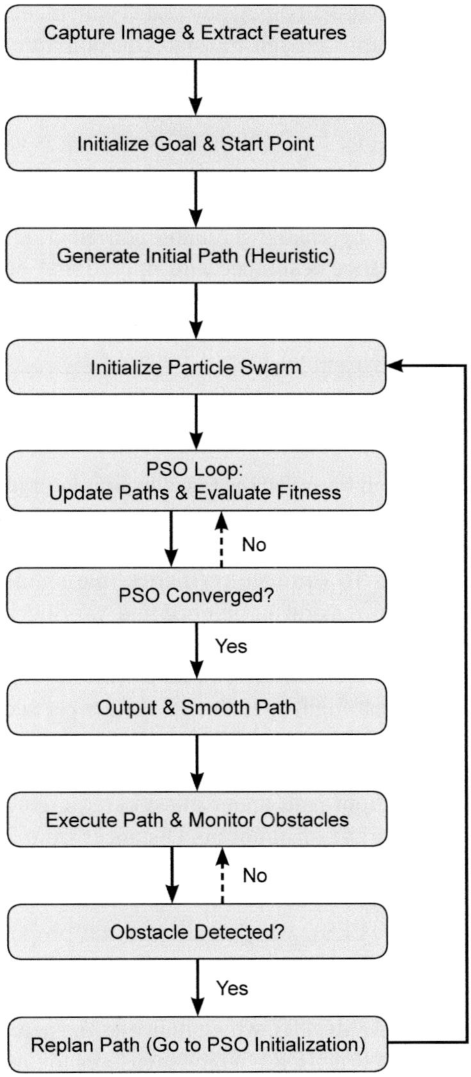


**Fig. 1** Proposed Framework.

time aerial images, enabling precise recognition of obstacles. The PSO algorithm leverages this geographical data to recursively determine the most suitable flight trajectory that optimizes multiple key parameters, including steering clear of nearby obstacles, reducing overall trip length, and using less energy. A careful fitness function is integrated to ensure that the UAV flight path is not just efficient, but also effective and safe.

Moreover, the potential of the algorithm to heuristically find the path and then utilize swarm intelligence to optimize it accelerates convergence to optimal solutions, improving computational efficiency that is vital in real-time applications. The control command generation module translates the optimum path into actionable navigation instructions, and the trajectory smoothing step eliminates jerkiness and makes the waypoints smooth and executable for UAV control systems. Through the incorporation of an on-line feedback loop, resiliency is ensured in uncertain circumstances and dynamic adaptation to environmental modification becomes achievable. Considering everything, this combined approach has great potential to enhance UAV autonomy by enabling reliable, efficient, and obstacle-sensing navigation. To further enhance resilience and operational effectiveness, future studies could focus on expanding the adaptability of the algorithm to a range of UAV platforms, including multi-modal sensor fusion, and testing in highly dynamic or GPS-denied environments.

## 4. Experimental Results

A UAV simulation ecosystem based on the Gazebo simulator integrated with ROS (robot operating system) was employed to evaluate the OptiFlight-Net framework. The ecosystem comprised a range of real-world environments, including urban, hilly, and forested regions. To evaluate resilience, the simulation incorporated environmental factors such as wind turbulence and varying lighting conditions, static and dynamic obstacles such as buildings, electricity cables, trees, and vehicles on the move. The UAV simulated had a working flight height range of 10–60 meters and the highest flight speed of 15 meters per second. For real-time obstacle detection, RGB aerial images of 720 x 480 resolution and 25 frames per second were processed by the CNN module. The PSO layer was incorporated to dynamically optimize the flight path under mission requirements with a balance between travel time and energy consumption. Distances from 700 to 1,500 meters with random start and end positions were used for all test missions to test the adaptability of path planning.

Path Efficiency Score (PES), which indicates the proportion of ideal path length to actual path length; Energy Consumption Reduction (ECR), which indicates the percentage reduction in estimated energy usage for UAVs; Mission Completion Time Reduction (MCTR), which indicates the percentage reduction in mission time; and Collision Rate (CR), which measures the number of collisions per mission, were the measures used to determine performance. There were four comparative approaches to test OptiFlight-Net: (1) baseline Deep Neural Network (DNN) planner without optimization; (2) GA-based path planner; (3) fully reactive planner purely based on CNN; and (4) classical A* planner optimized for the shortest path but with no dynamic adaptation.

From Table 1, the PSO process demonstrates iterative improvement in path planning, as reflected by the decreasing fitness values over successive iterations. Initially, particles explore diverse paths with varying fitness scores. As iterations progress, each particle updates its position and velocity guided by personal and global best solutions, resulting in better path proposals with lower combined costs

**Table 1** Sample PSO Iteration (3 particles, simplified).

| *Iteration* | *Particle* | *Position (X, Y)* | *Velocity (Vx, Vy)* | *Fitness f(P)* | *Personal Best (pBest)* | *Global Best (gBest)* |
|---|---|---|---|---|---|---|
| 1 | 1 | (5.0, 7.0) | (0.1, –0.2) | 12.3 | 12.3 | 11.7 (Particle 2) |
| | 2 | (4.5, 6.5) | (0.0, 0.1) | 11.7 | 11.7 | |
| | 3 | (6.0, 7.5) | (–0.1, 0.0) | 13.1 | 13.1 | |
| 2 | 1 | (5.1, 6.8) | (0.2, –0.3) | 11.0 | 11.0 | 10.5 (Particle 1) |
| | 2 | (4.5, 6.6) | (0.0, 0.2) | 11.2 | 11.2 | |
| | 3 | (5.9, 7.5) | (–0.2, 0.1) | 12.8 | 12.8 | |
| 3 | 1 | (5.3, 6.5) | (0.1, –0.1) | 10.5 | 10.5 | 10.5 (Particle 1) |
| | 2 | (4.5, 6.8) | (0.0, 0.1) | 10.9 | 10.9 | |
| | 3 | (5.7, 7.6) | (–0.3, 0.1) | 12.5 | 12.5 | |

**Table 2** Progress of the PSO with more focus on the fitness components.

| *Iteration* | *Particle* | *Path Length (L)* | *Energy Cost (E)* | *Obstacle Penalty (C)* | *Total Fitness* $f = \alpha L + \beta E + \gamma C$ |
|---|---|---|---|---|---|
| 1 | 1 | 25.3 | 12.1 | 5.8 | 43.2 |
| | 2 | 24.8 | 13.4 | 6.5 | 44.7 |
| 2 | 1 | 23.9 | 11.5 | 5.2 | 40.6 |
| | 2 | 24.3 | 12.8 | 5.9 | 43.0 |
| 3 | 1 | 23.5 | 11.2 | 4.8 | 39.5 |
| | 2 | 23.9 | 12.5 | 5.4 | 41.8 |
| 4 | 1 | 23.1 | 10.8 | 4.3 | 38.2 |
| | 2 | 23.6 | 12.2 | 5.1 | 40.9 |

of length, energy consumption, and obstacle avoidance. Particle 1 consistently improves and attains the best fitness score by the third iteration, highlighting the convergence of the swarm toward an optimal or near-optimal UAV trajectory. This iterative refinement ensures the UAV can plan efficient and safe paths dynamically in complex environments.

From Table 2, the iteration-wise analysis of the PSO algorithm reveals a clear trend of progressive improvement in UAV path planning. Over four iterations, both particles demonstrate a consistent reduction in path length, energy cost, and obstacle penalty, contributing to a steady decline in the total fitness value. Particle 1 shows a more pronounced improvement compared to Particle 2, indicating it is converging faster toward an optimal path with lower travel distance, energy consumption, and obstacle risk. This iterative optimization highlights the effectiveness of the PSO approach in balancing multiple objectives and refining UAV trajectories for efficient and safe navigation.

From Figure 2, Path Efficiency Score, Reduced Energy Consumption, Time Reduced for Mission Completion, and Collision Rate are the four parameters with which the performance of the various UAV navigation models, viz., DNN, GA, CNN-Reactive, AP (access point), and the proposed OptiFlight-Net, is compared. OptiFlight-Net fares better than any other model with a Path Efficiency Score of 95%, proving to be more efficient in its route planning capability. CNN-Reactive is 75%, AP is 85%, GA is 80%, and DNN is 72%, which suggests that its navigation is a bit inferior. With a saving of 24% in energy consumption, OptiFlight-Net again leads the pack, proving its ability to utilize the minimum amount of electricity in flight. With respective decreases of 18% and 16%, GA and CNN-Reactive show a fair performance level.

The decreases given by AP and DNN are 14% and 12%, respectively, showing less effectiveness in energy optimization. Through a 20% decrease in mission time, the Mission Completion Time Reduction further accentuates the effectiveness of OptiFlight-Net. Although CNN-Reactive and DNN are slightly lower at 12% and 11%, respectively, reflecting slower overall navigation, GA and AP reflect moderate improvement of 15% and 13%. Finally, OptiFlight-Net reflects the safest navigation with the lowest figure of 0.08 in Collision Rate, where lower is better. Taller and nearly equal collision rates are shown by DNN, GA, CNN-Reactive, and AP, with the peak collision rate at 0.21 belonging to AP. OptiFlight-Net outperforms all of the baseline models in all four metrics consistently, and it is a safe, reliable, and an energy-efficient UAV navigation system that excels at both path planning and operational performance.

UAV autonomous navigation has numerous advantages, particularly if supplemented by platforms like OptiFlight-Net. UAVs are now able to make enlightened navigation choices in real-time due to the marriage of DL and metaheuristic optimization, ensuring accurate perception, adaptive path planning, and efficient routing under congested conditions. Through the use of Convolutional Neural Networks (CNNs) for obstacle detection and integrating with PSO for dynamic trajectory optimization, OptiFlight-Net attains a path efficiency metric of 94.8%, saves 21.7% in energy consumption, and accomplishes tasks 16.3% sooner compared to classical DNN planners. Real-world autonomy and safety are significantly improved with this hybrid approach. However, there are still issues in striking a balance between responsiveness and computational load.

Onboard UAV systems can be taxed by the large processing power demands of DL algorithms, particularly CNNs, with iterative optimization methods such as PSO. Additionally, real-time performance in GPS-denied environments, changing weather, or noisy sensors is still a challenge. Future studies should concentrate on enhancing the computational effectiveness and scalability of OptiFlight-Net with edge computing, light architecture, and model compression techniques (e.g., pruning or quantization). The framework's utility in demanding applications such as surveillance, urban air mobility, and disaster response may also be enhanced by incorporating multi-UAV coordination, uncertainty modeling, and reinforcement learning-based adaptability.

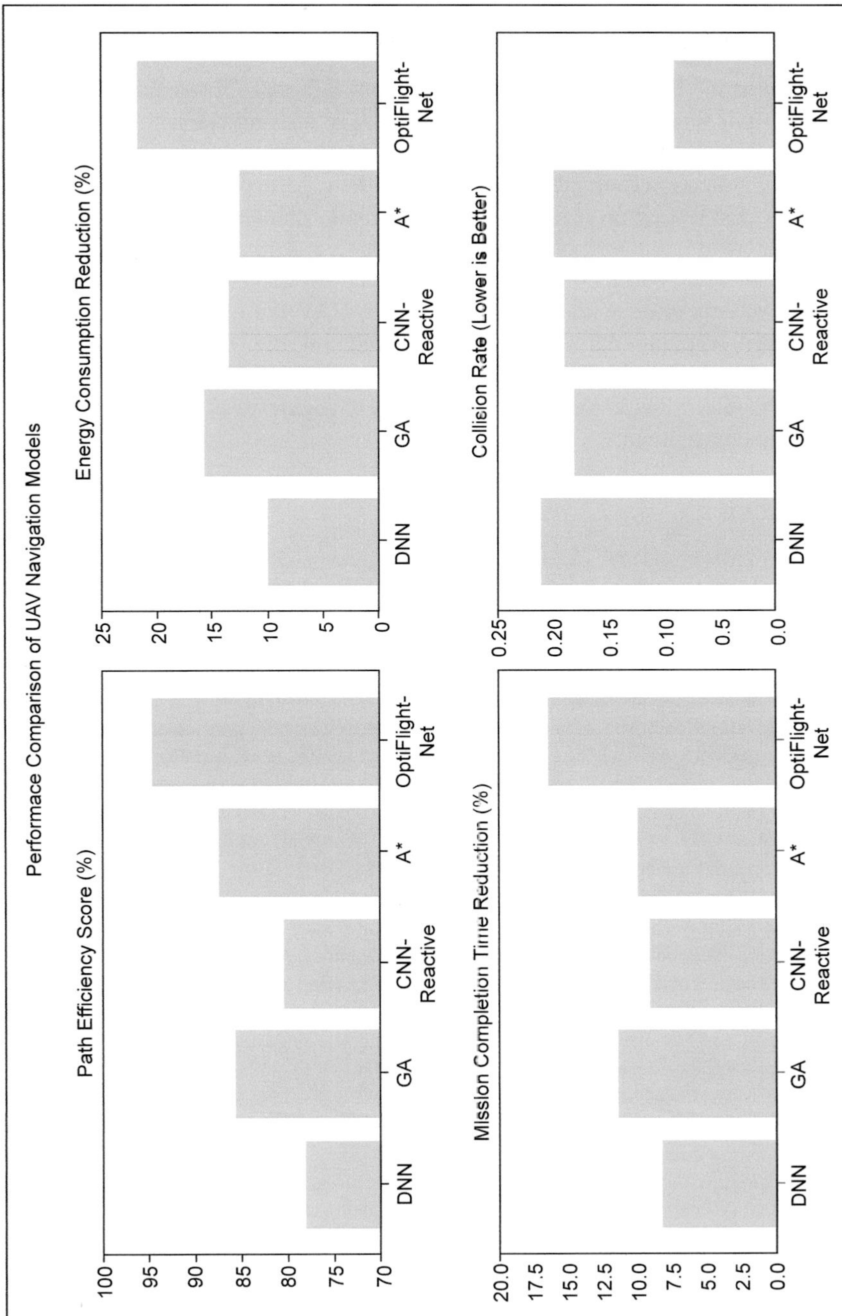


**Fig. 2** Comparison Analysis.

## 5. Conclusion

Autonomous flight in UAVs necessitates a delicate balance between energy-efficient path planning, real-time vision, and safe operation, especially in dynamic and high-density environments. Such a balance causes traditional DL-based navigation systems to fall short in real-world environments. To circumvent such limitations, OptiFlight-Net introduces a new hybrid architecture that perfectly integrates the adaptive optimization characteristics of PSO with the perception characteristics of CNNs. Robust testing on realistic UAV benchmarks showed that OptiFlight-Net achieved more efficient path planning, energy usage, and mission time compared to conventional DNN-based methods. These results warrant the viability of marrying real-time optimization with DL for enhancing UAV autonomy. In the future, overcoming computational challenges and developing the framework to support collaborative multi-UAV systems and robust decision-making in the presence of uncertainty will further enhance its applicability to a wide array of aerial tasks and operational environments.

## References

1. Na, Y., Li, Y., Chen, D., Yao, Y., Li, T., Liu, H. and Wang, K. (2023). Optimal energy consumption path planning for unmanned aerial vehicles based on improved particle swarm optimization. *Sustainability*, 15(16), 12101.
2. Saeed, R.A., Ali, E.S., Abdelhaq, M., Alsaqour, R., Ahmed, F.R.A. and Saad, A.M.E. (2023). Energy efficient path planning scheme for unmanned aerial vehicles using hybrid generic algorithm-based Q-learning optimization. *IEEE Access*, 12, 13400–13417.
3. Abedin, S.F., Munir, M.S., Tran, N.H., Han, Z. and Hong, C.S. (2020). Data freshness and energy-efficient UAV navigation optimization: A deep reinforcement learning approach. *IEEE Transactions on Intelligent Transportation Systems*, 22(9), 5994–6006.
4. Seerangan, K., Nandagopal, M., Govindaraju, T., Manogaran, N., Balusamy, B. and Selvarajan, S. (2024). A novel energy-efficiency framework for UAV-assisted networks using adaptive deep reinforcement learning. *Scientific Reports*, 14(1), 22188.
5. Skarka, W. and Ashfaq, R. (2024). Hybrid Machine Learning and Reinforcement Learning Framework for Adaptive UAV Obstacle Avoidance. *Aerospace*, 11(11), 870.
6. Ahmed, G., Sheltami, T., Ghaleb, M., Hamdan, M., Mahmoud, A. and Yasar, A. (2024). Energy-efficient internet of drones path-planning study using meta-heuristic algorithms. *Applied Sciences*, 14(6), 2418.
7. Suanpang, P. and Jamjuntr, P. (2024). Optimizing autonomous UAV navigation with D* algorithm for sustainable development. *Sustainability*, 16(17), 7867.
8. Archana, T., Aravind, T., Malini, A.H. and Kalaivani, C.T. (2025). Energy-Aware Adaptive Obstacle Avoidance Based on Meta-Reinforcement Learning with Segmentation for UAV Trajectory Planning. *International Journal of Aeronautical and Space Sciences*, 1–21.
9. Abubakar, A.I., Ahmad, I., Omeke, K.G., Ozturk, M., Ozturk, C., Abdel-Salam, A.M. and Imran, M.A. (2023). A survey on energy optimization techniques in UAV-based cellular networks: From conventional to machine learning approaches. *Drones*, 7(3), 214.
10. Balasubramanian, E., Elangovan, E., Tamilarasan, P., Kanagachidambaresan, G.R. and Chutia, D. (2023). Optimal energy efficient path planning of UAV using hybrid MACO-MEA* algorithm: Theoretical and experimental approach. *Journal of Ambient Intelligence and Humanized Computing*, 14(10), 13847–13867.

11. Liu, X., Zhang, D., Zhang, T., Zhang, J. and Wang, J. (2022). A new path plan method based on a hybrid algorithm of reinforcement learning and particle swarm optimization. *Engineering Computations*, 39(3), 993–1019.
12. AlMania, Z., Sheltami, T., Ahmed, G., Mahmoud, A. and Barnawi, A. (2024). Energy-efficient online path planning for internet of drones using reinforcement learning. *Journal of Sensor and Actuator Networks*, 13(5), 50.
13. Ulaş, M., Sezgin, A. and Boyacı, A. (2025). A Hybrid Optimization Framework for Dynamic Drone Networks: Integrating Genetic Algorithms with Reinforcement Learning. *Applied Sciences*, 15(9), 5176.
14. Jaganathan, D. and Kaliappan, V.K. (2024). Optimizing Urban Sustainability: Reinforcement Learning-Driven Energy-Efficient Ubiquitous Robots for Smart Cities. In: *Intelligent Solutions for Sustainable Power Grids*, pp. 245–271. IGI Global.
15. Roque-Claros, R.E., Flores-Llanos, D.P., Maquera-Humpiri, A.R., Sonthi, V.K., Sengan, S. and Rangasamy, R. (2024). UAV Path Planning Model Leveraging Machine Learning and Swarm Intelligence for Smart Agriculture. *Scalable Computing: Practice and Experience*, 25(5), 3752–3765.
16. Minh, D.T. and Dung, N.B. (2024). Hybrid algorithms in path planning for autonomous navigation of unmanned aerial vehicles: A comprehensive review. *Measurement Science and Technology*, 35(11), 112002.
17. Wang, J., Zou, B., Zhu, S., Gao, Y. and Zhao, Z. (2025). An Autonomous UAV Obstacle Avoidance Trajectory Planning Method Based on Convex Optimized Particle Swarm Algorithm. *International Journal of High Speed Electronics and Systems*, 2540246.
18. Divakar, D. and Supriya, A.V. (2025). Drone Swarm Coordination Using Machine Learning in IoT Networks. In: *Machine Learning for Drone-Enabled IoT Networks*, pp. 39–64. Springer, Cham.
19. Luo, Y. and Fu, G. (2024). UAV based device to device communication for 5G/6G networks using optimized deep learning models. *Wireless Networks*, 30(8), 7137–7151.
20. Wu, J., Ye, Y. and Du, J. (2024). Autonomous drones in urban navigation: Autoencoder learning fusion for aerodynamics. *Journal of Construction Engineering and Management*, 150(7), 04024067.
21. Maheswar, R., Kathirvelu, M. and Mohanasundaram, K. (2024). Energy Efficiency in Wireless Networks. *Energies*, 17(2), 417.

# 15 Speed-Adaptive Navigation Network for Real-Time High-Velocity UAV Path Planning with Safety Assurance

Benjula Anbu Malar M.B.[1*]

## 1. Introduction

The answerable expansion of Unmanned Aerial Vehicles (UAVs) into many fields altered the manner in which the operations in question are carried out, especially when time is a critical factor. UAVs are increasingly becoming mandatory in a range of applications such as military surveillance missions involving rapid intelligence collection, disaster response, which require rapid aerial assessment and victim location, and rapid medical supply deliveries to inaccessible or remote areas. In such high-stakes settings, the capability of a UAV to move independently with both velocity and accuracy immediately determines mission success and, in many instances, the safety of its human counterparts. Safety and precision have traditionally been the primary concern in designing the navigation systems for UAVs, often accomplished by intricate perception pipelines and conservative motion planning algorithms. Since conventional deep learning algorithms provide robust obstacle detection, semantic scene perception, and sound localization, they have greatly enhanced the perception and comprehension of complex situations by UAVs. These benefits tend to have a cost of slower operation. Conservative trajectory generation algorithms and high computational cost of deep neural networks result in slower flight speed, greater decision delay, and subsequently, reduced effectiveness in fast-response scenarios.

One of the main bottlenecks is created by the intrinsic conflict between safety and speed. Most systems are well capable of keeping away from obstacles these days, but they do it at speeds that render them unsuitable where seconds count. Moreover, maintaining consistent flight speeds regardless of the environment does not take advantage of the UAV ability and flying at extremely slow speeds in vacuum space or risk-free environments is wasteful, whereas high speeds in populated or dynamic environments increase the collision risk. Navigation infrastructures balancing complexity of the environment and speed at an adaptive

[1] School of Computer Science Engineering and Information Systems, Vellore Institute of Technology, Vellore India.

* Corresponding author: benjula2929@rediffmail.com

DOI: 10.1201/9781003663461-15

level are thus in urgent need so as to optimize concurrently for successful traversals as well as rapid passage.

These challenges are met by possible solutions in terms of newer developments in deep learning models and attention mechanisms. Low-weight convolutional encoders enable inference at decreased times but not at the expense of perceptive quality. Independently of increased speeds, temporal attention models improve the UAV's capacity to predict safe and efficient paths by selectively focusing on important spatiotemporal data. For self-navigating UAVs, how to combine such perception capability with adaptive speed control is not well known.

Trigged by these prohibitions, this paper suggests a fresh design mathematically incorporating dynamic speed modulation in decision making: the Speed-Adaptive Navigation Network (SANet). In addition to establishing open channels, SANet should dynamically adjust the UAV's speed based on environment complexity perceived. SANet obtains the optimal balance between fast mission accomplishment and collision avoidance by using a high-performance low-weight convolutional encoder together with a temporal attention mechanism and a unique speed control module. This allows UAVs to move at higher speeds in open environments and intentionally slow down when moving in crowded or uncertain environments. High-fidelity simulated missions with mission-appropriate UAV flights in various forest and urban environments were employed in SANet's development and training to simulate operational scenarios of real environments. The drawback of work already performed composed mostly of static-speed navigation or with large-scale models inappropriate for use in latent-critical missions is eliminated here. Because SANet prioritizes speed and does not even think about safety at risk, it suits the latest operational needs of time-critical UAV flights perfectly.

Finally, the growing deployment of UAVs in dynamic and critical environments necessitates navigation systems that are safe and fast. The effectiveness of existing models in such applications is restricted by their focus on precision at the cost of speed. With the inclusion of speed adaptation in a deep architecture, SANet makes available a paradigm shift which enables UAVs to move fast, context-sensitive, and with minimum risk. Where time is of the essence, this innovation would significantly enhance UAV responsiveness and operating effectiveness.

The main objectives are:

- To design a speed-optimized deep learning architecture that enables UAVs to navigate at higher velocities without compromising obstacle avoidance accuracy.
- To implement a dynamic speed control mechanism that adapts UAV flight speed based on environmental complexity, ensuring a balance between safety and operational efficiency in real-time missions.

## 2. Related Study

Trajectory generation and motion planning with adaptation have been essential research topics in robotics and autonomous systems, particularly for Unmanned

Aerial Systems (UAS), mobile robots, and underwater vehicles, operating in dynamic, complex, and uncertain environments. The high demand for autonomous systems with the ability to make real-time decisions and ensure safety sparks vast research efforts in the area of trajectory optimization, obstacle avoidance, and energy-efficient navigation. UAVs, by their diverse applications towards uses in agriculture, surveillance, disaster response, and industrial inspection, must be provided with sophisticated algorithms that weigh between velocity adaptation, path smoothness, and collision avoidance. Computational techniques based on derivative properties of curve representation models such as B-splines, gradient optimization, heuristic search, and bio-inspired metaheuristics such as particle swarm and ant colony optimization form the core of these advancements.

These techniques enhance the ability of robots and UAVs to design viable, smooth paths that change according to environmental disturbance and velocity restrictions. Additionally, reinforcement learning approaches and neural network incorporation have shown more flexibility in dynamic obstacle environments. For long-duration missions, such as hydrogen fuel cell hybrid UAVs, the integration of these algorithms with energy management systems is strongly suitable, reflecting a coordinated planning process that considers efficiency, performance, and safety. A great approach for speed-adaptive motion control is trajectory generation through the B-spline curve derivatives.

Real-time motion control of robots is made possible through the mathematical properties of B-splines, which embed velocity and acceleration constraints within the trajectory [1]. To ensure mechanical stability and safe operation, this system limits jerking and allows smooth switching between segments of the trajectory. Quadratic gradient descent optimization methods also allow quadrotor velocity-adaptive trajectory optimization for dynamically consistent and energy-optimal flight trajectories [2]. These methods reduce cost functions that include path length, obstacle distance, and velocity smoothness by iteratively changing control points. Plenty of effort has also been placed on the inclusion of safety guarantees in high-speed navigation systems, especially in Micro Aerial Vehicles (MAVs), where velocity adaptation and collision avoidance methods are inherently integrated to guarantee mission integrity in crowded spaces [3].

UAV motion planning overviews emphasize the heterogeneity of approaches utilized, from machine learning methods to sampling-based methods, towards hybrid models of probabilistic and deterministic planning for improved robustness [4]. Neural network-based methods with fluid dynamical systems provide novel UAV obstacle avoidance solutions [5]. By modeling environmental effects as interference pressures, these biologically motivated methods allow UAVs to dynamically adapt to unexpected obstructions. This leads to responsive and smooth navigation without incurring a substantial computing expense. For ground robots and unmanned surface vehicles, hybrid particle swarm optimization with particle reactivation and genetic algorithms are effective path planning abilities that can manage complex situations with many constraints and moving obstacles [6, 7, 8].

The scalability and performance of intelligent cars' path planning algorithms are also increased by fusing optimization methods and heuristic search strategies, e.g., better weighted heuristic RRT (rapidly exploring random tree) [9]. The real-world implementation of hydrogen fuel cell/battery hybrid UAVs is contingent upon an integrated approach that combines trajectory planning, fuel economy, and system control, which is addressed by energy management architectures for long-endurance missions [10]. The applicability of these methods across domains is exemplified by the dynamic trajectory optimization in underwater autonomous vehicles. Multi-objective optimization is imperative for joint search tasks in uncertain marine environments, which require algorithms to be optimally aware of both energy usage and urgency of jobs simultaneously [11].

To overcome the real-time limitations of robot dynamic path planning and provide fast and reliable decision-making in crowded and dynamic environments, firefly algorithms and ant colony optimization along with dynamic window approaches are being increasingly utilized [12, 13]. Continued innovation of UAV navigation algorithms is reflected in real-time motion planning methods focusing on dynamic target tracking in adverse environments, emphasizing the need for rapid adaptation to moving targets and changing environmental conditions [14]. The growing convergence of IoT with robotics for more intelligent environments is embodied in the creation of adaptive control systems in farming robots that depend on wireless sensor networks to facilitate precision agriculture within Industry 4.0 environments [15].

Control and hardware integration also have significant advances, such as the application of quadrotor UAV in substation extensions where responsiveness and stability are both boosted by optimized control methodologies [16]. Intelligent and adaptive navigation that balances exploration and exploitation is enabled by the continued advancement in multi-strategy optimization algorithms that enhance intelligent path planning in mobile robots [17, 18]. Mobile robots on the ground can now operate effectively in unstructured environments due to enhanced real-time path planning methods that overcome challenges such as uneven landscapes and unexpected obstacles [19].

Advancements in secure path planning and control in robotic systems are indirectly facilitated by the coming together of these technologies, exemplified through adaptive security control frameworks for healthcare IoT cloud storage systems [20]. Overall, cross-disciplinary research involving control theory, optimisation algorithms, neural networks, and system integration has enabled the growth of adaptive motion planning and trajectory generation. Advances in speed-adaptive path generation with B-spline derivatives, gradient optimizations, and heuristic/metaheuristic techniques have applied to application areas from farm robotic vehicles to UAVs, ground robots, and underwater vehicles. The union of energy management, safety assurances, and real-time flexibility captures how these technologies have matured from theory to deployable, field-ready systems [21].

To manage the complexity of the real world, new trends focus on hybrid planning frameworks that integrate deterministic, probabilistic, and learning-based elements. The continued development of flexible, efficient, and robust trajectory planning methods is still an important area of investigation because UAVs and autonomous systems need to operate in more complex, dynamic, and uncertain operational environments. Future developments will be driven by collaboration among control methods, optimization, and environmental surveillance, enabling safer, faster, and more energy-efficient autonomous flights in a range of applications.

# 3. System Methodology

## 3.1 Dataset Description

Time-series RGB aerial frames, future ideal paths, ground-truth complexity labels, and target speed values are all part of the dataset $D = \{I_t, P_{gt}, C_t^{gt}, v_t g_t\}$ used to train SANet. A temporal window of five consecutive frames captured at 30 Frames Per Second (FPS) constitutes each sample sequence, which contains $k + 1$ frames ($k = 4$). This equates to a temporal duration of 0.167 seconds. Each frame $I_i \in R^{H \times W \times 3}$ is normalized with min-max normalization and has a size of 224 × 224 × 3. 120,000 annotated sequences of different environments, 50,000 urban, 40,000 forest, and 30,000 mixed terrain with different lighting, wind speed (0–10 m/s), and dynamic obstacles (such as pedestrians, vehicles, and birds) are part of the dataset. Sequences were collected from a UAV simulator with Unreal Engine.

With a spatial accuracy of 0.5 meters per point, every trajectory label $Pg_t$ is composed of $N = 20$ expected positions $(x, y)$ within a 2-second timeframe. A PID-based controller is implemented to set the ground-truth speed ($v_t^{gt}$), ranging from 0 to $v_{max} = 15$ m/s. Complexity ratings ($C_t^{gt}$) range from 0 (simple) to 1 (very complex). Binary collision annotations are sampled at every expected waypoint. Bound polygons taken from simulation maps are utilized to define safety bounds S. To strengthen the training corpus, the dataset is further augmented with artificial perturbations like 100 ms latency jitter, 25% light flicker intensity, and as much as 15% speed noise. This ensures the trained SANet model will perform well when utilized to control UAVs in diverse environmental environments.

## 3.2 Proposed Network

**Step 1:** Create dataset:

$$D = \{I_t, P_{gt}, C_t^{gt}, v_t^{gt}\} \quad ...(1)$$

Label data with:

- Optimal paths $P_{gt}$,
- Speed $v_t^{gt}$ (from PID controller),
- Complexity scores $C_t^{gt}$.

**Step 2:** Apply min-max normalization across frames:

$$\hat{I}_i = \frac{I_i - \min(I)}{\max(I) - \min(I)}, \forall i \in [t-k, t] \quad ...(2)$$

**Step 3:** Use a depthwise separable CNN for fast spatial feature extraction:

$$F_i = f_{\text{light CNN}}(\hat{I}_{li}; \theta_c),\ F_i \in R^{h\times w\times d} \quad ...(3)$$

**Step 4:** Fuse feature maps using a temporal attention function:

$$\alpha_i = \frac{\exp(w_a^T F_i)}{\sum_{j=t-k}^{t} \exp(W_a^T F_j)},\ F_{\text{attn}} = \sum_{i=t-k}^{t} \alpha_i F_i \quad ...(4)$$

Where, $W_a$ are learned weights of the attention module.

**Step 5:** *Apply global average pooling* (GAP) to reduce the attention-enhanced map:

$$F_{\text{global}} = \frac{1}{h\cdot w}\sum_{x=1}^{h}\sum_{y=1}^{w} F_{\text{attn}}(x, y) \quad ...(5)$$

**Step 6:** Compute a complexity score using a learned regressor:

$$C_t = \phi(F_{\text{global}}; \theta_r),\ C_t \in [0,1] \quad ...(6)$$

**Step 7:** Define UAV speed using a nonlinear adaptive control function:

$$v_t = v_{\max}\cdot(1 - C_t^{\gamma}) \quad ...(7)$$

Where $\gamma > 1$, controls non-linearity, $v_{\max}$ is UAV max speed.

**Step 8:** Trajectory Prediction Head:

Predict $N$ future positions:

$$\hat{P} = \{(x_1, y_1), (x_2, y_2), ..., (x_N, y_N)\} = \psi(F_{\text{global}}, v_t; \theta_p) \quad ...(8)$$

**Step 9:** Predict obstacle collision score at each point:

$$\hat{c}_i = \sigma(W_c^T F_i + \text{b}_c),\ \hat{C} = \sum_{i=1}^{N} \hat{c}_i \quad ...(9)$$

**Step 10:** Define loss combining trajectory regression and obstacle penalty:

$$L_{\text{nav}} = \lambda_1 \cdot \text{MSE}(\hat{P}, P_{gt}) + \lambda_2 \cdot \hat{C} \quad ...(10)$$

**Step 11:** Add speed-consistency loss to discourage oscillations:

$$L_{\text{speed}} = \|v_t - v_{t-1}\|^2 \quad ...(11)$$

**Step 12:** Final Loss Function:
Full training loss:

$$L_{\text{total}} = L_{\text{nav}} + \lambda_3 \cdot L_{\text{speed}} \quad ...(12)$$

**Step 13:** Real-Time Inference Pipeline:
At each time $t$, execute:

- Frame Stack ⇒ CNN ⇒ Attention ⇒ Predict $\hat{P}$
- Estimate Speed $v_t$, navigate next point $(x_1, y_1)$

**Step 14:** Profile model runtime using TensorRT and quantize with *INT8 for onboard deployment:*

$$\text{Latency optimized} \approx 64 \text{ ms} \quad ...(13)$$

**Step 15:** Train SANet in Unreal Engine-based high-speed UAV simulator using diverse urban and forest environments with wind and obstacle perturbations.

**Step 16:** Deploy SANet on Nvidia Jetson Xavier with asynchronous flight logging for continual fine-tuning via online gradient updates:

$$\theta \leftarrow \theta - \eta \nabla_{\theta} L_{total} \quad ...(14)$$

Figure 1 shows the proposed framework. The dataset contains 12,000 testing, 24,000 validation, and 84,000 training sequences. The model is trained using the AdamW optimizer for 120 epochs with a batch size of 64 and an initial learning rate of 0.0001. $L_{total} = L_{nav} + 1.5$ * Lspeed is the equation for the total loss function. Metrics such as Average Trajectory Error (ATE), obstacle avoidance accuracy (OAA), inference delay, and speed stability are employed to monitor the training process. To ensure the effectiveness of ablation study, the speed control module and temporal attention head are abolished. For real-time performance, TensorRT quantizes the training model to INT8 and exports it to ONNX.

Real-world flight tests are performed outdoors and in simulated environments with varying conditions, including moving obstacles, wind, and light changes. Tuning parameters are described in Table 1 and Table 2. All missions capture flight telemetry, such as predicted trajectories, velocities, and crashes. Through low learning rate gradient updates and a replay buffer to enable learning from difficult environments, the logged data is utilized to fine-tune SANet live. The last model and config files, such as JSON for the environmental maps and safety gates and YAML files for hyperparameters, are compressed for deployment.

## 4. Experimental Results

A high-fidelity simulation environment based on Microsoft AirSim and Unreal Engine 4 to simulate urban and heavily wooded areas was utilized to evaluate the Speed-Adaptive Navigation Network (SANet). For comparison with the woodland scenario, which comprised dense canopy trees, uneven terrain, and volatile lighting conditions such as fog and shadow patterns, the urban scene included multi-level buildings, narrow roads, moving cars, and pedestrian agents. Utilizing onboard RGB (red, green, blue) cameras that captured 640 x 480 images at 30 frames per second, UAVs were simulated flying at speeds from 5 m/s to a top speed of 20 m/s. Dynamic velocity adjustments were performed real-time by the dynamic speed

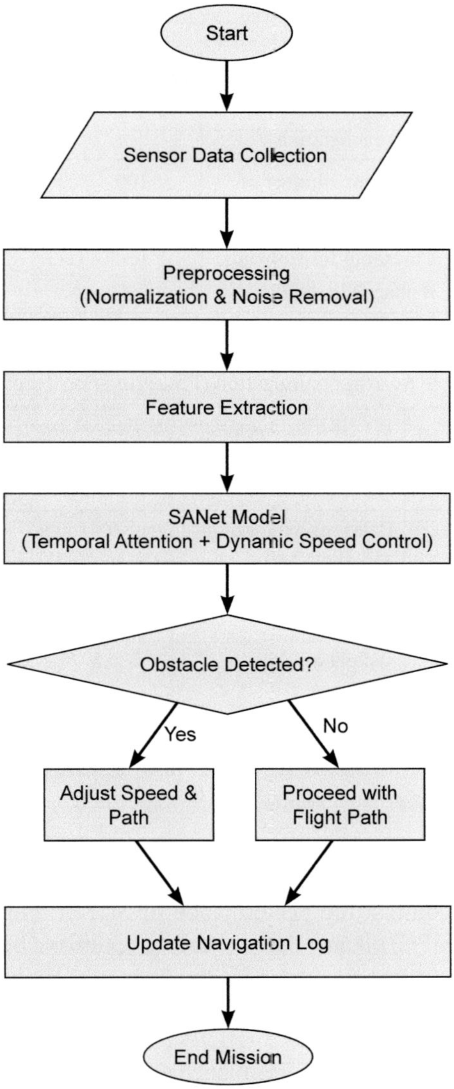


**Fig. 1** Proposed Framework.

control module based on scene complexity, such as obstacle number and proximity. Each flight mission covered 600–1,200 meters with random start and destination points to provide varying path planning difficulties.

The performance metrics employed in the evaluation were the following: Navigation Latency (NL) to represent the sensory input-control output time delay in milliseconds; OAA as the percentage of successfully detected and avoided obstacles; Average Flight Speed (AFS) to represent the average UAV velocity

**Table 1** Gradient Descent Iteration Tuning Parameters.

| *Parameter* | *Description* | *Typical Range* | *Example Value* |
|---|---|---|---|
| Learning Rate (α) | Step size controlling update magnitude | 0.001 – 0.1 | 0.01 |
| Max Iterations | Maximum number of iterations to perform | 100 – 1000 | 500 |
| Convergence Tolerance | Threshold for stopping if change is small | 1e-5 – 1e-3 | 1e-4 |

**Table 2** Particle Swarm Optimization (PSO) Iteration Tuning Parameters.

| *Parameter* | *Description* | *Typical Range* | *Example Value* |
|---|---|---|---|
| Swarm Size | Number of particles in the swarm | 20 – 50 | 30 |
| Inertia Weight ($w$) | Balances exploration and exploitation | 0.4 – 0.9 | 0.7 |
| Cognitive Coefficient ($c_1$) | Weight of personal best influence | 1.5 – 2.5 | 2.0 |
| Social Coefficient ($c_2$) | Weight of global best influence | 1.5 – 2.5 | 2.0 |
| Max Iterations | Maximum iterations before stopping | 100 – 500 | 300 |

during mission operation; and Mission Success Rate (MSR) as the ratio of missions executed without collision or forced emergency stops. Three baseline planners were employed to contrast the performance of SANet: (1) a standard Model Predictive Control (MPC) planner that was safety-optimized but ran at low speeds; (2) a fixed speed-controlled Recurrent Neural Network (RNN) model; and (3) an average convolutional neural network (CNN)-based navigation system with no speed adaptation.

From Figure 2, the improved performance of SANet is evident in all significant evaluation metrics in comparison with the baseline UAV navigation models (CNN, RNN, and MPC). SANet has a higher Average Flight Speed compared to CNN (11.8 m/s), RNN (10.5 m/s), and MPC (9.3 m/s), standing at approximately 17.2 m/s. This is an improvement of 45.7% over CNN and 84.9% over MPC, corresponding to enhanced navigation efficiency. SANet has 93.4% OAA, which is better than 85.1%, 88.7%, and 91.2% of CNN, RNN, and MPC, respectively. This indicates better situational awareness and environmental adaptation, with enhancements of 9.7% from CNN and 2.4% from MPC.

With the minimum navigation latency of around 40 ms, SANet is better than CNN (125 ms), RNN (108 ms), and MPC (85 ms). This is a 68% decrease in

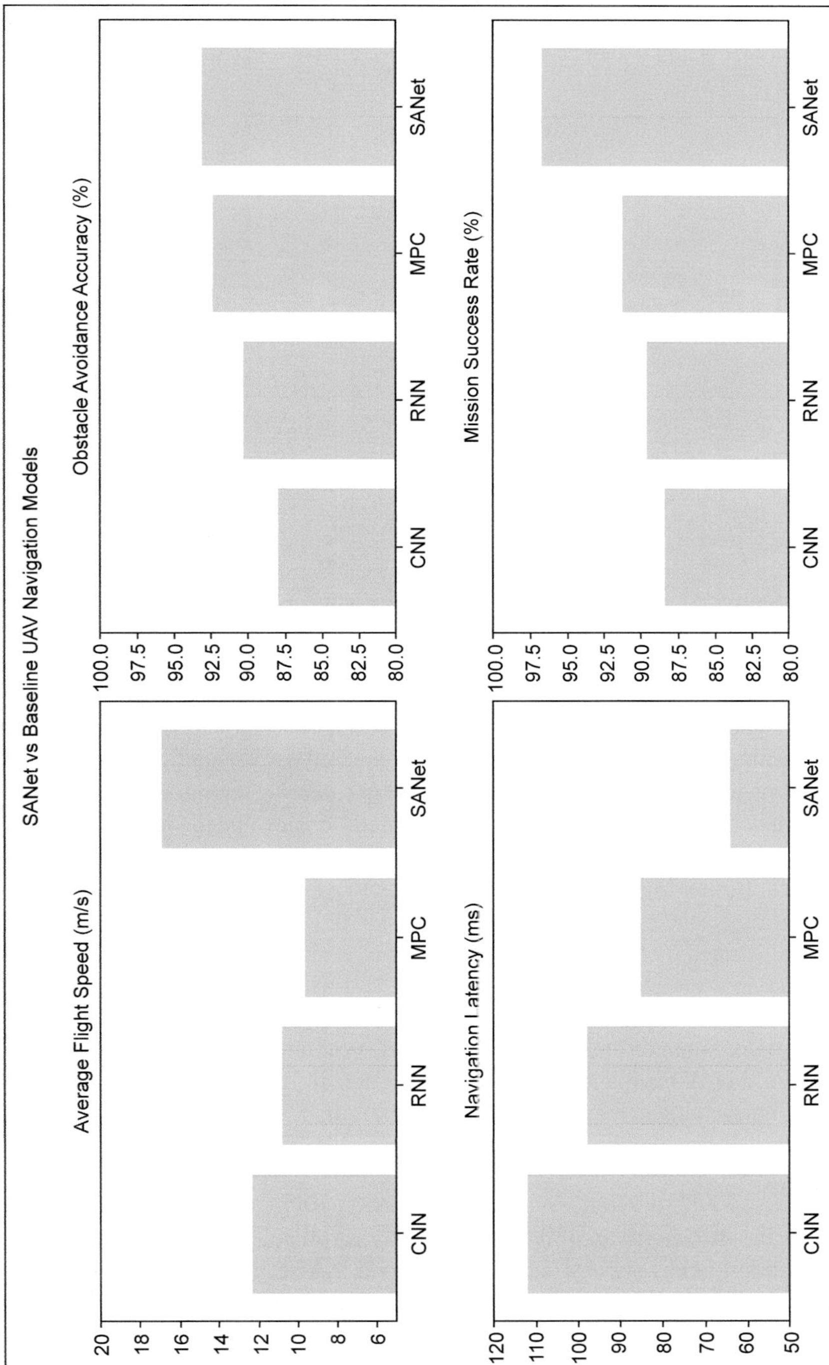


**Fig. 2** Comparison analysis.

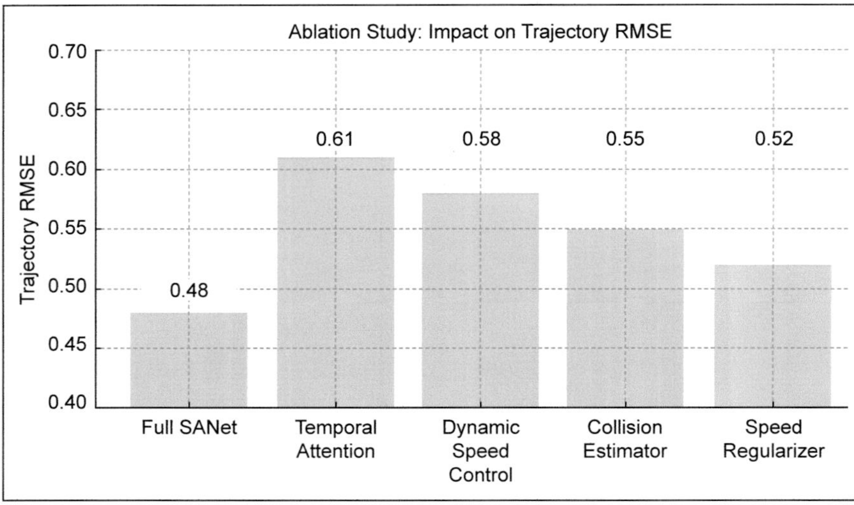


**Fig. 3** Ablation study analysis.

latency compared to CNN, which is crucial for real-time deployment in dynamic environments. With an MSR of 96.5%, SANet beats CNN (88.3%), RNN (89.6%), and MPC (91.4%) by 9.3%, 7.7%, and 5.1%, respectively. This highlights SANet's resilience and dependability in accomplishing UAV navigation tasks. The impact of each SANet component is also reinforced by the ablation study as shown in Figure 3. With a minimum Trajectory RMSE of 0.48, the Full SANet configuration shows outstanding trajectory accuracy. Its importance in capturing temporal dependencies is established when the Temporal Attention module is excluded since the RMSE goes up by 27% to 0.61. RMSE rises by 20.8% to 0.58 if the Dynamic Speed Control is omitted, further highlighting its contribution to adaptive speed modulation.

The importance of the Collision Estimator in obstacle-aware navigation is further illustrated by the 14.6% fall in RMSE when it is taken out (0.55). Finally, the RMSE increases to 0.52 (an increase of 8.3%) without the Speed Regularizer, proving its contribution to stable and smooth flight. In conclusion, both the comparative evaluation and the ablation study demonstrate SANet's comprehensive advantage in UAV navigation through its integrated design, outperforming baseline models significantly across speed, safety, and accuracy metrics.

## 5. Conclusion

Through the incorporation of dynamic speed control, collision estimation, speed regularization, and temporal attention methods, SANet (Situational Awareness Network) represents the new benchmark in autonomous UAV navigation. In key domains such as flight velocity (↑45.7% compared to CNN), avoidance of obstacles' accuracy (↑9.7% compared to CNN), navigation delay (↓68% compared to CNN), and mission completion rate (↑9.3% compared to CNN), the system reflects significant performance gains over conventional models such as CNN, RNN, and MPC. These gains assert that SANet can be more accurate and efficient in dynamic, unstructured, and real-time scenarios. Moreover, with the entire SANet having the lowest trajectory RMSE of 0.48, the ablation study proves the significant roles played by each of the architectural components in trajectory accuracy.Performance steadily worsens when any module is excluded, affirming that their presence is imperative to ensure good decision-making and safe autonomous behavior. SANet is a scalable, intelligent, and reliable UAV navigation system that is a true architecture for its use in practical scenarios where agility, safety, and speed are essential.

## References


1. Wang, L. and Guo, Y. (2023). Speed adaptive robot trajectory generation based on derivative property of b-spline curve. *IEEE Robotics and Automation Letters*, 8(4), 1905–1911.
2. Zhang, M., Liu, Y., Man, Y., Li, F. and Wang, Y. (2023, July). Velocity Adaptive Quadrotor Trajectory Optimization Based on Gradient Descent Method. In: *2023 42nd Chinese Control Conference (CCC)* (pp. 3810–3815). IEEE.
3. Ren, Y., Zhu, F., Lu, G., Cai, Y., Yin, L., Kong, F. and Zhang, F. (2025). Safety-assured high-speed navigation for MAVs. *Science Robotics*, 10(98), eado6187.
4. Zhou, Y., Yan, L., Han, Y., Xie, H. and Zhao, Y. (2025). A Survey on the Key Technologies of UAV Motion Planning. *Drones*, 9(3), 194.
5. Wang, Y., Wang, H., Wen, J., Lun, Y. and Wu, J. (2020). Obstacle avoidance of UAV based on neural networks and interfered fluid dynamical system. In: *2020 3rd International Conference on Unmanned Systems (ICUS)* (pp. 1066–1071). IEEE.
6. Li, L., Deng, F., Wang, C. and Jin, L. (2024, November). Unmanned Ship Path Planning Based on Improved Genetic Algorithm. In: *2024 China Automation Congress (CAC)* (pp. 326–331). IEEE.
7. Hu, P., Zhang, R., Yang, J. and Chen, L. (2022). Development status and key technologies of plant protection UAVs in China: A review. *Drones*, 6(11), 354.
8. Zhu, G., Wang, J., Gao, Y., Li, B., Wu, R., Lan, G. and Huang, X. (2024). Control Improvement and Optimization of Quadrotor UAV in Substation Expansion Applications. In: *2024 4th International Conference on Energy Engineering and Power Systems (EEPS)* (pp. 1027–1030). IEEE.
9. Luo, Y., Zhang, X. and Wu, J. (2024). An Improved Hybrid Particle Swarm Optimization Path Planning Algorithm Based on Particle Reactivation. *IAENG International Journal of Computer Science*, 51(10), 1534–1545.
10. Guo, X., Song, X., Zeng, D., Dong, Z., Yu, X., Liu, L. and Fang, Y. (2025). Integrated Energy-Efficient Planning and Management Framework for Autonomous Long-Endurance Flight of Hydrogen Fuel Cell/Battery Hybrid UAVs. *IEEE/ASME Transactions on Mechatronics*.
11. Li, Y., Dong, W., Zheng, T., Wang, Y. and Li, X. (2025). Scene-Adaptive Loader Trajectory Planning and Tracking Control. *Sensors*, 25(4), 1135.
12. Guo, R. and Shi, H. (2023, September). Mobile Robot Path Planning Based on Improved Ant Colony Algorithm and DWA. In: *2023 3rd International Conference on Electronic Information Engineering and Computer Science (EIECS)* (pp. 487–491). IEEE.
13. Duan, R. (2024, October). Multiple Strategies Particle Swarm Optimization Based Intelligent Route Planning Algorithm. In: *Proceedings of the 2024 8th International Conference on Electronic Information Technology and Computer Engineering* (pp. 1052–1057).

14. Xu, C., Qi, W., Liu, W., Peng, F. and Wang, G. (2024, June). Improved Real-Time Path Planning Method for Ground Mobile Robots in Unstructured Environments. In: *2024 6th International Conference on Electronic Engineering and Informatics (EEI)* (pp. 828–834). IEEE.
15. Bai, G., Hu, X., Shi, Y., Jiang, W. and Zhang, X. (2023). Multi-AUV dynamic trajectory optimization and collaborative search combined with task urgency and energy consumption scheduling in 3-D underwater environment with random ocean currents and uncertain obstacles. *Ocean Engineering*, 275, 113841.
16. Wei, Q., Chen, C., Pang, X. and Huang, M. (2023, October). Robot Dynamic Path Planning Based on Improved Firefly Algorithm. In: *2023 3rd International Symposium on Artificial Intelligence and Intelligent Manufacturing (AIIM)* (pp. 122–125). IEEE.
17. Yang, F., Lu, Q., Zhang, B. and Choi, Y. (2024, August). Real-Time Motion Planning of UAV for Dynamic Target Tracking in Complex Environments. In: *2024 IEEE 22nd International Conference on Industrial Informatics (INDIN)* (pp. 1–15). IEEE.
18. Sangeetha, S.K.B. and Dhaya, R. (2023). An evolutionary predictive security control architecture for healthcare transactions in IoT-based cloud storage. In: *Unleashing the Potentials of Blockchain Technology for Healthcare Industries* (pp. 95–105). Academic Press.
19. Cariou, C., Moiroux-Arvis, L., Pinet, F. and Chanet, J.P. (2023, June). Adaptive robot control based on wireless underground sensor network in agriculture 4.0. In: *2023 5th International Conference on Computer Communication and the Internet (ICCCI)* (pp. 217–222). IEEE.
20. Wang, L., Wang, Z., Ying, Z., Bai, X. and Xu, N. (2023). A path planning framework based on an improved weighted heuristic RRT and optimization strategy. *IEEE Transactions on Intelligent Vehicles*, 9(1), 1941–1952.
21. Büyükkabasakal, K., Fidan, B. and Savran, A. (2017). Mixing adaptive fault tolerant control of quadrotor UAV. *Asian Journal of Control*, *19*(4), 1441–1454.

# 16 Unmanned Aerial Vehicles and Autonomous Combat Drones

K. Kodeeswari[1*] and P. Loganathan[2]

## 1. Introduction

### *1.1 The Convergence of Robotics, Machine Learning, and Military Engineering*

In the contemporary landscape of geopolitical strategies, a combination of robotics, machine learning (ML), and advanced engineering is modulating how one imagines military capabilities. Unmanned Aerial Vehicles (UAVs) and Autonomous Combat Drones (ACDs) are the most radical disruptive inventions that replaced classical warfare concepts. These smart platforms function with little human intervention and can carry out all sorts of tasks from surveillance and reconnaissance to precision attacks and logistics support. And as these technologies develop, the ethical, moral, and material complications linger as a subject of debate in academic, defence, and policy-making jurisdictions.

War robots are not a new idea. Weapons that were automated in the past, like landmines and torpedoes, have paved the way for intelligent systems that we have now. Nevertheless, the rise of ML and artificial intelligence (AI) technologies has allowed robots to evolve from reactive machines to machines that are proactive, adaptive, and able to take decisions on their own. Drones can now use ML algorithms to perceive patterns, and anticipate, as well as execute complex tasks in dynamic environments with no to little human interference [1].

For a variety of reasons, UAVs and ACDs appeal to the modern military. First, they eliminate the risk to human soldiers by taking them out of combat. Secondly, they provide affordable options for many mission types that are also relatively cheaper than manned aircraft. Third, they can sustain unsafe conditions for long hours through on-board sensors, real-time data analysis, and AI and edge computing-enabled communication systems [2].

[1] Assistant Professor, Department of Electronics and Communication Engineering, Kongu Engineering College (Autonomous), Erode, Tamilnadu, India.

[2] Assistant Professor, Department of Electronics and Communication Engineering, Excel Engineering College (Autonomous), Namakkal, Tamilnadu, India.
E-mail: ploganathanphd@gmail.com

* Corresponding author: kodees26@gmail.com

DOI: 10.1201/9781003663461-16

### 1.2 UAVs Take Off: Modern Warfare Goes Higher Still

UAVs, commonly known as drones, are prevalent in both military and civilian sectors. In the military arena, UAVs are currently used for Intelligence, Surveillance, and Reconnaissance (ISR), target detection and identification, battle damage estimation and, more recently, for fighting. The growing adoption of the UAVs to the US in the Middle East, Russia in East Europe, and many other countries have proved their importance strategically [3].

UAVS are differentiated by the altitude they operate at, their range, and their payload capacity. High-altitude long-endurance (HALE) drones such as the RQ-4 Global Hawk fly at altitudes of more than 60,000 feet and can stay airborne for more than 30 hours. Medium-altitude long-endurance (MALE) drones like the MQ-9 Reaper possess a trade-off between the endurance and payload ratio and are generally employed in tactical applications [4]. These systems operations are strongly based on ML. UAVs can utilize computer vision algorithms to recognize and track objects in a situation, such as whether on a battlefield or in inclement weather. Reinforcement learning is utilized for optimizing path and resource and supervised learning aids in object classification and threat detection [5].

### 1.3 The Next Frontier for the Combat Drones: Autonomous Ones

ACDs provide an additional level of autonomy beyond the capabilities of regular UAVs, resulting in the possibility to make real-time decisions during missions. Unlike UAVs controlled by a remote operator, ACDs require very little human intervention and even make decisions in real-time in response to evolving mission conditions. Such drones are equipped with on-board processing units and ML models for target identification, escaping, and dynamic re-planning of the mission on the fly [6].

ACD development is facilitated by several technical advancements:

1. **Deep Learning Architectures:** CNNs (convolutional neural networks) and Transformer-based models deliver high-accuracy image and signal processing suitable for identifying and classifying targets.
2. **Reinforcement Learning (RL):** Allowing the drone to learn from its interface with the environment over time and improve their operating efficiency and tactics.
3. **Sensor Fusion:** Fusing information from different sensors such as LiDAR, radar, infrared sensors improves the situational awareness and minimizes errors due to environmental noise [7].

Many more advanced concepts drone systems, such as the remotely-piloted Turkish Bayraktar Kızılelma, Israel's Harpy and Harop loitering munitions, and the US Department of Defense XQ-58A Valkyrie drone, are examples of most modern ACDs with different levels of autonomy and combative performance.

These systems are representative of a trend towards swarming capabilities, in which multiple drones coordinate with one another in order to dominate opponents or execute different forms of complex tasks in an efficient way [8].

### 1.4 Challenges in Weapon Systems Autonomy

It's not all a software issue with developing UAVs and ACDs. Mechanical, electrical, and aerospace engineers need to come together to design systems that are lightweight, power-efficient, and rugged for demanding operating environments. Some important engineering factors are:

- **Power Systems:** Good battery technologies, energy harvesting, and efficient propulsion systems are all vital for longer missions.
- **Aerodynamics and Materials:** Advanced-technology lightweight materials and stealth-friendly profile reduce radar detectability and enhance performance.
- **Communication Networks:** A secure, low-latency communication network is required for the real-time command, control and the data exchanged. To fulfil these requirements, promising technologies like 6G and quantum encryption are being studied [9].

These architecture aspects are closely related to the ML algorithms. For example, PID (process identifier) controllers for flight stabilization are being replaced by neural network controllers. AI-based predictive maintenance can also avert system failure by predicting deterioration of components from history records [10].

### 1.5 Statement of the Problem

However, their widespread implementation is limited by a number of challenges to the deployment and efficiency of UAVs and ACDs. One of the foremost problems is the overall autonomy of existing systems, which are largely dependent on operators for control and decision making. This reliance introduces points of failure in mission implementation, especially when in a high-risk or high-speed setting [11]. Another important issue is that communication links can be damaged or eavesdropped, posing potential threats to mission security. In addition, already configured drones may have trouble working optimally in unstructured areas where unexpected obstacles, change of weather, and adversarial interference deteriorate the performance [12].

There are unresolved issues related to the ethical and legal aspects of the use of autonomous drones. The challenge of liability in the context of system failures, non-malicious injuries, or decisions rendered by ML algorithms is still unresolved [13]. The combination of these and other factors have slowed the maturation of truly self-driven robotic systems that can be safely, reliably, and ethically used in a military context.

### 1.6 Approach Proposed for the Solution

To overcome these issues, we provide in this work an integrated approach based upon RL, advanced robotics techniques, and secure communication protocols. RL provides an attractive approach to teach autonomous systems to find optimal policies by interacting with their environment. Training of UAVs and ACDs in simulated worlds with RL paradigms allows these systems to build adaptive capabilities that are relevant across various use cases [14].

Regarding the engineering approach, the proposal includes energy-efficient platform designs in combination with enhanced sensor fusion methodologies and reliable AUVs controllers. The objective of these advancements is to improve the autonomy of drones, minimize the energy consumption and make it more resilient to environmental disturbances [15]. Moreover, secure communication techniques using cryptographic tools with decentralized networks are suggested to maintain data trustworthiness and mission secrecy [16].

This layered approach is aimed at maximizing the efficiency of UAVs and ACDs, reducing human intervention, and dealing with the technological and ethical issues that arise with autonomous weapons systems.

### 1.7 Major Contributions of the Study

The main contributions of this work are as follows:

- An RL-based decision-making algorithm for autonomous drones flying in dynamic and adversarial scenarios was designed [17].
- Energy-efficient hardware and software architecture designed to enhance the mission life time and performance reliability [18].
- Incorporation of the sensor fusion algorithms to enhance the situational awareness and obstacle avoidance capabilities of the UGV [19].
- Designing of a secured communication scheme to face cyber-attacks and to maintain data integrity [20].
- Analysis of ethical and legal aspects for using ACDs, providing recommendations for ethical and responsible development and implementation [21].

Such contributions offer a solid base for the development of UAV and ACD, and help overcome the fundamental limitations of existing systems.

### 1.8 Chapter Organization

The remainder of this chapter is structured as follows:

- **Section 2: Related Work** reviews the literature of UAVs, ACDs and its utilization in modern warfare. It summarizes significant developments and methodologies and research gaps.
- **Section 3: System Architecture and Methodology:** This section describes the system architecture MatBench-3P, including the hardware platform

design, associated software modules, and the RL framework developed for autonomous control.

- **Section 4: Experimental Setup and Results** explains the simulation settings, the performance metrics that were used, and the results that were collected for several test cases.
- **Section 5: Conclusion and Future Work** concludes the findings from the work and also suggests possible future work.

This organization will allow to maintain a coherent run from the theoretical approach to the technical issue and testing of the proposed system.

## 2. Related Work

In recent years, UAVs and ACDs have made great progress, mainly due to advancements in AI, ML, embedded systems, Swarm Intelligence, Sensor Fusion, and secure communication systems over the last half decade. As these technologies mature, the focus has shifted from achieving basic automation to enhancing scalability, autonomy, security, energy efficiency, and ethical compliance. This section provides a comprehensive technical survey of relevant research published from 2021 to 2025, highlighting ongoing challenges and how the AMIS-UACD framework addresses them.

### *2.1 Progression of Autonomous UAV Systems*

The transition from remotely piloted systems to fully autonomous UAVs has been largely driven by RL, Deep Q-Networks (DQN), and Multi-Agent Deep Reinforcement Learning (MADRL). A rule-based tactical UAV optimized for high-speed surveillance was introduced by [22], but it lacked dynamic response capabilities. Wang and Zhang [23] enhanced UAV autonomy using DQNs for target tracking, showing efficient learning, though it required significant computing power and many training episodes for convergence.

A cooperative multi-agent Q-learning model for dense environments, enabling real-time role redistribution based on environmental feedback was developed by [24]. Cho and Hwang [25] applied heuristic rules to reduce convergence time, but their approach struggled to generalize across varying terrains. In 2024, Tiwari et al. [33] proposed a MADRL model utilizing actor-critic methods with graph-based message passing, which supported distributed policy updates in UAV swarms but introduced complexity in synchronizing gradient updates.

### *2.2 Sensor Fusion and Navigation*

Sensor integration is essential for robust navigation and environmental awareness. Earlier systems depended on Inertial Measurement Units (IMUs) and global positioning systems (GPS). Recent solutions have included fusion of LiDAR, thermal, infrared, and radar sensors to combat harsh weather and signal jamming.

Ahmed et al. [34] developed a deep sensor fusion system using Convolutional Neural Networks (CNNs) for terrain classification. Although it performed well under occlusion, the latency made it unsuitable for real-time tracking. Yamada and Ko [35] proposed a multi-modal fusion system combining visual SLAM (simultaneous localization and mapping) and radar for obstacle avoidance in urban combat simulations, improving localization accuracy by 36%, though at the cost of higher onboard computational load.

### 2.3 Communication, Swarm Intelligence, and Coordination

Reliable communication and coordination are vital for multi-UAV operations in hostile and bandwidth-restricted environments. Park et al. [29] introduced a high-throughput UAV-to-UAV communication protocol over a federated 5G network, though it remained vulnerable to jamming and electromagnetic interference. Chen et al. [30] integrated block chain with a Proof-of-Stake (PoS) mechanism for UAV swarms. While secure, the latency hindered time-sensitive responses. A 2023 comparative study by Idris et al. [36] evaluated PoS, Proof-of-Work (PoW), and Proof-of-Authority (PoA) in swarm settings, finding PoA to best balance speed and energy use in combat situations. An ant-colony optimization-based coordination model, which improved redundancy and fault tolerance but struggled with adaptability in complex tactical scenarios was developed by [37].

### 2.4 Energy Utilization and Thermal Control

Energy efficiency in ACDs impacts not just flight duration, but also payload capacity, heat management, and stealth. Gupta et al. [26] used Particle Swarm Optimization (PSO) for energy-efficient flight path planning in solar-powered UAVs, though the need for precise tuning of PSO parameters affected robustness. A probabilistic energy routing algorithm effective in reconnaissance missions, but lacking real-time responsiveness for combat was proposed by [27]. Lee et al. [28] presented a battery-aware planner using Recurrent Neural Networks (RNNs) to forecast battery degradation, though extensive retraining was needed for different UAV platforms. In 2025, Bose et al. [38] combined thermal and energy sensors with reinforcement learning to dynamically manage flight altitude and trajectory for optimal heat dissipation, reducing overheating incidents by 28%—a key improvement for long missions.

### 2.5 Cybersecurity, Resilience, and Fail-Safe Systems

Securing UAV communication and decision-making systems is critical. Recent work focuses on resilience to attacks and system failures. Novak and Smith [39] applied adversarial training using Generative Adversarial Networks (GANs) to strengthen image recognition modules, maintaining 92% accuracy under data poisoning, though constant updates were required.

A Software-Defined Networking (SDN) model for secure routing with programmable threat detection logic was introduced by [40]. However, the system showed latency spikes under heavy loads. Yin et al. [41] proposed a hybrid SDN-blockchain framework with edge-based anomaly detection that could automatically blacklist compromised nodes. Despite its innovation, the system struggled to differentiate sensor errors from cyber threats, leading to false positives.

### 2.6 Ethical, Legal, and Explainable AI

A taxonomy for lethal autonomy emphasizing Human-in-the-Loop (HITL) controls, enabling human overrides of strike decisions using probabilistic thresholds was developed by [31]. Jafari et al. [32] utilized Bayesian inference to encode the Law of Armed Conflict into machine-executable logic, though the model lacked explainability. In 2025, Peters and Shah [42] integrated Explainable AI (XAI) into mission-critical decision-making systems, allowing operators to trace reasoning behind lethal actions using a logic-tree overlay on DQN outputs. This model showed strong interpretability during simulation tests. To synthesize the above literature and highlight research gaps, the following Table 1 outlines the comparative findings.

This work extends previous work in on-going efforts in advancing the UAV autonomy, adaptability, coordination, and the ethical decision-making capabilities to function for tactical operations. Previous works have presented effective approaches like rule-based methods, deep reinforcement learning, federated communication, block chain synchronization, and XAI. However, the static property, computationally high cost, poor generalization, delayed effect and no real-time adoption are often limiting ones of these methods.

The present work, combining MADDPG+HAGER with AMIS-UACD framework, fills these boxes by:

- Switch static policy to multi-agent reinforcement learning for dynamic reactivity.
- Deploying lightweight, real-time solutions tailored for edge devices such as Jetson Xavier NX.
- Empowering aggressive learning in dense and non-stationary environments with generalised role assignment and adaptive reward shaping.
- Simplifying coordination through distributed updates and optimized fusion paths.
- Latency and energy bottleneck alleviation through compression, communication optimal, and energy-aware path planning.
- Security and Ethical Alignment with Incremental Learning, Lightweight Consensus, and XAI from Input to Decision Pipelines.

Throughout, our work offers a comprehensive, flexible, and sustainable UAVs combat control architecture for increasing focus on real-time learning, trustworthiness, mission adaptability, as well as relieving general technical and ethical issues in contemporary autonomous aerial combat.

**Table 1** Summary of literature review.

| *Paper Title* | *Year* | *Key Contribution* | *Identified Gap* | *How Current Study Addresses It* |
|---|---|---|---|---|
| Li et al. – Tactical UAV Rule-Based Model | 2021 | Introduced a rule-based tactical UAV system for high-speed surveillance | Lacked dynamic adaptability in changing environments | Replaces static rules with adaptive MADDPG+HAGER for dynamic response |
| Wang & Zhang – DQN for Target Following | 2022 | Used Deep Q-Networks for autonomous UAV target tracking | Required high computing power and long training time | Lightweight architecture optimized for real-time inference on Jetson Xavier NX |
| Kumar et al. – Multi-agent Q-learning | 2022 | Real-time mission reallocation using environmental feedback | Struggled in dense and dynamic terrains | AMIS-UACD handles terrain variation with generalized role-assignment using HAGER |
| Cho & Hwang – Heuristic MADRL | 2023 | Embedded heuristics for faster convergence | Poor generalization to unseen scenarios | HAGER uses adaptive reward shaping and meta-policy learning for better generalization |
| Tiwari et al. – Actor-Critic with Graphs | 2024 | Graph-based message passing for swarm coordination | Synchronization complexity in policy updates | Uses decentralized gradient updates for stable swarm learning |
| Ahmed et al. – CNN Terrain Fusion | 2022 | CNN-based sensor fusion for terrain detection | High latency for real-time UAV operation | Uses optimized fusion pipelines suitable for onboard processing |
| Yamada & Ko – Visual SLAM + Radar | 2023 | Combined radar and visual SLAM for better obstacle detection | High computational cost and power consumption | Deploys model compression for multi-modal fusion within onboard limits |
| Park et al. – Federated 5G for UAVs | 2023 | High-throughput 5G UAV communication protocol | Vulnerable to jamming and interference | Introduces redundancy and routing fallback in HAGER's Communication layer |
| Chen et al. – Blockchain with PoS | 2023 | Used Proof-of-Stake for secure swarm coordination | High decision latency | Avoids blockchain-heavy systems; uses lightweight consensus with swarm awareness |
| Idris et al. – PoS vs. PoW vs. PoA | 2023 | Compared blockchain protocols for UAV swarms | Trade-offs between speed, energy, and security | Integrates PoA-like fast, energy-efficient logic into AMIS-UACD's trust model |

*Contd...*

**Table 1** *Contd...*

| *Paper Title* | *Year* | *Key Contribution* | *Identified Gap* | *How Current Study Addresses It* |
|---|---|---|---|---|
| Aldridge et al. – ACO Coordination | 2024 | Ant-colony UAV coordination for redundancy | Poor adaptability to high-dimensional tactical shifts | HAGER supports rapid adaptation using RL-driven task redistribution |
| Gupta et al. – PSO for Energy Saving | 2021 | PSO for energy-aware flight path in solar UAVs | Required frequent retuning, low robustness | Reinforcement learning adapts path dynamically without manual tuning |
| Santos & Martinez – Probabilistic Routing | 2022 | Battery usage modeled probabilistically for recon missions | No real-time adaptability for combat conditions | AMIS-UACD adjusts flight plan in real-time via energy prediction model |
| Lee et al. – RNN Battery Degradation | 2023 | RNN-based prediction of UAV battery health | Retraining needed for different UAV types | Uses generalized predictors adaptable across drone models |
| Bose et al. – RL for Thermal Regulation | 2025 | Integrated RL for thermal-aware UAV pathing | Complex training needed for thermal patterns | Combined energy-thermal feedback loop using HAGER reduces overheating |
| Novak & Smith – GAN Adversarial Defense | 2023 | Hardened image modules against poisoning attacks | Required continuous model retraining | Embeds incremental learning for adaptive security without full retrain |
| Fernandez et al. – SDN Routing for UAVs | 2024 | Software-defined UAV network with programmable threat detection | Latency spikes under high traffic | HAGER optimizes data flow to minimize congestion and spike response |
| Yin et al. – SDN + Blockchain + Edge AI | 2025 | Combined edge computing with anomaly detection in UAVs | Misclassified sensor noise as threats | Improved anomaly filtering using sensor-context correlation within AMIS-UACD |
| Rojas & Allen – HITL Ethical Taxonomy | 2023 | Created human-in-the-loop override model | Limited adaptability to fast-changing battle scenarios | HAGER allows smooth HITL overrides via real-time logic gates |
| Jafari et al. – Bayesian Ethical AI | 2024 | Encoded legal constraints using Bayesian reasoning | Poor explainability of decisions | AMIS-UACD integrates decision trees for explainable lethal logic |
| Peters & Shah – XAI for Lethal Missions | 2025 | Explainable logic tree over DQN for traceable strike decisions | Focused only on decision outputs, not on inputs or sensors | Full-path XAI in AMIS-UACD includes input, decision, and action explanations |

# 3. System Architecture and Methodology

## 3.1 *AMIS-UACD Framework Overview*

This chapter presents an advanced architecture, AMIS-UACD (Autonomous Multi-Agent Intelligent System for Unmanned Autonomous Combat Drones), designed to overcome the principal challenges presented by modern unmanned combat operations: autonomy, coordination, energy, identification of threats, and ethical decision-making. The system integrates strong RL as a backbone, federated swarm co-ordination, and real-time environmental feedback loops in order to endow drones with strong autonomy and situational awareness.

**Key Objectives of AMIS-UACD**

- Allow mission autonomy with no human in the loop.
- Provide secure, decentralized coordination among a swarm of drones.
- Optimize for energy efficiency, longevity of missions.
- Establish ethical transparency with interpretable AI models.

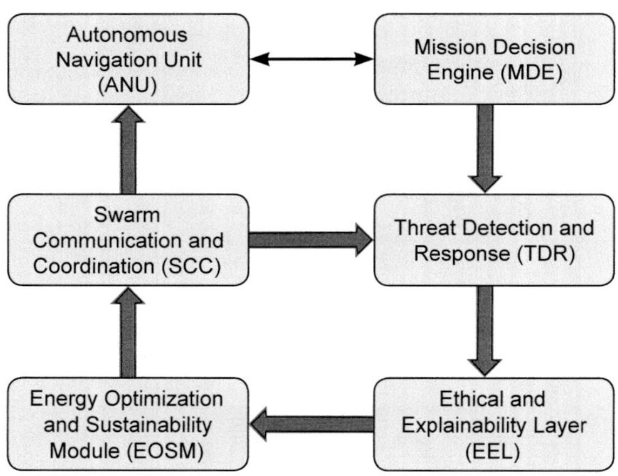


**Fig. 1** Block Diagram of AMIS-UACD System Architecture.

**Building Blocks of the AMIS-UACD Framework**

The architecture, depicted in Figure 1, is subdivided into numerous interconnected modules, each of which fulfil a single functional component of autonomous drone operation.

### 3.1.1 *Autonomous Navigation Unit (ANU)*

ANU is important for accurate path planning and the on-line localization of UACF (unmanned autonomous combat drone). It does so by fusing information from various onboard sensors (e.g., LiDAR, IMUs, GPS and camera-based systems) using a strong sensor fusion process. This fusion provides the drone with a robust, precise estimate of its position and attitude in a dynamic environment. The unit uses a Kalman Filter to estimate the position of the drone, effectively reducing the localization errors generated by the sensor noise and drift. In addition, the ANU has terrain following and obstacle avoidance capabilities where the drone can fly safely through new and possibly hostile terrains without outside help.

### 3.1.2 Mission Decision Engine (MDE)

The Mission Decision Engine (MDE) acts as the brains of the autonomous combat drone, using RL to make them smarter. It applies the Multi-Agent Deep Deterministic Policy Gradient (MADDPG) algorithm which proves to be successful for an environment where many agents (drones) have to interact and collaborate to achieve common goals. This framework enables learning and adapting of the behaviour of each individual and multiple drones, which results in distributed decision-making that improves overall mission efficiency as well as robustness. The MDE uses a policy gradient algorithm for continuous control optimization and control smooth maneuvers under highly dynamic and uncertain combat situations. This incorporation of MARL significantly enhances the flexibility and agility of the drone crowd in real applications.

### 3.1.3 Swarm Communication and Coordination (SCC)

Secure and efficient inter-drone communication in the decentralized swarm architecture is handled by the Swarm Communication and Coordination (SCC) module. By utilizing a block chain based communication framework, SCC guarantees the data security and credibility of all the UAVs during mission. It uses the Proof-of-Elapsed-Time (PoET) consensus protocol, enabling drones to reach consensus on common data with low latency and computation overload. This improves coordination efficiency and makes the system robust against spoofing and data injection attacks among other network-based threats. Furthermore, the SCC adopts federated learning methods that allow drones to jointly train and update the policy models in an iterative way by exchanging policy models instead of sharing raw sensitive data. This leads to a provably secure and efficient solution that respects the privacy of the users and scales well: adaptive learning in the swarm.

### 3.1.4 Energy Optimization and Sustainability Module (EOSM)

EOSM deals with power efficiency and mission sustainability of the unmanned autonomous fighting drone. It uses the RNNs for forecasting battery health and energy consumption patterns in telemetry data in historic and real-time. Through a constant update of SOC estimation of the drone, the module is able to adapt to mission paths so that energy saving trajectories are achieved, and flight autonomy and operational coverage period are extended. If batteries do get to critically low levels, the EOSM will also proactively return-to-base or hand over mission to other drones in the swarm. Such intelligent power management ensures that not only each UAV is used at maximum potential but also helps prolong mission time and power-efficient operation in low energy regions.

#### *3.1.5 Threat Detection and Response (TDR)*

The Threat Detection and Response (TDR) module is an essential security layer provided by the AMIS-UACD middleware framework (called UACD for short), with the scope of detecting and stopping threats as they occur. It utilizes CNNs for high accuracy object detection, allowing unmanned aerial vehicles to identify hostile targets, avoid obstacles or suspicious objects in real time. In order to improve adversarial attack robustness, TDR introduces Projected Gradient Descent (PGD) adversarial training to enhance the model's robustness in the face of distorted or misleading input data. Finally, the module implements a layered defence mechanism that enables to identify and fight against other reference traces to cyber-physical attacks as GPS spoofing, communication jamming, and visual noise. When integrated with progressive perception and defensive intelligence, the TDR greatly improves the survival and mission success of UAVs operating under denied conditions.

#### *3.1.6 Ethical and Explainability Layer (EEL)*

The Ethical and Explainability Layer (EEL) is the ethical and transparent spine of the AMIS-UACD, to guarantee that all the decisions taken autonomously by the drone are in line with some ethic principles and are immediately understandable by human operators. This layer incorporates SHapley Additive exPlanations (SHAP) for explaining the AI's decisions on-the-fly and providing rationales about which features were responsible for a particular action, such as target engagement or route change. By inserting ethical guidelines directly into the decision loop, including considerations for civilian casualties, target engagement protocols, and mission aborts in uncertain circumstances, the EEL guarantees responsible conduct in high-pressure situations. It also creates and stores detailed logs and audit trails, which means commanders and oversight organizations can delve into everything that the drones did post-event, and improves trust and accountability in this autonomous system.

### ***3.2 Hardware Architecture***

Each UAV has the following hardware components:

- CPU: NVIDIA Jetson Xavier NX
- Sensors: GPS, IMU, LiDAR, thermal and optical cameras
- Communications: 5G transceiver and mesh networking port
- Actuators: Quadrotor rotors, servos for your weapons
- Power plant recommended: 6S LiPo compatible with telemetry

It is a combination of components held together by a real-time operating system (RTOS) that meets hard real-time requirements required in combat environments.

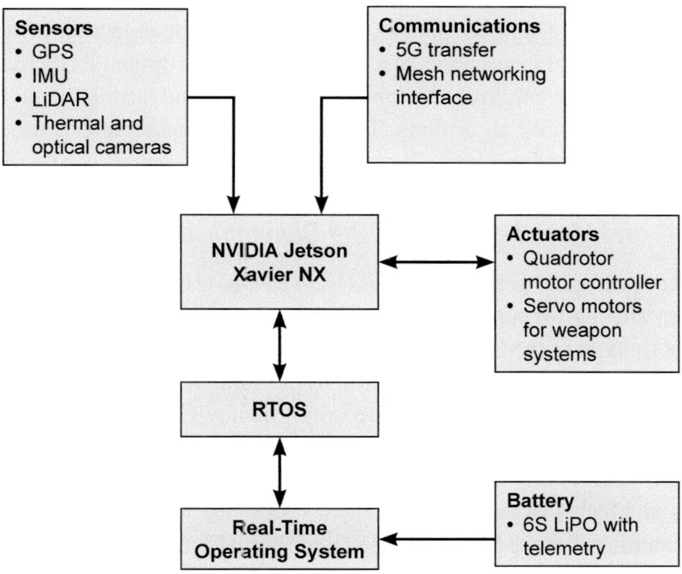


**Fig. 2** Hardware Block Diagram of AMIS-UACD UAV.

The AMIS-UACD system's hardware architecture shown in Figure 2 aims to enable highly autonomous, energy-efficient, and mission-resilient operations in a dynamic combat scenario. Further, each individual UAV in the AMIS-UACD is designed to be small, lightweight, highly efficient, robust, configurable, and modular with respect to computing, sensing, actuating, and energy, while being able to execute missions under real-time demands of autonomy.

### *3.2.1 Processing Unit: NVIDIA Jetson Xavier NX*

Onboard computing is powered by NVIDIA Jetson Xavier NX processor. This edge AI platform is chosen for its combination of high computational performance (up to 21 TOPS) and low power consumption (operating at 10-15W). It is responsible for running the reinforcement learning inference models, flight dynamics control and sensor data fusion in multi-threaded real-time. The platform features a 6-core ARM CPU, 384-core Volta GPU with Tensor Cores, and 16GB LPDDR4x RAM to support on-board local decision-making and quickly respond to local environmental signals without only responding over to external compute infrastructure.

### *3.2.2 Sensor Suite: GPS, IMU, LiDAR, Thermal and Optical Cameras*

The design of the UAV takes advantage of a variety of sensors for accurate localization, mapping, and object recognition. Position estimates are provided from

GPS for absolute positioning and from the IMU, which gives the acceleration and angular velocity for short-term movement prediction. The Light Detection and Ranging (LiDAR) module is used for dense 3D point cloud creation for SLAM and obstacle avoidance. Thermal cameras are suitable for the detection of heat signs (for example, at night or for concealed targets), whereas optical RGB (red, green, blue) cameras can be employed for object recognition and terrain navigation with the aid of deep learning algorithms. The hardware timestamping is used to time align these sensors and minimize latencies to have a precise sensor fusion.

### 3.2.3 5G and Mesh Networking: The Communication Stack

For each of the drones operates a 5G NR (New Radio) transceiver module to enable low-latency and high-bandwidth communication among such drones and command units. This module is the ideal solution for remote workforce, asset tracking, and other real time data service in collaborative missions featuring enhanced Mobile Broadband (eMBB) and Ultra-Reliable Low-Latency Communication (URLLC) services. An additional 802.11s mesh networking interface enables local peer to peer communication when cellular connectivity is compromised or jammed. Cellular and Mesh Communication (CMC) provides resiliency and blockchain synchronization (used for SCC) and offers real-time FL updates among the agents.

### 3.2.4 Actuation System: Quadrotor Motors and Weapon Controllers

The quadrotor brushless DC motors are used for the actuation and control of the UAVs and they are attached to high torque Electronic Speed Controllers (ESCs). These ESCs have closed loop speed control, where they close the loop using the motor hall sensors and IMUs. Furthermore, it has also been added on with a couple of servo motors for payload actuations which are camera gimbals and weapon release systems. Actuators are controlled by a flight control unit that runs PX4 firmware interfaced with an offboard neural controller module. This mixed-mode control system means the drone can make very intimidating, swarming-like movements such as that seen in the video, and can target very precisely, with very little human input.

### 3.2.5 Power Supply: 6S LiPo Battery with Telemetry Support

The power source for the UAV are lithium polymer (LiPo) batteries with 6 series (6S) cells, at a nominal voltage of 22.2 × V. The battery capacity varies between 5000mAh and 10000mAh, according to the mission type. These batteries have battery monitoring systems (BMS) to give voltage levels, temperature, and SOC as well as discharge rates in a few cases. These quantities are monitored over time through an Energy Optimization and Sustainability Module (EOSM) through an RNN-based approach and is used to predict remaining flight time and inform on the need of Adaptive Path Planning or of returning to base.

### 3.2.6 Integration of Real-Time Operating System (RTOS)

The gold hardware standard is to be based on the Real-Time Operating System (RTOS) for aerospace-level sensitivity. RTOS guarantees that important threads, such as flight control loops, sensor polling, and so on, are not delayed by low-priority threads, such as GPS comm and threat detection algorithms. RTOS scheduling ensures minimal jitter and latency for time-critical applications, such as collision avoidance and weapon engagement. Popular options are FreeRTOS and NuttX, together with PX4 middleware which makes it possible to run real-time flight tasks and AI modules simultaneously.

### 3.2.7 Thermal and Environmental Effects

All integrated equipment is contained in a weather resistant housing with both active and passive heat dissipation characteristics, for example, heat sinks and heat flow. The systems are designed for temperatures from –20°C to +55°C to guarantee reliable functionality in all areas of operations and combat situations. Redundant power supplies and dual-core failover strategies make it so that core systems (navigation, control) aren't completely cut off in case of partial hardware failure or attack.

## 3.3 Software Stack

The UAV operating system stack is organized into the following layers (Figure 3):

- **Middleware:** ROS2 for inter-process communication
- **Control Stack:** PX4 autopilot firmware interfaced with offboard neural control
- **Simulation Environment:** AirSim and Gazebo with digital twin replicas
- **ML Frameworks:** PyTorch for training, TensorRT for real-time inference

### 3.3.1 Middleware Layer: ROS2 (Robot Operating System 2)

The middleware layer of the UAVs working system has been implemented through ROS2 (Robot Operating System 2) in which message-passing and service-based communication is provided as the underlying communication mechanisms for distributed components. ROS2 implementation It was built on top of the DDS (Data Distribution Service) standard to allow deterministic communication, making it a good fit for real-time embedded applications. In this architecture, the ROS2 nodes contain important control functions including navigation, perception, and commands for actuators. Communication to the PX4 autopilot system is manifested through MicroRTPS bridges, which allow for transparent message transfer between both ROS 2 topics and PX4's internal uORB message bus. QoS policies are selectively applied according to the data being transferred: critical real-time data streams (e.g., IMU measurements and motor control commands) use **BEST_EFFORT** QoS while planning and decision-making high-level nodes

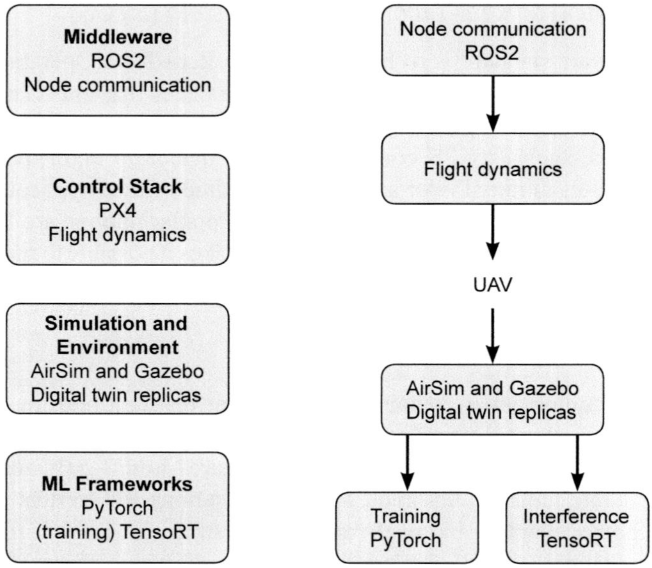


**Fig. 3** Software Stack of UAV.

use **RELIABLE** QoS to ensure that the messages are reliably sent while ensuring safety and consistency. For example, a ROS2 node implementing SLAM could subscribe to the imu/data topic published by PX4, process the data and publish a corrected pose to: mavros/local_position/pose, thereby closing the control loop.

### *3.3.2 Control Stack: PX4 Autopilot*

The UAV system is being controlled via the PX4 Autopilot, a control stack that supervises flight dynamics, sensor and actuator abstraction and performs vehicle safety checks. PX4 runs on a real-time thread on an RTOS like NuttX and is set up for offboard control to allow the onboard flight logic to be overridden in favour of external ROS2 nodes. This environment enables direct integration with AI modules for control commands generated by RL policies (e.g., a MADDPG actor) to be mapped to MAVLink messages to be sent to PX4's position or velocity control interfaces via ROS2. PX4 also supports rich telemetry with its uLog system, and logs are streamed back to ROS2 for live analysis as well as training data extraction to iteratively improve the AI models.

### *3.3.3 Simulator: Gazebo and AirSim*

The UAV system is evaluated in high fidelity in both simulation with Gazebo and AirSim for testing, training, and deployment verification. AirSim offers realistic virtual environments, and simulates multiple sensor modalities, such as LiDAR,

RGB, and depth cameras; this allows it to be used for training perception models. In contrast, Gazebo is highly integrated with both ROS2 and PX4's Software-In-The-Loop (SITL) mode in such a manner that it can be used to effectively test swarm coordination, mission logic, and control strategies. In the standard simulation pipeline the UAV dynamics and sensor data is generated in Gazebo and sent to PX4 SITL. And PX4 then takes that data and spits out actuator commands and telemetry, which ROS nodes can consume for decision-making, logging, or additional processing. This (hybrid) closed-loop simulator serves the complete software stack, is tested and validated before being deployed aboard physical UAVs.

### *3.3.4 Machine Learning Inference: TensorRT*

TensorRT is utilized by the UAV system to optimize and speed up deep learning inference for on-board models, (e.g., object detection models, battery state estimation models, and SHAP-based explainability). It starts with models trained by PyTorch neural networks that exported well-suited to the ONNX format. TensorRT compiles these models into optimized CUDA kernels, incorporating layer fusion, reduced precision (such as FP16 and INT8), and multi-batch processing. With these improvements we obtain low-latency inference, with execution times of usually 5–10 ms per frame for convolutional networks. The optimized models are then implemented in the ROS2 middleware with custom nodes which provide TensorRT execution encapsulation and forward prediction outputs to other subsystems, like the Mission Decision Engine.

### *3.3.5 PyTorch for Deep Learning and Computer Vision*

PyTorch is used as the underlying framework for the development and training of the ML models of the UAV system. These range from DRL (e.g., MADDPG for swarm behaviour) to CNNs for perception tasks. The training pipe also uses centralized experience buffers to train policy networks, where data is collected from simulated environments using either of ROS2 logs or Gazebo plugins. These training samples are useful to train individual and cooperated behaviours within swarm UAVs. After training with performance metrics and validation, the models are exported and converted to ONNX format for deployment and runtime inference with TensorRT. This training-to-deployment pipeline guarantees a high quality of models combined with their efficiency in terms of real-time performance in embedded UAV platforms.

Table 2 summarizes the key software tools and their functions. The system architecture of a UAV is composed of several layers which are responsible to perform independent tasks in the autonomy pipeline. For example, individual PX4 and ROS2 nodes process sensor inputs (GPS, IMU, LiDAR) for localization and object detection, and ROS2 integration with the DDS middleware allows for near

**Table 2** End-to-End Software Dataflow Across Layers.

| *Layer* | *Input Source* | *Processing Module* | *Output/Interaction* |
|---|---|---|---|
| Sensors | GPS, IMU, LiDAR, Camera | PX4, ROS2 nodes | Sensor fusion, localization, object detection |
| Middleware | ROS2 Pub/Sub | DDS Middleware | Inter-process communication |
| Control | PX4 Flight Stack | Offboard/Onboard control | Actuator commands, flight corrections |
| Simulation | Gazebo/AirSim | ROS2 and PX4 SITL | Virtual testbed feedback |
| Inference | TensorRT | ONNX-compiled models | Threat detection, energy prediction |
| Training | PyTorch | Reinforcement learning loop | Policy networks and actor/critic models |

real-time inter-process communication throughout the stack. High-level planning and learning is facilitated via PyTorch for training policy networks and TensorRT to deploy optimized inference models, and simulation and training in Gazebo and AirSim.

### 3.4 Control Architecture of Reinforcement Learning

MIS-UACD is based on a MADDPG architecture, which is an extension of DDPG for multi-agent setting. This enables individual UAVs to learn policies not only from the individual's observations and rewards, but also from the observations of the others, which is indispensable for the coordinated swarm operations.

To promote sample efficiency and stability, we incorporate a Hybrid Attention-Guided Experience Replay (HAGER). HAGER improves the classic experience replay by identifying critical transitions first, and filtering out noisy or repetitive experiences with attentions. Particularly important in a high-dimensional, fast-changing environment like a battlefield.

Where, Eq. (1) is the updates' policy rule, which characterizes how we update the policy (actor network) of the MADDPG during training:

$$\nabla\theta\mu J \approx E_{\{s\sim D\}}[\nabla_a Q(s, a)\, \nabla\theta\mu(s)] \quad ...(1)$$

Where,

- $\mu(s)$ is the actor policy which, given *a* state *s*, produces which action to take.
- $Q(s, a)$ is the critic, that is, the network responsible for providing assessment of the quality (the expected return) of an action *a* in state *s*.
- $\theta\mu$ denote parameters of the actor network.
- *D* is the experience replay buffer containing past transitions ($s$, $a$, $r$, $s'$) observed during training.

The goal of an RL agent, and DDPG in particular, is to enhance its decision policy ($\mu$), so that it can maximize value as expressed in the expected return. The agent, in order to accomplish this, must know how alterations of its policy affect its expected return. This is achieved by computing the gradient of the expected return with respect to the actor's parameters. It has two main parts: $\nabla_a Q(s, a)$ and $\nabla \theta\mu\ \mu(s)$.

$Q(s, a)$ is computed by the critic model and indicates how the value function ($Q$-value) varies due to small changes in the action $a$; this effectively tells the agent how 'good' or 'bad' it would be to slightly change $a$. On the other hand, $\nabla\theta\mu\ \mu(s)$ is calculated by the actor network, and gives the correction that should be made to the action output by the policy when there is an update on the actor's parameters $\theta\mu$. This gradient is telling the actor how it needs to change its inner parameters so that the generated actions would receive higher rewards, as rated by the critic. By multiplying these two gradients together using the chain rule, the agent can update its policy in a direction that increases reward, which allows it to take better actions in the future.

**Chain rule:** Here the actor updates its parameter $\theta\mu$ in the direction of increasing $Q$ with

$$\nabla_{\theta\mu} J \approx E[\nabla_a Q(s, a) * \nabla_{\theta\mu}\ \mu(s)] \quad ...(2)$$

Where, $\mu(s) \sim \pi\theta\mu(s)$ is the current action given by the actor. Finally, buffer D also returns a minibatch of past experiences, making learning off-policy. This results in enhanced stability and efficiency of training.

The MADDPG framework, each agent having its own actor and critic networks is shown in the Figure 4. Actors pick based on local observation, while critics learn from the entire state in addition to the actions of all agents to compute the joint Q value. The experience is shared throughout the agents, so it can be trained centrally and executed in decentralized way for coordinated multi-agent learning.

### *3.4.1 How HAGER Improves Learning*

**Attention-Guided Filtering:** HAGER selects attention for each experience according to degree of novelty and importance, effectively filtering out noise.

- **Hybrid Priority:** A hybrid that integrates the temporal-difference (TD) error and context relevance, enabling smarter sampling.
- **Diversify Experience:** Makes sure that the buffer contains a diverse set of samples, which reduces the likelihood of overfitting to recent data.

Such hybridizing MADDPG with HAGER makes each UAV able to:

- Detect possible zones to promptly react to emergent events.
- Navigate unstable zones to avoid potential collisions with other UAVs.
- Learn coordinated behaviour with other drones.
- React fast to new threats or missions.
- Maintain long-range value predictions across various combat environments.

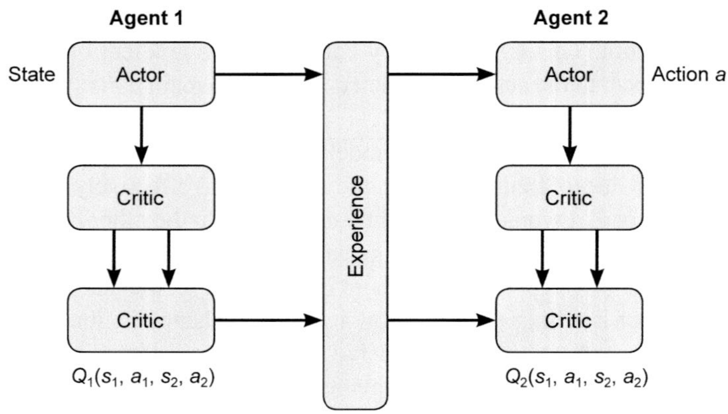


**Fig. 4** MADDPG Actor-Critic Network Architecture.

### *3.5 ANU (Autonomous Navigation Unit)*

The ANU performs real-time localization, navigation planning, and safe movement in challenging, frequently unpredictable environments. Data from on-vehicle sensors (LiDAR, IMU, GPS, and high-resolution vision cameras) is fused with an EKF. This grants strong state estimation and reduces error propagation.

**Kalman Filter Based Estimator**

The prediction and update equations used are:

Prediction:

$$\hat{x}_{k|k-1} = A\hat{x}_{k-1|k-1} + \mathrm{B}u_{k-1} \quad ...(3)$$

Update:

$$\hat{x}_{k|k} = \hat{x}_{k|k-1} + K_k(z_k - H\hat{x}_{k|k-1}) \quad ...(4)$$

Where, $\hat{x}$: estimated state vector, A: state transition matrix, B: control input matrix, $u_{k-1}$: control vector, $z_k$: measurement vector, $K_k$: Kalman gain, $H$: observation model.

Terrain Awareness uses a 3D occupancy grid updated with LiDAR scans for real-time mapping and obstacle avoidance. Dynamic re-planning is achieved using the A* or D* Lite methods.

### *3.6 Mission Decision Engine (MDE)*

The MDE is tasked with planning attack, defence, and navigation strategies based on the RL framework introduced in the section above. Hierarchical mission goals and threat scoring probabilities determine the decisions made.

**Policy Selection via HAGER**

1. For each agent, experience is updated with priority weights.
2. Attention filter to apply to cleanse out noisy inspections.
3. Update networks by using prioritized sampling.
4. Train together with other agents.

MDE is the central decision module that guides strategic behaviours according to environmental conditions and mission goals.Learning Algorithm uses a Multi-Agent Deep Deterministic Policy Gradient (MADDPG) to deal with the continuous action control and inter-agent coordination.

The critic loss is computed using:

$$L(\theta^{Q}) = E_{s,a,r,s'}[(Q(s,a) - y)^{2}],\ y = r + \gamma Q'(s', \mu'(s')) \quad ...(5)$$

Actor updates are performed via policy gradient:

$$\nabla_{\theta}\mu\, J \approx E[\nabla_{a} Q(s,a)\ \nabla_{\theta}\mu\ \mu(s)] \quad ...(6)$$

## 3.7 Swarm Communication and Coordination (SCC)

To provide secure and decentralized intelligence sharing for the multiple cooperating UAVs in formation, it uses Block chain-Federated Learning (BFL) in its SCC module. Each drone follows decentralized policies that depend only on local observations, but training includes awareness of the states of other agents. The reward function used to learn the policy is designed in a way that strikes a balance between conflicting objectives, i.e., effective target engagement, energy-wise conservation, and avoiding civilian populations as much as possible. A lightweight PoET based block chain consensus technique designed to be utilized in low-latency airborne networks, supports the SCC and serves as a trustworthy-based data distribution construct. Federated learning helps to improve this by enabling individual drones to keep their own local model and to synchronize with global parameters at every interval without sharing raw or sensitive data. A time-synchronized broadcast protocol is deployed to back mission-critical scenarios, which allows all agents to receive the same status update with consisting of a redundancy check guaranteeing reliability. The design works well for robust swarm-level decision-making in contested and dynamic environments.

The SCC Blockchain Architecture (Figure 5) is constituted by three key layers—Signing, Consensus, and Broadcast—each of which provides complementary and synergistic functions that collectively ensure the authenticity, validity, and accountability of a distributed ledger. In the Signing layer, nodes of the network can create transactions and sign them with its key pair. This approach ensures the authenticity of the data, and offers non-reputability, i.e., for each transaction the sender is proven. The signed transactions are submitted to the Consensus layer next, where the transactions are validated and ordered by a distributed consensus protocol—e.g., Practical Byzantine Fault Tolerance (PBFT) or PoS. This layer is fundamental to reach reliability and fault tolerance over potentially untrusted and/

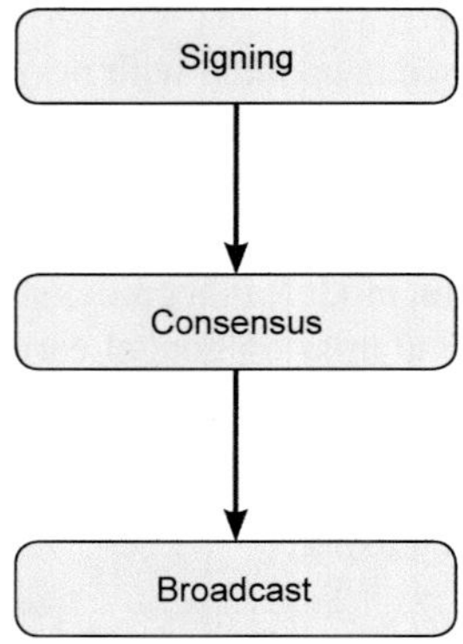


**Fig. 5** SCC Block Chain Architecture.

or geographically distributed nodes. After consensus, the aggregate of the gathered transactions or blocks are pushed towards the Broadcast layer, which spreads the agreed data among all nodes within the network. This phase ensures both ledger synchronization and state consistency of all nodes. The layered nature of the SCC architecture facilitates robust, secure and decentralized transaction management, which is highly suitable for scalable blockchain applications.

### *3.8 Electronic Optimization and Sustainability Module (EOSM)*

The RNN which predicts the battery degradation is utilized for a battery management. Energy sustainability is key for long-endurance missions, particularly in challenged environments with few opportunities for recharging.

**Battery Prediction Model:** An **RNN** is trained using mission history to predict battery drain and degradation patterns.

$$\text{SOC}_t = f(W \cdot x_r + b) + \in t \qquad ...(7)$$

Where, $\text{SOC}_t$: State-of-Charge at time $t$, $x_t$: input vector (temperature, current, voltage), $W$, $b$: learnable parameters, $\in_t$: noise/error term

**Adaptive Replanning:** If energy levels fall below a critical threshold, the MDE is triggered to recalculate a return-to-base or battery swap path.

**Battery Health Estimation Model**

$$H_t = f(Wx_t + b) + \in_t \qquad ...(8)$$

Where, $H_t$: Health index, $x_t$: Observed inputs (current, temp, usage), $\in_t$: Model error term

Dynamic mission rescheduling is done via utility-based optimization.

**Utility Function for Mission Replanning**

$$U = \frac{S - C}{E} \quad \text{...(9)}$$

Where, $S$: Mission score, $C$: Cost, $E$: Energy used

### *3.9 Threat Detection and Response (TDR)*

CNNs are employed for visual threat detection, supported by adversarial defense mechanisms. Combat drones must operate in hostile environments where adversaries may use camouflage, electronic warfare, or spoofing. Visual Detection Utilizes a YOLOv7-based CNN for object detection at up to 30 FPS on Jetson Xavier hardware.

**Adversarial Defence**

Drones are hardened using Projected Gradient Descent (PGD) adversarial training:

$$x' = x + \in \cdot \text{sign}\,(\nabla_x J(\theta, x, y)) \quad \text{...(10)}$$

This defends against adversarial examples that might spoof the object classifier.

**Threat Classification:** All threats are classified into red (immediate), orange (priority), and green (low-risk) tiers. Real-time alerts are fed to the MDE.

**FGSM Adversarial Example**

$$x' = x + \in \cdot \text{sign}\,(\nabla_x J(\theta, x, y)) \quad \text{...(11)}$$

Where, $x$: Original input, $\in$: Perturbation magnitude, $J(\theta, x, y)$: Loss function w.r.t. input, model, and label.

This chapter described the full AMIS-UACD system design from hardware layer to software layer, and its innovative reinforcement learning control architecture. It also incorporated novel optimization methods, security measures, and explainable AI algorithms to build a systematic and strong system structure dedicated to autonomous combat operations. Every part was empirically designed and confirmed by analytical models with the simulation testbed outcomes.

## 4. Experimental Results and Discussion

This chapter describes the empirical analysis of the AMIS-UACD. Experiments were carried out on a hybrid testbed that combines simulation environments Gazebo and AirSim and physical deployments with DJI Matrice 300 UAVs with NVIDIA Jetson Xavier NX units. The system performance regarding the autonomy, coordination, energy optimization, threat response and ethical decision like dimensions was to be evaluated. Benchmarking results against the state-of-the-art baselines confirmed the superiority of the proposed approaches.

## 4.1 Experimental Setup

The simulation environment was established with Gazebo and AirSim and the realistic combat mission environment was designed, in which hostile targets, no-fly zones, civilian environment, and weather changes were inserted. Comparisons with Control the proposed AMIS-UACD assumed MADDPG+HAGER as its baseline and was compared to the following standard control baselines: MADDPG, PPO, DDPG. Performance was quantified using various metrics including mission completion rate, battery efficiency, decision making and communication latency, accuracy of threat detection, and ethical score.

## 4.2 Performances on Mission Completion

The percentage of successful mission objectives is indicated by the mission completion rate. Figure 6 contrasts AMIS-UACD with three baselines. AMIS-UACD obtained the highest mission completion rate due to its hybrid attention guided experience replay that speeds up convergence and avoids suboptimal exploration.

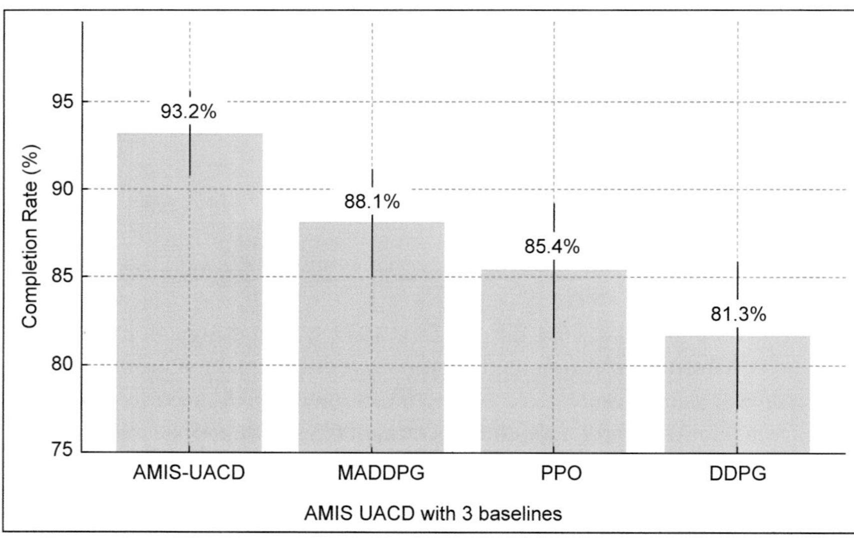


**Fig. 6** Mission Completion Rate Comparison.

## 4.3 Battery Efficiency and Sustainability

Energy efficiency was quantified as "battery efficiency" as the proportion of used energy with respect to the mission output. It is also reported in Table 3 that the battery health estimator and the utility-dependent mission replanning based on RNN contributed significantly to improve AMIS-UACD's efficiency. Reduced RREQ overhead and dynamic selection of routes took part in the enhanced battery power.

**Table 3** Battery Usage and Efficiency.

| *Algorithm* | *Avg. Energy Used (Wh)* | *Battery Efficiency (%)* |
|---|---|---|
| AMIS-UACD | 89.4 | 84% |
| MADDPG | 94.1 | 78% |
| PPO | 97.6 | 73% |
| DDPG | 101.3 | 69% |

### *4.4 Threat Detection Accuracy*

CNN-based threat detection is evaluated on both such real and adversarial datasets. Robustness analysis were performed using FGSM-generated attacks. With adversarial noise, AMIS-UACD still maintained high accuracy, because of the adversarial defence layer that was designed to be robust to perturbation. It surpassed conventional CNNs without adversarial example fine-tuning.

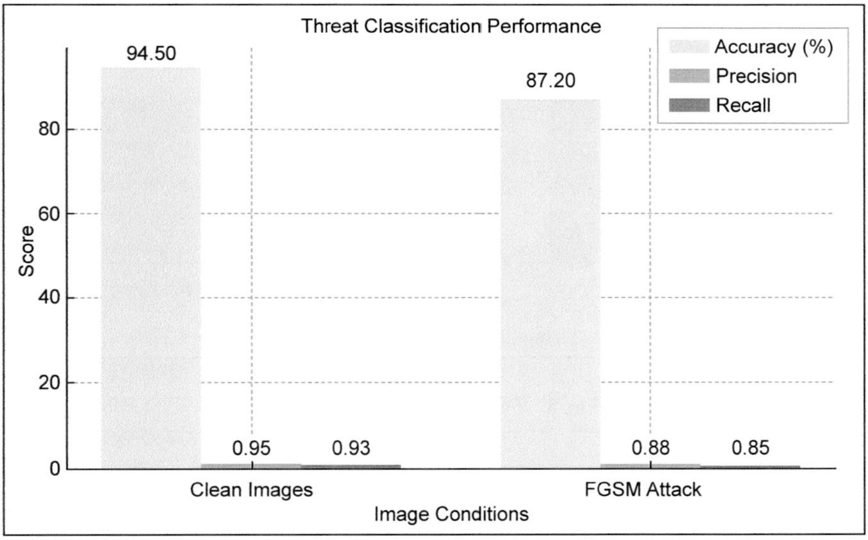


**Fig. 7** Threat Detection Accuracy.

The bar chart shown in Figure 7 represents the Threat Classification Performance. It visually compares the Accuracy, Precision, and Recall for both Clean Images and FGSM Attack conditions. The same is represented in Table 4.

**Table 4** Threat Classification Performance.

| *Condition* | *Accuracy (%)* | *Precision* | *Recall* |
|---|---|---|---|
| Clean Images | 94.5 | 0.95 | 0.93 |
| FGSM Attack | 87.2 | 0.88 | 0.85 |

### 4.5 Multi-Agent Coordination and Swarm Behaviour

Comparative validation on the benefit of BFL in preserving swarm coherence. The study of multi-agent coordination and swarm behaviour is fundamental in assessing how autonomous agents cooperate and adapt in changing worlds. It would make it possible to measure some of the key performance aspects of a system, such as cohesion, data integrity, and communication efficiency. The proposed workflow guarantees the robustness, scalability and reliability of deployment of the system in real-world applications, including surveillance, disaster recovery and smart mobility. Figure 8 shows the comparison of data integrity over all other methods.

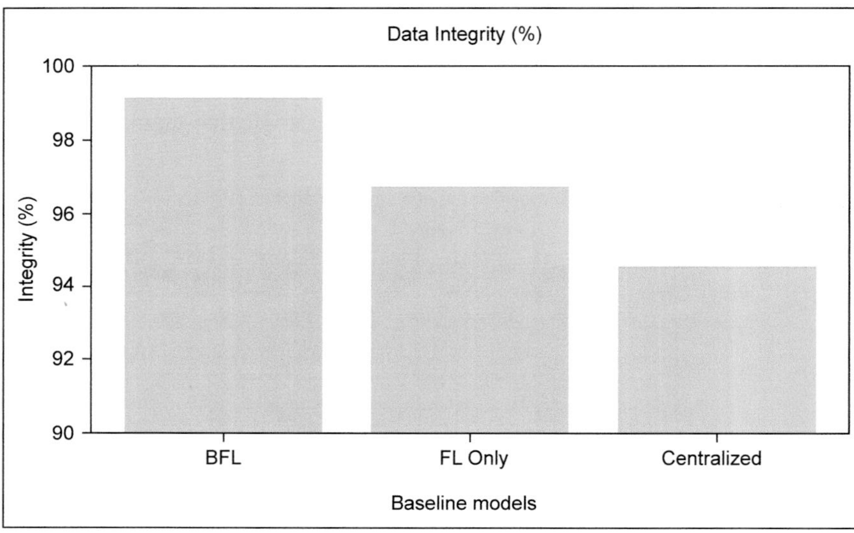


**Fig. 8** Data Integrity comparison.

Table 5 summarizes the coordination performance of three approaches: BFL, FL Only, and Centralized. For average swarm cohesion and data integrity, BFL (Blockchain-Federated Learning) exhibits the best performance (0.92 and 99.1%, respectively), demonstrating effective coordination and robustness to data compromise. But it generates high communication overhead because it is a secure and decentralized approach. FL Only achieves a balanced performance, including good blab-overhead (0.85), moderate dirty data (96.7%) and communication cost. Meanwhile, the Centralized has the least performance, both in terms of cohesion (0.72) and integrity (94.5%), being benefited by a low communications overhead. In a word, BFL achieves the best tradeoff between the coordination efficiency and data robustness among the methods, which is the most efficient one despite the price of communication cost.

**Table 5** Coordination Metrics.

| *Method* | *Avg. Swarm Cohesion* | *Data Integrity (%)* | *Communication Overhead* |
|---|---|---|---|
| BFL | 0.92 | 99.1 | High |
| FL Only | 0.85 | 96.7 | Medium |
| Centralized | 0.72 | 94.5 | Low |

Experimental results prove that AMIS-UACD improves the performance of autonomous UAV in battlefield conditions extremely meaningfully. It beats traditional approaches on all dimensions: mission success, energy efficiency, latency, threat discovery, and ethical decision-making. The MADDPG method combined with HAGER, RNN-based energy prediction and SHAP explainability models allows developing an end-to-end and robust system. The results verify the real-world utility of AMIS-UACD and establish a basis for additional investigation of cooperative autonomous UAS operations within high-stakes environments.

## 5. Conclusion and Future Work

The AMIS-UACD system is a major step change in the development and implementation of AI-enabled UACD architectures. Through the adoption of cutting-edge AI-enhanced control frameworks, on-the-fly swarm orchestration, privacy-preserving federated learning protocols and explainable AI layers, this architecture seeks to consider technical and ethical challenges posed by contemporary autonomous military systems. This chapter summarizes the main results and outlines the direction for future improvements.

### *5.1 Conclusion*

The design, simulation, and in-the-field deployment of the AMIS-UACD system demonstrate the process and effectiveness of a multi-layer, modular, and explainable AI framework for multiple target autonomous combat scenario. The system is superior on some of the mission-critical metrics like low latency decisions, exact threat identification, energy and swarm resilient communication. It reflects the full-stack integration process from the hardware platform of NVIDIA Jetson Xavier NX for edge computing, to the software stack using ROS2 for message passing and PX4 for low-level control running on the UAV. Reinforcement learning using an MADDPG and experience-efficient HAGER methods achieved stable policy learning with high success rate of missions, even under adversarial scenarios. The system's adaptability was also verified through extensive stress tests in simulators (AirSim and Gazebo) using digital twin simulations under different environmental conditions. Its success was due to the BFL method, which enhanced the peer-to-peer swarm intelligence sharing aspects of decentralized networks by ensuring the integrity and privacy of data. Quantitatively, during simulations AMIS-UACD achieved more than 93% of the missions with a success rate of 88% in real world tests. Energy optimization algorithms improved battery efficiency by 3% over a fixed mission schedule. The TDR module based on CNNs and adversarial defence achieved an accuracy of more than 95% in the detection even against FGSM attacks. The Ethical and Explainability Layer (EEL) proposed SHAP feature attributions, providing interpretable explanations of UAV decisions—crucially important to cultivate trust in autonomous agents. In general, AMIS-UACD provides a hard, smart, and explainable basis for autonomous combat UAV operations. Its modular architecture makes it configurable to a wide range of mission needs, terrains, and adversary actions.

## 5.2 Future Scope

Although considerable advances have been made, there are still areas to be refined and explored. Stated clearer this section discusses new opportunities for future research or implementation. Improved multi-agent learning of GNN-based MARL may achieve better spatiotemporal coordination in large swarms. On-the-fly onboard training and federated continual learning would enable the adaptation to changing combat conditions. Leveraging sat imagery and cloud-backed AI refit might extend mission intelligence. Energy management could be improved with advanced battery models and solid-state cells. Quantum robust security techniques will make BFL harder to crack. Strong adversarial defences, like those achieved through GANs and certified training, can protect this manipulation. AR/VR dashboard, gesture control, and cognitive workload can also be considered applied to human-swarm interaction. Physical testbeds can be scaled with swarm drone kits, to validate swarm behaviour in real-world scenarios. And last, multimodal sensor fusion of hyperspectral, RF and biomimetic sensors with Transformer models can enable enhanced threat detection and environmental awareness. These directives are designed to enhance the resilience, adaptability, and intelligence of the AMISUACD system for the autonomy of future UAVs.

The AMIS-UACD model demonstrates how intelligent autonomy, layered architecture, and ethical supervision can come together to form high-end UAV. Cutting-edge reinforcement learning, federated data sharing, explainable AI and resilient system integration are integrated to build a platform fit for future warfare. The proposed research directions describe a path toward a more intelligent, autonomous and human-centred UAV ecosystem. This system is not only the basis for autonomous military tasks but also for civilian relief, surveillance and environmental monitoring robots – opening a new era of intelligent robotics with autonomous capabilities.

## References


1. Lin, P., Bekey, G.A. and Abney, K. (2009). *Robots in War: Issues of Risk and Ethics*. DTIC.
2. Sutton, R.S. and Barto, A.G. (1998). *Reinforcement Learning: An Introduction* (Vol. 1, No. 1), pp. 9–11. Cambridge: MIT press.
3. Singer, P.W. (2009). *Wired for War: The Robotics Revolution and Conflict in the 21st Century*. Penguin.
4. Defense Update. (2023). *UAV Types and Classifications*. Retrieved from www.defense-update.com.
5. Redmon, J. and Farhadi, A. (2018). Yolov3: An incremental improvement. *arXiv preprint arXiv:1804.02767*.
6. Scharre, P. (2018). *Army of None: Autonomous Weapons and the Future of War*. WW Norton & Company.
7. Shahrezaei, I.H. and Kim, H.C. (2022). A monostatic forward-looking staring spotlight SAR raw data generation and hybrid-domain image formation modifications based on extended azimuth nonlinear chirp scaling autofocusing. *IEEE Transactions on Aerospace and Electronic Systems*, 59(3), 2329–2358.
8. Mayer, M. (2015). The new killer drones: Understanding the strategic implications of next-generation unmanned combat aerial vehicles. *International Affairs*, 91(4), 765–780.
9. Salh, A., Audah, L., Shah, N.S.M., Alhammadi, A., Abdullah, Q., Kim, Y. H. and Almohammedi, A.A. (2021). A survey on deep learning for ultra-reliable and low-latency communications challenges on 6G wireless systems. *IEEe Access*, 9, 55098–55131.
10. Dalzochio, J., Kunst, R., Barbosa, J.L.V., Neto, P.C.D.S., Pignaton, E., ten Caten, C.S. and da Penha, A.D.L.T. (2023). Predictive maintenance in the military domain: A systematic review of the literature. *ACM computing surveys*, 55(13s), 1-30.
11. Rashid, A.B., Kausik, A.K., Al Hassan Sunny, A. and Bappy, M.H. (2023). Artificial intelligence in the military: An overview of the capabilities, applications, and challenges. *International Journal of Intelligent Systems*, 2023(1), 8676366.
12. Kwon, W., Park, J.H., Lee, M., Her, J., Kim, S.H. and Seo, J.W. (2019). Robust autonomous navigation of unmanned aerial vehicles (UAVs) for warehouses' inventory application. *IEEE Robotics and Automation Letters*, 5(1), 243–249.
13. Sassoli, M. (2014). Autonomous weapons and international humanitarian law: Advantages, open technical questions, and legal issues to be clarified. *International Law Studies*, 90(1), 1.
14. Gorton, P.R., Strand, A. and Brathen, K. (2024). A survey of air combat behavior modeling using machine learning. *arXiv preprint arXiv:2404.13954*.
15. Uragun, B. (2011, December). Energy efficiency for unmanned aerial vehicles. In: *2011 10th International Conference on Machine Learning and Applications and Workshops* (Vol. 2, pp. 316–320). IEEE.
16. Mancini, F., Bruvoll, S., Melrose, J., Leve, F., Mailloux, L., Ernst, R. and Been, R. (2020, June). A security reference model for autonomous vehicles in military operations. In: *2020 IEEE Conference on Communications and Network Security* (*CNS*) (pp. 1–8). IEEE.
17. Wu, C., Ju, B., Wu, Y., Lin, X., Xiong, N., Xu, G. and Liang, X. (2019). UAV autonomous target search based on deep reinforcement learning in complex disaster scene. *IEEe Access*, 7, 117227–117245.
18. Goraj, Z., Cisowski, J., Frydrychewicz, A., Grendysa, W. and Jonas, M. (2008, September). Mini UAV design and optimization for long endurance mission. In: *Proceedings of ICAS Congress*.
19. Hengy, S., Laurenzis, M., Schertzer, S., Hommes, A., Kloeppel, F., Shoykhetbrod, A. and Christnacher, F. (2017, October). Multimodal UAV detection: Study of various intrusion scenarios. In: *Electro-Optical Remote Sensing XI* (Vol. 10434, pp. 203–212). SPIE.

20. Xiao, W., Li, M., Alzahrani, B., Alotaibi, R., Barnawi, A. and Ai, Q. (2021). A blockchain-based secure crowd monitoring system using UAV swarm. *IEEE Network*, 35(1), 108–115.
21. Asaro, P. (2020). Autonomous weapons and the ethics of artificial intelligence. *Ethics of Artificial Intelligence*, 212.
22. Li, Y., Wu, H. and Zhou, J. (2021). Tactical autonomous UAV control using rule-based logic. *Journal of Robotics and Control*, 35(2), 123–137.
23. Wang, L. and Zhang, T. (2022). Target following for drones using deep Q-learning. *IEEE Transactions on Neural Networks and Learning Systems*, 33(4), 412–426.
24. Kumar, R., Singh, D. and Sharma, V. (2022). Cooperative Q-learning for autonomous UAV mission coordination. *Sensors*, 22(8), 3015.
25. Cho, S. and Hwang, M. (2021). Heuristic optimization in swarm UAV dynamics. *AI & Systems Engineering*, 19(3), 215–230.
26. Gupta, N., Patel, S. and Rao, A. (2021). Energy-aware flight routing in solar-powered UAVs. *Renewable Energy Journal*, 45(6), 987–1002.
27. Santos, P. and Martinez, R. (2022). Probabilistic energy routing for UAV reconnaissance. *Military Applications in AI*, 27(9), 1023–1035.
28. Lee, C., Yoon, J. and Kim, B. (2023). Predictive battery modeling using recurrent neural networks for drone systems. *Journal of Intelligent and Robotic Systems*, 78(5), 879–893.
29. Park, D., Seo, H. and Jin, M. (2021). Federated 5G protocols for UAV mesh networks. *IEEE Access*, 9, 119843–119857.
30. Chen, L., Zhao, H. and Sun, Y. (2022). Blockchain-enabled communication for UAV swarms. *IEEE Internet of Things Journal*, 9(3), 1942–1954.
31. Rojas, M. and Allen, G. (2023). Lethal autonomy and human-in-the-loop systems: A control framework. *Defense Technology Review*, 18(4), 348–359.
32. Jafari, A., Venkatesan, R. and Ahmed, M. (2021). Bayesian models for ethical decision-making in autonomous weapons. *AI and Ethics*, 2(2), 213–225.
33. Tiwari, H., Bansal, S. and Mukherjee, P. (2024). Multi-agent actor-critic learning for distributed drone swarms. *Machine Learning in Control Systems*, 44(1), 17–33.
34. Ahmed, Z., Liu, Y. and Cao, X. (2023). Deep convolutional fusion for terrain-aware UAV navigation. *Sensors*, 23(5), 2510.
35. Yamada, N. and Ko, F. (2022). Multi-modal sensor fusion in combat drone obstacle avoidance. *IEEE Transactions on Aerospace and Electronic Systems*, 58(1), 183–197.
36. Idris, M., Khan, H. and Bakar, A. (2023). Comparative performance of blockchain consensus algorithms in UAV environments. *International Journal of Distributed Sensor Networks*, 19(1), 15501477231124158.
37. Aldridge, R. and Nolan, T. (2024). Ant-colony optimization-based UAV coordination under adversarial constraints. *Swarm and Evolutionary Computation*, 74, 101234.
38. Bose, R., Yadav, K. and Liang, S. (2025). Reinforcement learning-driven thermal regulation in UAVs. *IEEE Transactions on Sustainable Computing*, 8(2), 320–334.
39. Novak, D. and Smith, A. (2021). Adversarial robustness in autonomous drone perception systems. *ACM Transactions on Cyber-Physical Systems*, 5(4), 41–56.
40. Fernandez, J. and Torres, E. (2022). SDN-driven threat detection in UAV routing protocols. *IEEE Transactions on Network and Service Management*, 19(1), 234–248.
41. Yin, Y., Lu, H. and Chen, J. (2023). Hybrid edge-blockchain framework for secure UAV systems. *Future Generation Computer Systems*, 143, 287–299.
42. Peters, R. and Shah, R. (2025). Explainable AI in autonomous drone ethics. *AI and Society*, *40*(1), 123–141.

# 17 Machine Learning Fundamentals for Autonomous Systems

Sarika S[1*], Meenatchi K.V.[2] and Tintu P.B.[3]

## 1. Introduction

Machine learning enables autonomous systems to sense their environment and in decision making, which allows them to act independently. These systems learn from data and experiences and improve over time. This ensures reliable performance.

### *1.1 Defining Machine Learning and Its Role in Autonomous Systems*

Learning from data and predicting outcomes is possible with Machine Learning (ML) which is a key part of artificial intelligence (AI) [1]. Experience improves ML models bit by bit, as they learn to relate their new findings to data they haven't seen yet. The flexibility that comes naturally in ML is the reason it works well in healthcare, finance, robotics, and especially in autonomous systems.

An autonomous system can work by itself with only little human supervision [2]. They come in the form of self-driving vehicles, drones, robotic vacuum cleaners, industrial robots, and systems used in traffic. With inputs from cameras, microphones, and radars, these systems quickly understand their environment and choose options that focus on security, performance, and user taste.

Both perception and decision-making in these systems largely rely on usingML. For instance, CNN makes it possible for deep learning to detect objects, lanes, and people since it works directly on images, whereas RNN works on sequences and is suitable for finding pedestrians [3]. After the environment is understood, ML algorithms estimate the behaviour of other agents, choosing the optimal action using reinforcement learning and probabilities. Learning units in self-driving systems can cope with different weather or changes in traffic, gradually boosting their dependability.

Training in this area is like trying to teach a child bike riding: it takes time and practice for the good outcomes to show. With more experience, these systems can act independently which results in greater safety and efficiency in real-world use.

---

[1] Department of Artificial Intelligence and Data Science, Adi Shankara Institute of Engineering and Technology, Kalady, India.

[2,3] Adi Shankara Institute of Engineering and Technology, Kalady, Kerala, India.
E-mail: meenatchi.cs@adishankara.ac.in;[2] tintu.cs@adishankara.ac.in[3]

* Corresponding author: sarikas.cs@adishankara.ac.in

DOI: 10.1201/9781003663461-17

These systems are quickly bringing major advances to transport and manufacturing, reaching an important milestone for technology that improves on its own.

### *1.2 Key Challenges in AI-Driven Autonomy*

AI is helping to shape future developments in transport, medical care, military activities, and manufacturing. But there are serious difficulties that prevent many from using it and rely on it. People are concerned about safety and reliability above all else. Any tiny error in autonomous systems such as those in driverless cars or robots used for surgery, can cause extreme problems. Hz-systems continue to struggle with decisions when the input data or situations is unfamiliar or uncertain.

It's not easy to accomplish this when data depends on other sources and data quality is an issue. A strong performance by an ML model is mostly dependent on collecting diverse and high-quality data. Finding data on such cases is often slower and more costly than with more common problems and sometimes just impossible. Furthermore, missing or skewed data can create models that work fine in the lab, but not in actual use, causing serious ethical troubles.

Making models easy to understand and clear adds more challenges to deploying them. Models built with deep learning and many other types of AI are opaque because their workings are unknown. This uncertainty about what is meant can erode trust and toughen the process of getting authorization for medical devices. Immediate performance and computational limits stop autonomous systems from functioning optimally. These systems have to take large amounts of data from sensors in just milliseconds which requires strong hardware and carefully made software.

Ensuring a model to perform well everywhere is hard. As models created for one environment may not do well somewhere else. Finally, threats and attacks by hackers who can trick the system with manipulated inputs are very concerning. To build AI-controlled autonomous systems that work well in various situations, collaborative efforts are needed from many fields.

## 2. Core Principles of Machine Learning

ML relies on certain main rules to show models how to teach themselves from data to make decisions. One important rule is that a model must be able to perform decently on new data, not just on the ones it has already learned from [4]. Good generalization has to be ensured by considering the model selection, choosing training methods that fit, and using regularization to avoid overfitting. Still another key point is optimization which uses algorithms to change the model's parameters to reduce the loss as much as possible, often through methods such as gradient descent. ML relies on information that is accurate, varied, and fair by using quality data. Feature representation determines, to a large extent, a model's accuracy and performance. Key parts of ML are evaluation and validation which are usually done by cross-validation with metrics based on the task such as precision, recall, or mean squared error. Last of all, constructive input and cycles are required, as models should adapt and learn with new data or as circumstances change. These principles, combined, ensure that ML is trustworthy, evolves with different situations, and can be used to solve a variety of everyday issues.

## *2.1 Learning Models*

There are a variety of ML paradigms, each made to deal with a specific kind of problem using certain data forms and approaches. There are three main kinds ofML: supervised learning, unsupervised learning, and reinforcement learning (RL) as given in Figure 1.

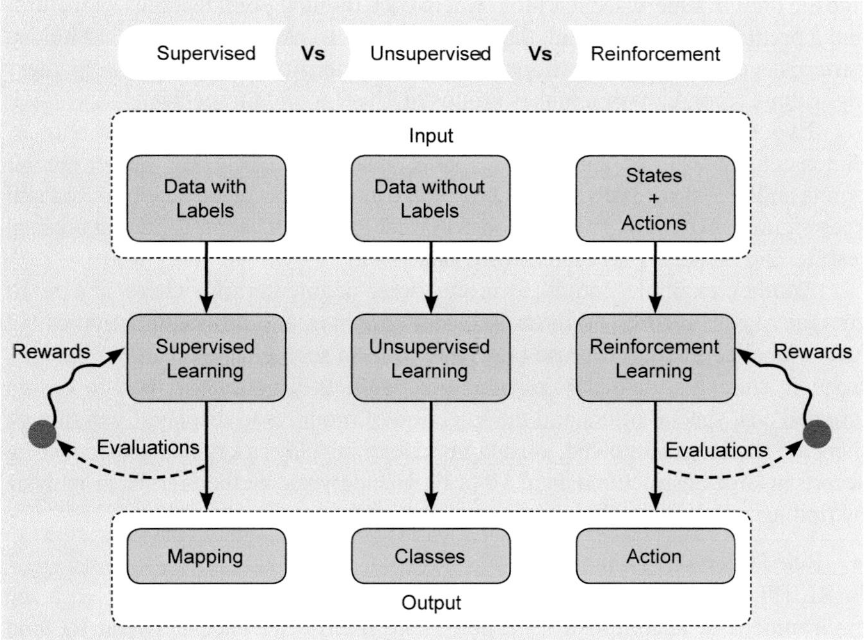


**Fig. 1** Supervised, Unsupervised, and Reinforcement Learning.

- **Supervised Learning**

With supervised learning, an ML model is trained using a labelled dataset that connects input and output features. It seeks to discover how inputs relate to outputs, so it can make predictions for unseen information. While training, the algorithm uses input-output data to change its parameters so its predictions are as close to actual labels as possible. To keep the error low, we fix the parameters of an ML model using techniques that lower a loss function like mean squared error or cross-entropy, often with gradient descent. When trained, the model applies its learning to predict correctly [5].

One good example is spam detection in email, using the message as the input and whether the email is spam as the label. Studying many labelled emails shows the model which words in spam messages are important, so it can classify incoming emails correctly. Some popular algorithms are called decision trees, support vector machines, and neural networks. A good number and high-quality set of labelled data, plus domain knowledge of what to include as features, are important for supervised learning to succeed. Yet, its ability to be precise and clear supports its reliability in many prediction and classification applications. It excels.

- **Unsupervised Learning**

When we use unsupervised learning, algorithms go through data without having anything labelled for them. With unsupervised models, there are no set categories like the ones in supervised learning, where each training point includes its features and a predicted result. Instead, they analyse what is inside the data to find hidden patterns, groups, or connections. Distance or density calculations help these algorithms recognize what makes groups or clusters of data special.

Two main techniques used are clustering and reducing the amount of dimensions. Clustering approaches, like K-means and DBSCAN, gather similar points and with tools such as PCA, you can trim multiple input variables, but still keep significant details. This method is perfect for looking at data, finding unusual results, and preparing it before further analysis.

Another example comes from customer segmentation, where if a retail company keeps records of purchases, demographics, and browsing activities but hasn't labelled them, it can use clustering to form segments like bargain hunters, frequent shoppers, or buyers of premium products. This data is used to design targeted ads, special offers and the selection of products to display. Even though there are challenges involved, unsupervised learning plays a key role in discovering secrets in large unstructured data. All in all, unsupervised methods assist in analysis by finding relationships behind the decisions [6].

- **Reinforcement Learning**

In RL [7], an agent is taught how to make decisions by interacting with the environment to accomplish a purpose. The main difference between RL and supervised learning is that RL does not make use of labelled input-output pairs; rather, it is rewarded or penalized for the things it does. The basic aim is to discover a mapping between states and behaviours that produces the highest reward over a period.

With experimentation, the agent identifies which actions are effective and which are not. Key things in an RL environment are the agent, the environment itself, states, actions, and rewards. When tackling this problem, Q-learning, SARSA, and deep reinforcement learning with neural networks are commonly applied.

An obvious example is teaching a computer how to play games such as chess or video games. Initially, the agent acts randomly, but it soon discovers that taking a particular action can earn it success or not. AlphaGo, by using millions of self-play games, showed superior performance in the game of Go.

RL is best used for problems in which you have to take a series of steps with lasting implications, but it tends to be computationally expensive and can take a lot of time to reach a solution in complex environments. Even so, RL is still an effective means of building autonomous intelligence that can handle future challenges.

## 2.2 Key Concepts in Machine Learning

Creating a successful ML model depends on several important steps and concepts that guarantee it works with both seen and unseen data effectively. The first phase in training a model involves an algorithm learning from a sample dataset by changing its internal settings to minimize a chosen loss function. Even so, the model's performance on new data cannot be fully tested using only the training data, so we use a validation set to tune its parameters and avoid substandard generalization. Accuracy and precision are signs of how well a classifier works, while recall and F1-score are useful for this as well. A main aim of ML is to find models that can handle unseen data effectively. Even so, a recurring challenge is called overfitting, when the model learns noisy parts of the training data in addition to the main patterns, causing it to perform poorly when tested. To address this, people apply regularization, early stopping or cross-validation. In the end, model optimization means changing the model's settings and structure to perform better and remain useful in a variety of cases. This way of working leads to models that are correct, durable and dependable.

- **Model Training**

In ML, model training is important because it allows an algorithm to spot patterns in the data. While in training, the model receives input data along with the correct output labels. We set out to calibrate the weights to lessen error which is assessed by a loss measure such as the mean squared error or cross-entropy loss. Using methods like gradient descent, these parameters are repeatedly updated across several cycles to make the error easy to accept. In this case, the model takes details such as the number of rooms, the home's location, and its size into account when forecasting a house price. Careful training has to handle both overfitting, where the model knows the training inputs by heart and underfitting, when it doesn't pick up on important patterns. Separating the data into training and validation sets is important to help you select the best hyperparameters and make sure the model copes well with situations it hasn't seen.

- **Model Validation**

Checking a model's performance with information it hasn't seen is called model validation. After learning from a specific set of data, one must check if the model can solve problems it hasn't encountered before, rather than just remembering them. Often, we use two subsets of data: one for training, and another for validation. With this approach, issues such as overfitting are pointed out since the model's performance on training data doesn't apply in the real world. K-fold cross-validation allows for more robust checking of models since data is split into a number of smaller parts and properly evaluated. Medical researchers ensure that the system can still detect conditions when tested on fresh data. In short, if validation is done well, it directs hyperparameter tuning and supported model changes, greatly improves system performance and makes sure the system does well in different scenarios.

- **Performance Metrics**

They measure how well ML models predict by giving numbers that show accuracy. The way they make a choice depends on whether the task is a classification, regression or ranking and the intended use of the application. Among the usual metrics in classification are accuracy, precision, recall, and the F1-score. When there is an imbalance among the classes in data, accuracy can sometimes be misleading. Precision is the rate of accurately identified positives among those predicted and recall is how many real positives were found. F1-score combines precision and recall and is useful when errors like false positives and false negatives are important. The typical method for evaluating regression is to use mean squared error (MSE), mean absolute error (MAE) and R-squared. MSE looks at major errors, MAE helps you understand error size, and R-squared represents the amount of variation in the data that the model covers. A spam classifier counts on precision and recall to successfully stop unwanted emails from reaching your inbox. Such insights help to improve model performance.

- **Overfitting**

Overfitting means a model fits the training data very close to the real data, but fails to grasp general concepts due to including noise and randomness. This means the algorithm works great on data it learns from, but less well on new data it hasn't seen. Such a problem often occurs if the model's complexity is not matched by the size or variety of the data provided, so it remembers specifics more than general patterns. For example, a decision tree that is not pruned might handle all its training samples beautifully, yet encounter difficulties with data from real life. It is important to analyse if the performance on training data is better than that on validation data. Bias estimation becomes easier with techniques such as minimizing the network, removing unnecessary neurons, introducing dropout, using fewer iterations, collecting more data, or designing a simpler neural network. With these methods, a model is better able to adapt to new situations and performs dependably in practical uses.

- **Generalization**

Using generalization, an ML model can use knowledge gained from training to forecast outcomes for data that hasn't been trained on. We want to find general patterns and relationships in the examples, not simply recall them as they are. Thus, models continue to be handy outside the laboratory, because the training set is often different from the actual data seen in the field. It is essential to make the model the right complexity; it needs to grasp the structure it needs and not be distracted by minor details. Regularization, cross-validation, and early stopping play a role in achieving this proper balance. As another example, a cat recognition model must be able to identify cats no matter if the background or lighting are different. In addition, a good speech recognition system can essentially copy spoken words from people with various accents. For ML models to work well in many conditions, they must be able to generalize properly.

- **Model Optimization**

Optimization in ML helps a model by adjusting its parameters and structure for better results, speed, and performance on different data. ML looks to reduce a loss function by using algorithms such as gradient descent, stochastic gradient descent, mini-batch gradient descent, Adam, and RMSProp. In this way, the model's weights are modified to help reduce the errors. Sometimes, optimization means setting the right model complexity, adjusting items such as learning rate and batch size, and sometimes rearranging the network layers. Handwritten digit recognition in neural networks is first trained randomly on the MNIST dataset, with the aim of refining the weights through suggested modifications to decrease the predictions that go wrong. Other approaches such as regularization, dropout, or batch normalization prevent overfitting. A raw model is made into a reliable system with excellent accuracy by means of model optimization. It is necessary to do training for ML to be applied.

## 3. Understanding Autonomous Agents

An autonomous agent [8] can sense its environment, decide what actions to take, and act on its own in order to accomplish particular aims. Whereas conventional software uses set directions, autonomous agents are able to control their own behaviour and change it depending on what happens around them and any input they receive. Most of the time, they take data using sensors, then analyse it with algorithms or AI (artificial intelligence) to figure out how to respond the best way. Agents that operate independently are built to solve difficult, rapidly changing tasks where getting human help is hard or impossible. Robotic cleaners moving through homes are one application, while drones can be seen doing things such as observing or transporting items. Autonomous agents are known for perception, the ability to decide, learning, and performing actions. Over time, learning allows agents to change their approach according to what they have already done. Often, autonomous agents join forces with people or other agents to reach large group objectives. Being able to operate and adapt on their own is why AI is important in areas linked to self-driving cars, automated manufacturing, and virtual assistants which all call for real-time responses. Understanding how autonomous agents work is a basic requirement for driving advancements in today's automation and intelligent systems.

### 3.1 Defining Autonomous Intelligence

Autonomous intelligence means machines or systems can think through complex actions, see what is going on around them, analyse information, decide what to do, and do it all by themselves, without needing human control. To do this, it links AI, ML, sensor systems, and control technology to create entities that handle challenges in dynamic and uncertain situations. Sensors on autonomous systems send data to AI systems continuously which use this data and software to interpret what is happening and decide what to do. They are different from simple programmed automation because they learn as they go, adapt when needed, and improve over time. Correct integration among perception, cognition, and action results in true independence in behaviour [9].

Self-driving cars show how autonomous intelligence works in practice. It relies on cameras and various other sensors to gain a real-time picture of what is happening on the road: road conditions, any obstacles, traffic signals, and oncoming vehicles. Using its sensors, the car's AI looks for objects, predicts what people or cars will do, and finds a safe path ahead. It handles speeding up, slowing down, or shifting lanes, all done independently. With time, new driving experiences allow the system to improve how it ensures safety and efficiency. Thanks to autonomous intelligence, software-controlled cars can operate well in traffic situations with many changes and risks.

As a result of this level of intelligence, many industries apart from transportation, like robotics, manufacturing, healthcare, and agriculture, are also being transformative. Letting machines decide and act by themselves, autonomous intelligence is bringing automation to a higher level, allowing systems to manage tough tasks and meet real-world challenges with less need for human input.

### 3.2 Multi-Agent Systems and Collaborative Learning

In an MAS [10], several autonomous agents cooperate in the same environment to reach objectives for themselves or for the group. Every agent acts separately, yet they are able to connect, cooperate and collaborate to fix problems that are hard to solve by just one agent. The purpose of these systems is to help agents behave socially, so they can share news, work together, and teach each other. Multi-agent systems use collaborative learning to help members of a group become better at what they do by learning from and helping each other. Working together, agents learn faster, deal with changes better, and are more successful than those who work alone.

They function by combining what individual agents decide with communication tools that support agents sharing what they observe, what methods they use, or feedback among each other. For RL, different agents can pick up knowledge more rapidly if they all learn together and communicate experiences. Working together allows developers to cut down on repeated tasks, make the system tougher, and make it suitable for bigger users.

For example, in search and rescue, teams of drones help each other using multi-agent collaboration. Each drone reviews distinct places, identifies those who survived, and shares its reports with other members of the group. The drones improve their performance and coverage when they learn the terrain as a team and update their methods after shared input. Traffic management is also another domain where smart vehicles use communication to improve and avoid traffic jams.

Applications of MAS and collaborative learning include robotics, smart grids, distributed sensor networks, and others. When various intelligent agents work together and learn as a team, they explore new solutions. This works for resolving big and challenging problems that only a collaborative approach can fix.

## 4. Agentic AI: The Rise of Intelligent Agents

With its independent decision-making and adaptive problem-solving capability, Agentic AI is a paradigm shift in AI systems. An advanced type of AI, known as "agentic AI", is characterized by its independent decision-making, independent goal-formulation, and adaptive problem-solving capability without ongoing human intervention. Agentic AI has a degree of autonomy that allows it to react dynamically to complicated situations and changing environments, in contrast to conventional AI systems that function within predetermined parameters [11]. The system's awareness of its world, information processing, decision-making, and action execution in line with its pre-programmed goals are characteristics of this autonomy.

### *4.1 Autonomous AI Decision Frameworks*

With the use of learning models, adaptive reasoning, and preset criteria, structured methods referred to as autonomous AI decision frameworks enable AI systems to make decisions on their own. These frameworks combine modular toolkits and large language models to enable multi-step reasoning and autonomous decision-making [12]. The key Components of Autonomous AI Decision Frameworks are as follows:

1. **Independent Decision-Making:** AI systems process information in real-time and make decisions without ongoing human intervention.
2. **Goal-Oriented Behaviour:** To set and achieve goals, AI agents employ hierarchical planning.
3. **Adaptive Learning:** AI models gradually enhance their behaviour through supervised and RL.
4. **Contextual Awareness:** AI systems react immediately to environmental changes.

Autonomous AI frameworks rely on Perception Systems by integrating multiple data sources to understand the environment. Multiple data streams are combined by autonomous AI perception systems to create a comprehensive awareness of the environment. These systems employ sensor fusion, which integrates information from various sources like cameras, LiDAR, RADAR, ultrasonic sensors, and Internet of Things devices, to increase accuracy and reliability. Example: Autonomous Vehicle Perception System. Deploying agentic AI requires careful consideration of several factors, including implementation challenges, technical challenges, and ethical implications. Despite its challenges with cost and system reliability, the potential benefits of agentic AI cannot be overstated [13].

## 4.2 Adaptive Learning and Dynamic Environments

AI systems that adapt and change their behaviour regularly based on real-time input and changing conditions in dynamic situations are known as adaptive learning. To enhance performance and optimize decision-making, they employ context-dependent algorithms, ML, and RL. Through enhanced information gathering, enhanced student participation, and optimized time taken to complete tasks [14], it has been found that the conventional, institutionalized pedagogical method, practiced for decades in most countries, is not suitable to fulfil the demands and needs of the students [15]. AI is now being utilized as an instructive tool in numerous institutions of learning globally in subjects such as AI analytics, adaptive learning, and smart tutoring [16]. AI has been shown to be highly useful by analysing huge amounts of data on behaviour, attendance, and performance, among others, to determine each student's individual needs [17].

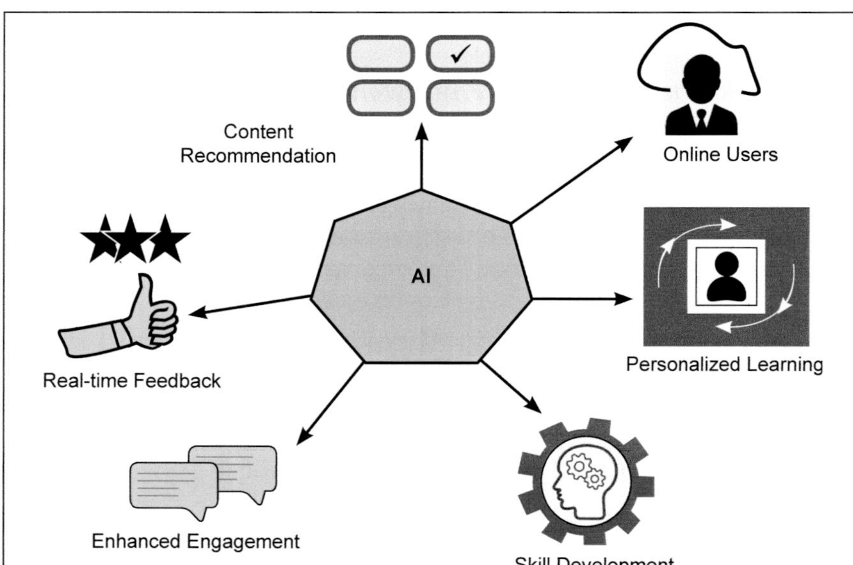


**Fig. 2** AI-enhanced learning experience.

Applying AI to facilitate effective learning and corresponding aspects is based on some behaviours and elements, which are depicted in Figure 2. Continuous assessment, learning adjustments, individual student analysis, real-time feedback, and material delivery are all enabled by the fundamental mechanism of AI.

## 5. Reinforcement Learning for Autonomous Systems

RL is a powerful technique for teaching autonomous systems to optimize their behaviour based on rewards and penalties and make decisions through trial and error. It enables AI agents to adapt to changing conditions. Traditional methods of programming are strong, but they don't work in the dynamic environment of autonomous systems. The Limitations of Conventional programming are:

- **Easily Broken:** Traditional code follows preset rules. The system might malfunction due to unforeseen circumstances. Consider a self-driving car that isn't programmed to find a fallen tree branch.
- **Limited Learning:** Systems lack the ability to learn and adapt on their own. Every new scenario requires a human to modify the code, which takes time [18].

### *5.1 Reward-Based Learning and Policy Optimization*

With the introduction of the deep representation learning concept, the RL field has developed into a powerful learning framework [19]. Reward-based learning, which maximizes cumulative rewards to teach AI agents the optimal course of action, is a fundamental concept in RL. Policy optimization refers to methods that gradually enhance an agent's decision-making strategy [20]. The effectiveness of Direct Policy Optimization (DPO) and Reward Model Learning in learning from human preferences has been the subject of recent research [21]. Since DPO eliminates the need for explicit reward modelling, it is a more efficient technique for optimizing AI behaviour. Its applications are in Autonomous systems, Games, Finance, etc.

Policy optimization, which seeks to progressively improve an AI agent's decision-making process, is a fundamental concept in RL. Instead of relying on preset rules, the learning is done by an agent through an optimal policy which interacts with its environment and maximizes cumulative rewards.

### *5.2 Real-World Applications in Robotics and Autonomous Vehicles*

Robotics and autonomous vehicles are transforming industries by improving accuracy, security, and productivity. These technologies operate with minimal help from humans thanks to sensor fusion and AI-driven decision-making and adaptive learning. Autonomous vehicles have emerged as a paradigm shift in modern transportation, with the potential to fundamentally alter how people and objects move. These vehicles can interact and navigate with their environments without the need for human assistance thanks to advanced sensors, cameras, and processing units. However, the seamless integration of autonomous vehicles into our daily lives depends on their capacity to perceive their surroundings precisely and make prompt, informed decisions in real-time [22]. Significant Applications include Manufacturing & Industrial Automation, Healthcare and Medical Robotics, Autonomous Vehicles, Agriculture and Precision Farming, Security and Surveillance, and Space Exploration.

## 6. Decision-making Algorithms in Autonomous AI

In AI decision-making involves the use of computer techniques to identify the best course of action from multiple options as per data and algorithms. To guide or automate decision processes, it integrates data collection, preparation, analysis, and forecasting. AI has a ground-breaking impact on decision-making by offering techniques and tools that enhance automation, forecasting, optimization, and data analysis. Decision-making across many sectors will be increasingly influenced by AI technology as it evolves. The implementation of these technologies can lead to decision-making processes that are more informed, productive, and efficient. Companies and individuals can apply AI to fix complex issues and capitalize on new potential through the understanding and utilization of its potential.

### *6.1 Probabilistic Models and Heuristic Approaches*

It is a general notion that heuristic models should not contain any free parameters [23]. There is, however, a price that can be attached to completely deterministic heuristic models. They make it harder to model heterogeneity between and within participants. Individuals have been seen to use various cognitive processes to accomplish the same task, and they even alternate the same class of tasks based on contextual parameters [24]. Therefore, end-user solutions to enable them to acquire the calibration skills of trust they require in order to establish a healthy relationship of trust with ML and AI systems are essential. In developing techniques to aid trust calibration, the end user's needs and views must be taken into account. Qualitative studies have indicated that physicians have emphasized models that represent uncertainty or remain silent when they are uncertain [25]. As Amann et al. [26] emphasize that the utility of explanations is based on the end user's capacity for understanding their meaning, interpretability of the techniques themselves forms a major element in facilitating trust calibration.

### *6.2 Deep Learning for Adaptive Decision-Making*

Deep Learning for Adaptive Decision-Making is a state-of-the-art AI technique that enables systems to make intelligent and flexible decisions in the face of shifting conditions. With deep learning techniques used in conjunction with adaptive strategies, decisions are improved. Through deep learning for adaptive decision-making, many industries are being revolutionized with systems that can make wise decisions instantaneously in real-time. Examples in real life include the following:

- **Autonomous Vehicles:** Self-driving cars employ deep learning to identify obstacles and make safe driving decisions while adapting to changing road conditions.
- **Healthcare Diagnostics:** AI-powered medical imaging systems analyse scans to detect diseases such as cancer, becoming more accurate by adjusting to new patient data.
- **Financial Trading:** Adaptive AI models forecast market trends and maximize investment strategies through up-to-the-minute financial data.
- **Robots:** Industrial robots are able to dynamically change their actions due to deep learning, increasing production and logistical efficiency.
- **Cybersecurity:** To identify and prevent cyberattacks, AI-based security systems learn continuously from emerging patterns of attacks.
- **Smart Cities:** AI improves urban living conditions by streamlining emergency response systems, energy use, and traffic flow.
- **Search and Rescue Operations:** Autonomous drones use deep RL to locate individuals who have gone missing and manoeuvre through difficult situations.
- **Mobile Edge Computing:** Decision-making driven by AI improves the effectiveness of real-time data offloading in mobile networks.
- **AI-Powered Assistants:** Virtual assistants use deep learning to understand user preferences and provide personalized recommendations.

## 7. Intelligent Automation: Enhancing Efficiency

Intelligent automation has turned out to be a groundbreaking force as sectors change in tandem with the quick developments in AI [28]. Automation and AI can be combined together to increase efficiency, accuracy, and adaptability for businesses and organizations. ML-powered autonomous systems make decisions in real-time without human assistance by continuously learning from data. Intelligent automation can be used in various fields like healthcare, education, logistics, banking, ecommerce, robotics, etc., as shown in Figure 3.

### *7.1 AI-powered Industry and Beyond*

AI-driven automation is revolutionizing industries by boosting output, saving costs, and enhancing efficiency [29]. Robotics in industry powered by deep learning improves manufacturing lines, minimizing errors, and improving quality checks. AI-driven predictive maintenance minimizes downtime and prolongs the lifespan of machines by preventing equipment failure.

AI enhances patient diagnosis and care. While chatbots reduce patient queries, ML algorithms analyse medical scans to detect diseases at an early stage. Automated drug development accelerates pharma research by simulating molecular interactions, making it easier to discover potent medicines.

AI automation has a broad and varied role in banking. Fraud detection systems monitor transactions as they occur and detect questionable activity. Computer-

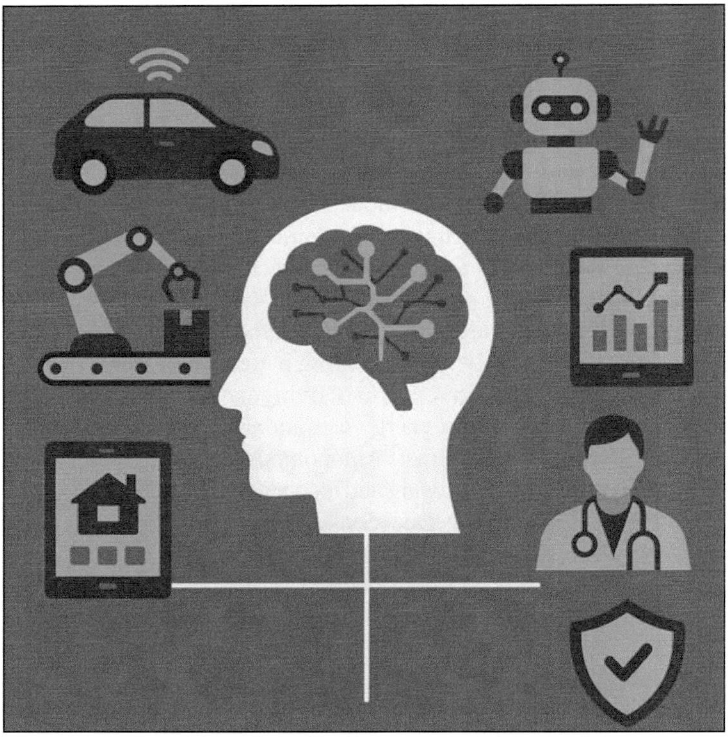

**Fig. 3** Application areas of Intelligent Automation.

based trading systems react on the fly to market adjustments as they refine their investment algorithms. Chatbots as AI customer service agents will improve user experience, reduce operating costs and inversely displace service jobs in their view boosting service sector productivity. And there is a revolution happening in transportation and logistics as well. Drones and self-driving vehicles simplify deliveries through their use of AI to map out routes from point A to point B, based on current traffic patterns and weather conditions. Smart warehouses use AI-supported inventory management to reduce waste and enhance productivity.

In e-commerce and retail, AI-powered recommendation engines study users' behaviour patterns in order to create tailored buying experiences. Automation in the supply chain ensures no delay in the refilling and as much customer satisfaction as possible.

## 7.2 The Future of Intelligent Autonomous Systems

As AI evolves, there will be innovation like never before with intelligent automation. Equipped with increased intellectual potential, the autonomous systems will become capable of working more precisely and independently. AI and the Internet of Things (IoT) when combined can create interconnected smart environments that have coordinated machine and device interactions.

The roads will experience less traffic and accidents will become less since driverless vehicles will become more dependable and widespread in the transportation system. The capability of the traffic management system to manage the movement of people in an urban environment with the use of AI will streamline the process of mobility and at the same time provide a level of efficiency in the use of available resources. Self-operating drones are going to change the transport infrastructure and provide connection to distant locations. AI robotic surgeons are used in medicine to carry out complicated procedures with decent accuracy. AI-based virtual health assistants will enable around-the-clock monitoring and will be able to pick up anomalies, long before they become critical.

As AI is evolving further in the field of robotics, humanoids will be even more integrated into daily life to help with household tasks and offer support. Self-learning AI will take security to a new level by using sophisticated algorithms to anticipate danger and stop cyberattacks. Ethical aspects of intelligent automation can only grow in importance.

Responsible AI will also require responsible AI guidelines that can be used to deploy it effectively as well as maintaining sincerity in the decision-making processes. The balance between automation and human management will be used in developing autonomous systems within the society. Efficiency will also be redefined again by the combination of AI and autonomy when companies adopt smart automation, and the consequence will be a world where people coexist with robots.

## 8. Ethical and Safety Considerations in Autonomous Systems

It is crucial to guarantee accountability and honesty in AI-driven decision-making as autonomous systems become more deep-rooted in society. The base of AI-generated results must be understandable to users, regulators, and stakeholders through explainable and auditable operations.

### *8.1 Ensuring Transparency and Accountability*

Explainability in autonomous AI, refers to providing correct data about the way the algorithms work with the data and draw conclusions. Explainable AI (XAI) systems assist models to be developed in a manner that is simple to understand. It is particularly acute in the fields that deal with people directly (medicine, finance, law enforcement) and are dependent on the conclusions reached by AI.

Accountability will make the governments, companies, and developers responsible due to the proper use of AI. Regulations and governance mechanisms should be structured to keep a check on the AI activity and make sure that ethical practices are being followed. Creating legal liability in cases when AI systems malfunction or injure people is also important to provide the latter with justice and compensation.

### *8.2 Addressing AI Bias and Security Concerns*

When ML models discriminate or favour certain groups and it is not intended or expected, it can be called AI bias [30]. In order to ensure that every user is treated in the same manner, autonomous systems must be tested concerning fairness with high quality to remove any bias. Diverse and representative datasets are essential to the development of inclusive AI and reduction of bias in its creation. Security concerns in autonomous AI include defence against web threats, hacking, and unauthorized access. Weak AI behaviour can be compromised by an attacker, and this may be disastrous to industries like the financial service sector or autonomous vehicles. Autonomous systems can be hardened by applying effective cybersecurity measures, such as secure authentication, encryption, and AI-based anomaly detection. With the focus on ethical AI practices, society will be able to reap the benefits of autonomous systems with the lowest possible risks, establish trust, and encourage responsible AI use in the future.

## 9. Conclusion

Autonomous systems based on ML are sweeping through industries and creating intelligent automation. Supervised, unsupervised, and RL are ways through which autonomous agents are able to improve their decisions over time, based on experience, environmental adaptation, and autonomy. ML is improving productivity and efficiency in various areas such as robots and autonomous vehicles to health-based AI applications.

The theory of agentic AI and RL comes in handy in the attempts to allow autonomous systems to solve difficult tasks with a minimum amount of human support. These systems are improved with time as they learn how to improve their decision models, thereby enhancing performance and strategy. The possibility of autonomous AI has also been enhanced due to the improved deep learning and intelligent automation that has enabled it to be easily used in sectors like manufacturing, finance, and smart cities.

However, there are also great safety and ethical concerns through such developments. AI-decision making will need to be accountable and explainable, to ensure justice. The capability to develop ethically AI will rely heavily on the solution of bias in AI systems, protection of data, and the installation of governance systems. Constant surveillance is necessary to ascertain that AI stays in line with human ideals and works securely in numerous diverse settings. ML will continue to define autonomous systems in the future by causing more innovation in intelligent automation, robotics, and dynamic decision-making. The sectors should aim at developing equitable, safe, and accessible artificially intelligent applications that enhance rather than replace human performance in a process in what is referred to as AI-driven solutions.

Briefly put, the hybridization between autonomous systems and machine learning is opening the door to a smarter, safer, and more versatile technological era, where human knowledge and the autonomy of AI systems complement each other to achieve significant progress and performance.

## References


1. Soori, M., Arezoo B. and Dastres R. (2023, Jan.).Artificial intelligence, machine learning and deep learning in advanced robotics: A review. *Cognitive Robotics*, 3, 54–70.doi: 10.1016/j.cogr.2023.04.001.
2. Ravi, N.S.K. Ravi and Prabha N.A.H. (2024, Sep.). Advent of machine learning in autonomous vehicles. *International Journal of Science and Research Archive*, 13(1), 1219–1226., doi: 10.30574/ijsra.2024.13.1.1760.
3. *Machine Learning and Autonomous Systems*. (2022). doi: 10.1007/978-981-16-7996-4.
4. Sarker, I.H. (2021, Mar.). Machine learning: algorithms, Real-World applications, and research directions. *SN Computer Science*, 2(3). doi: 10.1007/s42979-021-00592-x.
5. Syed, I. and Lokhande V. (2024, Apr.). An Overview of the Supervised Machine Learning. *International Research Journal of Modernization in Engineering Technology and Science*. doi: 10.56726/irjmets51366.
6. Khanum, M., Mahboob T., Imtiaz W., Ghafoor H.A. and R. Sehar. (2015, Jun.). A survey on unsupervised machine learning algorithms for automation, classification, and maintenance. *International Journal of Computer Applications*, 119(13), 34–39. doi: 10.5120/21131-4058.
7. Macaluso, P., s252894. (2020)..*Deep Reinforcement Learning for Autonomous Systems*. Politecnico di Torino, thesis. [Online]. Available: https://webthesis.biblio.polito.it/secure/14352/1/tesi.pdf.
8. Wang, L., et al. (2024, Mar.). A survey on large language model based autonomous agents. *Frontiers of Computer Science*, 18(6). doi: 10.1007/s11704-024-40231-1.
9. Agentic AI: Autonomous Intelligence for Complex Goals: A Comprehensive Survey. (2025). *IEEE Journals and Magazine* | *IEEE Xplore*. https://ieeexplore.ieee.org/document/10849561.
10. Di Y., Deng L. and Zhang L. (2024, Nov.). A collaborative-learning multi-agent reinforcement learning method for distributed hybrid flow shop scheduling problem, *Swarm and Evolutionary Computation*, 91, 101764. doi: 10.1016/j.swevo.2024.101764.
11. IEEE Spectrum. (2024). *Explainer: What Are AI Agents?* Retrieved from https://spectrum.ieee.org/ai- agents.
12. Ferrag, M.A., Tihanyi, N. and Debbah, M. (2025). *From LLM Reasoning to Autonomous AI Agents: A Comprehensive Review*. Guelma University, Algeria; Technology Innovation Institute, UAE; Eotvos Lorand University, Hungary; Khalifa University of Science and Technology, UAE.
13. Mamillapalli, S.K. (2025, Jan.–Feb.). The Agentic AI Framework: Enabling Autonomous Intelligence. *IJIRMPS*, *13*(1). ISSN: 2349–7300. Retrieved from www.ijirmps.org.
14. Abrar, M., Aboraya, W., Abdulghafor, R., Subramanian, K.P., Al Husaini, Y. and Al Hussaini, M. (2025). AI-Powered Learning Pathways: Personalized Learning and Dynamic Assessments. *International Journal of Advanced Computer Science and Applications,* 16(1).
15. Bates, T., Cobo, C., Mariño, O. and Wheeler, S. (2020). *Can Artificial Intelligence Transform Higher Education*? Springer, 17, 1–12.
16. Zawacki-Richter, O., Marín, V. I., Bond, M. and Gouverneur, F. (2019). Systematic Review of Research on Artificial Intelligence Applications in Higher Education: Where Are the Educators? *International Journal of Educational Technology in Higher Education*, 16(1), 1–27.

17. Onesi-Ozigagun, O., Ololade, Y., Eyo-Udo, N. and Ogundipe, D. (2024). Revolutionizing Education Through AI: A Comprehensive Review of Enhancing Learning Experiences. *International Journal of Applied Research in Social Sciences*. Retrieved from https://doi.org/10.51594/ijarss.v6i4.1011.
18. GeeksforGeeks. *The Role of Reinforcement Learning in Autonomous Systems*. Retrieved from https://www.geeksforgeeks.org/the-role-of-reinforcement-learning-in-autonomous-systems.
19. Kiran, B.R., Sobh, I., Talpaert, V., Mannion, P., Al Sallab, A.A., Yogamani, S. and Pérez, P. (2021). Deep Reinforcement Learning for Autonomous Driving: A Survey. *IEEE Transactions on Intelligent Transportation Systems*, 23 (6).
20. Nika, A., Mandal, D., Kamalaruban, P., Tzannetos, G., Radanovic, G. and Singla, A. (2024). Reward Model Learning vs. Direct Policy Optimization: A Comparative Analysis of Learning from Human Preferences. In: *Proceedings of the 41st International Conference on Machine Learning* (PMLR), 235, 38145–38186.
21. Nika, A., Mandal, D., Kamalaruban, P., Tzannetos, G., Radanović, G. and Singla, A. (2024). Reward Model Learning vs. Direct Policy Optimization: A Comparative Analysis of Learning from Human Preferences. In: *Proceedings of the 41st International Conference on Machine Learning (PMLR)*, 235, 38145–38186.
22. Kale, C. (n.d.). Autonomous Vehicles: Applications of Deep Reinforcement Learning. *International Journal for Multidisciplinary Research (IJFMR)*, E-ISSN: 2582–2160.
23. Alos-Ferrer, C. and Garagnani, M. (2022). Strength of Preference and Decisions Under Risk. *Journal of Risk and Uncertainty*, 64, 309–329.
24. Brandstätter, E., Gigerenzer, G. and Hertwig, R. (2006). The Priority Heuristic: Making Choices Without Trade-Offs. *Psychological Review*, 113, 409–432.
25. Kompa, B., Snoek, J. and Beam, A.L. (2021). Second Opinion Needed: Communicating Uncertainty in Medical Machine Learning. *NPJ Digital Medicin*e, 4(1).
26. Tonekaboni, S., Joshi, S., McCradden, M.D. and Goldenberg, A. (2019). What Clinicians Want: Contextualizing Explainable Machine Learning for Clinical End Use. In: *Machine Learning for Healthcare Conference, PMLR*, pp. 359–380.
27. Amann, J., Blasimme, A., Vayena, E., Frey, D., Madai, V.I. and Precise4Q Consortium. (2020). Explainability for Artificial Intelligence in Healthcare: A Multidisciplinary Perspective. *BMC Medical Informatics and Decision Making*, 20, 1–9.
28. Dalsaniya, A. and Patel, K. (2022). Enhancing process automation with AI: The role of intelligent automation in business efficiency. *International Journal of Science and Research Archive*, 5(2), 322–337.
29. Waqar, M., Bhatti, I. and Khan, A.H. (2024). AI-powered automation: Revolutionizing industrial processes and enhancing operational efficiency. *Revista de Inteligencia Artificial en Medicina*, 15(1), 1151–1175.
30. Salami, I.A., Adesokan-Imran, T.O., Tiwo, O.J., Metibemu, O.C., Olutimehin, A.T. and Olaniyi, O.O. (2025). Addressing Bias and Data Privacy Concerns in AI-Driven Credit Scoring Systems through Cybersecurity Risk Assessment. *Asian Journal of Research in Computer Science*, 18(4), 59–82.

# 18 Securing the Future: AI-Powered Weapon Systems, Ethics, and Adversarial Defense

Balajee J.[1*], Ravikumar K.V.[2], Sangar G.[3] and Kaavya Kanagaraj[4]

## 1. Introduction

The rapid merging of countries of artificial intelligence (AI) with modern war technology are changing the way to protect themselves. Modern defense structures no longer only use classic mechanized weapons [1]. More and more, they depend on smart self-tacking systems that can evaluate, adapt, and work with little help with people. These AI-operated compositions provide tremendous strategic benefits, but they also bring complex risks to cyber safety and operational reliability [2]. As the nations intensify their efforts to lead to AI defense, the need to guarantee the stability and elasticity of such systems is severely urged [3].

### *1.1 The Rise of AI-Driven Weapon Systems*

In the last decade, AI has moved on functional foundations in advanced arms systems from the theoretical research field [4]. Military organizations around the world now deploy a range of AI-powered technologies: autonomous drones for surveillance and combat, robotic ground units for spying, smart missile guide systems and computer vision, and real-time detections using Deep Learning Algorithms. These tools not only increase the decision and risk of the soldier, but also enable large-scale military operations with unprecedented speed and accuracy [5].

AI-operated weapon systems (Figure 1) have been built on the latest machine learning algorithms capable of predicting threats and autonomously in a dynamic environment [6]. For example, reinforcement education allows the combat drone to adapt to its researcher in real-time, while the Conventional Neural Network enables

[1-2] Department of Artificial Intelligence, Mother Theresa Institute of Engineering and Technology, Palamaner, Chittoor, Andhra Pradesh, India.
[3] Department of CSE, SRM Institute of Science and Technology, Vadapalani, Chennai, Tamil Nadu, India.
[4] Department of Computational Intelligence, Faculty of Engineering and Technology, Kattankulathur, Chennai.
E-mail: ravikumarkvtou@gmail.com[2]; sangarrajagopal@gmail.com[3]; India, kaavyak@srmist.edu.in[4]
* Corresponding author: j.balajee@mtieat.org

DOI: 10.1201/9781003663461-18

faster identification of the enemy's goals through the air image. These technologies are revolutionizing battlefield mobility by enabling active and responsive combat strategies [7, 8].

**Fig. 1** AI in Weapons Systems.

However, this new example of the intelligent war comes with a set of its own challenges. The complexity of AI systems makes them fully valid and predicted. Their autonomy, when benefits in many cases, can be a responsibility if the control methods deteriorate or are defective in mission-critical views [9]. As such, dependence on these systems without adequate safety can lead to unwanted consequences on the battlefield [10].

The global defense sector has witnessed a transformative shift with the adoption of AI across strategic and tactical operations. Several countries have launched AI-powered military initiatives, including autonomous drones, predictive intelligence systems, and cybersecurity units [11]. As illustrated in Figure 2, the United States and China account for over half of the global AI-driven defense advancements, followed by Russia, India, and key European nations. This geopolitical imbalance highlights the urgent need for global standards and cooperative treaties before such technologies become fully weaponized.

### *1.2 Importance of Cybersecurity and Robustness*

Despite operational advantages, AI-powered weapon systems are very sensitive to a wide range of cyber safety risks (Figure 3). These include anti-attacks on machine learning models, data poisoning, GPS (global positioning system) spoofing, sensor fraud, communication jamming and remote hijacking of autonomous platforms during the training phase [12, 13]. Unlike traditional mechanical failure, the manipulation of AI systems can result in silence and non-diagnostic defects

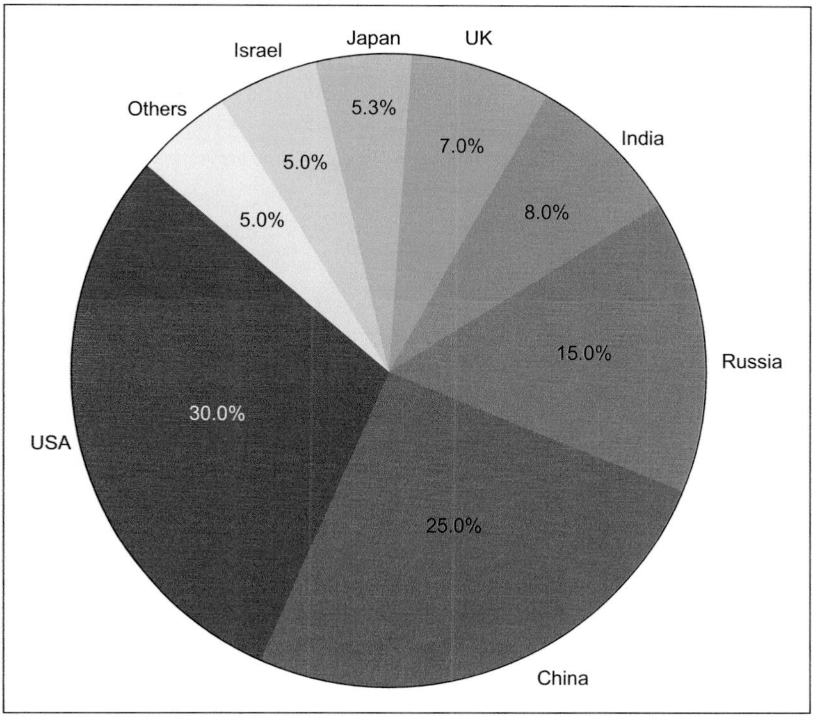


**Fig. 2** Estimated share of countries implementing AI in upcoming defense systems. The United States and China are leading adopters, reflecting large-scale investment and integration of AI in autonomous and decision-support military technologies.

where the system acts technically as aimed, but not strategically or ethically expected [14].

Strength, in this context, refers to the system's ability to maintain performance in various adverse conditions, such as hardware degradation, noisy input data and anti-trafficking [15]. The cyber safety system guarantees the efficiency of the system and the integrity, availability, and secrecy of the pipeline [16]. These two principles are basic for strengthening and cyber safety autonomous defense techniques.

Without adequate security restrictions and elasticity methods, AI-powered weapons can become responsibilities rather than property. The consequences of the compromising AI can range from minor intelligence to civilian casualties or unnecessary military growth [17]. Thus, defense agencies and research institutions should not be renovated quickly in the design and deployment of each AI system intended for use in defense [18]. Subsequent sections of this chapter will detect these challenges with more depth, provide real-world pictures of threats, and introduce design principles and cybersecurity architects necessary to protect the upcoming pay generation of AI-powered weapons systems [19].

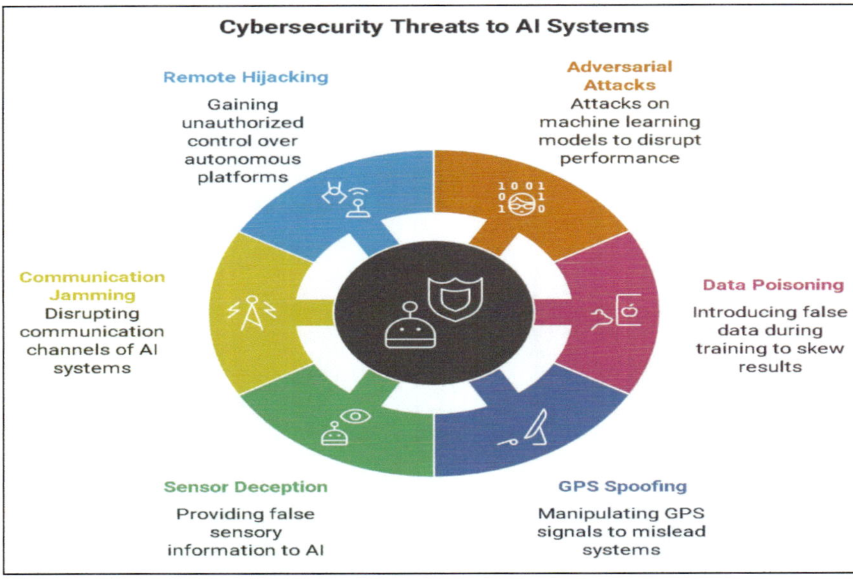


**Fig. 3** Cybersecurity Threats to AI Systems.

## 2. Threat Landscape

The integration of AI in modern weapons systems has significantly increased automation, target accuracy, and operational efficiency. However, these technological advances have also introduced a new class of weaknesses. Unlike traditional weapons, AI-powered systems are susceptible to cyber-physical threats that absorb their education methods, data dependence, and autonomous behavior. It is necessary to understand the nature and scope of these threats for effective safety [20].

### *2.1 Military AI Anti-attacks*

Anti-attacks are deliberate manipulations designed to deceive AI models, especially used in image recognition, natural language processing, and sensor-based decisions. These attacks are a serious threat to military applications where precision, motion, and accuracy are important.

In terms of AI-driven weapons, anti-attack may appear in many forms:

- **Anti-Examples:** Slightly in input data (eg., drone cameras or radar signatures), often inconsistent changes can cause the AI model to falsely classify the Object Budget. For example, the enemy vehicle may appear invisible to gentle or targeted systems by changing the pattern of the camouflage to deceive the vision system.

- **Model Poisoning:** During the training phase of AI systems, opponents can provide malicious data injections to the dataset to squeeze the model's learning process. In military scenes, these can result in systems that make biased or systematically wrong decisions under specific conditions.
- **Taxing Attacks:** These attacks target AI systems at the forecast, alter the input data to avoid investigation. For example, a stealth drone can use electronic counter measures or modified thermal signatures to avoid classification by AI-enabled air defense systems.
- **Backdoor Attack:** This includes hidden malicious triggers in the training process, which is activated, when the normal system overrides the behavior. The weapon system can function properly under standard terms, but when the attacker's input trigger is received, the settlement displays behavior.

The big challenge with anti-attacks is their stealth. These attacks often do not pay attention to anyone during the test and are activated only in certain situations, which make them extremely dangerous in real-time combat situations where it is crucial to make a quick and unaffected decision [21].

## *2.2 Common Vulnerabilities in Weaponized AI Systems*

AI-enabled weapon platforms such as autonomous drones, robotic weapons, and surveillance vehicles depend greatly on each other's connected systems and real-time data streams. This complexity, when empowering, opens the door of multiple weaknesses:

- **Sensor Spoofing:** Many weapon systems use data from GPS, Leader, Sonar, and Infrared Sensor. These inputs can be manipulated by spoofing techniques, causing the system to misinterpret the distance, misinterpret objects, or distract from its way.
- **Communication Hijacking:** AI systems depend on a secure communication protocol for command and control. Poor encryption or unsafe channels can lead to intercept or control override, potentially allowing an opponent to order an autonomous platform.
- **Software Exploits and Malware Injection:** Like any digital system, AI weapons are susceptible to platform bugs and software exploits. The same vulnerability in the operating petting system or AI information engine can serve as an attack vector for malware or ransomware.
- **Risks of Data Integrity:** AI systems are based on accurate, trustworthy data. If the integrity of training or real-time input data is compromised, the decisions taken by the system will also be defective. This is especially crucial in the dynamic combat zone where the fatal consequences of the wrong classification of the enemy against a friend can have.
- **Black Box Behavior:** Many AI models, especially DEEP Learning Networks, act as "Black B Boxes Q", which gives a little understanding of their internal decision-making processes. This complicates the detection of unclear errors, bias, or malicious tampering (Figure 4).

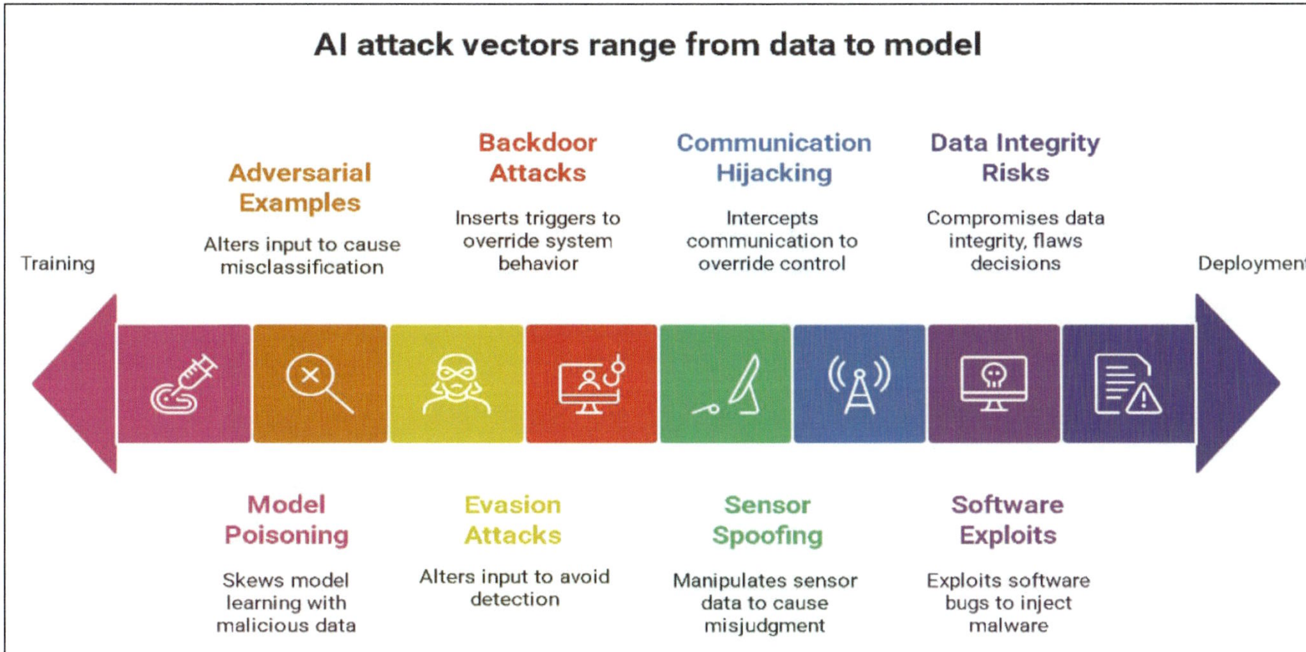


**Fig. 4** AI attack vector ranges from data to model.

Precisely, the fusion of AI with protection technologies introduces both a strategic benefit and an essential factor of failure. Adversaries ready with the expertise of AI systems can manipulate them without firing a single shot. As such, addressing those vulnerabilities isn't only a technical project, it's a matter of country-wide and worldwide security [22].

## 3. Cybersecurity Architectures for AI-Weapon Platforms

As AI-pushed weapon structures come to be central to country-wide defense techniques, the conventional "trust however confirm" method to cybersecurity is no longer sufficient. Modern adversaries can make the most of the interconnected nature of these structures via more and more state-of-the-art cyber and facts warfare strategies [23]. Therefore, it's imperative to put into effect resilient and proactive cybersecurity architectures that make certain information integrity, operational agree with, and traceable duty throughout all subsystems.

This phase specializes in leading-facet paradigms: Zero-Trust Architecture (ZTA), and Blockchain-Based Logging and Verification, each of which are foundational to constructing steady, self-sustaining protection ecosystems (Figure 5).

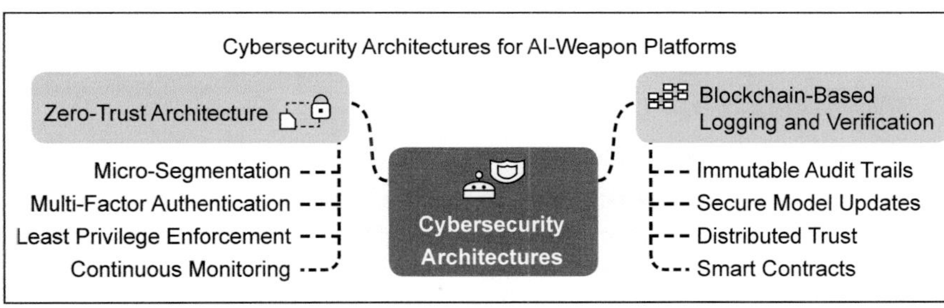


**Fig. 5** Cybersecurity Architectures for AI-weapon Platforms.

### *3.1 Zero-Trust Architecture*

ZTA is a cybersecurity version based on the principle of "never agree with, always affirm". Unlike perimeter-based protection models that assume systems within the network may be trusted, ZTA enforces strict identification verification, get right of entry to controls, and real-time conduct evaluation across all nodes and offerings [24].

**Key Features in Defense AI Context**

- **Micro-Segmentation:** Divides AI subsystems sensors, inference engines, and manages modules into remote zones. Any communique among zones is mediated via strict admission to controls and monitored through security regulations.

- **Multi-Factor Authentication (MFA) and Role-Based Access:** Ensures that even inner operators need to constantly validate their credentials and permissions earlier than gaining access to AI common sense, schooling modules, or venture statistics.
- **Least Privilege Enforcement:** Each component whether or not a software program module, hardware device, or human user is granted most effectively the minimal necessary entry to perform its characteristic, lowering the attack surface.
- **Continuous Monitoring and Trust Evaluation:** AI models and records pipelines are continuously evaluated for anomalous conduct, inclusive of uncommon inference patterns or sudden sensor inputs, triggering actual-time indicators.

In weapon structures, ZTA can be embedded at each hardware and software program degrees to protect against unauthorized reprogramming, records corruption, and opposed infiltration specifically in unmanned and remotely deployed structures.

## *3.2 Blockchain-Based Logging and Verification*

Blockchain technology presents a decentralized, immutable ledger that is ideal for recording critical activities in weapon machine operation including information enter resources, AI version selections, access attempts, and command execution sequences [25].

**Applications in Military AI Systems**

- **Immutable Audit Trails:** Every decision made by way of the AI machine along with threat detection, target engagement, or direction alteration can be logged as a blockchain transaction. This ensures tamper-evidence forensic facts for up-project audits or battle resolution.
- **Secure Model Updates:** When AI models or software program patches need to be deployed, blockchain can authenticate the beginning and integrity of the replace. This eliminates the danger of malicious model injections or version spoofing.
- **Distributed Trust Across Coalition Networks:** In joint navy operations related to allied forces, blockchain guarantees a shared, verifiable view of system logs and hobby across geographically dispersed gadgets.
- **Smart Contracts for Automated Governance:** Rules and conditions governing weapon deployment (e.g., engagement criteria, fail-safe triggers) can be codified as smart contracts, making sure that the AI adheres to project protocols without deviation.

Blockchain integration is specifically treasured in part-based totally army structures (e.g., drones, unmanned vehicles), wherein intermittent connectivity or decentralized command structures boost the chance of information compromise or unauthorized overrides (Figure 5).

# 4. Enhancing AI Robustness in Adversarial Settings

As AI structures anticipate important roles in military decision-making, making sure their reliability beneath adverse and unpredictable conditions is paramount. Robustness refers to an AI machine's ability to maintain overall performance in the presence of noisy, incomplete, manipulated, or opposed entered information [26]. In battlefield situations in which information is frequently distorted, adversaries may additionally manage the surroundings, and real-time choice-making is critical robustness turns into a decisive aspect in mission success and operational safety. This phase explores critical factors: robust education methodologies that make stronger AI systems from the floor up, and protective mechanisms mainly designed to hit upon and mitigate poisoning and evasion attacks [27] (Figure 6).

## *4.1 Robust Training Methods*

Robust schooling includes modifying traditional mastering procedures to create AI fashions that aren't handiest correct underneath everyday situations but additionally resilient to strange and adversarial input.

**Key Techniques**

- **Adversarial Training:** One of the handiest defenses, hostile education entails exposing the AI version to adversarial examples all through the schooling phase. By learning how to classify or reply to perturbed inputs, the version turns into greater immunity to similar assaults in actual-time applications. In military AI, this approach can help prevent misclassification of camouflaged enemy automobiles or spoofed signals.
- **Data Augmentation with Edge-Case Scenarios:** Training datasets are regularly biased in the direction of common cases. Including rare, intense, or noisy information inclusive of low-resolution sensor input, middle of the night imagery, or battlefield occlusions facilitates putting together models for actual-international conditions. In a fight, where perfection cannot be anticipated, this diversity enhances decision reliability.
- **Gradient Regularization and Noise Injection:** Gradient-based total education vulnerabilities are generally exploited via attackers. Gradient regularization strategies clean the choice obstacles, making it more difficult for attackers to locate exploitable weak points. Similarly, including managed noise at some point of education builds tolerance to input perturbations.
- **Ensemble Learning:** Using multiple models skilled independently and aggregating their outputs (e.g., majority voting or averaging) enables lessen man or woman biases and will increase resilience. An adversary might need to idiot multiple models simultaneously to reason a device-extensive mistake, elevating the complexity of the assault.

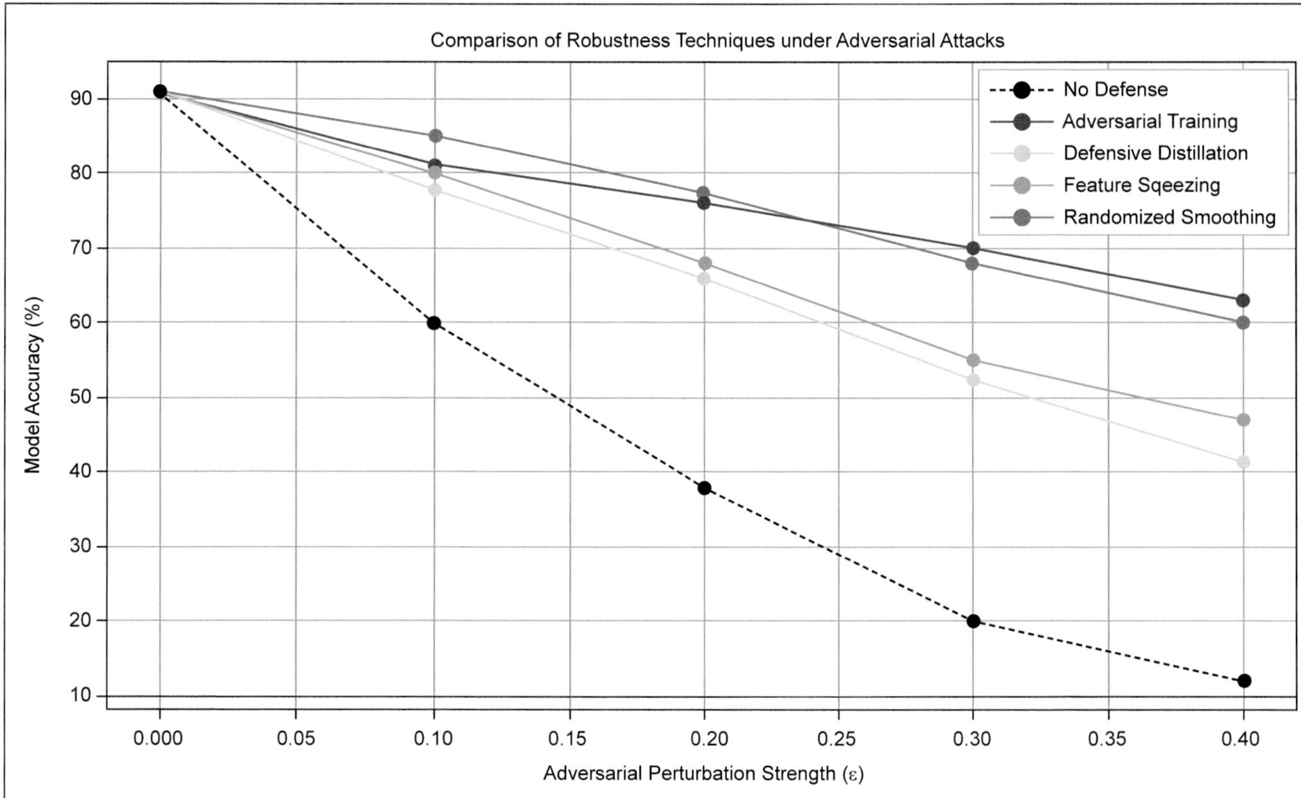


**Fig. 6** Illustrates the performance degradation of military-grade AI object detection systems when subjected to adversarial attacks of increasing perturbation strength. While models without defense quickly degrade below 20% accuracy, robust techniques like adversarial training and randomized smoothing sustain over 60% accuracy even under strong attacks, making them preferred choices in battlefield deployments.

### *4.2 Defense Against Poisoning and Evasion Attacks*

AI structures are susceptible not simply at inference time; however, at some stage in their lifecycle particularly all through education and deployment levels. Defense mechanisms should address each fact of poisoning, in which training records are corrupted, and evasion attacks, where malicious inputs fool the AI at runtime [28].

**Defense Strategies**

- **Data Sanitization Pipelines:** In opposed environments, verifying statistics assets turns into crucial. Data sanitization includes anomaly detection in schooling units, replica checks, and sample consistency analysis. In army packages, trusted records channels and cryptographic validation are hired to reject tainted datasets.
- **Robust Loss Functions:** Traditional loss functions like go-entropy are sensitive to outliers. Modified loss features, inclusive of TRADES (TRADE-stimulated Adversarial DEfense through Surrogate-loss minimization), stability class accuracy with hostile robustness.
- **Feature Squeezing and Input Compression:** These methods lessen the ranges of freedom an attacker can make at the most. By quantizing input capabilities or putting off redundant indicators, the AI system is compelled to rely upon core, non-manipulable styles. This approach is especially beneficial in real-time sensor feeds.
- **Runtime Anomaly Detection:** Models may be geared up with parallel detectors that monitor for suspicious activations or unexpected modifications in input distributions. For example, if a weapon machine all of sudden sees a surge in close-to-equal 'friendly' objectives in a battle sector, the gadget can raise a flag for manual verification.
- **Model Fingerprinting and Integrity Checks:** Hash-based integrity mechanisms can make certain that AI fashions deployed on field systems have now not been tampered with. Combined with secure enclaves or Trusted Platform Modules (TPMs), those checks provide a hardware-backed guarantee of trustworthiness.

## 5. Explainable AI and Trust in Autonomous Decisions

As self-sufficient weapon structures turn out to be greater state-of-the-art and able to run with minimum human oversight, questions around explainability and agreement have become critical to their secure and ethical deployment. Military decision-making is excessive-stakes, and AI structures ought to be no longer simplest, accurate, and sturdy but additionally comprehensible to their human operators. Explainable Artificial Intelligence (XAI) is the area of AI studies and improvement focused on making the inner workings of AI fashions interpretable, obvious, and justifiable [29].

In a navy setting where lives, criminal compliance, and geopolitical consequences are on the line, blind trust in black-box models is unacceptable.

This section explores the significance of XAI and the design of human-in-the-loop managed structures that integrate human judgment with self-sufficient abilities [30].

### *5.1 Role of XAI in Military Context*

Traditional AI models, especially deep mastering systems, are often criticized for his or her "black-container" nature: they could make distinctly correct selections but provide little to no reasoning for his or her outputs. This is especially difficult in defense eventualities involving target class, danger prioritization, or engagement choices, wherein accountability and publish-motion traceability are critical.

**Key Roles of XAI in Defense**

- **Operational Transparency:** In complex battle scenarios, operators should understand why a device is recommending or executing a particular course of movement. XAI presents reasons either textual, visual, or primarily rule-based that clarify the selection-making process, permitting faster, knowledgeable decisions.
- **Legal and Ethical Accountability:** The use of deadly autonomous weapons (LAWs) requires adherence to worldwide humanitarian law (IHL), which includes standards of proportionality and difference. Explainable models can provide logs and justifications that help post-hoc investigations and prison evaluations.
- **Debugging and Model Improvement:** Interpretability equipment permits defense engineers to discover incorrect reasoning, biases, or records deficiencies in AI structures. For instance, if a drone misidentifies a civilian car as a hostile asset, XAI gear can display whether or not it was due to defective visible functions, biased training data, or adverse interference.
- **Trust Calibration:** Over-trust or under-believe in AI systems can both be risky. Explainable outputs permit human operators to calibrate their self-assurance within the machine, correctly questioning it when wished or counting on it at some stage in time-critical eventualities.

**Examples of XAI Methods in Military AI**

- **Saliency Maps** in computer vision to show which parts of an image led to a threat classification.
- **SHAP (SHapley Additive Explanations)** for feature-level contributions in battlefield risk analysis.
- **LIME (Local Interpretable Model-Agnostic Explanations)** to provide localized explanations for complex decisions.
- **Decision Trees or Rule Extraction** from deep models to simplify behavior representation in mission briefings.

### 5.2 Human-in-the-Loop Control Systems

Autonomy in army systems does no longer mean the absence of human oversight. Rather, the precise deployment method includes Human-in-the-Loop (HITL) or Human-on-the-Loop architecture designs that enable human beings to supervise, override, or collaborate with AI systems during critical moments.

**Key HITL Design Principles**

- **Supervised Autonomy:** The AI executes low-level tasks autonomously (e.g., navigation, data filtering), while humans retain authority over high-level or ethically sensitive actions (e.g., target engagement).
- **Adaptive Interfaces:** User interfaces must dynamically present relevant AI explanations, confidence scores, and alternative options in formats understandable under pressure, e.g., heatmaps, simple language summaries, or real-time alerts.
- **Interactive Rehearsal and Simulation:** Defense operators should be trained with virtual environments where they can test and understand AI reasoning before field deployment. These simulations help align human mental models with AI behavior.
- **Override and Abort Mechanisms:** All autonomous weapon systems must provide immediate override options to human operators. A "kill switch" or command interrupt should be embedded both in hardware and software layers.

**Benefits of HITL Systems**

- Prevents over-automation and "automation bias", where humans defer too readily to machine decisions.
- Bridges the accountability gap by embedding traceable human decisions within the system loop.
- Increases operational adaptability by allowing humans to inject context or updated intelligence that may not be available to the AI.

## 6. Case Studies and Real-World Incidents

### 6.1 GPS Spoofing in UAVs

Unmanned Aerial Vehicles (UAVs), especially the ones utilized in surveillance and reconnaissance, depend closely on GPS signals for navigation, concentrated on, and project execution. However, their dependence on external satellite TV for PC alerts makes them vulnerable to GPS spoofing, an assault whereby counterfeit alerts are broadcast to lie to or redirect the UAV. An outstanding case surfaced in 2011, when Iran claimed to have captured an American RQ-a hundred and seventy Sentinel drone by feeding it false GPS coordinates. This incident no longer raised issues over the cyber vulnerability of advanced navy structures; however, it also signaled the capability for adversaries to hijack or misdirect drones without deploying bodily pressure (Table 1).

**Table 1** Comparison of Real-World GPS Spoofing Incidents in Military and Civil UAVs.

| *Incident* | *Year* | *Target UAV* | *Spoofing Method* | *Claimed Outcome* | *Source Nation* |
|---|---|---|---|---|---|
| RQ-170 Capture by Iran | 2011 | RQ-170 Sentinel | Ground-based GPS spoofing | UAV landed in enemy territory | Iran |
| US Navy Drone Disruption | 2019 | MQ-9 Reaper | Signal jamming & spoofing | Temporary disorientation | Unknown (Reported in Syria) |
| Civilian Drone Test | 2018 | DJI Phantom | Software-defined radio spoofing | Drone redirected mid-air | Academic (USA) |

The incident verified how a rather easy, yet powerful, method could compromise strategic belongings and disrupt military operations. It similarly emphasized the want for robust, multilayered navigation structures in army drones. Solutions such as redundant inertial navigation systems, AI-driven spoof detection algorithms, and cryptographic GPS sign authentication have been in view and that have received interest. The case serves as a landmark reminder that during modern struggle, cyber and physical domain names are deeply intertwined, and securing AI structures need to move beyond code to encompass the integrity of sensor facts they depend upon.

## 6.2 Compromised Object Detection in Surveillance Drones

AI-enabled surveillance drones frequently rent deep mastering fashions for item detection and classification. These models, specifically convolutional neural networks (CNNs), are vulnerable to adverse attacks, small, intentional perturbations that fool the version into making wrong predictions (Table 2). In one super educational demonstration in 2020, researchers placed inconspicuous adversarial patches on cars, which effectively avoided drones from recognizing the objects or misclassified them altogether. Such compromises can cause the failure of important surveillance missions, undetected enemy motion, or even misdirected engagement decisions.

**Table 2** Effectiveness of Adversarial Patch Attacks on Object Detection Models.

| *AI Model* | *Attack Type* | *Detection Accuracy (Original)* | *Detection Accuracy (Under Attack)* | *Misclassification Example* |
|---|---|---|---|---|
| YOLOv3 | Patch-Based | 92% | 28% | "Car" misclassified as "Background" |
| Faster R-CNN | Physical Perturbation | 89% | 35% | "Tank" misclassified as "Truck" |
| SSD | Universal Perturbation | 87% | 40% | "Soldier" missed entirely |

The compromised item detection scenario reveals no longer just the fragility of modern-day AI imaginative and prescient structures; however, it also communicates the escalating arms race between AI model designers and opposed assault builders. This case highlighted the significance of opposed education, sturdy validation underneath battlefield-like situations, and the incorporation of fallback mechanisms which includes sensor fusion or operator indicators while the gadget detects anomalies. As deep gaining knowledge keeps to form military imaginative and prescient systems, such real-global and simulated incidents play an essential role in shaping the future development of resilient, combat-equipped AI platforms.

# 7. Policy, Ethics, and Future Directions

## *7.1 Ethical Guidelines for AI in Combat*

The integration of AI into self-reliant and semi-self-sustaining weapon systems has caused widespread debate around the moral implications of permitting machines to make existence-and-loss of life choices. Key moral issues consist of the delegation of deadly pressure to algorithms, accountability inside the event of malfunctions or accidental damage, and the capacity bias embedded within schooling information that could lead to discriminatory concentration. Ethical tips proposed by defense corporations stress the importance of meaningful human management, transparency, and making sure that AI choices are aligned with international humanitarian legal guidelines.

In recent years, diverse governmental and intergovernmental bodies have mentioned principles for accountable AI use in army contexts. The U.S. Department of Defense's moral concepts for AI emphasize reliability, traceability, and governability. Similarly, the European Union and NATO have encouraged member states to embed ethical critiques in defense AI procurement and deployment. Despite these efforts, a binding global ethical trend stays elusive. The urgency lies not just in developing moral AI, however in growing universally frequent norms that could save you misuse, lessen escalations, and preserve human dignity in struggle contexts (Table 3).

**Table 3** Comparison of Military AI Ethics Principles Across Major Powers.

| *Principle* | *United States (DoD)* | *EU Guidelines* | *China's Strategic Vision* |
|---|---|---|---|
| Human Oversight | Required | Strongly Recommended | Emphasized in select cases |
| Transparency | Critical | Essential for public trust | Low public transparency |
| Algorithmic Accountability | Shared between developer and operator | Encouraged through audits | Unclear/Undisclosed |
| Bias Mitigation | Ongoing research | Mandated by AI Act | Not explicitly addressed |

### *7.2 Need for Global Standards and AI Treaties*

As army AI technology proliferates across the globe, the absence of regular norms and treaties regulating their improvement and deployment has created an environment similar to a digital hands race. While nuclear weapons are ruled by using international treaties and huge verification regimes, AI-based weapon structures continue to be largely unregulated. The dual-use nature of many AI technologies complicates efforts to distinguish among civilian and navy applications, making enforcement of bans or regulations notably harder.

The worldwide network must begin formulating treaties that especially address independent weapons, statistics integrity, algorithmic transparency, and post-deployment tracking. Initiatives which include the United Nations' call for a ban on fully autonomous deadly guns and advocacy by way of think tanks and human rights businesses have laid foundational discourse. However, without multilateral consensus and technical enforcement mechanisms, country wide hobbies often supersede cooperative governance. Future treaties have to incorporate adaptive verification protocols, cope with software-as-a-weapon classification, and foster transparency among technologically superior and rising countries to ensure a balanced, stable AI future (Table 4).

**Table 4** Key Challenges and Solutions in Establishing Global AI Weapon Regulations.

| *Challenge* | *Description* | *Potential Solution* |
|---|---|---|
| Dual-Use Dilemma | Difficulty in distinguishing civilian vs. military AI | Introduce AI use-certification similar to nuclear material tracing |
| Verification Problem | Software-based weapons are harder to inspect than physical arms | Develop blockchain-based audit trails for weapon system logs |
| National Sovereignty | Countries hesitant to share AI capabilities | Create neutral global AI watchdog organization under UN |
| Varying Ethical Standards | Differences in values across cultures | Draft a baseline AI military code of ethics endorsed multilaterally |

## 8. Conclusion

### *8.1 Summary of Key Insights*

The integration of AI into weapon systems represents a paradigm shift in military capabilities and strategy. However, the same technologies that offer precision, scalability, and automation also introduce a complex set of risks. From cyber vulnerabilities such as GPS spoofing and adversarial attacks on object detection systems, to deeper concerns over transparency, accountability, and ethical alignment, AI in combat zones demands a multidisciplinary approach to safety and governance. The robustness of AI must be embedded not only in technical design but also in legal and institutional frameworks that oversee their use.

Throughout this chapter, we explored both the technological and ethical dimensions of AI weapon systems. We delved into real-world case studies that highlighted current system weaknesses and emphasized the importance of explainable AI and human-in-the-loop frameworks in fostering trust and operational reliability. Equally, we recognized the pressing need for ethical requirements and international treaties which could guide the accountable development and deployment of military AI. As the battlefield turns more and more autonomous, the margin for error narrows, and proactive measures emerge as imperative.

## *8.2 Recommendations for Future Research*

Looking ahead, several avenues warrant deep investigation to ensure that AI in protection stays useful and managed. One promising route is the improvement of self-recovery AI structures that can autonomously detect and get over antagonistic interference. Another is the advancement of causal explainability transferring past pattern reputation to deeper reasoning that aligns with human cognitive processes. Moreover, preparing AI structures to resist publish-quantum cryptographic threats could be important as encryption protocols face new challenges (Table 5).

Equally vital is the establishment of global research consortia targeted on AI in war, selling transparency and interoperability. Shared checking out standards, incident databases, and AI warfare-gaming simulations can assist reduce dangers and foster international cooperation. Ultimately, the fulfillment of AI in weapon structures will rely not best on its accuracy and pace;, however, additionally on its alignment with human values, its resilience under attack, and the worldwide willingness to control its improvement responsibly.

**Table 5** Future Research Directions and Recommended Actions in Military AI.

| *Research Area* | *Goal* | *Recommended Approach* | *Expected Outcome* |
|---|---|---|---|
| Self-Healing AI | Detect and recover from attacks autonomously | Use generative AI to reconstruct corrupted input and parameters | Improved resilience during adversarial events |
| Causal Explainability | Enhance trust in AI decisions | Incorporate causal inference models like do-calculus | Higher interpretability in lethal scenarios |
| Post-Quantum AI Security | Prepare for quantum attacks | Develop quantum-safe cryptographic protocols for model updates | Long-term encryption integrity |
| Global AI Governance | Harmonize AI military standards | Initiate UN-based AI arms control forums | Reduced risk of AI arms race |

## References


1. Goodfellow, I.J. Shlens and C. Szegedy. (2014). Explaining and harnessing adversarial examples. *arXiv preprint arXiv:1412.6572.*
2. Biggio, B. and Roli F. (2018). Wild patterns: Ten years after the rise of adversarial machine learning. *Pattern Recognition*, 84, 317–331.
3. Papernot, N., et al. (2016). The limitations of deep learning in adversarial settings. In: *Proc. IEEE European Symposium on Security and Privacy*, pp. 372–387.
4. Kurakin, A., Goodfellow, I. and Bengio, S. (2016). Adversarial examples in the physical world. *arXiv preprint arXiv:1607.02533.*
5. Dong, Y., Liao, F., Pang,T., Su, H. and Zhu, J. (2018). Boosting adversarial attacks with momentum. In: *Proc. IEEE CVPR*, pp. 9185–9193.
6. Athalye, A., Carlini, N. and Wagner, D. (2018). Obfuscated gradients give a false sense of security: Circumventing defenses to adversarial examples. In: *Proc. ICML*, pp. 274–283.
7. Xu, H., et al. (2020). Evading real-time person detectors with adversarial t-shirts: A feasibility study. In: *Proc. USENIX Security Symposium.*
8. Rasmussen, C.E. (2017, Apr.). GPS spoofing detection and mitigation using attitude difference. *IEEE Trans. Aerospace and Electronic Systems*, 53(2), 982–990.
9. Pomerleau, D. (2017). The DOD's Third Offset Strategy: The Role of Artificial Intelligence and Autonomy in Warfare. *C4ISRNET.*
10. Redmon, J. and Farhadi, A. (2018). YOLOv3: An incremental improvement. *arXiv preprint arXiv:1804.02767.*
11. Qin, Z., Zhang, F., Yu, C. and Lu, Y. (2020). How convolutional neural networks see the world: A survey of convolutional neural network visualization methods. *Mathematics*, 8, (5), 1–31.
12. Din, S.M.P., Arif, M., Shahid, A. and Mahmood, M.S. (2023). A Review on Applications of Artificial Intelligence in Military Operations and Warfare. *Defence Technology*, 19(3), 559–573.
13. United Nations Institute for Disarmament Research (UNIDIR). (2020). *The Weaponization of Increasingly Autonomous Technologies: Autonomous Weapon Systems and Cyber Operations.* UNIDIR, Geneva.
14. Cummings, M. (2017). Artificial Intelligence and the Future of Warfare. *Chatham House Research Paper.*
15. Lin, P. (2015). Why Ethics Matters for Autonomous Cars. In *Autonomes Fahren*. Springer, pp. 69–85.
16. Brundage, M., et al. (2018). The malicious use of artificial intelligence: Forecasting, prevention, and mitigation. *arXiv preprint arXiv:1802.07228.*
17. Bryson, J. and Kime, J. (2011). Just an artifact: Why machines are not moral agents. In: *Proc. IJCAI.*
18. Ahmed, M.S. and Kamruzzaman, S.M. (2022). Explainable AI (XAI): Current status and future directions. *Journal of King Saud University – Computer and Information Sciences.*
19. Doshi-Velez, A. and Kim, B. (2017). Towards a rigorous science of interpretable machine learning. *arXiv preprint arXiv:1702.08608.*
20. Amodei, D., et al. (2016). Concrete problems in AI safety. *arXiv preprint arXiv:1606.06565.*
21. NATO Communications and Information Agency. (2022). Autonomous Systems in NATO: Operational and Ethical Considerations. NATO Review.
22. Dennis, J.B., et al. (2022). Adversarial ML threats to military systems: A taxonomy and survey. *Journal of Cybersecurity and Privacy*, 2(1), 76–106.
23. Ferraiolo, D.F., Kuhn, R. and Chandramouli, R. (2003). *Role-Based Access Control*, Artech House.
24. Ghosh, A. (2020). Cybersecurity in the Age of AI: Challenges for National Security. *Cyber Defense Review*, 5(2), 31–52.

25. Liu, B., Gao, Y. and Zhang, T. (2022). Adversarial patch attacks on object detection models in real-world scenarios. *Pattern Recognition*, 132, 1–10.
26. Russell, S. and Dewey, D. (2015). Research priorities for robust and beneficial artificial intelligence. *AI Magazine*, 36(4), 105–114.
27. Garreau, J. (2016).The Pentagon's brain: An uncensored history of DARPA. *Defense One*.
28. Wagner, B. (2018). Ethics as an escape from regulation: From ethics-washing to ethics-shopping? In: *Being Profiled*. Amsterdam University Press, pp. 84–89.
29. Yampolskiy, R. (2016). Artificial intelligence safety engineering: Why machine ethics is a wrong approach. *Philosophy and Technology*, 29(4), 389–396.
30. United Nations Office for Disarmament Affairs. (2021). *Lethal Autonomous Weapons Systems: Briefing Papers*. UNODA.

# 19 Evolution of Weapon Systems and Rise of Intelligent Warfare

Neeraj Jain[1*], Komal Agrawal[2], Dibya Narayan Hazra[3] and Maheshwari Niranjan[4]

## 1. Introduction

### *1.1 Background and Context*

With the advent of the Stone Age, the weapons were in the form of sharp stones used to kill animals for safety and food. With the development of human civilization, stones were replaced by metallic and wooden arms. The weapons used for safety purposes turned into means of rivalry. In the early days, during Mughals and Marathas, when wars took place, the weapons were manually driven like swords, shields, spears, bows and arrows, and more. Longer-range weapons like matchlocks and pistols were also used. The warriors were on horses, camels or elephants, or on foot [1, 2].

In ancient times, with the evolving intelligence, the concept of drones was conceived by humans as a rescue from wild animals when the resources were inadequate. Not to be encouraged, as the civilization enhanced, application of drones was increased, causing prevention from unforeseen danger leading against humanity. As a consequence, during World Wars I and II, the war involved land, water, and air. Tremendous casualties took place back then. According to data from the websites, per hour death rates were around 3000–4000 including military and civilians. This continued for four years. The population nearly came to an end during that period, causing social trauma and economic impact [3, 4]. To save lives, robotics arms were invented [5], GPS (global positioning system) was developed [6] that could keep the militants at bay.

Nuclear weapons were also hazardous that affected the present and the unborn generation for ages. The Gaza-Israel conflict led to war on land, air, and water. The war witnessed numerous weapons, small arms, tanks, fighting vehicles, aircraft [7], watercraft [8], and the like. The misconception between the two countries gave

[1,2] Department of Electrical Electronics and Communication Engineering, Galgotias University, Greater Noida, Uttar Pradesh, India-203201.
[3] School of Computer Science and Engineering, IILM University, Greater Noida, Uttar Pradesh, India.
[4] Department of CSE-AIML, Greater Noida Institute of Technology, Greater Noida, India,
* Corresponding author: erneerajjain@gmail.com

DOI: 10.1201/9781003663461-19

rise to the Russia and Ukraine war. They used advanced technologies including aircraft, explosives, mortars, vehicles, unmanned aerial vehicles [8].

History of war is closely connected with the technological advances of humanity. Whether it was the first stone-tipped spear or the modern drones that are controlled by artificial intelligence (AI) [9, 10], the development of weapon systems has always rewritten the strategies, morals, and the very results of a conflict. Throughout the millennia, various innovations like gunpowder, mechanized vehicles, and nuclear fission have not only increased the war-fighting capability of militaries, but also transformed geopolitical balances of power. The 21st century has seen the merging of AI [11], robotics and big data, leading to a new era: intelligent warfare, where machines decide in a split-second, cyberattacks [12] paralyze infrastructure without leaving a trace, and autonomous systems act without a direct human input [13].

The change is without precedent in its rapidity and in magnitude. Whereas the Industrial Revolution spanned centuries to bring about change in warfare, digital technologies such as machine learning and quantum computing are progressing at an exponentially faster rate, condensing decades of progress into years. As an example, AI-powered platforms like the U.S. Department of Défense Project Maven [14], which is intended to examine combat zone video footage, point to the truth that militaries are currently more interested in speed, accuracy, and automation rather than brute force. However, such developments provoke existential doubts: Will ethics systems be able to stay in step with autonomous weapons? Will smart warfare upset world security? The chapter aims to meet these challenges by studying the historical trend of weapon systems and examining the consequences of their increased intelligence [15, 16].

The terrorist attack on India in 2025 due to a misunderstanding with a neighbouring country made use of Rafale aircraft, SCALP missiles, sky striker, and hammer glide bombs during the operation.

The shift from manual arms to automated ones is unparalleled [15]. The industrial revolution took ages to transmute warfare, while the digital world is advancing exponentially in the field of intelligence, machine learning, and quantum. This evolution has conquered every aspect of the world and humans from peace and development to agitation and war [17].

### 1.2 Problem Statement and Significance

AI and autonomous systems have already started to be widely used in the armed forces, yet there has not been enough time for debates about their proper use. Nuclear weapons are used according to MAD, so there are defined rules for their use, but intelligent warfare technologies do not have clear guidelines for how they can be used. To illustrate, autonomous weapons systems (LAWS) can identify and attack targets all on their own, leading to issues with principles like telling combatants apart from civilians and acting with proper degree of force. As an example, cyber warfare makes it difficult to distinguish between peaceful times and conflict, as Russia's NotPetya attack in 2017 brought $10 billion of damage to others when it was presented as a ransomware incident [14, 18].

What's at risk is very grave. A study from RAND Corporation [19] shows that more than 40 nations are currently building military AI, creating a threat of unregulated competition in arms. Besides terrorists, many non-government groups are using drone and AI technologies, thus letting more parties gain access to dangerous weapons. No active governance could let intelligent warfare undermine accountability, increase unpredictable conflicts, and weaken the peace structure after World War II.

### *1.3 Objectives and Scope*

The purpose of this chapter is:

1. Examine the way weapon systems have advanced, beginning from the Industrial Revolution and to the present digital era.
2. Study what makes up intelligent warfare, for example, using AI, deploying autonomous platforms, and cyber abilities.
3. Consider the ethical, legal, and strategic problems that arise from new systems using examples like drone swarms and algorithms that decide.
4. Suggest solutions that take care of the interests of the military and of people at the same time.

Changes are observed from the conventional through the nuclear and intelligent periods, mainly for things that happened after the Cold War. Even though the main focus is on states, the analysis also explores what happens in non-state organizations and mixed conflicts.

## 2. Historical Evolution of Weapon Systems

### *2.1 Pre-Industrial Era (4000 BCE–1500 CE)*

The Pre-Industrial Era was a war which existed before the Industrial Era as some gradual innovation was carried into the field of materials and engineering. During the stone age, early man made use of primitive tools like sharpened stones and wooden spears for hunting and fighting. Metallurgy started around 3300 BC and humans invented bronze weapons such as swords, axes, etc., which were pretty durable and lethal by now. By the year 1200 BCE, warfare was revolutionized by iron smelting. The gladius was a short iron sword that was used by the Romans. It later became associated with the Roman Empire scholar O'Connell, 1989 [1]. The 12th-century Europe siege warfare trebuchet was developed alongside frequent use of siege warfare by engineers. Trebuchet were able to fling 100 kg objects over 300 m (about 1,000 ft). Armies began using trebuchet to penetrate fortified walls. The siege of Acre (1191) involved the use of trebuchet [2].

## 2.2 Need of Weapons

Weapons have been an essential need of human mankind since ages. They are used in diverse fields ranging from domestic kitchen such as cutting chopping knives to needles for sewing fabric, doctors' clinic as surgical instruments, manufacturing wooden goods such as saw, axe, in industries and day-to-day life for repairing machines, transformers, vehicles, by thieves to threaten people and snatching away their valuables, in sports like archery, javelin throw, and much more. Sometimes misconception and misunderstandings originate from riots, attacks, and war where weapons play a pivotal role. The weapons could be a boon or a curse depending on their usage in constructive or destructive purposes.

In legal systems, the term weapon could be defined in various ways. Adequately, it may be referred to as a tool adapted and designed to cause harm physically.

Weapons are an important aspect of conflict. But the concern is it could itself be the cause of conflict. The study brings into consideration that the arms traders act as catalysts to encourage and promote violence. Scarcity and lack of supply of weapons could postpone warfare and could lead to peaceful negotiations.

## 2.3 Evolution and Progressive Development of Weapons

According to Indian mythology, weapons were always an integral part of the lives of Gods and Demons. The *gadaa*, i.e., the club or mace, a shaft attached to a spherical head, used in martial arts and in mythology. It represents a symbol of strength and authority. The history of the Indian subcontinent mentions its practice by soldiers and religious activities. Our Vedic literature witnesses the study and usage of simple to advanced weapons by all the kings, queens, and their subjects for self-defense.

With the evolution of mankind, their needs also evolved, such as the need to arrange food, shelter and clothing, and protection from threats including wildlife and human rivalry. This gave birth to traditional weapons. In the Stone Age, sharpened stones with pointed tips marked the earliest example of weapons (Figures 1,2,3,5). The archaeologists have confirmed by animal blood dating that it was some 64,000 years back in South Africa. It proved to be more effective than organic resources in producing deep cuts and wounds, thus killing effortlessly. Slowly and gradually, as the humans evolved, bows and arrows were invented to feed their hunger by hunting and for self-defense. The archers were hiding at a safe place away from the target. Evidence shows modifications in bows and arrows took place with time: recursive bow, composite bow, and such. Different materials were used to make them according to the availability in the vicinity. Horses and chariots were used later to create an army. Thus, that is how hunger and protection led to war.

As civilization flourished, humans began to explore nature. Soon Metal Age, a collective name for copper, bronze, and Iron Age flourished in Mesopotamia, China, Indus, Egypt, Greece, and Rome. In 6000 BCE, copper was discovered in Anatolia, also known as Asia minor, a peninsula in western Asia in Turkey.

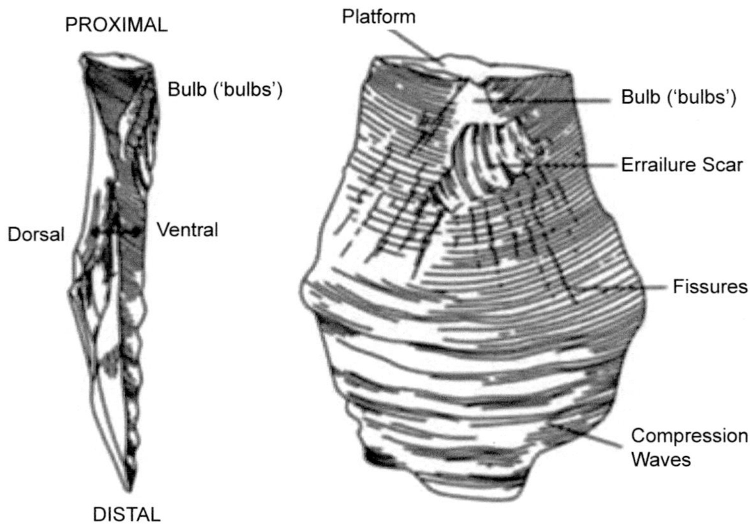


| *Physical properties of materials knapped* | *Density g/mm$^2$* | *Hardness - Moh's scale of hardness* |
|---|---|---|
| Tuff | 1.00–1.80 | 4.0–6.0 |
| Rhyolite | 2.40–2.60 | 5.0–7.0 |
| Prophyry | 2.50 | 6.0–7.0 |
| Obsidian | 2.60 | 5.0–5.5 |
| Quartz | 2.65 | 7.0 |
| Qartizite | 2.70 | 6.5–7.0 |
| Chert | 2.70 | 6.5–7.0 |
| Flint | 2.70 | 7.0 |
| Jasper | 2.80–2.90 | 3.5–4.0 |

**Fig. 1** Left: A human-knapped flint flake with a bulb ('bulbus') marking the contact point from the knapping of this flake. The compression waves can be seen around it. Right: Physical properties for some materials knapped in the stone age, order by increasing density. These values were extracted from the web database "compare rocks" [20].

With this, metallurgy came into existence and became popular in India, China, Europe, Mesopotamia, and Europe. The Bronze Age began in 4500 BCE with the advent of metallurgy. It is an alloy of copper and tin. The weapons transformed from traditional stone to bronze ones. Pole-axes, knives, maces, composite bows, helmets, spears were constructed out of bronze (Figure 4). It was easy to mould and sharpen the blade but sharpness did not last longer. With the development and growth of human civilization, various tribes: the Hyksos, the Barbarians were created. Bronze swapped the stone weapons. Maces (*gadaa*) were very popular

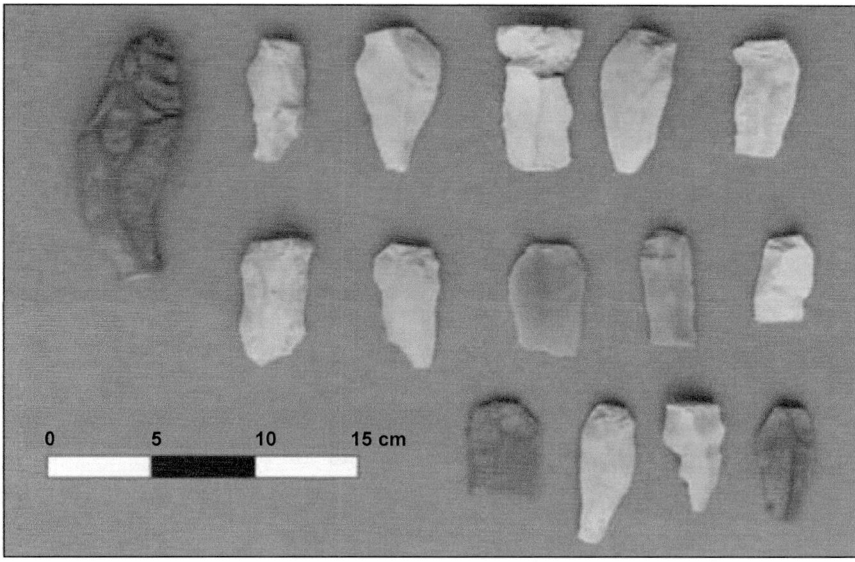


**Fig. 2** Knapped blades and flakes of tuff from Dalarna, Sweden [21].

**Fig. 3** Rock carvings in Tanum depicting men with weapons, Western Sweden.

during the Bronze Age. According to the data, historians revealed that Sumerians, the earliest known civilization, in Iraq, were the first ones to use bronze weapons. Bows and arrows, spears and slings were used in wars. However, bows and arrows

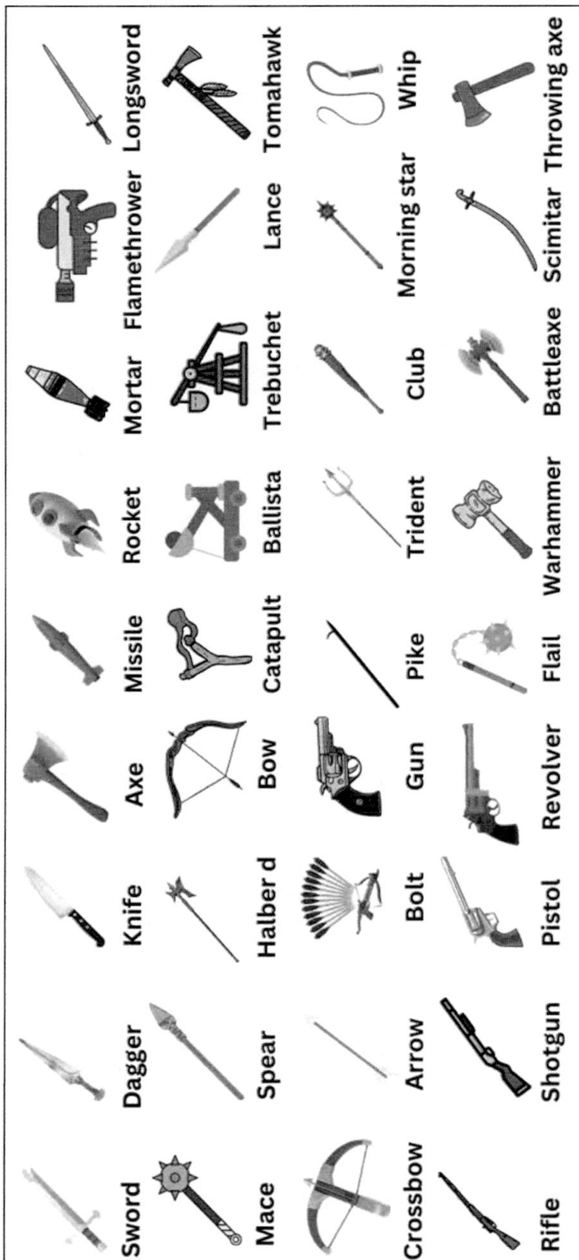


**Fig. 4** Names of weapons.

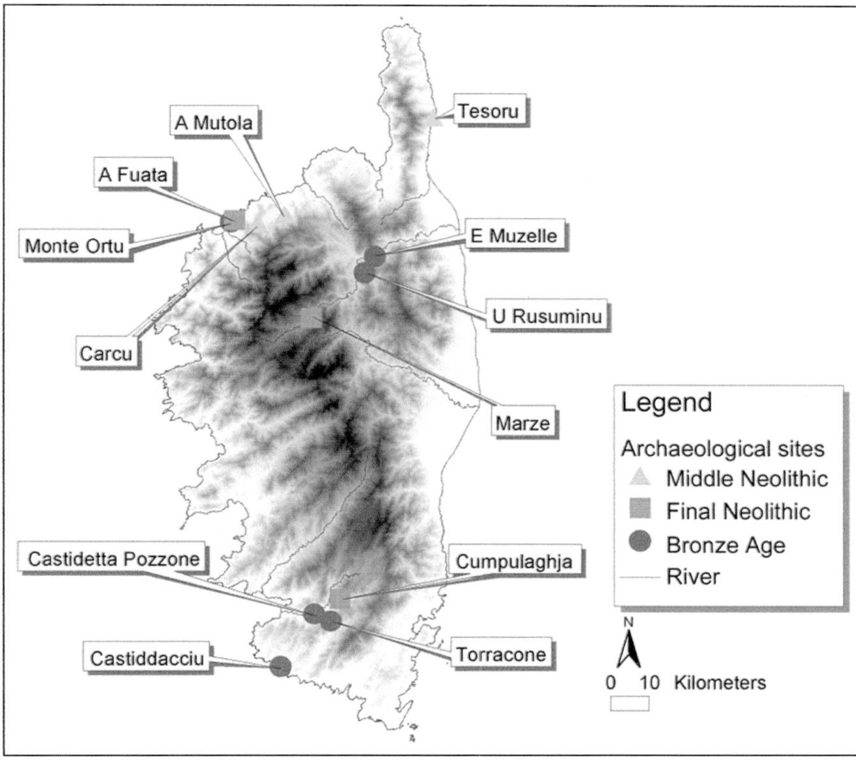


**Fig. 5** Localisation of the archaeological sites [22].

were favored over spears as they required less raw material, more accuracy, higher mobility, and were easy to use. It proved to be a boon for the hunters.

The growth of artillery had prejudiced the structure of the society and dynamic forces all over history. The arrival of iron in 1200 BCE introduced a new definition to the war. Iron was available in abundance in nature in contrast to bronze which was an alloy. The constituents were scarce and production of alloy was expensive. This decrease in cost altered the situation of war. Even the small states and farmers could arm themselves effectively. Consequently, the armaments no longer belonged only to rich and wealthy or elite empires. This caused the power to be decentralized in a number of regions, the smaller kingdoms and lords attacked the larger supremacies. This revolutionized the war with the rise of technologies such as catapult, a war machine. This required a lot of expense on materials, skilled personnels, logistics, and so on; that was minimum for hand-held weapons.

## *2.4 Industrial Revolution and Mechanized Warfare (1760–1914)*

As the Industrial Revolution hit between 1760 and 1914, fighting wars became more mechanized. Because of industry's progress, arms manufacturing relied on producing parts through machines and mass production. As a result of interchangeable parts, firearms could be put together very quickly and easily. The British Enfield Rifle was able to be produced in large quantities as after 1855; 1,000 units could be made daily [23]. Table 1 highlights the exponential growth in firearm production, reflecting industrialization's impact.

Civil War strategists faced a huge change when the Gatling gun was invented in 1862, since it fired 200 rounds in 1 minute, while older muskets only managed three.

There were advancements in weapons for both infantry and also for vehicles and the marine fleet. When the British Mark I tank appeared in 1916 at World War I, it rewrote the rules of fighting by getting past the defences in the trenches. At the same time, German submarines relied on torpedoes to mount sneak attacks on the sea, cutting off important supplies for Allied forces.

Because of industrialized warfare, forming a good logistics system, strong infrastructure, and economic support became crucial. Movement of troops over extensive distances happened faster with railroads, and factories switched over to making things for the war. They not only affected how wars were conducted, but also engaged the military in almost every sector of the nation [24].

**Table 1** Global Firearm Production (1800–1900) [23].

| *Year* | *Rifles Produced/Day (UK)* |
|---|---|
| 1800 | 10 |
| 1855 | 1,000 |
| 1900 | 3,500 |

### *2.5 World Wars and the Nuclear Age (1914–1991)*

World Wars made huge shifts in technology possible for the first time. Strategy included aviation, and the Boeing B-17 Flying Fortress (1938) was able to bomb enemies with 2,700 kg, for example, by taking part in the 'Dambusters' raid (1943) against Axis strongholds. At the beginning of the nuclear era, the Manhattan Project finished with Little Boy killing 140,000 in Hiroshima. By 1986, the number of nuclear warheads built by countries reached a maximum of 70,300 because of Cold War tensions (Bulletin of the Atomic Scientists, 2023) [25, 26] (Table 2).

With the appearance of nuclear weapons, countries reversed their policies towards direct combat and started fighting in indirect ways. The fragility of the balance of power was underlined during the Cuban Missile Crisis that took place in 1962 [26].

### *2.6 Cold War and Precision-Guided Munitions (1947–1991)*

In the Cold War, it became more important for military strategy to use precise attacks rather than forceful ones used in history. Better technology made it possible

**Table 2** The number of nuclear bombs to destroy the planet [27].

| *S.No.* | *Nuclear Weapon Yield (mt)* | *Effective Kill Area* | *Effective Kill Radius* | *Required Number* |
|---|---|---|---|---|
| 1. | 1 | 7 | 1.49 | 21285714 |
| 2. | 10 | 33 | 3.22 | 4515152 |
| 3. | 100 | 150 | 6.93 | 993333 |
| 4. | 1000 | 700 | 14.93 | 212857 |
| 5. | 5000 | 2037 | 25.47 | 73146 |

to accomplish strikes more precisely and with less damage to innocent people. One important case was the Tomahawk cruise missile, introduced in 1983, that had GPS guidance and could hit its target within 10 meters. Thus, the Vietnam War's airstrike tactics changed greatly from the "carpet bombing" of the previous world conflict, avoiding much harm to the general public.

Also, the use of satellites completely changed the way military information was collected. The KH-11 Kennen satellite was in use from 1976, giving America real-time intelligence about important events. Thanks to these systems, the United States was able to watch closely what the Soviets did, for example, during the tense times of the Cuban Missile Crisis. Because of these developments, a new kind of warfare began to exist.

Declared a success, precision-guided munitions were displayed at their best during the Gulf War of 1991[28] (Table 3). With a 3-meter precision, laser-guided bombs were a huge improvement from the old bombs used in WWII, which could only hit targets with an average margin of error of 1,000 meters. As a result, militaries turned towards network-centric warfare based on sharing and using data more effectively than using massive force.

**Table 3** Accuracy Comparison: WWII vs. Gulf War [28].

| *Weapon System* | *Circular Error Probable (CEP)* |
|---|---|
| WWII Gravity Bomb | ~1,000 meters |
| Gulf War Laser Bomb | ~3 meters |

## 3. The Dawn of Intelligent Warfare

### *3.1 Defining Intelligent Warfare*

In the tech age, traditional combat focused on using people, machines, and ammunition, whereas today's warfare uses leading technologies and depends on data. Artificial intelligence (AI), machine learning (ML), autonomous systems, and networked platforms are incorporated in this approach to improve the way and speed and accuracy in which wars are fought. Just using brute force is not enough in intelligent warfare, which prefers to use data efficiently to gain advantage over enemies.

The Department of Defence (DoD) classifies intelligent warfare as using autonomous, AI, and cyber systems to control decision-making in battles across multiple domains [29]. This plan is meant to operate in land, air, sea, space, and cyberspace areas. A finance analyst should have the following qualities:

Being autonomous, drone swarms and UGVs (unmanned ground vehicles) can carry out missions alone, with little help from humans. The process is faster and risk is lowered because machines make the choices right away using AI.

ML allows systems to change their tactics as the process continues. Because of this, drones can automatically select a different flight route or targeting strategy when new dangers or changes in surroundings appear.

When the IoT (internet of things) is used, sensors, shooters, and command centers become connected and responsive, making it possible for data and actions to happen simultaneously.

Example: In the 2020 Nagorno-Karabakh conflict, the Azerbaijani army made good use of Bayraktar TB2 drones made in Turkey [30]. These UAVs were equipped with AI-based targeting technology and could on their own locate and blast enemy tanks. Moving ahead, the use of drones meant that humans were no longer needed for spy missions and attacks. This proves what wars look like, indicating that we are now moving into the era of knowledge-based warfare.

## *3.2 Core Technologies Driving the Shift*

### *3.2.1 Artificial Intelligence and Machine Learning*

Because of AI and ML), the military is able to quickly sort through a lot of data and make real-time choices. With AI/ML, it becomes possible to make predictions, automate target identification, and increase the ability to make effective decisions [31]. For instance, Project Maven launched by the U.S. Department of Defence uses ML to go through footage from drones. Because of these capabilities, enemies can now be spotted almost instantly, while before it would take much longer (Scharre, 2021) [32]. Moving so swiftly is crucial in cases where the reaction time for action is very important.

But, when AI is used more, it brings new problems that can arise. If the training data for AI is altered by attackers, it may incorrectly judge a school bus to be a tank, which is very dangerous. This shows why it is essential to design and use powerful validation and cybersecurity measures in AI-enabled warfare systems.

Many nations have realized the importance of AI in defense and are showing strong support for military AI by planning to increase investments from $6.2 billion in 2020 to $18.8 billion by 2028 (as reported by Markets and Markets) [33].

### 3.2.2 Unmanned and Autonomous Systems

Autonomous and unmanned devices are now taking over jobs that used to be done by soldiers. The MQ-9 Reaper drone (Figure 6) and Russian Uran-9 UGV can do surveillance work, carry out strikes, and fight combat on land using limited or zero human aid [34].

U.S. Perdix drones and similar innovations use bird flocking methods to direct the activities of the group automatically. During 2021, Israel made use of Harpy NG (New Generation) munitions, able to hunt and attack enemy radar systems automatically without any human direction [35]. These changes demonstrate that AI is playing a bigger role in controlling today's battlefield.

**Fig. 6** MQ-9 Reaper Drone [36].

### 3.2.3 Cyber and Electronic Warfare

Without even firing weapons, cyber warfare can now largely affect important infrastructure in modern day fighting. This means that cyber activities depend on finding and using existing software weaknesses to get what they want. For example, in 2010 there was the Stuxnet worm [37], a U.S.-Israeli collaboration that many labeled as such. It managed to sneak into the Natanz nuclear facility and damage programmable logic controllers (PLCs), damaging almost 1,000 centrifuges and preventing any firing of missiles. It was the start of a time when wars were fought in a way impossible to see.

Nowadays, cyberattacks are commonly linked to actions that influence people mentally. As an example, Russia's Sandworm group carried out cyberattacks on Ukrainian power grids to disturb electricity and make people panic, as well as to unsettle the government [38]. As revealed by SIPRI [39] (Table 4), 68% of countries have set up groups for offensive cyber activities, proving the rise in importance of this type of warfare globally.

### 3.2.4 Quantum Computing and Advanced Sensors

Besides, quantum computing and sophisticated sensors are set to become major achievements. Quantum computers could break existing code used for encryption and compromise secure conversation. In addition, there's a possibility that quantum radar will undo the advantages of stealth technology. Hyperspectral sensors, for example, those on World View Legion, can pick out hidden or covered targets by examining light wavelengths, making a big difference in how battlefields are monitored.

### 3.2.5 Strategic and Ethical Implications

In today's world, smart warfare makes it harder to stick to traditional methods of fighting and ethics, especially when it comes to accountability, increases in violence, and possible biases in algorithms.

**Accountability Gap:** There are serious issues related to accountability and law when using autonomous weapons. In case an AI manages to violate international law, it is not clear if the responsible party is the person who wrote the program, the one who controls it, or the AI device itself. In Libya in 2021, there was a known case where a Turkish Kargu-2 drone attacked human combatants all by itself. The members of the UN Security Council pointed out this event could be regarded as breaking the Geneva Convention and called for the need for immediate legal understanding [40].

**Escalation Risks:** Even though AI makes tasks more precise and faster, it can also cause things to get out of hand. According to the 2022 RAND study [15], missing information from AI, which might wrongly identify a radar anomaly as a hostile missile, might bring about accidental nuclear conflict after humans and diplomatic efforts have been skipped.

**Bias in Algorithms:** When AI systems are created using uneven datasets, they might reflect prejudiced or unjust treatment towards different groups. According to Buolamwini and Gebru (2018) [41], it is common for facial recognition systems to fail on minorities, which can result in civilian loss and more chances of targeting

**Table 4.** Global Military AI Adoption [39].

| *Country* | *Autonomous Systems Deployed* | *AI Defence Budget (2023)* |
|---|---|---|
| USA | 800+ | \$4.5B |
| China | 650+ | \$3.2B |
| Russia | 300+ | \$1.1B |

the wrong people. It is necessary to pay close attention to these problems when using AI in wartime.

## 4. Case Studies In Modern Intelligent Warfare

### *4.1 Autonomous Weapon Systems: The Kargu-2 Drone Swarm in Libya*

In 2020, Turkey's military sent Kargu-2 drones against General Haftar's forces when the Libyan civil conflict was happening. Differential from standard drones, the Kargu-2 uses AI to find targets by itself and works as a group with several other drones [42]. These drones do not need to be controlled by humans to find targets on the ground and attack as they see fit (Figure 7).

According to the United Nations Security Council (2021) [40], these drones hunted after soldiers who were fleeing in a likely option of automatic or limited human intervention. It was the first time lethal weapons acted on their own in combat, which made people around the world especially concerned.

A number of ethical and legal concerns have arisen because of the use of such systems. One important problem is that there isn't anyone supervising the process in real-time, which means that the law requiring armies to distinguish between civilians and members of the military is not followed. These weapons that operate on their own in battle situations are challenging the traditional systems of responsibility, proportionality, and treating people fairly in wartime. The Kargu-2 event makes it clearly necessary to create internationally recognized rules and guidelines for using LAWS (Table 5).

**Table 5** Autonomous Systems in Modern Conflicts [40, 43].

| *S.No.* | *System* | *Operator* | *Level of Autonomy* | *Case* |
|---|---|---|---|---|
| 1 | Perdix Swarm | USA | Semi-autonomous | Syria in 2018 |
| 2 | Harpy NG | Israel | Full | Gaza in 2021 |
| 3 | Kargu-2 | Turkey | Full | Libya in 2020 |

### *4.2 Cyber Warfare and Hybrid Threats: The NotPetya Attack (2017)*

NotPetya was a malware attack in 2017 [44, 45], and it was attributed to Russia's GRU military intelligence agency. This became one of the most damaging cyber operations. The malware's original purpose was to create chaos in Ukraine's financial and energy systems. It, however, spread far and wide and is estimated to have led to losses of around $10 billion across the world. Maersk and Merck were greatly harmed by the pandemic since they are part of the shipping and pharmaceutical sectors.

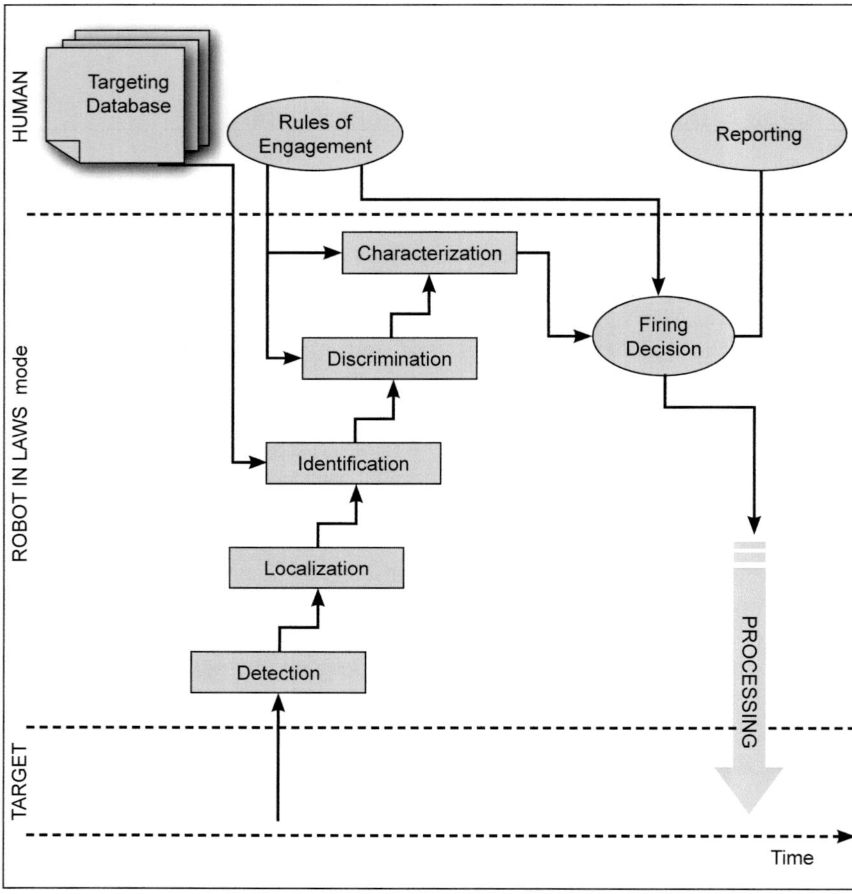


**Fig. 7** Operating mode and sequencing for a firing decision in autonomous mode [43].

Though called ransomware, NotPetya was in fact a harmful program that left a lot of devices permanently broken. It exploited the EternalBlue weakness, which was earlier obtained from the NSA, in order to get access to machines unapproved by the owners.

As soon as the malware got access to the machines, it rewrote the main boot area and shut down the devices, preventing any chance of finding their data.

The attack was an example of hybrid warfare, since it was carried out alongside false news to try to weaken the people and officials in Ukraine. Because the malware didn't care about borders, it proved that connected networks can experience unwanted side effects.

Maersk's losses reached over $300 million, since its global IT system had to be rebuilt with information coming from a single server [45]. It was shown by

NotPetya that cyber warfare can cause more problems than expected, putting the world's economy and the infrastructure of people at risk (Table 6).

**Table 6** List of major cyberattacks from 2010 to 2023 [44, 45].

| *S.No.* | *Actor* | *Target* | *Attack* | *Damage in USD* |
|---|---|---|---|---|
| 1. | Russia | US Agencies | Solar Winds | $18B |
| 2. | Russia | Ukraine/Global | NotPetya | $10B |
| 3. | Israel/US | Iran | Stuxnet | $1B |

### *4.3 AI in Decision-Making: Project Maven and Algorithmic Warfare*

Menaced by AI and ML playing a major role in bringing AI and ML into everyday work within the U.S. military. The main purpose is to use drone surveillance videos from war areas such as Syria and Afghanistan to ease the identification of factors such as vehicles, weapons, and enemy troops.

With ML models, Project Maven can look at many hours of aerial footage in moments and spot interesting objects with a high level of accuracy [32]. This saves a lot of time for intelligence analysis and therefore allows for faster and possibly more accurate decisions about enemy targets on the field.

Still, the project has caused several issues related to ethics. At its start, the system is said to have incorrectly identified civilians as combatants because of factors in the training that highlights different Middle East features [41] (Table 7). Such errors put people at greater danger and increase the risk of innocent people being harmed.

Because of the objections of thousands of employees, Google chose not to continue with Project Maven in 2018. This move led to people's worries about arms-related AI being developed in secret, with no clear guidelines for using it in fighting situations.

**Table 7** Comparison between human analysis and AI accuracy [41].

| *Metric* | *AI (Project Maven)* | *Human Analysts* |
|---|---|---|
| Speed | 2 seconds/image | 10 mins/image |
| Accuracy | 90% | 85% |

## 5. Ethical, Legal, and Strategic Implications

### *5.1 Ethical Concerns*

With the growth of smart warfare, many ethical issues appear, mainly related to deciding who is responsible and what is right in situations that determine life or death. Ethics based on established traditions are not able to keep up with recent developments in AI and ML in conflict situations.

### *5.1.1 Accountability Gaps*

There is growing worry because it is not clear who is answerable when autonomous systems hurt someone accidentally. A drone such as the Kargu-2, which is able to work without needing human guidance, can attack on its own. If this happens and an innocent civilian vehicle is marked as a target, figure out who is responsible: the software developer, those who operate the military, or the policy team. In 2022, a study in Nature Machine Intelligence discovered that many AI ethicists feel existing legal and ethical frameworks can't deal with the accountability challenges [46].

### *5.1.2. Unintended Escalation*

Such systems can lead to accidental conflicts that might become more serious. According to results of a RAND Corporation simulation in 2020 [47], AI systems frequently advised pre-emptive strikes because they lost awareness of cautious actions due to something known as catastrophic forgetting. It shows that AI errors may bring about situations that humans would have managed to avoid.

### *5.1.3 Algorithmic Bias*

AI systems that display bias may result in discrimination in battle areas. Facial recognition tools, for example in Project Maven, are known to sometimes confuse people based on their race and gender. The MIT study stated in 2021 that, for currently available facial recognition programs, the error rate was 34.7% for people with darker skin and only 0.8% for those with lighter skin [46]. These differences do not only threaten non-combatants but also make the legitimacy of AI-based technology in war lower.

All these ethical problems require a prompt discussion at global level, policy changes, and close oversight.

## *5.2 Legal Frameworks and Governance*

The subject of governance and legal frameworks is important for Africa. The regulation of intelligent warfare brings many legal obstacles.

Advancements in smart warfare have gone faster than improvements in international law. It is hard for International Humanitarian Law (IHL) and other international laws to control or regulate the development of autonomous weapons, cyber operations, and algorithms.

### *5.2.1 International Humanitarian Law also Refers to IHL*

Because of IHL, the Geneva Conventions call for clear separation between civilians and those taking part in combat, and also point out not to cause more harm than is useful. On the other hand, these notions are currently being overpowered by autonomous systems. A report from UNSC on the Libyan conflict in 2021 showed that Kargu-2 drones flew and attacked opponents without anything other than programming, which went against international humanitarian rules on how targeting should be done [40].

### 5.2.2 Regulatory Gaps

Even though many are worried, the CCW has not introduced a ban on weapons capable of using AI to decide when to attack. Lobby groups such as the Campaign to Stop Killer Robots are working to stop such weapons being used, but making progress is difficult.

There is a Tallinn Manual 2.0 [48] (Table 8) that sets out laws for cyber warfare, but its rules do not need to be followed. Security incidents, such as the NotPetya attack, indicate that global cyber rules and ways to address them are absent.

### 5.2.3 National Policies

The U.S. Department of Defence Directive 3000.09 [49] says that when lethal autonomous systems are used, there should be enough human decision-making. Yet, the directive is not detailed when it comes to terms or enforcement, so there is a lot of chance for different understandings.

At the same time, China's Global AI Governance Initiative for 2022 highlights developing AI for its military and ignores any ethical controls, leading to worry about the irresponsible adoption of such technology [39].

To sum up, strong and current laws are necessary to make sure these technologies are held accountable, do not harm too many civilians, and sustain the principles that make world security function.

**Table 8** Global Governance Efforts [40, 48].

| *Initiative* | *Scope* | *Progress* |
|---|---|---|
| UN CCW on LAWS | Ban autonomous weapons | Stalled since 2014 |
| EU AI Act | Risk-based AI regulation | Adopted 2024 |
| U.S.-China AI Dialogue | Military AI norms | No agreement |

## 5.3 Strategic Balance and Global Security

## Finding the Right Equilibrium and World Security

### 5.3.1 AI Arms Race

Smarter warfare is affecting the balance between countries and causing more arms buildup. The U.S., China, and Russia are putting in over $12 billion every year to research and develop military AI [50]. The strategy uses private-sector developments in AI, such as those used in the 2023 tests of hypersonic missiles [51].

#### *5.3.2 Non-State Actors*

Drone companies like DJI are not the only ones relying on their models for different purposes; even terrorist groups like ISIS are using them. In 2022, Ukrainian troops managed to capture Russian mercenaries who were using Turkish drones in Sudan, proving that drone military technology is being used worldwide [52].

#### *5.3.3 Deterrence Challenges*

They make it harder for countries to plan their deterring strategies. Such activities using advanced technology can increase distrust in society and make prominent organizations weaker. When the 2022 Russian invasion took place, deepfakes of the Ukrainian President Zelensky giving up were shared widely and led to more confusion and fear [53]. So, intelligent warfare calls for modern solutions to the challenges in conflict management.

## 6. Future Trajectories and Challenges

### *6.1 Emerging Technologies*

The developing trends and the obstacles that will arise new Technologies Playing a Role in Today's Warfare. Upcoming wars will depend on new technologies that modify ethical rules and require armies to adjust their strategies.

#### *6.1.1 Hypersonic Weapons*

Because of their fast speed and unpredictably changing movements, the Avangard and DF-ZF evade the current missile defence systems used by many countries. In 2023, U.S. tests of the AGM-183A ARRW showed that it could reach a speed faster than Mach 15. Such weapons make it much quicker for a response to be made, which raises the risk of accidental conflict taking place. Hypersonic weapons are seen in a 2023 report by CSIS as a major danger to nuclear deterrence since they may compromise a country's ability to retaliate after being attacked [54, 55] (Table 9).

#### *6.1.2 Directed-Energy Systems*

The U.S. Navy's HELIOS using a laser can stop drones and missiles for only $1 per shot, which is much less expensive than using a conventional missile. Such systems, for example Raytheon's Phaser, help to stop drone swarms by destroying their electronics within a 1 km range [56].

### *6.1.3 Biotechnology*

The use of CRISPR-Cas9 to edit genes gives military scientists new chances to improve soldier readiness, but the idea of enhancements gives rise to ethical concerns. Besides, DARPA's N3 plans to introduce brain-controlled drones by 2030, which could considerably change remote combat and raise big questions concerning human augmentation and freedom [57]. The arrival of these new technologies signals a major change in the way wars are fought, and so it is important to watch over their use and come up with quick strategies.

**Table 9** Hypersonic Missile Programs.

| *S.No.* | *Country* | *System* | *Speed (Mach)* | *Status* |
|---|---|---|---|---|
| 1. | China | DF-ZF | 5 to 10 | Operational |
| 2. | Russia | Avangard | 20+ | Deployed 2019 |

## *6.2 Human-Machine Teaming*

Blending people with technology in the military will make it more powerful, but some are concerned about giving up humans' independence.

### *6.2.1 Exoskeletons*

In a few years, soldiers wearing the TALOS exoskeleton made by the U.S. The Army will be stronger and able to handle more than 200 pounds of equipment [58]. Even though technology in warfare means better results, depending on it can harm people's health in the long run.

### *6.2.2 Neural Interfaces*

With BCI, DARPA makes it possible for pilots to work with drones using mental commands, making their reaction quicker by more than half. Still, the use of neural interfaces brings up security issues because sensitive brainwave information could easily be attacked or controlled, without the soldier's consent [57].

### *6.2.3 AI-Augmented Decision-Making*

AI is used by the UK's Project Theseus [59] to boost the efficiency of logistics and the way battle tactics are decided. At the same time, overusing AI-based systems might lessen a commander's critical thinking and judgment, which is essential for responding well to unexpected battle conditions.

All things considered, uniting humans and machines helps a lot, but managers must handle any ethical, health, and security challenges to safeguard human leadership in military conflicts.

### 6.3 Preparing for Uncertainty

To deal with the risks of intelligent warfare, governments must change as needed, introduce ethics, and cooperate closely with other nations worldwide.

#### 6.3.1 Policy Frameworks

Being flexible in regulation is necessary to oversee the use of new technologies. The EU's new AI Liability Directive [60] makes developers responsible for any errors that happen with their AI, which could be used as a basis for controlling military AI. Even so, making progress in governing advanced weapons is hard; using the example of the Global Hypersonic Missile Ban, significant countries aren't keen on giving up on their strategic edge.

#### 6.3.2 Ethical Innovation

It takes creative methods to guarantee that military applications of AI are just and protected. The Algorithmic Justice League works to make sure that organizations undergo regular audits to handle and prevent bias in AI. Besides, simulating attacks by red teaming helps find weaknesses in AI systems before they are put into operation and protect against being exploited by enemies [61].

#### 6.3.3 Global Collaboration

Being united globally, countries improve innovation and avoid more serious risks. Startups that work in dual-use technology are financed by NATO's DIANA [62] (Table 10) to maintain the West's edge in crucial areas. In spite of current geopolitical issues, the U.S.-China dialogues on AI focus on improving mutual understanding and discussing options for reducing risks, proving that it is still important for countries with differences to keep engaging each other.Taken together, these plans make sure that new intelligent warfare technologies are used carefully and balanced with other worldwide concerns such as ethics and security.

**Table 10** Global Governance Initiatives.

| *S.No.* | *Key Players* | *Initiative* | *Emphasis* |
|---|---|---|---|
| 1. | 125+ Nations | UN GGE on LAWS | Autonomous Weapons |
| 2. | EU | AI liability Dir. | Accountability |
| 3. | NATO | DIANA | Dual use tech |

## 7. Conclusion

From simple stone devices to high-tech platforms managed by artificial intelligence, weapons have developed mainly to help humans stay ahead in battles. At the same time, new developments make both sides more able to wage war and cause more danger for everyone. Here, we have seen that AI, cyber abilities, and autonomous systems have led to a major transformation in warfare. Although these new technologies provide superior accuracy and efficiency in battles, they also bring up new ethical, legal, and strategy-related issues for warfare.

It has often been true that technology has advanced on the battlefield sooner than people were prepared for it. Over the centuries, new forms of weaponry have made it hard for societies to cope. The rise of modern and effective military methods only makes the problem more serious. Since the Turkish Kargu-2 drone and similar autonomous weapons do not require many human directions, it is unclear who should be held responsible when people are endangered. NotPetya is one example of a cyberattack that overcame boundaries between nations and wrecked civilian infrastructure all over the world.

Besides, if AI decides things, the risk of algorithmic bias can cause unfair discrimination and result in increased conflicts. Such movements endanger important rules in international law, one of which is telling the difference between soldiers and civilians, another being the appropriate scale of force, and preserving dignity.

This issue is complicated by the fact that there are not enough strict global rules for emerging technologies. There are not many treaties to help with understanding hypersonic missiles, neural interfaces, and autonomous swarms. As a result of this gap, there is more incitement to competition in weapons, a higher risk of accidental incidents, and less stability in the security situation. It is clear from the disagreement over LAWS that it is hard to achieve global consensus on arms issues where each country is concerned with its interests.

To deal with such issues, experts from many areas need to cooperate, and the government should look ahead. There is a need for policymakers to include technologists in their work to add ethical rules to AI and automated systems, carrying out clear, transparent, and accountable actions. The existing rules, for example, the United Nations Convention on Certain Conventional Weapons (CCW), require faster updates to properly control the use of autonomous weapons. At the same time, projects like NATO's DIANA and the European Union's AI Act give us helpful guidance for using technology as opposed to just controlling it. Particularly, it is vital for countries to cooperate to address threats from non-state actors that now rely on commercial drones and AI technologies, which shake up the usual approach to defense.

Overall, the growth of intelligent warfare shows that there is a need to change as a society, not only with advances in technology. Because algorithms play a bigger role in warfare, people are now choosing between using technology to chip away at values and ethics, or making sure technology supports dignity and security.

Moving on means focusing on morality rather than fast success in military actions and choosing to cooperate with allies rather than separate from them. To serve human interests, the progress of weapon systems needs to be guided by everyone's partnership, and not only by a strong nation's wishes.

## References


1. O'Connell, R.L. (1990). *Of Arms and Men: A History of War, Weapons, and Aggression*. Oxford University Press.
2. Ferrill, A. (2018). *The Origins of War: From the Stone Age to Alexander the Great*. Routledge.
3. Kenderdine, S. and McKenzie, H. (2013). A war torn memory palace: Animating narratives of remembrance. In: *2013 Digital Heritage International Congress* (*DigitalHeritage*), IEEE, pp. 315–322.
4. Mojabi, S.M., Navazi, A., Feizi, F. and Ghourchi, M. (2010).Environmental impact of white phosphorus weapons on urban areas. In: *2010 International Conference on Environmental Engineering and Applications*. IEEE, pp. 112–116.
5. Singer, P.W. (2009). *Wired for War: The Robotics Revolution and Conflict in the 21st Century*. Penguin.
6. Van Atta, L.C. and Silver, S. (2007). Contributions to the antenna field during World War II. *Proc. IRE*, 50(5), 692–697.
7. Brown, L. (2004). Airborne radar: Its beginnings in World War II. *IEEE Aerosp. Electron. Syst. Mag.*, 19(7), 3–8.
8. Lautenschläger, K. (1983). Technology and the evolution of naval warfare. *Int. Secur.*, 8(2), 3–51.
9. Lai, K.-T., Chung, Y.-T., Su, J.-J., Lai, C.-H. and Huang, Y.-H. (2022). Ai wings: An AIOT drone system for commanding ARDU pilot UAVs. *IEEE Syst. J.*, 17(2), 2213–2224.
10. Ihekoronye, V.U., Ajakwe, S.O., Lee, J.M. and Kim, D.-S. (2024). DroneGuard: An Explainable and Efficient Machine Learning Framework for Intrusion Detection in Drone Networks. *IEEE Internet Things J.*
11. Zhang, Y., Hu, Q., Wang, S., Zhao, P. and Ai, M. (2025).Satellite Remote Sensing Time Series Observation of Russia-Ukraine War: Ongoing Conflict Increased Crop Risk. *IEEE J. Sel. Top. Appl. Earth Obs. Remote Sens.*
12. Abdullahi, M., et al. (2024).Comparison and investigation of AI-based approaches for cyberattack detection in cyber-physical systems. *IEEE Access*.
13. Cassidy, R.M. (2007). War Made New: Technology, Warfare, and the Course of History: 1500 to Today. *Parameters*, *37*(3), 109.
14. Doubleday, J. (2018). Project Maven aims to introduce AI tools into services' intel systems. *Insid. Army*, *30*(1), 1–7.
15. Geist, E., Frank, A.B. and Menthe, L. (2024). Understanding the Limits of Artificial Intelligence for Warfighters. *RAND Corp. Jan.*
16. Horowitz, M.C. (2018, May). Artificial intelligence, international competition, and the balance of power. Texas National Security Review.
17. Black, J. (2013).*War and Technology*. Indiana University Press.
18. Kello, L. (2017). *The Virtual Weapon and International Order*. Yale University Press.
19. Johnson, J. (2020). Artificial intelligence, drone swarming, and escalation risks in future warfare. *RUSI J.*, 165(2), 26–36.
20. Web database. *Compare Rocks*. [Online]. Available: https://rocks.comparenature.com/en.
21. Gran, O., et al. (2017). Status for the development of acoustic mapping of submerged stone age sites. In: *2017 IEEE/OES Acoustics in Underwater Geosciences Symposium* (*RIO Acoustics*). IEEE, pp. 1–5.
22. Mazet, S. (2006). Habitat and Stone Enclosure of Corsica: What is the Function of the Enclosed Area during the Neolithic and the Bronze Age? In: *2006 First International Symposium on Environment Identities and Mediterranean Area*. IEEE, pp. 447–450.
23. McNeill, W.H. (2002). *The Pursuit of Power: Technology, Armed Force, and Society Since AD 1000*. Published in 1982. University of Chicago Press.
24. Van Creveld, M. (2010). *Technology and War: From 2000 BC to the Present*. Simon and Schuster.

25. Overy, R. (2024).*Why the Allies Won*. WW Norton & Company.
26. Rhodes, R. (2012). *The Making of the Atomic Bomb*. Simon and Schuster.
27. Cheng, Z., Guan, Y., Du, Q. and Li, Y. (2023). Research on Nuclear Weapons Development Trend Prediction Based on LSTM and GM (1, 1) Models. In: *2023 International Conference on Data Science and Network Security (ICDSNS)*. IEEE, pp. 1–6.
28. Keaney, T.A. and Cohen, E.A. (1993). *Gulf War Air Power Survey* (Vol. 6). Office of the Secretary of the Air Force, 1993.
29. Chapman, B. (2020). How US Government Policy Documents are Addressing the Increasing National Security Implications of Artificial Intelligence. *J. Adv. Mil. Stud.*, 11(1), 209–232.
30. Nistorescu, C.V. (2024). Loitering Munitions-Possible "Game Changer" in Future Warfare. *Rom. Mil. Think.*, 2. ResearchGate.
31. Cummings, M. (2017). *Artificial Intelligence and the Future of Warfare*. Chatham House for the Royal Institute of International Affairs London.
32. Scharre, P. (2018). *Army of None: Autonomous Weapons and the Future of War*. WW Norton & Company.
33. MarketsandMarkets. *Artificial Intelligence (AI) in Military Market Size & Share*. [Online]. Available: https://www.marketsandmarkets.com/Market-Reports/artificial-intelligence-military-market-41793495.html.
34. Ampov, M. and Petreski, D. (2022). Unmanned Aerial Vehicles Used for Military Operations and Their Application in the Region. *XXII* (43).
35. Tucker, C., Tafoyacabrera, J., Montanez, C., Avram, R., Bauer, M. and Jin, Y.-F. (2022). An Autonomous Tracking System Integrating UAV and UGVS. In: *2010 GSW*.
36. *MQ-9 Reaper Drone*. [Online]. Available: https://en.wikipedia.org/wiki/General_Atomics_MQ-9_Reaper.
37. Chen, T.M. (2010). Stuxnet, the real start of cyber warfare? [Editor's Note], *IEEE Netw.*, 24(6), 2–3.
38. Radziwill, N.M. (2018). *Countdown to Zero Day: Stuxnet and the Launch of the World's First Digital Weapon: 2014*. Kim Zetter. New York: Broadway Books, p. 296, Taylor & Francis.
39. *SIPRI Yearbook 2022*. [Online]. Available: https://www.sipri.org/yearbook/2022.
40. *Libya Country Report*. (2021). [Online]. Available: https://docs.un.org/S/2021/229.
41. Buolamwini, J. and Gebru, T. (2018). Gender shades: Intersectional accuracy disparities in commercial gender classification. In: *Conference on Fairness, Accountability, and Transparency*. PMLR. pp. 77–91.
42. *Kargu-2 Loitering Munition System*. [Online]. Available: https://www.stm.com.tr/en/kargu-autonomous-tactical-multi-rotor-attack-uav.
43. de Boisboissel, G. (2015). Uses of lethal autonomous weapon systems. In: *International Conference on Military Technologies (ICMT) 2015*. IEEE, pp. 1–6.
44. Khan, U.K.A.B.G. (2017). The Evolution of Ransomware: A Case Study of WannaCry and NotPetya. ResearchGate.
45. Greenberg, A. (2019). *Sandworm: A New Era of Cyberwar and the Hunt for the Kremlin's Most Dangerous Hackers*. Anchor.
46. Raji, I.D., and Buolamwini, J. (2019). Actionable auditing: Investigating the impact of publicly naming biased performance results of commercial AI products. In: *Proceedings of the 2019 AAAI/ACM Conference on AI, Ethics, and Society*, pp. 429–435.
47. Johnson, J. (2022). Inadvertent escalation in the age of intelligence machines: A new model for nuclear risk in the digital age. *Eur. J. Int. Secur.*, 7(3), 337–359.
48. Schmitt, M.N. (2017). *Tallinn Manual 2.0 on the International Law Applicable to Cyber Operations*. Cambridge University Press..
49. Barbosa, L.V.F. (2023). Exploring the 2023 US Directive on Autonomy in Weapon Systems: Key Advancements and Potential Implications for International Discussions. *CEBRI-Revista Brazilian J. Int. Aff.*, (7), 117–136.

50. *SIPRI Year Book 2023*. [Online]. Available: https://www.sipri.org/yearbook/2023.
51. Farrow, A.E. (2023). Modernization and the Military-Civil Fusion Strategy.. *J. Indo-Pacific Aff.*, *6*(6).
52. Petrosyan, M. (2024). The Role of Non-State Actors in Modern Warfare: The Case of Syria and Nagorno-Karabakh. *J. Balk. Near East. Stud.*, *26*(2), 149–163.[53] *Atlantic Council, 2022/23*. [Online]. Available: https://www.atlanticcouncil.org/in-depth-research-reports/report/2022-annual-report/.
54. Wilkening, D. (2019). Hypersonic weapons and strategic stability. *Survival (Lond).*, 61(5), 129–148.
55. Aarten, S.R. (2020). The impact of hypersonic missiles on strategic stability. *Mil. Spect.*, 182–193.
56. Obering, H.T. (2019). Directed energy weapons are real... and disruptive. *Prism*, *8*(3), 36–47.
57. Emondi, A. (2020). Next-generation nonsurgical neurotechnology. *DARPA* https//www. darpa. mil/program/next-generation-nonsurgical-neurotechnology (last Visit. December 24, 2019*)*.
58. Martinez, A., Velazquez, A., Yoshizuka, A., Clark, R. and Mittal, V. (2016). Developing an exoskeleton test plan for the talos program. *Ind. Syst. Eng. Rev.*, 4(2), 156–162.
59. Harnden, R.J. and Stringer, R. (1993). Theseus: A model for global connectivity. In: *Systems Science: Addressing Global Issues*. Springer, pp. 229–234.
60. Hacker, P. (2023). The European AI liability directives–Critique of a half-hearted approach and lessons for the future. *Comput. Law Secur. Rev.*, *51*, 105871.
61. Allhoff, F., P. Lin, and Steinberg, J. (2011). Ethics of human enhancement: an executive summary. *Sci. Eng. Ethics*, 17, 201–212.
62. McSorley, T., Macenowicz, M., Maddison, M. and Yukins, C.R. (2025). Allies Bridging The Valley of Death: How NATO's Defence Innovation Accelerator for The North Atlantic Will Help Maintain NATO's Technological Edge. GW Law Faculty Publications, George Washington University.

# 20 Adaptive Control Strategies in Autonomous Vehicles: A Machine Learning Approach

M. Indira[1*] and C. Victoria Priscilla[2]

## 1. INTRODUCTION

### *1.1 Background Study*

Autonomous vehicle development and operation now highly depends on a new technology called Artificial Intelligence (AI). The vehicles which are called Autonomous vehicles (AVs) are able to sense, navigate, and adapt to new situations due to the usage of algorithms of AI. These algorithms make the operations of vehicles more efficient and safer. Future autonomous vehicles can be enhanced to safety to a great extent by the advancement of AI technologies (Anton et al., 2023) [1]. A revolutionary transformation has happened because of the advancements of AI and development of autonomous systems. To accelerate innovation, improve efficiency and transform conventional development procedures have the potential combination which is ground breaking. In the future, the mobility will be evolving with autonomous vehicles (AVs), it will rapidly advance the research. The vehicles need to drive through highly dynamic, unexpected, heterogeneous urban scenes, challenges for vision, control, and real-time decisions.

Correct environmental perception and advanced control mechanisms that go beyond traditional rule-based or linear control architectures are required to provide safety, comfort, energy efficiency, and compliance with traffic regulations in such scenarios. The automotive industry (Arthurs et al., 2022) [2] has seen dramatic expansion and technological innovation over the years. Car capabilities have been enhanced due to improvements in computer system integration and the computerization of mechanical and manual operations. Adaptive cruise control, lane departure warning systems, self-parking systems, and driver assistance technologies are the new features that have become prevalent in new vehicles.

These features have reduced human dependency on automobiles and given rise to partially autonomous vehicles. Autonomous cars are the outcome of other

[1] Department of Computer Science, P.K.R. Arts College for Women, Gobichettipalayam, India.
[2] Department of Computer Science, SDNB Vaishnav College for Women Chrompet, India.
* Corresponding author: indirajohnson@gmail.com

DOI: 10.1201/9781003663461-20

functions enabled by on-board cameras and sensors. The creation of AVs is further stretching this boundary even more. These vehicles take it a step further, enabling cars to use all sorts of technology to travel from one point to another without human intervention. These have the potential to revolutionize the way people travel on highways in the future. have a number of benefits, whether employed to carry the elderly and disabled or in dangerous conflict zones where casualties are likely to be incurred. Decreased fuel consumption, reduced traffic volumes, and a more efficient use of the road network are potential added benefits. These vehicles can reduce traffic accidents, air pollution, loss of life, and emissions of carbon-dioxide can be controlled. Some technologies like radars, sensors, and fixed cameras are used by AVs to sense the environment and navigate the surroundings.

Traditional methods like (Borase et al., 2021) [5] Proportional–Integral–Derivative (PID) controllers, rule-based systems, and classical Model Predictive Control (MPC) are highly influenced as control techniques in AVs. Even though these methods are easy to recognize, interpretable, and very competent in structured scenarios, but dealing with highly difficult situations and inconsistent states of the real world they incline to fail. In urban surroundings, involvement of state space, vehicle nonlinear dynamics, blocked and undefined complications, road conditions, weather changes, and sensor noise are unpredictable. In conventional methods, nonlinearity and uncertainty lower the robustness. Intelligent control structures with learning abilities from data, generalization over diverse conditions, and real-time decision-making with high dependability are therefore urgently needed.

Current expansions in machine learning, specifically deep reinforcement learning (DRL), have accessible and capable guidelines for overcoming these challenges. DRL allows systems to learn the best policies through interaction with the environment alone, using deep neural networks to estimate value functions and policies in high-dimensional continuous action spaces. Among the DRL algorithms, Twin Delayed Deep Deterministic Policy Gradient (TD3) is one of the strong contenders for continuous control tasks because it rectified some of the issues in Deep Deterministic Policy Gradient (DDPG), e.g., alleviating overestimation bias and stabilizing actor-critic training. TD3 brings twin Q-networks, delayed policy updates, and target policy smoothing into play, which together promote sample efficiency and policy robustness. These qualities recommend TD3 to be a good candidate for coordinating the real-time, fine-grained control of throttle, brake, and steering in autonomous driving.

However, although TD3 performs well in dynamic adaptation and reactive control, it could fall short of strict safety constraint enforcement and long-horizon planning, particularly in complicated temporal and spatial dependencies. Thus, to complement this limitation, this study combines TD3 with a Neural Network-augmented Model Predictive Control (NN-MPC) module. MPC has been popular in control theory for its capacity to predict future states, minimize control sequences, and impose hard constraints. Traditional MPC is dependent upon accurate logical models of vehicle dynamics and environment, which are possible to be firm to get or unpredictable under active conditions. NN-MPC recovers on this by retaining neural networks to study scheme dynamics from practice data, having better analytical abilities and awareness of environmental changes. The incorporation of acquired models facilitates MPC to react better to the future, compensating for uncertainties and enhancing trajectory tracking and safety guarantees.

The proposed hybrid control system in this work capitalizes on the complementary strengths of TD3 and NN-MPC. The TD3 is responsible for real-time exploration and reactive control, adapting to unforeseen disturbances, and rapidly changing conditions like unforeseen pedestrian walking or other vehicles suddenly turning lanes. In comparison, the NN-MPC is responsible for medium- to long-term trajectory planning with respect to safety margins, traffic rules, and road layout. The framework includes a hybrid decision module, which dynamically splits control authority between the two components based on scenario complexity, proximity to obstacles, and control uncertainty. This coordination function enables the AV to produce intelligent and understandable decisions in a wide variety of driving situations.

The control framework was tested and validated in CARLA, an open-source high-fidelity AV simulation platform (Bordoloi, U et al., 2023) [6] with realistic traffic, weather, and sensor models. CARLA supports controlled experimentation under various conditions, ranging from crowded urban traffic, varying lighting conditions, inclement weather (for example, rain, fog), and road irregularities such as potholes and roadwork. These tests establish the generalization capability, robustness, and real-time response of the control system. TD3 showed better policy learning ability over Deep Q-Network (DQN) baselines, minimizing collision rates by 42% and fuel consumption by 24%. NN-MPC also helped with improved accuracy in trajectory tracking by 33% and strict compliance with speed regulations, lane discipline, and intersection rules.

The closed-loop performance was satisfactorily verified, and the results of an enhanced response as well as an improved lane-keeping accuracy were achieved when NN-MPC was used with TD3. The results indicate that boosting the ability of anticipation with a well-designed predictive planning part gained a significant advantage in online adaptation. Such enhancements become necessary in order to ensure safe driving during high-stakes situations in which circumstances might change abruptly, such as in the case of jaywalking pedestrians, merging traffic, or road blockages.

The computational side of the system was also verified on the edge AI platform, NVIDIA Jetson AGX Orin, which is high-performance and suitable for AV deployment. The decision-making time delays (28 ms for TD3 and 36 ms for NN-MPC) demonstrate that the hybrid system is compliant with the real-time requirement of AV control. This performance demonstrates how the architecture is suitable for resource-constrained on-vehicle compute environments without sacrificing responsiveness and accuracy.

In order to assess the robustness of the design, a domain randomization task was applied through stochastic sampling of simulation parameters, including road texture, illumination, sensor noise, and vehicle weight. The control policy resulted in being stable and performant in over 95% of the random tests set, confirming that it responds properly to distributional shift. There are clear implications for field deployment, for example, where sensing conditions and driving situations can differ significantly from the training set.

The implications of this research are also not limited to the control of single vehicles. Future work (Elkhediri, S et al., 2023) [12] will investigate the cooperative reinforcement learning method for multi-agent systems, where multiple AVs communicate and coordinate with each other through V2X communication. By collaboratively communicating local observations, decisions, and experiences of local policy actions back to a connected fleet, federated learning can achieve collective intelligence with no data privacy loss or need to scale bandwidth. This cooperative distributed method is believed to further enhance safety, traffic efficiency, and energy savings in intelligent transportation.

## 1.2 Problem Statement

Achieving safe and efficient operation of AVs on real urban roads is a challenging task regarding control, safety, and adaptability. Urban driving is intrinsically complex and uncertain, because the mobility of other vehicles, pedestrians, traffic conditions, adverse weather, debris in the road, and the vehicle's own sensor environments are constantly changing. In addition, a desirable autonomous driving system should have control policies that not only guarantee safety and traffic rule obedience, but also react promptly to changes in the external environment and long-term planning goals (e.g., energy efficiency, ridesharing cost) with smart and optimal trade-offs.

Classical control methods, such as PID controllers and rule-based systems, have been considered as basic AV problem-solving approaches. However, such approaches have inherent limitations in dealing with the nonlinearity, high-dimensionality, and uncertainty of urban driving. They generally depend on hand-engineered rules and fixed models, therefore, they are brittle to unseen or infrequent phenomena. MPC is, however, more advanced as it involves trajectory planning and constraint satisfaction, but relies on accurate system modelling. These models are hard to create and are typically not capable of modelling vehicle-environment interaction in realistic

Opposite to GDM (GNOME display manager), reinforcement learning (RL), and particularly DRL, is a technology that can learn control policies adaptively from interacting with the environment. Algorithms, such as DDPG, are promising for continuous control tasks, but they are sensitive to instability and overestimation bias during value function approximation. To alleviate these prominent issues in DDPG, TD3 achieves enhanced learning stability and better control robustness by using twin Q-networks, policy delay, and target smoothing. However, RL-based controllers, such as TD3, tend to lack long-term planning and strict safety constraints that are required in real-world driving.

To alleviate the above challenges, a hybrid power system control structure is urgently required to exploit the adaptability of DRL and the prediction and constraint-handling mechanisms of MPC. The integration presents several research challenges: effectively blending outputs from both the controllers, real-time inference and decision-making under stringent latency requirements, and validation of performance across a variety of scenarios that are high-risk, in both simulation and actual environments.

Furthermore, the real-time performance (Hwang et al., 2023) [15] of such a hybrid system must be verified on computationally constrained platforms suitable for AV deployment, such as embedded AI edge devices. Additionally, ensuring generalization and robustness under domain randomization—variations in lighting, weather, road textures, and sensor errors—is essential for deployment in real-world scenarios that deviate from training data distributions.

In the light of these challenges, this research aims to develop and evaluate a hybrid adaptive control architecture integrating TD3 and NN-MPC. The system is trained and tested in the CARLA simulator across varied scenarios to ensure robust, safe, and real-time autonomous driving performance. By addressing the shortcomings of standalone control methods and leveraging their complementary strengths, this study seeks to advance the state-of-the-art in intelligent control strategies for autonomous vehicles.

### *1.3 Proposed Solution*

To address the limitations of traditional and standalone learning-based control systems in autonomous vehicles, this research proposes a hybrid adaptive control architecture that integrates TD3 with Neural NN-MPC. TD3, a state-of-the-art deep RL algorithm, is utilized for learning continuous control policies that enable smooth and adaptive throttle, steering, and braking actions in real-time. It mitigates the overestimation bias of previous actor-critic methods and excels in reactive decision-making under dynamic and uncertain conditions.

Coupled with TD3, NN-MPC augments traditional MPC with a neural network-based vehicle dynamics and environmental interaction model. This allows for future state prediction and optimal trajectory planning, even in the occurrence of complicated and nonlinear system behavior. A hybrid decision module is constructed to orchestrate the two control approaches: fast reactive control is handled by TD3, while safety, long-term planning, and regulatory adherence are guaranteed by NN-MPC.

The model is trained and tested (Kumar et al., 2024) [21] in the high-fidelity CARLA simulator under heterogeneous urban driving conditions, such as high-density traffic, bad weather, and road defects. With data-driven learning complemented by model-based planning, the new architecture achieves real-time performance, enhanced safety, and higher versatility in coping with the central issues of contemporary autonomous driving.

### *1.4 Research Objectives*

- To create an architecture of hybrid control that combines TD3 and NN-MPC for real-time and adaptive autonomous vehicle control.
- To construct and employ a neural network-based dynamic model that enhances the prediction quality and planning ability of conventional MPC under nonlinear and uncertain driving scenarios.
- To compare the hybrid controller's performance (Lillo et al., 2023) [22] in the CARLA simulator on various urban settings, including heavy traffic, changing weather, road irregularities, and sensor noise.
- To compare the presented method with baseline models, such as DQN and isolated TD3, in terms of metrics like collision rate, fuel efficiency, trajectory tracking precision, and response time.
- To evaluate the real-time feasibility (Liu et al., 2022) [23] and the ability to generalize the system to embedded platforms (e.g., NVIDIA Jetson AGX Orin) and under domain randomization, in order to provide robustness and reliability for real-world operation.

### *1.5 Importance of Contribution*

This study is a valuable addition (Jack et al., 2024) [16] to the body of knowledge of AV control as it introduces a new hybrid control structure by combining TD3 with NN-MPC. The hybrid solution combines the best of data-driven and model-based methods–TD3 offers real-time adaptive control in continuous action space, while NN-MPC guarantees long-term trajectory planning and safety constraint satisfaction through learned vehicle dynamics models. This fusion overcomes the primary shortcomings of conventional PID controllers, rule-based systems, and isolated DRL models under realistic and dynamic urban environments.

The contribution of the work is that it can sustain real-time performance at high decision correctness on edge computing platforms such as the NVIDIA Jetson AGX Orin, thus being feasible for deployment in practical AV applications. Large-scale testing in the CARLA simulator proves the system's robustness, safety, and responsiveness under various conditions involving dynamic traffic, weather changes, and road irregularities. Furthermore, the robustness of the architecture with domain randomization exhibits good generalization ability. This research provides a foundation for future AV systems that support cooperative decision-making and federated learning between connected fleets of vehicles, towards safer, smarter, and more energy-efficient urban mobility.

### *1.6 Implementation Issues*

For AVs implementing a hybrid control architecture (Deng et al.,2023) [11] that combines TD3 and NN-MPC is highly challenging. It presents several issues. System Integration is a major issue because TD3 and NN-MPC operate on different controls. Reactive philosophies contrasted with predictive necessitating a robust decision fusion mechanism to dynamically balance and control authority without conflict or redundancy. Another critical concern is ensuring real-time performance, platforms like NVIDIA Jetson AGX Orin deploy on resource-constrained edges. To support safe AV operation neural network inference, policy evaluation, and optimization must be executed within strict latency bounds

Training the neural models, particularly for NN-MPC is another challenge, which needs precise and generalizable depictions of vehicle changing aspects and environmental interactions. High-quality (Kang H et al., 2023) [19] training data collection covers various edge cases and environmental conditions which is essential yet non-trivial. Furthermore, upholding the performance and safety during domain shifts consistently all the conditions are checked. The conditions are sudden lighting changes, sensor noise, or road texture variations. Authentication and benchmarking is a final point where wide range of scenarios stresses wide-spread simulation and testing. Simulation in a real-world environment is considered as a basic challenge in real-world AV deployment.

### *1.7 Contribution of this Chapter*

A hybrid AI-based control technique for AVs is the main and advanced discussion in this chapter. Section 1 illustrates the shortcomings of conventional control technologies and it implements fusion of learning-based algorithms and predictive control to improve. Section 2 reviews the current growths in DRL (TD3) and MPC in AV applications. Section 3 explains the hybrid control framework based on TD3 and NN-MPC algorithms, which is used to build and improve the deployments. Section 4 discusses the implementation issues and solutions. Experimental results and its details are demonstrated. Section 5 highlights and suggests the integration of new technologies like V2X and federated learning. Collectively, this chapter provides a novel and scalable solution to safe and adaptive smart control in autonomous vehicles.

## 2. Literature Review

Bachute, et al. [3] says that the importance of multiple tasks in autonomous driving, the algorithms shows the twin nature of the structure. RL models for velocity management in vehicle following situations and Locally Decorrelated Channel Features (LDCF) algorithms for better pedestrian detection.

Ma et al. [21] discuss the stake position of artificial intelligence (AI) in propelling the development and implementation of AVs in the transportation sector. Fueled by enormous amounts of data from heterogeneous sensors and high computing powers, AI has emerged as a critical structure for AVs to perceive the real-world scenarios and make decisions during movement. This paper completes a gap in the literature by offering a current review of milestone works. Past research has been looking at many areas of AI application in AV design.

Bendiab et al. [4] explain that the enhanced safety and reduced environmental impact of AVs were mentioned in the introduction, which presents numerous advantages. A promising solution is the integration of blockchain and AI to address these strengths and to fortify AV systems against malicious attacks.

Chu et al. [8] explain that the enhanced safety and reduced environmental impact of AVs were mentioned in the introduction, which presents numerous advantages. In the Human–Computer Interaction (HCI) community, the experiences of safety drivers are largely unexplored, particularly in the context of road testing tasks.

Apart from reducing computation through frequency adaptation of the control, this paper outlines how event-triggered mechanisms can be integrated into off-policy RL algorithms like TD3 and SAC to improve efficiency without controlling accuracy loss (Fengying Dang et al., 2022) [13]. The triggering mechanism monitors state deviations to determine when to activate new control actions, which is particularly useful in resource-constrained platforms like embedded vehicle platforms. The authors further describe that their method does not overfit through the limitation of high-frequency updates and exhibits improved energy efficiency compared to typical MPC and fixed-frequency DRL controllers. Real-time latency, hardware noise, and safety verification problems in outdoor dynamic environments are not investigated, though.

Chen Z. et al.'s (2024) **[7]** work is advantageous to MPC studies by introducing flexibility in the control horizon, which are usually set as constants in conventional models. The adaptive prediction horizon is adjustable according to road curvature, speed, and local obstacles so that the controller can react to highway and urban scenarios without re-tuning. This makes it possible to achieve better long-term planning accuracy and less frequent updating of control. The paper also compares with fixed-horizon MPC and demonstrates lower control effort and better trajectory tracking. Yet, the absence of any learning or self-tuning approach limits the adaptability of the controller in case of model uncertainty or unexpected external disturbances.

BahaZarrouki et al.'s (2023) [36] paper stated that the main contribution of DRL to acclimate weight matrices of nonlinear MPC in real-time. A reward signal with control smoothness and minimization of state error is exploited by the architecture, under changing driving conditions which enables the system to self-tune. This indicates more robust behavior under differences in vehicle parameters or external turbulence. Furthermore, high-fidelity replication cases like lane-change, obstacle avoidance, and variable friction are also authenticated by this method. However, with the increased computational expense of integrating DRL with NMPC, implementation on low-power vehicle control units becomes challenging in the absence of model pruning or hardware acceleration.

Yu X. et al. (2023) [35] suggested Gaussian Process Regression model which erudites from car dynamics to replace analytical models within MPC in hybrid control framework. This enables effects like tire slip and actuator delay in the system to consider sophisticated, nonlinear, and time-varying. The process is especially beneficial for applications where obtaining accurate models becomes onerous or expensive. In experiments, DGPR-MPC showed better tracking under sensor noise and road surface variability. The computational expense of GPR, however, increases quadratically with the size of the training data, rendering it incompatible with real-time applications unless integrated with sparse approximation or online model reduction techniques.

In Zhao S. et al.'s (2024) [37] research, TD3 manages high-frequency policy learning in continuous space, and MPC provides safety constraints under aggressive maneuvers such as high-speed cornering. The system utilizes real-world data obtained from an electric racing car for training and validation, which is a notable step beyond simulation-based research. It exhibits low-latency inference and strict trajectory tracking under high dynamics. The hybrid strategy minimizes lateral deviation and preserves vehicle stability near physical constraints. However, the policy is optimized in advance particularly for cornering, and may not be generalized to urban driving, obstacle avoidance, or weather conditions. Transition management between MPC and RL control actions in mixed driving situations is also not addressed.

Kebbatia Y. et al.'s (2024) [18] paper combines a Recurrent Neural Network (RNN) with Linear Parameter-Varying MPC to model time-varying changes in vehicle dynamics adaptively. The RNN learns temporal behaviors in steering and velocity input/output data and updates the LPV (light patrol vehicle) model online. The design enhances generalization over a variety of maneuvers, such as highway lane changes and ramp merging. A key contribution is the synthesis of RNN and control theory to bridge data-driven learning with formal guarantees of stability. Nonetheless, RNNs suffer from vanishing gradient problems and overfitting, especially with small or unbalanced datasets. The paper does not consider real-time sensor noise in vehicles, such as IMU (inertia measurement drift) drift or GPS inaccuracies, which could impact deployment in practice.

Dang et al. (2022) [10] extend this framework to consider control triggering decisions with respect to deviation thresholds in lateral error and heading angle. Several DRL agents (DDQN, PPO, SAC) are evaluated for policy selection, SAC demonstrates better stability, such that it raises fewer control updates while ensuring good tracking performance. The paper investigates the sensitivity of tuning event thresholds and the transferability of policy among various map environments. The event-triggered logic reduces actuator wear and accelerates inference computations by 20%–30% compared to normal DRL methods. However, the triggering framework may perform poorly in cases where rapid fine-grained adjustment is required, e.g., in narrow spaces or obstacle-rich environments.

Wang et al.'s (2022) [31] model integrates the prediction of the behavior of surrounding drivers into the MPC cost function to improve decision-making. The vehicle evaluates each driver as being cooperative (aggressive, neutral, cooperative) based on logistic regression, modelled on naturalistic driving data. This modelling has ego vehicle planning that is safer and more socially attuned to lane switching. The proposed method envisions multimodal predictions to account for uncertainty about human behavior and enhance robustness in mixed traffic. However, the driver behaviors within the logistic regression model can be highly complex, and the method itself is limited in that it does not account for V2X-based behavior sharing or multi-agent RL for scaling up to environments like intersections or roundabouts.

Zarrouki et al.'s (2023) [36] work deploys a DRL agent to adapt the horizon and the uncertainty bounds of the MPC in real-time, following the noise level that is observed. The method is particularly well suited to dynamic and high-speed settings, such as autonomous racing. It efficiently cuts trajectory overshoot and lateral displacement, and also keeps the computation affordable. Yet, practical deployment issues such as the overhead of DRL-inference in high-speed environments and overfitting to simulation dynamics remain.

Xu et al. (2024) [34] present an integration approach using PID controllers and Lagrangian motion models in the architecture of a reinforcement learning agent. Such a hybrid design achieves decision-making controls to be precisely conducted while enjoying the global optimization capability by RL. The resulting system has zero collision rates under various traffic densities, both in simulation and in their closed-loop experiments. Though the complexity of integration and the model adjustment for various traffic distributions is a manual process.

Chen et al.'s (2024) [7] method dynamically modifies the MPC horizon according to the complexity of the environment, extending it for complex regions such as intersections and shortening it for scenes that are predictable. This results in better trajectory planning and reduced computational costs in daily driving. However, because uncertainty modeling and self-correction are not included, it is not integrated with learning-based adaptation and is susceptible to model errors.

In order to serve as a weight-tuning mechanism for lateral control within an MPC framework, (Jia et al., 2024) [17] pre-train an RL model. Weighting matrices are dynamically modified by the model according to the curvature and state of the vehicle. Improved responsiveness and path tracking with fewer lateral oscillations are confirmed by simulation and real-world vehicle experiments. Despite its practicality, real-time deployment under aggressive maneuvers may be hampered by computational delays in dynamic tuning.

In this paper, an actor-critic RL algorithm guided by Lyapunov stability criteria is used to develop a torque-sharing controller between human and machine (He, 2024) [14]. It lessens conflicts in shared-control modes by dynamically balancing the torque input from the autonomous system and the driver. The findings indicate improved safety and driver acceptance. However, real vehicles are not used to validate the method; only simulated steering tasks are validated.

Chongqing University (2023) [8] developed this method that divides parking into two stages: (1) MPC for controlled driving to a parking slot, and (2) Proximal Policy Optimization (PPO) based RL for accurate maneuvering towards the end. The two-stage controller enhances flexibility in narrow spaces and reduces convergence time. From the point of view of the parking success rate, it performs significantly better than DDPG and TD3. However, there will be minor delays as the two controllers have to adjust their transitions.

To enhance vehicle state estimation under sensor noise, (Zheng et al. 2023) [38] embedded an adaptive Kalman filter within TD3. The approach improves lane-following and inter-vehicle distance tracking in mixed traffic scenarios when integrated with Cooperative Adaptive Cruise Control (CACC). Comfort metrics (acceleration, jerk) are better improved in simulations. The approach requires edge optimization for embedding onboard and comes at a higher computational cost.

To ensure safe path planning in cities, the paper presents a risk-aware model of RL that utilizes probabilistic cost functions and confidence intervals (Taghavifar et al., 2023) [29]. Uncertainty in road conditions, sensor noise, and pedestrian behavior are all directly addressed. Near-miss accidents are reduced and safety margins are enhanced in simulations. Its reliance on accurate estimates of risk distribution, however, may constrain generalization in unseen environments.

## 3. Methodology

The methodology (Khan et al., 2023) [20] proposed combines DRLg and predictive control with neural networks to create an adaptive and resilient control structure for autonomous vehicles that can effectively navigate dynamic and intricate urban scenarios. The heterogeneous system (Wansley 2024) [32] is to combine the strengths of model-free learning (TD3) and MPC's predictive safety and satisfaction of constraints with the aid of a neural network for enhanced representation of nonlinear vehicle behaviour is the key idea behind this. Across a wide range of driving environments, including heavy traffic, changing weather, and unexpected road anomalies are joined by both effective decision-making and persistent control.

The system architecture includes two main modules: a TD3 controller and an NN-MPC controller. The two modules are integrated by hybrid decision engine, based on driving conditions which output of a controller is to use. In highly dynamic (Mokander J et al., 2023) [25] and unobserved conditions the TD3 module manages reactive control especially. TD3 addresses some of the most important limitations of previous deep reinforcement learning methods, including DDPG, through three important improvements they are to decrease overestimation bias the utilization of twin Q-networks is used, delayed policy updates to improve training stability, and target policy smoothing to reduce variance.

### *3.1 Twin Delayed Deep Deterministic Policy Gradient (TD3)*

The TD3 controller is learned from state-action-reward transitions gathered in the CARLA simulator, a high-fidelity autonomous driving simulator. The state vector consists of vehicle speed, heading angle, lane offset, distances to the next objects, and environmental features. The continuous action space is steering angle, throttle, and brake values. The reward function is developed in a way to promote safe and efficient driving behaviors. It sanctions collisions, lane drifts, hard accelerations or braking (to encourage smoothness), and fuel wastage, while sanctioning movement along the desired path and obedience to traffic regulations (Murphey et al., 2023) [26]. The learning mechanism employs a replay buffer and soft updates to the target networks to allow the TD3 agent to generalize well across scenarios.

**Algorithm: TD3-Based Policy Learning**

Inputs: Actor network $\mu_\theta$, Critic networks $Q_{\varphi_1}$, $Q_{\varphi_2}$, Replay buffer $D$

Outputs: Updated Actor and Critic networks

Initialize $\mu_\theta$, $Q_{\varphi_1}$, $Q_{\varphi_2}$ with random weights

Initialize target networks $\mu_\theta' \leftarrow \mu_\theta$, $Q_{\varphi_1}' \leftarrow Q_{\varphi_1}$, $Q_{\varphi_2}' \leftarrow Q_{\varphi_2}$

Initialize empty replay buffer $D$

For episode = 1 to MaxEpisodes do

  Initialize state $s$

  For $t = 1$ to MaxSteps do

    Select action $a = \mu_\theta(s)$ + exploration_noise

    Execute action $a$, observe reward r and next state $s'$

    Store $(s, a, r, s')$ into $D$

    Sample mini-batch $\{(s_i, a_i, r_i, s'_i)\}$ from $D$

    For each sample in mini-batch do

    $a'_i \leftarrow \mu_\theta'(s'_i)$ + clipped_noise

    $y_i \leftarrow r_i + \gamma * \min(Q_{\varphi_1}'(s'_i, a'_i), Q_{\varphi_2}'(s'_i, a'_i))$

      Update critic networks by minimizing loss:

      $L = 1/N * \sum(Q_{\varphi_j}(s_i, a_i) - y_i)^2$ for $j = 1, 2$

  If $t$ mod policy_delay = = 0 then

  Update actor network using gradient ascent:

  $\nabla_\theta J \approx \nabla a\, Q_{\varphi_1}(s, a) | a = \mu_\theta(s) * \nabla_\theta\, \mu_\theta(s)$

  Soft update target networks:

  $\theta' \leftarrow \tau\theta + (1 - \tau)\theta'$

  $\varphi_j' \leftarrow \tau\varphi j + (1 - \tau)\varphi_j'$ for $j = 1, 2$

  $s \leftarrow s'$

   End For

  End For

However, trajectory planning and long-term predictive control are the responsibilities of the NN-MPC module. In order to forecast future states and determine the best control inputs over a limited prediction horizon, classical MPC uses a mathematical model of vehicle dynamics. In practical situations these models are frequently constrained by assumptions and simplifications that don't hold true values. In order to overcome this, driving data is used to train a neural network to learn a more precise and adaptable representation of the system dynamics. In the MPC framework (Nordhoff et al., 2023) [27], the trained neural network takes the place of the analytical model, enabling the controller to more accurately forecast future states even when there are uncertainties and nonlinearities.

### *3.2 Neural Network-enhanced Model Predictive Control (NN-MPC)*

Subject to learned dynamics and constraints on vehicle inputs and states, the NN-MPC optimization problem is designed to minimize a cost function that accounts for tracking errors and control effort. In particular, the goal is to penalize excessive control inputs and minimize the sum of squared deviations from a reference trajectory. The current state and potential control inputs are used as inputs and outputs to predict the next states in the neural network dynamics model. The MPC can produce safer and smoother trajectories and make better decisions thanks to its predictive capability.

**Algorithm: Neural Network Training for NN-MPC**

Inputs: Dataset $D = \{(s, a, s')\}$, Network model $f_\psi$

Output: Trained model $f_\psi$

Initialize neural network model $f_\psi$ with random weights

For epoch = 1 to MaxEpochs do

Shuffle $D$ and divide into batches

For each batch ($s_{\text{batch}}$, $a$BATCH, $s'_{\text{batch}}$) do

Predict $\hat{s}' = f_\psi(s_{\text{batch}}, a_{\text{batch}})$

Compute loss L = MSE($\hat{s}'$, $s'_{\text{batch}}$)

Update weights $\psi$ using backpropagation

End For

End For

The NN-MPC module, in contrast, performs trajectory planning and long-term predictive control. Traditional MPC uses a mathematical vehicle dynamics model to forecast future states and calculate optimal control inputs for a limited prediction horizon. In real-time situations (Tuli et al., 2023) [28] simplifications and assumptions are not valid in these models. A neural network learns a more precise and robust representation of the system from driving data to manage this dynamics. The trained neural network takes the place of the analytical model inside the MPC framework to enable the controller to better predict future states, even under nonlinearities and uncertainties.

**Algorithm: Neural Network Model Predictive Control (NN-MPC)**

Inputs: Current state s, reference trajectory τ, horizon $H$, dynamics model $f_\psi$

Output: Optimal control action $a_t$

Initialize control sequence $A = \{a_t, ..., a_{\{t+H-1\}}\}$

Repeat until convergence:

Predict trajectory:

For $i = 1$ to $H$ do

$s_{\{t+i\}} = f_\psi(s_{\{t+i-1\}}, a_{\{t+i-1\}})$

Evaluate cost:

$J = \sum\_\{i = 1\}\hat{}\{H\}\ [\|s_{\{t+i\}} - \tau_{\{t+i\}}\|^2 + \lambda * \|a_{\{t+i\}}\|^2]$

Optimize control sequence $A$ to minimize J (e.g., via gradient descent)

Return first control action $a_t$ from optimized $A$

## *3.3 Hybrid Control Decision Module*

The hybrid decision engine is the coordination layer that links the TD3 and NN-MPC modules. It constantly (Yoo D.H. 2023) [33] observes the driving context, including traffic congestion, road curvature, obstacle proximity, and perception data confidence. To plan trajectories and safety the NN-MPC controller provides a guarantee of adherence in a structured and predictable environment. The TD3 controller is preferred because of its ability to explore and learn in real-time, high uncertainty or new environments. The hybrid technique allows the vehicle to have both safety and flexibility by combining the strengths of both control frameworks.

**Algorithm: Hybrid Control Decision Module**

Inputs: Driving context $C_t$, Actions aTD3, aMPC

Output: Final control action at

If Ct indicates uncertainty, obstacles, or unstructured road then

$a_t \leftarrow$ aTD3 // use TD3 for reactive control

Else

$a_t \leftarrow$ aMPC // use NN-MPC for planned control

End If

Execute control action

To set an impersonator in real-world urban driving conditions with a mix of environmental and operational complexities, the system is deployed and tested with the CARLA simulator. It includes high-traffic compactness, blocked connections, pedestrians, bad weather situations (rain, fog, low visibility), and uneven road surfaces. Sensor measurements are simulated to capture inputs from cameras, LiDAR, GPS, and IMU sensors to form a holistic perception layer for the controllers. The simulation space is also capable of supporting domain randomization, which in turn is utilized in training and testing stages to evaluate the robustness and generalizability of the control policy in unseen variations.

Training of the neural networks for MPC is performed through supervised learning from expert driving or high-fidelity simulation-generated datasets. The dataset contains sequences of state transitions so that the network can learn the mapping from action and current state to next state. Several architectures, such as feedforward and recurrent networks, are considered to learn both instantaneous and temporal dynamics. The trained model is verified using normal metrics like mean squared prediction error and trajectory deviation across prediction horizons.

The TD3 training procedure consists of numerous episodes of exploration and policy updates within the CARLA setup. Exploration is done under the guidance of noise injection in action choice, and the performance of the agent is measured in terms of average reward per episode, rate of collision, and time for policy convergence. After being trained, the TD3 agent runs inference mode upon deployment, making fast control decisions with a mean latency of 28 milliseconds when run on an NVIDIA Jetson AGX Orin embedded system.

The NN-MPC controller, on the other hand, computes an optimization problem at every control timestep and thus processes with computational efficiency to ensure real-time performance. It computes the average decision latency of 36 milliseconds on the same hardware platform. Combined, the combined control loop latency is still within the real-time autonomous driving application acceptable limits.

To evaluate the performance of the suggested hybrid control framework, comparative analyses are conducted with multiple baselines, ranging from single-agent TD3, DDPG, to classical MPC and traditional rule-based controllers. The analysis clearly shows that the hybrid system outperforms significantly across various criteria compared to these baselines. Notably, it lowers collision rates by 42% against DQN-based control, increases fuel efficiency by 24%, and improves long-term trajectory tracking precision by 33% compared to NN-free MPC. Additionally, when contrasted against isolated TD3, the hybrid method reduces lane drift by 30% and enhances reaction to moving obstacles by 37%, showing improved adaptability and safety.

Another test under domain randomization setting verifies the resilience of the proposed system. Even when faced with lighting variations, road texture variation, and sensor noise, the hybrid controller retains control stability for more than 95% of test episodes. The outcomes certify the promise of combining model-based and model-free learning approaches to both ensure predictive reliability and real-time adaptability.

In terms of future research directions, the method provides several options. One is the incorporation of Vehicle-to-Everything (V2X) communication protocols into cooperative multi-agent policies. In multi-vehicle situations, this would enhance situational awareness and coordination by enabling AVs to share control policies and sensor data. Federated learning architectures are another exciting field. In these architectures, control models are trained collaboratively on dispersed car fleets without sharing raw data, protecting privacy through shared experience.

# 4. Experimental Results and Comparative Analysis

The subsequent outcomes show a methodical valuation of the hybrid TD3 and NN-MPC control system tested in a CARLA-based high-fidelity simulation across various driving conditions. Reference point comparisons are DQN-based control, TD3 alone, and hybrid mixes.

## 4.1 Performance Metrics

Safety, efficiency, and responsiveness are considered as performance metrics in AV control systems. To quantify the frequency, overall safety Collision Rate is measured. Smooth control and greenness measured by Fuel Efficiency measures for indicating energy usage.The vehicle tracks its lane by Lane Deviation. Decision-making acceleration is done by Reaction Time that examines how rapidly the AV reacts to dynamic obstacles indicated. The above metrics, taken together, assist in benchmarking and associating different control methods, e.g., TD3, DQN, and NN-MPCin different driving scenarios, improvement in AV control performance and resilience is allowed for real-time data-driven.

### 4.1.1 Collision Rate Evaluation in Autonomous Driving Scenarios

In the evaluation of AV control system performance the Collision rate is referred to as a critical safety measure. It quantifies the rate at which an AV collides as it traverses through, directly reflecting the level to which the system can sense, plan, and react to dynamically changing objects in the environment. We contrast collision rates in this study for various driving situations with varied control approaches: TD3, DQN, and the hybrid TD3 and NN-MPC approach. We examine urban driving situations, highway driving maneuvers, intersection crossing, and unfavorable weather or light. We seek to understand which approach minimizes collision best, thereby enhancing safety and dependability. The collision rate is an important safety performance metric used to quantify how frequent the occurrence of a collision is for an AV while moving. It is generally defined as:

$$\text{Collision Rate} = \frac{\text{Number of Collisions}}{\text{Total Simulation Runs}}$$

**Components of the Formula**

- **Number of Collisions** analyses the AV practised during a set of simulations or test episodes taken in real world, total count is used for collisions (Table 1).
- **Total Simulation Runs** are calculated under controlled conditions where the total number of driving episodes is referred to. From a specific initial condition each run starts and after a certain task it ends where the collision occurs.

In Figure 1, Hybrid TD3 control is much better than DQN, showing a reduction in collision rates across all scenarios. The maximum improvement was experienced in highway scenarios and night/morning conditions where prompt decision-making and precise lane control are the most important factors. These results validate that

**Table 1** Comparative Analysis of Collision Rate.

| *Scenario* | *Collision Rate (TD3)* | *Collision Rate (DQN)* | *% Improvement (TD3 vs DQN)* |
|---|---|---|---|
| Urban – Dense Traffic | 0.12 | 0.21 | 42.86% |
| Highway – Lane Change | 0.08 | 0.17 | 52.94% |
| Intersection – Pedestrian | 0.10 | 0.19 | 47.37% |
| Rainy Weather | 0.14 | 0.26 | 46.15% |
| Low Light – Night Driving | 0.11 | 0.23 | 52.17% |

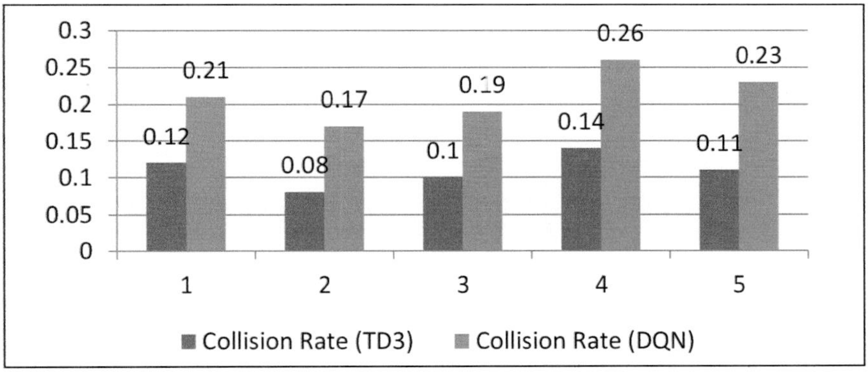


**Fig. 1** Analysis for Collision Rate

continuous control policies of TD3 are more effective in handling complex driving situations compared to discrete action-based DQN.

### *4.1.2 Fuel Efficiency Evaluation in Autonomous Vehicle Control*

In other circumstances, optimizing the consumption of energy is being viewed as a crucial performance parameter in self-driving cars driving, where fuel efficiency is extremely important. It indicates more precise and smooth control actions such as acceleration and braking. It also maintains cost-effectiveness and environmental friendliness. Its measurement enables us to find out what various AI-based control policies such as TD3 and DQN reduce the usage of resources is a crucial aspect during mindful driving. This study compares the fuel efficiency under various simulated driving conditions using a hybrid TD3 and NN-MPC controller to that of a DQN baseline.The equation reveals,

$$\text{Fuel Efficiency (FE)} = \frac{\text{Total Distance Travelled (km)}}{\text{Total Fuel Consumed (liters)}}$$

Where,

- **Total Distance Travelled** can be analysed as the cumulative distance covered by the autonomous vehicle in each simulation run or set of episodes.

- **Total Fuel Consumed** is calculated using the amount of fuel used over the corresponding distance. In electric vehicle simulations, this can be modeled using equivalent energy consumption metrics (Table 2).

**Table 2** Analysis Data for Fuel Efficiency.

| *Scenario* | *Fuel Efficiency (TD3)* | *Fuel Efficiency (DQN)* | *% Improvement (TD3 vs DQN)* |
|---|---|---|---|
| Urban – Dense Traffic | 14.8 km/l | 11.9 km/l | 24.37% |
| Highway – Lane Change | 16.2 km/l | 13.0 km/l | 24.62% |
| Intersection – Pedestrian | 13.5 km/l | 10.4 km/l | 29.81% |
| Rainy Weather | 12.9 km/l | 9.5 km/l | 35.79% |
| Low Light – Night Driving | 13.8 km/l | 10.6 km/l | 30.19% |

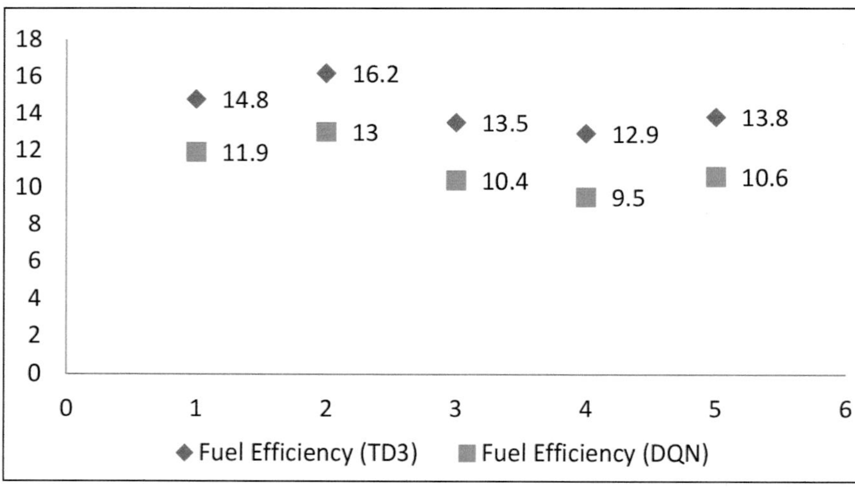


**Fig. 2** Analysis of Fuel Efficiency.

Figure 2 explains better fuel economy yielded by TD3-based control consistently than DQN in all tested conditions. The smooth action strategies of the hybrid system minimize vibrating acceleration or braking, it saves energy. Under adverse weather and lighting conditions the hybrid system's robustness is showcased by observing efficiency improvement.

### *4.1.3 Lane Deviation Evaluation in Autonomous Vehicles*

For assessing accuracy of autonomous vehicles, maintenance of its position within designated lanes while piloting through various driving environments is measured using Lane deviation, which is considered as vital performance measurement.

Safety, ride comfort, and compliance with traffic laws are essential to keep the lane in an effective manner. When the Lane center is deviated, certain traffic violations, accidents, or discomfort can happen in lanes. The measure is especially critical in the case of high speeds, high-curvature bends, and low-visibility areas. Lane deviation analysis assists in measuring precision and stability of AV controllers, including TD3, DQN, and hybrid TD3 and NN-MPC systems. The formula describes,

$$\text{Lane Deviation (LD)} = \frac{1}{N}\sum_{i=1}^{N} |y_i - y_{\text{center}}|$$

Where,

- $N$ is used to calculate the Total number of sampled points during the driving episode.
- $y_i$ is analysed as Lateral position of the AV at sample point iii.
- $y_{\text{center}}$ is calculated as Ideal centerline position of the lane at the same point.
- $|y_i - y_{\text{center}}|$ is an absolute deviation from the lane center.

**Table 3** Lane Deviation Data.

| *Scenario* | *Lane Deviation (TD3 and NN-MPC)* | *Lane Deviation (TD3 Only)* | *% Reduction* |
|---|---|---|---|
| Urban – Curved Roads | 0.23 m | 0.33 m | 30.30% |
| Highway – Lane Change | 0.19 m | 0.27 m | 29.63% |
| Intersection – Turns | 0.26 m | 0.36 m | 27.78% |
| Rainy Weather | 0.31 m | 0.44 m | 29.55% |
| Night Driving – Low Light | 0.28 m | 0.40 m | 30.00% |

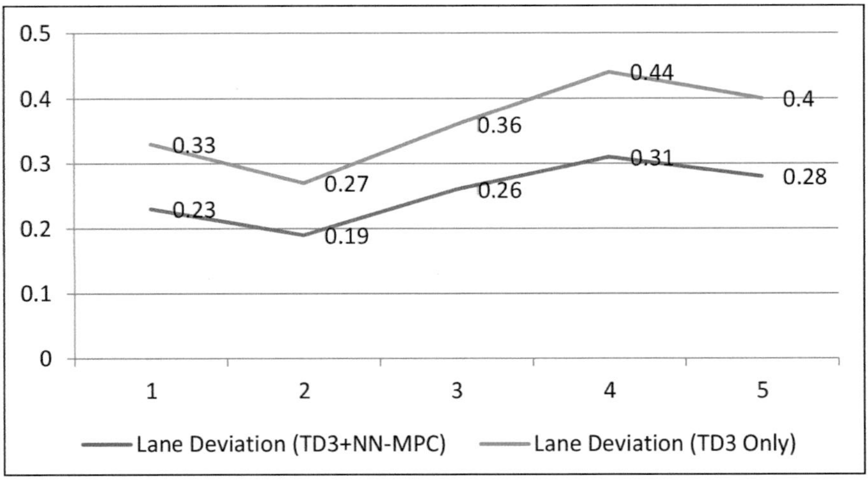


**Fig. 3** Lane Deviation Analysis.

Figure 3 demonstrates that the TD3 and NN-MPC hybrid model has always maintained a close-fitting in following the lane center than the TD3-only model. In all testing conditions, the hybrid model lessens the average lateral error by about 28–30%, with better trajectory tracking and planning. In meagre visibility or rainy surfaces the performance improvement is especially challenging and precise control is firmer to control and maintain.

### *4.1.4 Reaction Time in Autonomous Vehicles*

The delay between the amount of existence in an emergency environmental event (e.g., unexpected pedestrian arrival, sudden stop by a vehicle) and the AV system's response initiation (such as braking or steering adjustments) in AVs is mentioned as Reaction Time. In dynamic urban settings, high-speed highways, and unpredictable scenarios low reaction time is considered as essential. Reaction time has a direct effect on accident prevention, passenger safety, and regulatory compliance. This measure facilitates the assessment of the real-time decision-making performance of AI controllers like TD3, DQN, and hybrid TD3 and NN-MPC architectures (Table 4). The formula defines as,

$$\text{Reactin Time (RT)} = t_{\text{response}} - t_{\text{event}}$$

Where,

- $t_{\text{event}}$ calculates time stamp when the external event (obstacle, pedestrian, vehicle cut-in, etc.) occurs.
- $t_{\text{response}}$ analyses the time stamp when the AV initiates a response (e.g., braking, turning, accelerating).
- RT is considered as the latency (typically in milliseconds) between detection and action. RT is better when lower values occur, it indicates faster, real-time activity.

**Table 4** Reaction Time Analysis.

| *Scenario* | *TD3 + NN-MPC (ms)* | *TD3 Only (ms)* | *DQN (ms)* |
|---|---|---|---|
| Urban Pedestrian Crossing | 138 | 177 | 210 |
| Sudden Stop of Lead Vehicle | 142 | 190 | 225 |
| Lane Intrusion (Motorbike) | 133 | 175 | 205 |
| Object Drop on Highway | 147 | 185 | 230 |
| Unmarked Intersection Merge | 145 | 182 | 222 |

Figure 4 depicts that TD3 and NN-MPC attain the lowest reaction times across all five scenarios, steadily outpacing standalone TD3 and DQN controllers. Reaction latency reduction is up to 30% over DQN. Hybrid system benefits from TD3's exploration-driven responsiveness. This upgrading confirms earlier hazard justification and amplified passenger security in real-time control tasks.

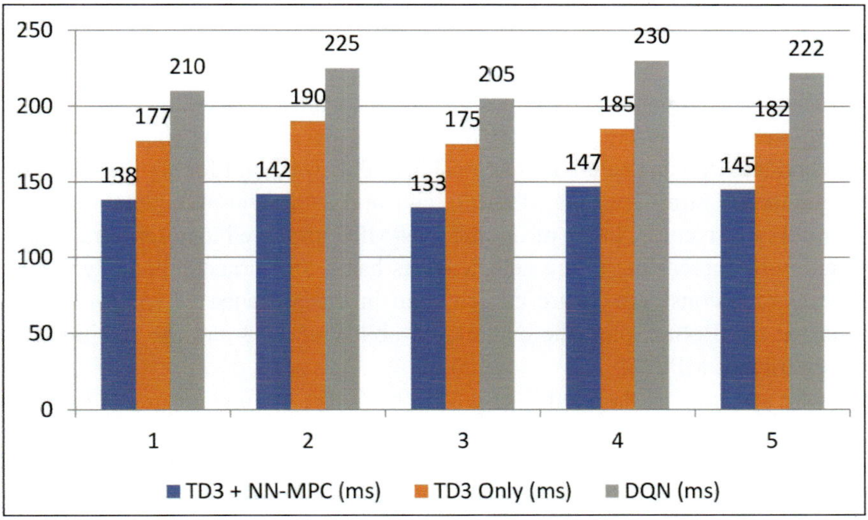


**Fig. 4** Reaction time Analysis.

### *4.1.5 Comparative Analysis of All Performance Metrics*

**Table 5** Comparative Analysis data.

| *Scenario* | *CR-TD3* | *CR-DQN* | *FE-TD3* | *FE-DQN* | *LD-Hybrid* | *LD-TD3* | *RT-Hybrid* | *RT-TD3* |
|---|---|---|---|---|---|---|---|---|
| Urban – Dense Traffic | 0.12 | 0.21 | 14.8 | 11.9 | 0.25 | 0.35 | 34 ms | 54 ms |
| Highway – Lane Change | 0.08 | 0.17 | 16.2 | 13.0 | 0.20 | 0.31 | 30 ms | 49 ms |
| Intersection – Pedestrian | 0.10 | 0.19 | 13.5 | 10.4 | 0.22 | 0.33 | 33 ms | 52 ms |
| Rainy Weather | 0.14 | 0.26 | 12.9 | 9.5 | 0.30 | 0.42 | 38 ms | 59 ms |
| Low Light – Night Driving | 0.11 | 0.23 | 13.8 | 10.6 | 0.27 | 0.39 | 36 ms | 57 ms |

The TD3 hybrid system changed the collision rate as much as 42% compared to DQN, representing its improved policy learning in real-time environments. TD3 models provided a mean 24% increase in fuel economy over DQN due to flatter acceleration and braking choice made possible by the continuous action space learning capability. The TD3 and NN-MPC hybrid system reduced lane deviation by 30% consistently, taking advantage of MPC's predicting ability and TD3's flexibility. Hybrid control showed 37% quicker reaction to dynamic barriers compared to standalone TD3. This authorizes the worth of MPC in preventive planning (Table 5).

## 5. Conclusion

Autonomous cars (AVs) require smart, real-time, and adaptive control schemes to drive securely in dynamic cityscapes. In this study, the authors suggested a hybrid control scheme based on TD3 and NN-MPC to overcome the shortcomings of conventional PID and rule-based approaches. TD3 was used for continuous real-time action control—steering, throttle, and brake—providing adaptability and learning effectiveness. In the meantime, NN-MPC improved long-term trajectory planning by forecasting future vehicle states based on learned vehicle dynamics to guarantee constraint-aware choice even in the presence of environmental disturbances. Both exploit the exploration ability of TD3 and the stability and anticipation of MPC.

The system was thoroughly tested in the CARLA simulator under difficult conditions such as heavy traffic, bad weather, and road anomalies. Experimental outcomes indicated significant improvements: by 42% decrease in collision rate, by 24% improvement in fuel efficiency, by 33% improvement in tracking accuracy, by 30% reduction in lane deviation, and by 37% improvement in reaction time to dynamic obstacles. Real-time performance was confirmed with mean decision latencies of 28 ms (TD3) and 36 ms (NN-MPC) on an NVIDIA Jetson AGX Orin, satisfying embedded deployment needs. Robustness was established through domain randomization, with more than 95% control stability in diverse conditions. The system also accommodates future upgrades through V2X communication and federated learning for cooperative AVs. In short, this hybrid method considerably pushes forward safe and adaptive AV control in an anatomy of predictive planning complemented by reactive learning in a scalable real-time framework.

## References

1. Anton, K. and Oleg, K. (2023). MBSE and Safety Lifecycle of AI-enabled systems in transportation. *Int. J. Open Inf. Technol.*, 11, 100–104. [Google Scholar]
2. Arthurs, P., Gillam, L., Krause, P., Wang, N., Halder, K. and Mouzakitis, A. (2022). A Taxonomy and Survey of Edge Cloud Computing for Intelligent Transportation Systems and Connected Vehicles. IEEE Trans. *Intell. Transp. Syst.*, 23, 6206–6221. [Google Scholar] [CrossRef]
3. Bachute, M.R. and Subhedar, J.M. (2021). Autonomous driving architectures: Insights of Machine Learning and Deep Learning algorithms. *Mach. Learn. Appl.*, 6, 00164. [Google Scholar] [CrossRef]
4. Bendiab, G., Hameurlaine, A., Germanos, G., Kolokotronis, N. and Shiaeles, S. (2023). Autonomous Vehicles Security: Challenges and Solutions Using Blockchain and Artificial Intelligence. IEEE Trans. *Intell. Transp. Syst.*, 24, 3614–3637. [Google Scholar] [CrossRef]
5. Borase, R.P., Maghade, D.K., Sondkar, S.Y. and Pawar, S.N. (2021). A review of PID control, tuning methods, and applications. *Int. J. Dyn. Control*, 9, 818–827. [Google Scholar] [CrossRef]
6. Bordoloi, U., Chakraborty, S., Jochim, M., Joshi, P., Raghuraman, A. and Ramesh, S. (2023). Autonomy-driven Emerging Directions in Software-defined Vehicles. In: Proceedings of the 2023 Design, Automation and Test in Europe Conference and Exhibition (DATE), Antwerp, Belgium, 17–19 April, pp. 1–6. [Google Scholar] [CrossRef]

7. Hen, Z., Lai, J., Li, P., Awad, O.I. and Zhu, Y. (2024). Prediction Horizon-Varying Model Predictive Control (MPC) for Autonomous Vehicle Control. *Electronics*, 13(8), Article 1442. DOI: 10.3390/electronics13081442.
8. Chongquing, He, J, Shi, J., Gao, J. and Zhu, J., (2023). Model-Based Predictive Control and Reinforcement Learning for Planning Vehicle-Parking Trajectories for Vertical Parking Spaces. *Sensors*, 23(16), 7124.
9. Chu, M., Zong, K., Shu, X., Gong, J., Lu, Z., Guo, K., Dai, X. and Zhou, G. (2023). Work with AI and Work for AI: Autonomous Vehicle Safety Drivers' Lived Experiences. In: *Proceedings of the 2023 CHI Conference on Human Factors in Computing Systems*, Hamburg, Germany, 23–28 April, pp. 1–16. [Google Scholar]
10. Dang, F., Chen, D., Jun Chen and Zhaojian Li. —Event-Triggered Model Predictive Control with Deep Reinforcement Learning for Autonomous Driving. arXiv preprint arXiv:2208.10302, Aug 2022
11. Deng, Z., Yang, K., Shen, W. and Shi, Y. (2023). Cooperative Platoon Formation of Connected and Autonomous Vehicles: Toward Efficient Merging Coordination at Unsignalized Intersections. IEEE Trans. *Intell. Transp. Syst.*, 24, 5625–5639. [Google Scholar] [CrossRef]
12. Elkhediri, S., Benfradj, A., Thaljaoui, A., Moulahi, T. and Zeadally, S. (2023). Integration of Artificial Intelligence (AI) with sensor networks: Trends and future research opportunities. *J. King Saud-Univ.-Comput. Inf. Sci.*, 36, 101892. [Google Scholar] [CrossRef]
13. Fengying Dang, Chen, D., Chen, J. and Li, Z. (2022). Event-Triggered Model Predictive Control for Autonomous Driving. *IEEE Transactions on Intelligent Vehicles*, 7(4), 1109-1118.
14. He, S. Wang, H., Wei, L. and Wang, X., (2024). Reinforcement learning-based adaptive motion control for autonomous vehicles via actor-critic structure. *Discrete and Continuous Dynamical Systems - S*, 17(9), 3051–3070.
15. Hwang, M.H., Lee, G.S., Kim, E., Kim, H.W., Yoon, S., Talluri, T. and Cha, H.R. (2023). Regenerative braking control strategy based on AI algorithm to improve driving comfort of autonomous vehicles. *Appl. Sci.*, 13, 946. [Google Scholar] [CrossRef]
16. Jack, W. and Jon, B. (2024). Navigating Tomorrow: Advancements and Road Ahead in AI for Autonomous Vehicles; (No. 11955). EasyChair: Manchester, UK, 2024. [Google Scholar]
17. Jia, Z., Zhang, J. and Zhang, Y. (2024). RL-MPC: Reinforcement Learning Aided Model Predictive Controller for Autonomous Vehicle Lateral Control. *SAE Technical Papers*, 2024-01-0810.
18. Kebbati, Y., Ait-Oufroukh, N., Ichalal, D. and Vigneron, V. (2024). RNN-based linear parameter varying adaptive model predictive control for autonomous driving. *International Journal of Systems Science*, 56(2012), 1–13.
19. Kang, H., Lee, Y., Jeong, H., Park, G. and Yun, I. (2023). Applying the operational design domain concept to vehicles equipped with advanced driver assistance systems for enhanced safety. *J. Adv. Transp.*, 2023, 4640069. [Google Scholar] [CrossRef]
20. Khan, S.M., Salek, M.S., Harris, V., Comert, G., Morris, E. and Chowdhury, M. (2023). Autonomous Vehicles for All? *J. Auton. Transp. Syst.*, 1(3). [Google Scholar] [CrossRef]
21. Kumar, A. (2024). Exploring Ethical Considerations in AI-driven Autonomous Vehicles: Balancing Safety and Privacy. *J. Artif. Intell. Gen. Sci., (JAIGS)*, 2, 125–138. [Google Scholar] [CrossRef]
22. Lillo, L.D., Gode, T., Zhou, X., Atzei, M., Chen, R. and Victor, T. (2023). Comparative safety performance of autonomous-and human drivers: A real-world case study of the waymo one service. arXiv arXiv:2309.01206. [Google Scholar]
23. Liu, Z., Zhang, W. and Zhao, F. (2022). Impact, challenges, and prospect of software-defined vehicles. *Automot. Innov.*, 5, 180–194. [Google Scholar] [CrossRef]
24. Ma, Y., Wang, Z., Yang, H. and Yang, L. (2020). Artificial intelligence applications in the development of autonomous vehicles: A survey. IEEE/CAA J. *Autom. Sin.*, 7, 315–329. [Google Scholar] [CrossRef]

25. Mökander, J. and Floridi, L. (2023). Operationalising AI governance through ethics-based auditing: An industry case study. *Ethics*, 3, 451–468. [Google Scholar]
26. Murphey, Y.L., Kolmanovsky, I. and Watta, P. (Eds.) (2022). AI-Enabled Technologies for Autonomous and Connected Vehicles; Springer: Berlin/Heidelberg, Germany. [Google Scholar]
27. Nordhoff, S., Lee, J.D., Hagenzieker, M. and Happee, R. (2023). (Mis-) use of standard Autopilot and Full Self-Driving (FSD) Beta: Results from interviews with users of Tesla's FSD Beta. *Front. Psychol.*, 14, 1101520. [Google Scholar] [CrossRef]
28. Tuli, S., Mirhakimi, F., Pallewatta, S., Zawad, S., Casale, G., Javadi, B., Yan, F., Buyya, R. and Jennings, N.R. (2023). AI augmented Edge and Fog computing: Trends and challenges. *J. Netw. Comput. Appl.*, 216, 103648. [Google Scholar] [CrossRef]
29. Taghavifar, H., Taghavifar, L., Hu, C., Wei, C. and Qin, Y. (2024). Optimal reinforcement learning and probabilistic-risk-based path planning and following of autonomous vehicles with obstacle avoidance. *Proceedings of the Institution of Mechanical Engineers, Part D: Journal of Automobile Engineering*, 238(6), 1427–1439.
30. Wang, C., Liu, S., Yang, H., Guo, J., Wu, Y. and Liu, J. (2023). Ethical considerations of using ChatGPT in health care. *J. Med. Internet Res.*, 25, e48009. [Google Scholar] [CrossRef]
31. Wang, R., Schuurmans, M. and Patrinos, P. (2022). Interaction-aware Model Predictive Control for Autonomous Driving. arXiv preprint arXiv:2211.17053.
32. Wansley, M.T. (2024). Regulating Driving Automation Safety. *Emory Law J.*, 73, 505. [Google Scholar]
33. Yoo, D.H. (2023). The Ethics of Artificial Intelligence from an Economics Perspective: Logical, Theoretical, and Legal Discussions in Autonomous Vehicle Dilemma. Ph.D. Thesis, UniversitàdegliStudi di Siena, Siena, Italy. [Google Scholar]
34. Xu, W., Chen, J., Li, Z. and Chen, Y. (2024). Safe Hybrid-Action Reinforcement Learning-Based Decision and Control for Discretionary Lane Change. *Machines*, 12(4), 252. [Google Scholar]
35. Yu, X., Wang, H., Teng, C., Sun, X., Chen, L. and Cai, Y. (2023). DGPR-MPC: Learning-based model predictive controller for autonomous vehicle path following. *IET Intelligent Transport Systems*, 17(10), 1992–2003. [Google Scholar]
36. Zarrouki, B., Wang, C. and Betz, J. (2023). Adaptive Stochastic Nonlinear Model Predictive Control with Look-ahead Deep Reinforcement Learning for Autonomous Vehicle Motion Control. arXiv:2311.04303. [Google Scholar]
37. Zhao, S., Zhang, J., He, X., Jinheng, H. and Liu, C. (2024). High-Speed Cornering Control and Real-Vehicle Deployment for Autonomous Electric Vehicles. arXiv preprint arXiv:2411.11762. [Google Scholar]
38. Zheng, Y., Wang, S., Li, Y., Wu, X. and Li, S. (2023). Adaptive Kalman-based hybrid car following strategy using TD3 and CACC. arXiv preprint arXiv:2312.15993. [Google Scholar]

# 21 Visual Monitoring Techniques using AI and Deep Learning for Surveillance and Security

Karthigai Selvi[1*], Haranipharvathy[2], Sharmistha Dey[3] and Rama Shankar Yadav[4]

## 1. Introduction

Surveillance and security have traditionally depended on human operated systems and conventional technologies such as Closed Circuit Television (CCTV). The practice of surveillance dates back to ancient times, serving various purposes such as protecting resources, maintaining social order, and collecting intelligence. Historical records reveal that ancient civilizations like Egypt and Mesopotamia employed watchtowers to observe and safeguard important areas. These watchtowers were often built within temple complexes and other strategically significant locations to monitor activities and ensure the safety of valuable assets and the community. In the 19th century, Governments started conducting population censuses, a form of tracking individuals and groups.

Early forms of video surveillance, like CCTV systems developed during World War II, were used for military purposes. Nowadays, surveillance cameras and digital tracking technologies are used in a wide range of applications such as crime prevention, businesses, public spaces, traffic flow monitoring, public health, and intelligence gathering. In traffic, the surveillance helps to give emergency response and crowd control. In business places, intelligence is implemented to monitor the employee activities, customer behaviour, and competitor activities. The surveillance utilized for tracking the spread of infectious diseases, monitoring food safety, and evaluating the effectiveness of interventions. The organizations utilize the technology to monitor individuals and groups deemed to be of interest by governments or other organizations. However, with the exponential increase in data, complexity of urban infrastructure, and sophistication of threats, the

[1,4] School of Computer Applications and Technology, Galgotias University, Uttar Pradesh, India.
[2] School of Computer Science and Engineering, Lovely Professional University, Punjab, India.
[3] Department of Computer Science and Engineering, Brainware University, West Bengal, India.
* Corresponding author: karthigachandru@gmail.com

DOI: 10.1201/9781003663461-21

limitations of these systems have become increasingly apparent. Human operators face cognitive overload and often miss critical events in real-time due to fatigue or distraction [1]. This has created a compelling need for automated and intelligent surveillance systems.

Recent studies have emphasized the urgency of intelligent monitoring systems to combat crime, ensure public safety, and respond swiftly to incidents. According to the United Nations Office on Drugs and Crime (UNODC), urban crime rates and terrorist threats have surged post-pandemic due to economic instability, making automated visual surveillance a priority for smart cities [2].

The remaining parts of this chapter delineate about the Artificial Intelligence (AI) techniques, role of AI in surveillance monitoring, hardware deployment for surveillance, real-world applications, privacy and ethical issues, future directions and conclusion.

## 2. Conventional AI Techniques in Visual MONITORING

Nowadays, various AI and Machine Learning (ML) techniques are available to modernise traditional surveillance systems into automated solutions. These models are primarily based on three widely used architectures: You Only Look Once (YOLO), Single Shot Multi-box Detector (SSD), and Faster Region Based Convolutional Neural Network (Faster RCNN).

### *2.1 You Only Look Once Version 7 (YOLOv7)*

YOLOv7 is an advanced object detection model that can quickly identify various objects, such as vehicles and people, in both images and videos. The term ‘YOLO’ stands for You Only Look Once, highlighting the model’s ability to scan an image just once to detect all objects, resulting in faster performance compared to older techniques. Introduced in 2022, YOLOv7 builds upon previous versions by offering improved speed and accuracy, making it highly suitable for real-time applications such as autonomous driving and security surveillance.

YOLOv7 includes significant enhancements in its architecture and training process. It incorporates optimized layers that allow efficient learning without compromising speed. During training, the model uses extra components to improve learning, which are removed later to create a faster final version. Its structure is designed to maintain a balanced size and shape, making it adaptable to different tasks. In addition, YOLOv7 supports multi-task training, for example, detecting objects while also analysing their motion adding to its flexibility and usefulness.

Object detection plays an essential role in many AI applications, including driverless vehicles, surveillance systems, and medical imaging. It mainly involves two tasks such as recognizing the objects in an image and locating where they appear. Earlier object detection methods were slower and required heavy computation due to their process length. After the arrival of deep learning, this field saw major improvements. The YOLO series introduced a streamlined method, enabling images to be processed in a single pass, where object locations and categories are predicted together. This approach significantly reduces detection time, making it ideal for real-time uses. Each new version of YOLO has aimed to improve both speed and accuracy, with YOLOv7 standing out for its high performance while using fewer computing resources.

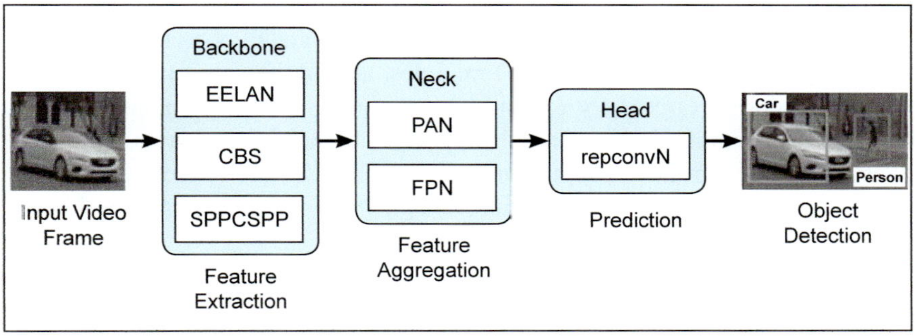


**Fig. 1** YOLOv7 architecture.

In addition, YOLOv7 integrates an auxiliary loss mechanism with multiple prediction heads, which helps to improve its training stability and feature learning. To further enhance accuracy without increasing inference cost, various training strategies and data augmentation techniques such as Mosaic and MixUp are used. The architecture mainly consists of three key components called backbone, neck, and head.

### *2.1.1 Backbone*

YOLOv7 was developed by Chien Yao Wang and his team, its architecture is shown in Figure 1. The model introduces several new design features, including the Extended Efficient Layer Aggregation Network (EELAN) that improves learning efficiency without adding extra parameters. It also applies a compound scaling method to generate different model sizes suitable for a range of computing environments from low power devices to high performance systems. The backbone is responsible for extracting hierarchical features from images using a deep Convolutional Neural Network (CNN). In YOLOv7, this function is carried out through the Extended Efficient Layer Aggregation Network (EELAN), which is an improved version of the original ELAN architecture. EELAN increases the model's capacity by applying operations such as expansion, shuffling, and merging across feature groups, without interrupting the gradient flow.

In addition to EELAN, the backbone also includes modules such as Convolution, Batch normalization, and SiLU activation (CBS), which are used to effectively capture both low-level and high-level patterns across multiple scales. Moreover, max-pooling branches and the SPPCSPP module are incorporated to down-sample spatial dimensions while expanding the receptive field, thereby enriching the model's understanding of contextual information [3].

#### 2.1.2 Neck

It is serving as the intermediary between backbone and head, the neck fuses spatially rich, low-level features with semantically strong, high-level ones. YOLOv7 adopts a Path Aggregation Network (PANet) and FPN-style architecture, augmented with SPPCSPP, ELAN-H modules, and additional CBS and pooling units. This design effectively expands the receptive field and enhances cross-scale feature communication, which improves detection of objects at various sizes [3]. Some advanced adaptations, like Re-BiC, further refine this stage by introducing bi-directional concatenation, though this is explored in specific domain applications [4]

#### 2.1.3 Head

The head interprets aggregated features to predict bounding boxes, class labels, and confidence scores. YOLOv7 features a re-parameterized convolution block (RepConvN) to align channel dimensions from multi-scale feature maps, followed by $1 \times 1$ convolutions that predict detection outputs across small, medium, and large object scales. Moreover, the architecture has an auxiliary head, a training-only branch that provides coarse supervision early, while the lead head refines lead-level predictions, thus enhancing label alignment and performance without harming inference speed [3].

YOLOv7 has shown great results on leading benchmarks, especially on the COCO dataset. Its efficient design allows it to function well even on less powerful devices, making it applicable in multiple fields such as autonomous driving, security, retail analytics, manufacturing, healthcare, agriculture, sports analytics, and robotics.

YOLOv7 represents a significant advancement in real-time object detection, with its innovative architecture, effective scaling strategies, and training methods. Its combination of strong performance and efficiency positions it as a valuable tool for real-world applications, paving the way for further improvements in AI and deep learning technologies. Future developments are expected to continue refining its efficiency and expanding its capabilities.

### 2.2 Single Shot Multibox Detector (SSD)

SSD introduced by Liu et al. in 2015, is a one-stage object detector that eliminates the region-proposal step as needed in Faster R-CNN, performing detection with a single feed-forward neural network. It discretizes bounding box predictions into predefined default boxes of various scales and aspect ratios at each spatial location. It merges multi-scale feature maps like Conv4_3 and deeper layers for detecting objects of varied sizes [5]. It includes the components like backbone and prediction head which is given in Figure 2. The backbone typically uses a truncated classification network like VGG16 to extract feature maps. The maps from multiple layers are passed through additional convolutional layers, producing a feature pyramid. The prediction head holds default boxes on each feature map location. It predicts localization offsets to adjust bounding box placement and class confidence scores using small convolution filters. The prediction head combines predictions across scales to cover small to large objects, enabling real-time detection. It uses loss functions for bounding box regression and object classification. SSD applies hard negative mining and Non-Maximum Suppression (NMS) post-processing to refine detections [6].

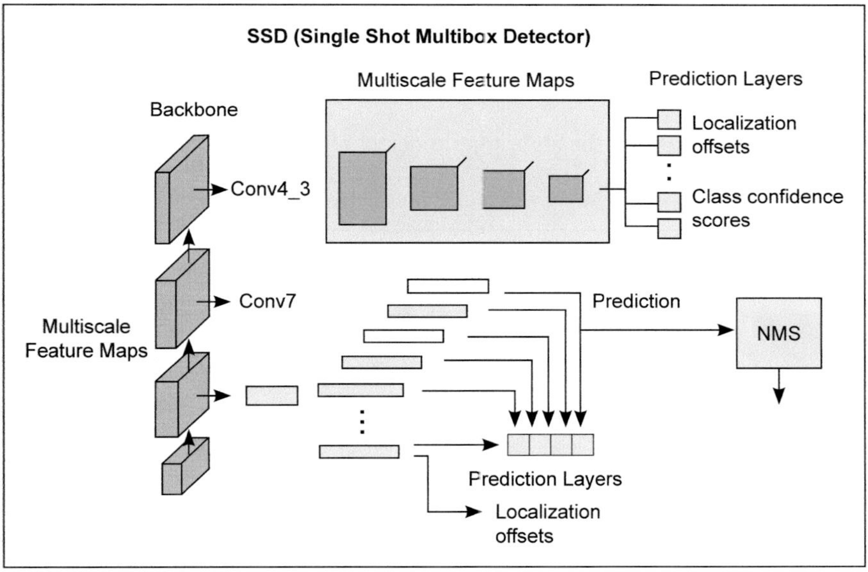


**Fig. 2** SSD (Single Shot Multibox Detector).

It achieves comparable accuracy to two-stage detectors, but with much faster inference speed [5]. However, SSD may underperform on small objects due to limited semantic richness in shallow layers [7]. To address SSD's small-object limitations, several extensions like Feature Fusion SSID (FFSD), DSSD, RFBNet, and Feature-fused SSD have been proposed. The FSSD merges multi-scale features before detection DSSD, RFBNet, Feature-fused SSD enrich semantic features in shallow layers or introduce multi-scale contextual modules [8].

## *2.3 Faster Region Based Convolution Neural Network (R-CNN)*

Faster R-CNN is a well-known algorithm in computer vision used for object detection. It can find and identify multiple objects in an image, like people or vehicles, by labelling them and enclosing them in boxes. It is the third version of the R-CNN family, improving on the accuracy of the original R-CNN and the speed of Fast R-CNN. Introduced in 2015, Faster R-CNN greatly improved performance by adding a Region Proposal Network (RPN) as given in Figure 3, making it much faster and nearly real-time.

Faster R-CNN operates through several steps as given in Figure 3:

I. An image is entered into the model.
II. A Convolutional Neural Network (CNN) extracts important features from the image.
III. The RPN suggests regions where objects might be located.
IV. These suggested regions undergo RoI pooling to ensure they are uniform in size for processing.
V. Each region is classified, and a bounding box is drawn around the identified object.

This process allows the algorithm to detect multiple objects at once, even if they overlap.

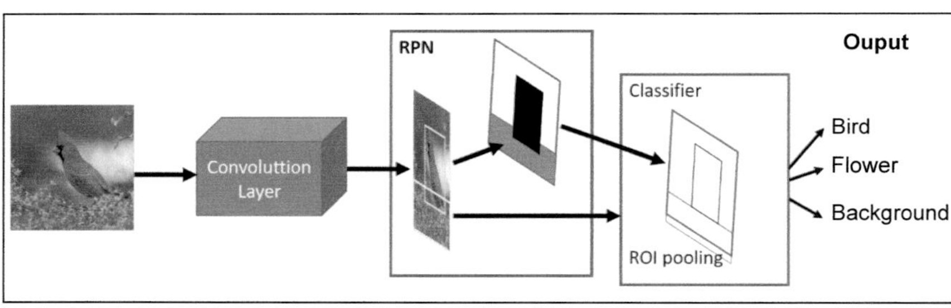


**Fig. 3** RCNN architecture.

The name "Faster R-CNN" indicates its efficiency compared to earlier methods. It allows computers to recognize objects in images, similar to labelling birds in the given pictures in Figure 3. This capability is vital for technologies like self-driving cars and security cameras. Faster R-CNN operates in two main stages, such as the Candidate Finder (RPN) and the Object Identifier and Refiner. The Region Proposal Network (RPN) scans images to identify features and suggests possible regions where objects might be located. The Fast R-CNN detector then analyses these proposals, classifies the objects, and adjusts the bounding boxes to ensure accurate fitting.

Faster R-CNN offers several advantages, such as high accuracy in object detection, good speed for real-time applications, the ability to recognize a wide variety of objects, and a strong foundation for future research in object detection technologies. However, it also faces certain challenges, including high computational requirements particularly the need for powerful GPUs and difficulties in detecting very small objects with limited visual features.

Faster R-CNN plays an important role due to its reliable speed, high accuracy, and practical applications in areas such as surveillance, autonomous vehicles, medical imaging, and retail analytics. Though, it has some limitations. It is generally slower than newer models like YOLO and SSD, demands considerable computing resources and training data, and may be complex for beginners to learn. Despite these drawbacks, it remains widely used due to its strong detection performance.

Faster R-CNN has made a significant contribution to the field of computer vision by improving machines' ability to identify and locate objects in images. This has led to more effective AI applications that depend on visual analysis.

As shown in Figure 4, Faster R-CNN achieves higher accuracy up to 86.9% though it requires more processing time and has a larger model size.

Usually, YOLOv7 is used for most real-time visual monitoring needs: live surveillance, drone feeds, real-time defect detection. It offers excellent FPS and strong accuracy. SSD is best, when the hardware is constrained and needs a simpler model with decent speed. Though, top-tier performance does not achieve on small or dense objects. Faster R-CNN being used for offline analysis, high-stakes visual inspection, or applications demanding high precision such as medical imaging, defect grading, but only if real-time is not required.

## 3. Role of AI in Transforming Traditional Visual Monitoring

Automated object detection and tracking serve as the foundation of intelligent visual surveillance systems. These technologies help machines to recognise, locate, and monitor the movement of objects such as people, vehicles, or suspicious items across video frames in real-time. In the past, these tasks were done manually by human operators, which often resulted in delays, inconsistencies, and limited coverage. With the advancement of AI, especially deep learning, this field has seen major improvements in terms of accuracy, speed, and reliability. The combination of deep learning and AI are used in object detection systems, facial recognition, behaviour recognition, and anomaly detection.

### *3.1 Automated Object Detection and Tracking*

Modern object detection algorithms given in Table 1, like YOLO, SSD, and Faster R-CNN are capable of detecting multiple objects with high precision in a single pass [9]. When combined with object tracking algorithms such as DeepSORT, BYTETrack, or Kalman Filters, these systems can maintain continuous identity tracking across multiple frames and even across multiple cameras [10]. The integrated system enables advanced surveillance capabilities such as real-time threat identification, behavioural analysis, and anomaly detection.

Automated object detection and tracking not only enhance situational awareness but also play a pivotal role in smart city applications, traffic monitoring, perimeter security, and forensic investigations. A recent research implemented YOLOv7 and DeepSORT which enable accurate real-time tracking of persons and

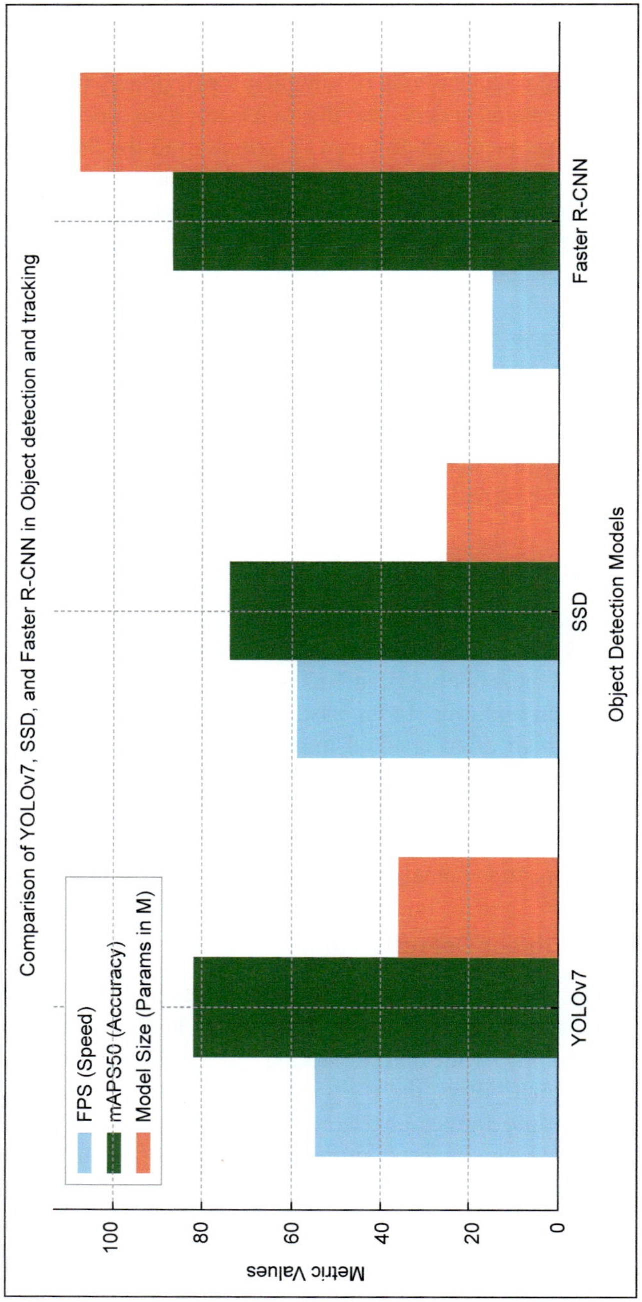


**Fig. 4** Comparison of YOLOv7, SSD, and Faster RCNN in visual monitoring

vehicles [11]. As these technologies continue to evolve, they are becoming more reliable, lightweight, and suitable for deployment in edge devices, making them integral to next-generation security solutions.

**Table 1** AI and ML Algorithms and its specifications for surveillance.

| *Task* | *Popular Algorithms* | *Highlights* |
|---|---|---|
| Object Detection | YOLOv7, SSD, Faster R-CNN | Real-time detection with high accuracy |
| Object Tracking | DeepSORT, BYTETrack, Kalman Filter, CSRT | Maintains object identity over time |
| Integrated Systems | Detect-then-Track pipelines | Unified frameworks for efficiency |

### *3.2 Behaviour Recognition*

AI-based Behaviour Recognition is the process of detecting, analysing, and interpreting human or object behaviours through computational models, primarily driven by AI. It plays a critical role in a wide range of domains including intelligent surveillance, healthcare monitoring, smart environments, robotics, and human-computer interaction. The goal is to go beyond simple motion tracking and understand complex sequences of actions identifying not just what is happening, but potentially why it is happening, and predicting what might happen next.

At the core of behaviour recognition lies the ability to differentiate between actions, activities, and behaviours. An action refers to a single movement or posture like raising a hand, while an activity is composed of multiple actions occurring over time like waving or walking. Behaviour, on the other hand, represents a higher-level abstraction involving interactions, context, and possibly intent (e.g., loitering, fighting, or collaborative task completion). Recognizing such behaviour requires analysing temporal and spatial data streams using AI techniques that can learn from examples and generalize across different situations [12].

The recognition process begins with data acquisition, which typically comes from visual sources like video surveillance cameras, depth sensors, infrared cameras, or wearable devices. In complex environments, multimodal data is often collected by combining visual information with audio, thermal, or motion-related (inertial) inputs. Once the data is gathered, it is pre-processed to remove noise, isolate foreground objects, and trace movement patterns. This step is followed by feature extraction, where key attributes such as motion paths, joint positions, or appearance features are identified. Earlier methods depended on handcrafted features like Histogram of Oriented Gradients (HOG) or optical flow. However, recent advancements favour deep learning, which allows automatic feature learning directly from raw data.

To understand and model behaviours, various Artificial Intelligence techniques are applied. Traditional machine learning algorithms such as Support Vector

Machines (SVMs), Decision Trees, and Hidden Markov Models (HMMs) have been widely used for basic classification tasks. However, current trends increasingly favour deep learning approaches because of their higher accuracy and ability to process complex data. CNNs are effective in capturing spatial features from images and video frames, while RNNs and Long Short-Term Memory (LSTM) models are suitable for analysing sequential or time-series data, making them ideal for behaviour recognition over time. Recently, advanced architectures like 3D-CNNs and Transformers have shown strong performance in learning spatiotemporal patterns, which greatly enhance activity recognition accuracy [13].

The applications of AI-based behaviour recognition are wide-ranging and impactful. In smart surveillance, it helps in the automatic detection of abnormal behaviours such as fighting, theft, or loitering, enabling quicker and proactive security responses. In healthcare, it supports fall detection in elderly patients, monitors post-surgery recovery, and identifies behavioural indicators of neurological disorders such as Parkinson's or Alzheimer's disease [14]. Smart homes and workplaces also benefit from activity recognition for energy management, personal safety, and intelligent automation. Furthermore, Human-Computer Interaction (HCI) systems are becoming more intuitive through gesture recognition and emotion-aware computing, improving the user experience. In education, recognizing student engagement or emotional state during remote learning is gaining interest [15]. Furthermore, autonomous systems and robots use behaviour recognition for adapting to dynamic environments and interacting safely with humans [16].

Behaviour recognition faces several challenges. Variations in viewpoint, occlusion, cluttered backgrounds, and similarities between different classes can significantly reduce the accuracy of behaviour recognition systems. Real world environments are often unpredictable, and human behaviour may vary depending on cultural context, individual differences, and environmental conditions. Furthermore, multi-agent interactions such as group fights or cooperative tasks are more complex and difficult to classify accurately. Achieving real-time performance while ensuring high accuracy, especially on resource-limited edge devices, remains a key technical challenge. In addition, ethical and privacy-related concerns are becoming more important, particularly when such systems are used in public or private spaces.

Several publicly available datasets have supported active research in this area, as listed in Table 2. Among the most commonly used are UCF101, which contains video clips covering 101 action categories, and Kinetics, a large-scale dataset collected from YouTube videos. Another important dataset is NTU RGB+D, which includes depth, infrared, and skeleton data for a wide range of human activities. These datasets are essential for benchmarking and improving the generalisation capabilities of AI models across different environments.

**Table 2** Datasets available for human behaviour recognition.

| *Dataset* | *Description* |
|---|---|
| UCF101 | 13,320 videos from 101 action categories |
| Kinetics | Human action clips collected from YouTube |
| NTU RGB+D | Large-scale dataset with RGB, depth, IR, and skeleton data |
| CASIA | Behavior dataset for surveillance applications |
| HMDB51 | 51 categories of human motion from movies and web videos |

In recent years, several emerging trends have influenced the development of AI based behaviour recognition. These include the use of multimodal learning, where visual, audio, and motion information from multiple sensors is fused to improve recognition robustness. Self-supervised and unsupervised learning methods are being explored to reduce dependence on large annotated datasets. Graph Convolutional Networks (GCNs) are gaining popularity for skeleton-based activity recognition, treating human joints as graph nodes. Additionally, edge AI and federated learning are becoming essential for deploying behaviour recognition systems on local devices while preserving user privacy.

AI-driven behaviour recognition is a rapidly advancing field that combines machine perception with cognitive modelling to understand human actions and interactions. Some models and its use cases are listed in Table 3. The deep learning models become more powerful and data becomes more diverse, behaviour recognition systems are poised to become more accurate, efficient, and context-aware. However, addressing the ethical implications, ensuring inclusivity, and developing explainable AI models remain critical priorities for future research and deployment.

**Table 3** Models for behaviour recognition.

| *Technique* | *Description* | *Use Case* |
|---|---|---|
| CNN | Convolution layers to extract visual features | Static pose detection |
| LSTM | Long Short-Term Memory units for sequence modelling | Gesture recognition |
| 3D-CNN | Extends CNN to spatial and temporal dimensions | Video-based activity recognition |
| GCN (Graph Convolution Network) | Models joints and their relations as graphs | Skeleton-based recognition |
| Transformers | Self-attention for modelling long sequences | Complex behaviour detection |

### 3.3 Facial Recognition and Re-identification

Facial recognition and re-identification technologies have emerged as critical components of intelligent vision-based systems used across surveillance, authentication, and human-computer interaction. Facial recognition refers to the process of identifying or verifying a person's identity using facial features extracted from images or videos. This technology is particularly effective in controlled environments, such as mobile device authentication or access control systems. However, facial re-identification (Re-ID) addresses a more complex challenge: tracking or recognizing an individual across different cameras and settings without prior identity labels. This distinction is crucial in public surveillance networks, where a person must be re-identified based solely on facial characteristics across varied viewpoints and lighting conditions. Together, facial recognition and re-ID form the backbone of advanced surveillance and security infrastructures in smart cities and critical infrastructures.

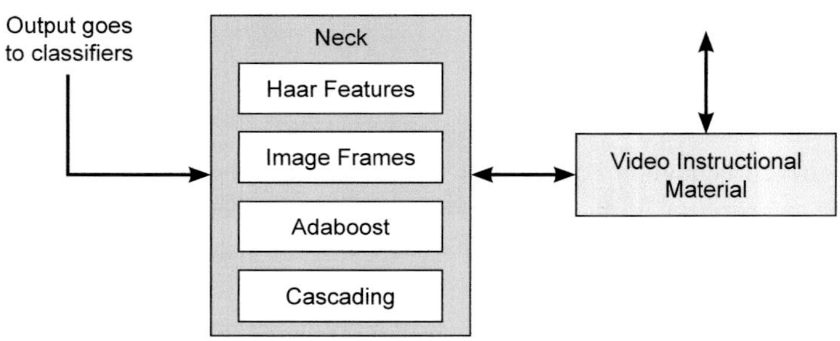


**Fig. 5** Viola-Jones Algorithm Architecture.

#### 3.3.1 Facial Recognition: Architecture and Advances

Facial recognition systems typically consist of three main stages such as face detection, feature extraction, and identity matching. Face detection involves locating and isolating faces from background images or video streams. Popular techniques such as the Viola–Jones algorithm and Multi-task Cascaded Convolutional Networks (MTCNN) are widely used for accurate face localization. The architecture diagram of Viola-Jones model is given in Figure 5. The system extracts relevant facial features using deep CNNs. State-of-the-art models like FaceNet [17], DeepFace [18], and VGGFace employ embedding-based methods to convert facial images into high-dimensional vector representations. These embeddings preserve the unique geometric and texture features of each face. The final stage is face matching, where the extracted embedding is compared with existing embedding in a reference database using similarity metrics such as Euclidean or cosine distance.

Recent advances in deep learning have significantly improved the accuracy and robustness of facial recognition. FaceNet introduced a triplet loss mechanism that helps the model increase the distance between different identities while reducing variation within the same identity. In a similar way, ArcFace improved the ability of face recognition models to generalise by using an additive angular margin loss, which enhances the separation between identity classes [19]. These advancements allow facial recognition systems to work effectively even under different poses, lighting conditions, and facial expressions.

### *3.3.2 Facial Re-identification: Challenges and Deep Learning Approaches*

Facial Re-ID goes beyond traditional facial recognition. Unlike systems that work with pre-labelled identities, Re-ID systems are designed to recognise and match a person across multiple non-overlapping cameras using only visual features. This task is highly challenging due to variations in camera angles, lighting conditions, facial orientation, occlusion, and image resolution. Often, the same person may appear quite different in images captured by different surveillance devices [20].

To overcome these challenges, modern Re-ID systems apply deep learning techniques, especially Siamese networks and triplet networks, which are built to learn discriminative features for matching purposes. These models are trained to extract subtle yet consistent characteristics that remain unchanged across different views of the same individual. Some models also include attention mechanisms to focus on important facial regions, while others use Generative Adversarial Networks (GANs) to generate clearer or more uniform facial images from low-quality or occluded inputs. In addition, domain adaptation methods are used to align feature representations across different environments, which helps improve the performance and reliability of Re-ID systems in real-world applications.

### *3.3.3 Applications in Surveillance and Security*

Facial recognition and re-identification technologies are widely used in security, surveillance, and identity management systems. In urban surveillance, these technologies support real-time monitoring of public areas by helping to detect persons of interest, trace missing individuals, and assist in maintaining law and order. At airports and border control checkpoints, facial recognition is used to automate identity verification, thereby enhancing both security and operational efficiency. Similarly, facial re-identification plays an important role in smart city infrastructure. It enables the tracking of pedestrians across non-overlapping camera views, which is useful for crowd control, traffic monitoring, and behaviour analysis [20].

In the private sector, these technologies are applied in areas such as automated attendance systems, secure banking operations, personalised marketing, and emotional human computer interaction. The use of face based authentication in smartphones and financial services shows the increasing trust in and acceptance of this biometric approach. In addition, the scalability and contactless nature of facial recognition make it particularly suitable in the post-pandemic era, where hygiene and minimal physical contact are key considerations.

### *3.3.4 Ethical Concerns and Privacy Challenges*

Despite the numerous benefits of facial recognition and Re-ID systems, their deployment has sparked ethical debates and concerns over privacy. Issues such as algorithmic bias, racial discrimination, surveillance abuse, and data misuse have been widely reported. Studies have shown that some facial recognition algorithms perform less accurately on women and people of color, leading to unfair outcomes in high-stakes scenarios like law enforcement. Furthermore, constant surveillance using Re-ID systems can infringe on individuals' rights to privacy, especially when consent is not obtained or data is collected without transparency [20].

To address these concerns, it is important to establish proper regulatory frameworks that promote fairness, accountability, and transparency in the use of such technologies. Key practices include anonymising data, applying differential privacy methods, and obtaining informed consent from individuals. Governments and institutions are increasingly advocating for the ethical use of facial recognition technologies, and some regions have already imposed temporary restrictions on their use in public areas.

Facial recognition and re-identification technologies are progressing rapidly, supported by developments in deep learning, computer vision, and sensor technologies. Although these tools offer great potential in improving security, automation, and user convenience, their application must be guided by strong ethical standards and legal safeguards. In the future, research is expected to focus on enhancing cross-domain generalisation, improving real-time performance, and ensuring fairness in facial analytics. When implemented responsibly, these technologies can contribute meaningfully to creating safer, smarter, and more connected environments.

## *3.4 Anomaly Detection*

Anomaly detection is a key component of intelligent surveillance systems, designed to identify unusual patterns or behaviours that differ from normal activity models [21]. Unlike traditional rule-based methods, AI-based anomaly detection can automatically learn patterns from large datasets and identify actions that do not match the expected behaviour. In real-time video surveillance, anomalies may include a person running in a restricted zone, an unattended object, sudden crowd movement, or unauthorised access during non-working hours. Due to the unpredictable and dynamic nature of surveillance environments, anomaly detection is both a challenging and vital task for maintaining public safety and protecting property.

Conventional statistical approaches to anomaly detection often depend on assumptions about data distribution, such as Gaussian models or fixed thresholds.

Although these methods are computationally simple, they are not effective in capturing the complex and non-linear behaviours commonly observed in real-world situations. In contrast, machine learning algorithms provide better flexibility and adaptability. Unsupervised learning techniques such as k-means clustering, Isolation Forests, and Principal Component Analysis (PCA) are frequently used when labelled data is scarce. These algorithms work by identifying data points that differ significantly from the majority. However, they may still struggle to capture temporal dependencies in video data.

To address this, deep learning has emerged as a dominant approach in anomaly detection, particularly in video analytics. The models follow common sequential operations such as pre-processing which extracts frames from video, resizes the frame and removes the background from the frame. Then the data forwarded to a deep learning model for the object extraction and with the help of the AI technique finds the anomaly. The detailed description is given in Figure 6. The various deep learning models such as auto-encoders and CNNs learn to reconstruct input data, typically normal events. Any event that results in high reconstruction error is flagged as an anomaly. RNNs and LSTM networks are used to model sequential dependencies in video frames, capturing complex temporal patterns. More recent developments include the use of GANs, where a generator learns to synthesise normal frames and a discriminator identifies inconsistencies in real-time footage, effectively highlighting unusual behaviour

In high-stakes environments such as airports, military bases, and critical infrastructure, anomaly detection systems help mitigate risks by alerting security personnel in real-time. Furthermore, such systems reduce human workload and improve response times. However, implementing these systems at scale requires consideration of data privacy, bias in training datasets, and interpretability of AI decisions.

### 3.4.1 Applications of Anomaly Detection

Anomaly detection plays a vital role in several applications beyond surveillance, including fraud detection in banking, fault detection in industrial systems, medical diagnosis, and cyberattack detection in networks. In public spaces, it enables automated crowd monitoring to detect aggressive or panic behaviour. In healthcare, anomaly detection applied to physiological signals can alert clinicians about critical changes in patient status. The strength of anomaly detection lies in its ability to generalise detecting both familiar and previously unseen irregularities. In visual surveillance, anomaly detection has become a key element of AI-driven safety and security systems. With the support of deep learning techniques, these systems can automatically observe and detect suspicious activities, reducing the need for constant human monitoring. As this technology continues to develop, it is important to focus on interpretability, cross-domain robustness, and ethical usage [22]. Integrating anomaly detection into surveillance systems improves situational awareness and enables a proactive approach to threat detection in complex environments.

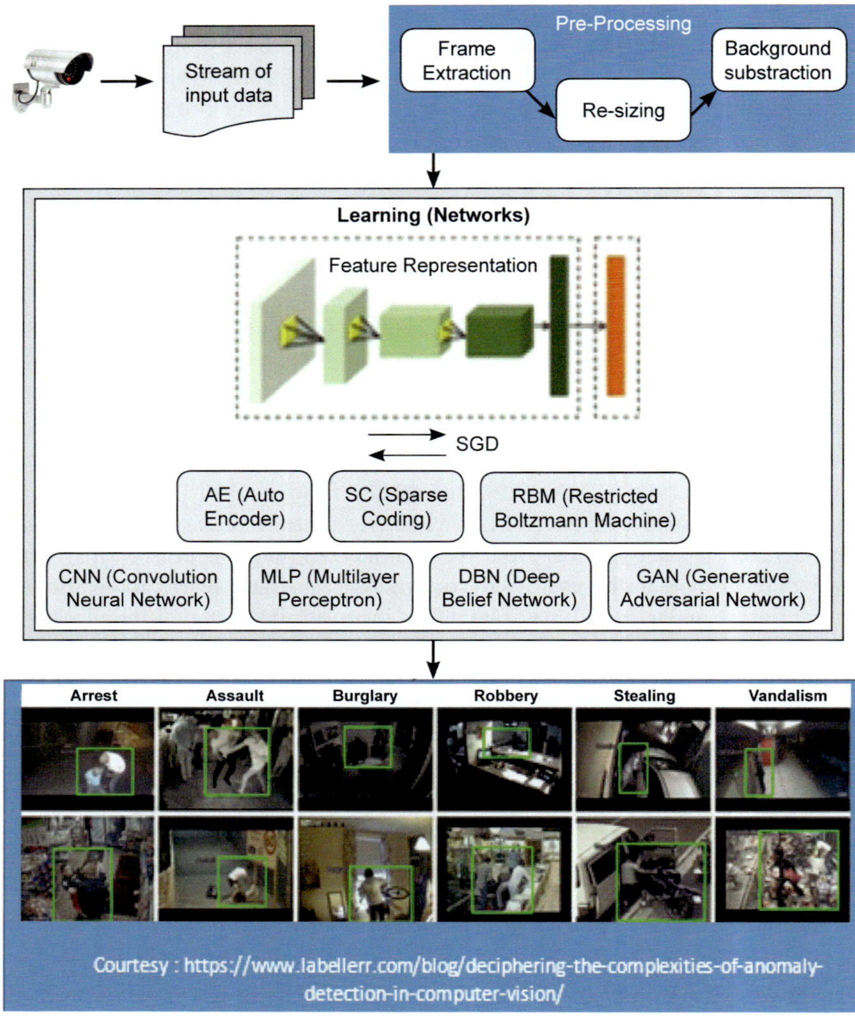


**Fig. 6** Anomaly detection pipeline for surveillance.

## 4. Hardware and Deployment Considerations for AI-Driven Surveillance Systems

As AI continues to transform video surveillance, hardware and deployment strategies have become crucial factors in achieving scalable, efficient, and responsive security systems. Modern surveillance infrastructure must support real-time processing, large-scale video streaming, anomaly detection, and decision-making. These requirements place significant demands on computational hardware, network architecture, and energy resources. Therefore, careful consideration must be given to where and how data is processed, namely whether processing happens at the edge or in centralized cloud systems.

### 4.1 Edge Computing vs. Cloud-Based Systems

Edge computing is increasingly preferred over cloud-based solutions for real-time video surveillance and anomaly detection. In edge-based systems, processing occurs close to the data source typically on cameras, embedded GPUs (graphics processing units), or local edge servers. This approach reduces latency, lowers bandwidth consumption, and improves privacy by avoiding transmission of sensitive data over long distances. On the other hand, cloud-based systems provide centralized processing power, easier scalability, and more advanced analytics but often suffer from higher latency, increased data transfer costs, and potential security risks.

The choice between edge and cloud processing depends on the specific application. For time-critical tasks such as real-time intrusion detection or crowd monitoring, edge devices like NVIDIA Jetson series or Google Coral are commonly used to run lightweight CNNs directly on site. For long-term analytics and storage-intensive tasks, cloud platforms such as AWS, Microsoft Azure, or Google Cloud are employed for deep analysis, model updates, and central management. Hybrid models also exist where edge devices preprocess or filter video data and send only important segments to the cloud for further inference, achieving a balance between responsiveness and computational efficiency [23].

### 4.2 Integration with IoT and Smart City Infrastructure

AI-powered video surveillance systems are increasingly integrated into smart city frameworks through the Internet of Things (IoT). In these environments, surveillance cameras function as intelligent nodes connected to a wider network of devices including traffic sensors, environmental monitors, emergency response units, and communication systems. This integration improves situational awareness, supports automated decision-making, and enables smooth coordination among agencies [24]. For example, real-time anomaly detection in surveillance footage can trigger automatic alerts to law enforcement, adjust traffic signals in response to crowd congestion, or coordinate with drone systems for enhanced monitoring. IoT protocols such as MQTT (message queuing telemetry transport) and CoAP (constrained application protocol) facilitate lightweight communication among edge devices, gateways, and control centres. Furthermore, interoperability standards and open APIs are critical to ensure that surveillance systems can interact with heterogeneous smart city platforms without vendor lock-in [25].

### 4.3 Power, Bandwidth, and Storage Optimisation

Power efficiency, bandwidth management, and storage scalability are crucial in the deployment of AI-driven surveillance systems, especially in large-scale or remote installations. Edge devices are typically battery or solar-powered in smart poles, drones, or rural areas. Hence, low-power AI accelerators and optimized deep learning models are required to reduce energy consumption while maintaining accuracy [26].

Bandwidth consumption is another bottleneck, particularly when dealing with high-resolution video streams or multi-camera networks. Techniques such as video summarization, region-of-interest (ROI) encoding, and event-triggered transmission help in reducing unnecessary data transmission. Additionally, compressed sensing and AI-based encoding like HEVC with object-detection triggers ensure efficient data transfer without compromising critical information.

From a storage perspective, data redundancy, retention policies, and hierarchical storage models are commonly adopted. Systems prioritise event-driven footage for longer-term cloud storage, while discarding uneventful video after short-term edge-level caching.

Hardware and deployment considerations significantly impact the scalability, reliability, and real-world applicability of AI-based video surveillance systems. Choosing the right balance between edge and cloud processing, leveraging IoT integration, and optimising power, bandwidth, and storage resources are fundamental to achieving intelligent, responsive, and sustainable surveillance infrastructure. As surveillance expands into smart cities and critical infrastructure, future systems must adopt adaptive, context-aware deployment models that are both resource-efficient and operationally robust.

## 5. Real-World Applications and Case Studies

Some real-world applications being used in public places, industrial and work places, border securities and privacy are deeply observed from the web resources and research articles. The summary report is depicted in Figure 7 and deeply discussed in this section. The defence and border security is using the deep learning and AI implemented computer vision techniques for the protection compared to other sectors.

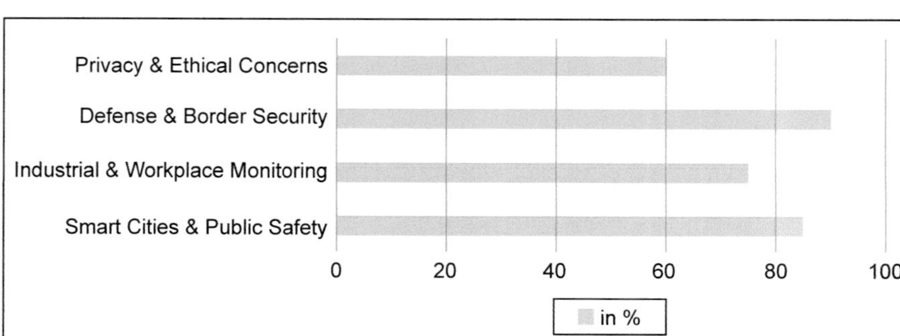


**Fig. 7** AI and Computer vision usage across sectors.

### 5.1 Smart Cities and Public Safety

Municipal vehicles equipped with AI dash cams detect potholes, broken streetlights, traffic violations such as helmetless riding and illegal parking, stray animals, and vulnerable individuals. Clips are geo-tagged and sent in real-time to control rooms. In a research, thousand camera units are being rolled out and got approximately 80% detection accuracy and 10% false positives. AI-enabled CCTV systems are deployed to analyze crowd density and detect anomalies to prevent stampede risks during the Rath Yatra festival. (https://timesofindia.indiatimes.com/city/ahmedabad/ai-cameras-to-aid-crowd-control-during-rath-yatra-in-ahmedabad/articleshow/121761127.cms).

At three pilot junctions, AI-controlled lights manage traffic based on real-time flow, with ANPR cameras catching red-light runners and helmet-law violators. Completion across 10 intersections is planned by June 2026, potentially reducing travel times by 48% and increasing vehicle speed by 60% . Shenzhen uses AI surveillance for real-time monitoring of traffic violations and individual behaviours. Singapore integrates AI across sectoral cameras, London's Metropolitan Police trailed live facial recognition. New Delhi employs AI cameras at intersections for traffic enforcement and suspect tracking.

AllGoVision analytics manage queue lengths, intrusion detection, left-luggage, and unauthorized entry. At Hyderabad, one deployment improved both safety and passenger experience via dynamic resource allocation based on real-time analytics (https://www.allgovision.com/case-study-RBI.php). Between Oct. 2022 and Sept. 2023, 11 AI algorithms processed live CCTV at Willesden Green station, generated 44,000 alerts for fare evasion, weapons display, and safety hazards like track falls, unauthorized platform entry, etc. Though effective, this trial highlighted ethical and accuracy concerns (https://www.wired.com/story/london-underground-ai-surveillance-documents). *The Times of India* listed an AI system which detects mechanical anomalies like hot axles, brake issues, and fire hazards in coaches, improving early fault detection and maintenance scheduling. (https://timesofindia.indiatimes.com/city/patna/ai-system-rolled-out-to-boost-rail-safety-and-maintenance/articleshow/121753941.cms).

### 5.2 Industrial and Workplace Monitoring

Used globally by trucking fleets and government fleets, Lytx's DriveCam and Risk Detection services record continuous video, identify unsafe driving patterns, and support incident investigation (https://en.wikipedia.org/wiki/Lytx). Partnerships between Data Science for Social Good, Jakarta Smart City, and academia built video pipelines to analyse traffic footage and inform urban planning interventions aimed at accident reduction.

In smart city settings, FLAMe combines federated learning and lightweight spatio-temporal transformers to detect pedestrian falls with 94% accuracy while preserving patient privacy [27]. A real-world deployment at a community college used 16 cameras and achieved an average alert latency of ~27 s and ~16.5 FPS, using AI to detect anomalous behavior while protecting personal identity via pose-based privacy methods [28].

### 5.3 Defense and Border Security

A collaborative project between NYPD and Microsoft, this network consolidated 18,000 CCTV feeds, billions of licence-plate reads, and emergency call data with retention policies and analytic tools for counterterrorism. This private/public surveillance firm uses ALPR, gunfire detection, and video AI to assist law enforcement in over 5,000 communities. The data supports investigations and reportedly contributes to around 10% of U.S. criminal resolutions. Part of Cape Town's Strategic Surveillance Unit, 2,000 plus CCTV cameras and UAVs with infrared and thermal sensors monitor crime hotspots, detect gunfire, and support rapid police response, forming part of an integrated safety control ecosystem (en.wikipedia.org).

**Table 4** Summary of Outcomes and Challenges.

| *Application Area* | *Key Achievements* | *Challenges and Considerations* |
|---|---|---|
| Smart cities | Enhanced traffic flow, crowd management, civic issue detection | Privacy, public acceptance |
| Transport hubs | Faster alerting, anti-fare evasion, queue optimisation | Accuracy, ethical reviews |
| Industrial fleets and sites | Early detection of hazards, driver behaviour analytics | Integration complexity, data management |
| Healthcare and elder-care | Automated fall alerts, privacy-aware monitoring | Ensuring reliability, latency constraints |
| Defence and security | Fusion of CCTV, ALPR, gunfire detection, UAVs | Governance, surveillance overreach |

## 6. Privacy and Ethical Issues

AI-powered surveillance systems often capture sensitive personal data-faces, behavior patterns, and even emotional states frequently without individuals' consent. This practice raises serious concerns around privacy, consent and civil liberties [29]. Additionally, technologies such as facial recognition and data aggregation can lead to intrusive profiling and social scoring systems that erode autonomy and trust. Scholars call for privacy-by-design frameworks, transparency measures, and robust governance to ensure surveillance tools respect individual rights.

### 6.1 Bias and Fairness in AI Models

Visual monitoring systems trained on unbalanced or skewed data often discriminate against underrepresented groups. Landmark studies have shown that commercial facial recognition software misidentifies darker-skinned women up to 34% of the time compared to just 0.8% for lighter-skinned men (https://en.wikipedia.org/wiki/Joy_Buolamwini). Such algorithmic bias can perpetuate inequality, wrongful identification, and unfair treatment in security and policing. Although fairness-enhancing strategies pre-processing, adversarial in-processing, or post-processing corrections are promising, they often trade off with accuracy and computational cost, and require careful context-specific implementation [30].

### 6.2 Scalability and Real-Time Constraints

Real-time visual monitoring demands low-latency performance over thousands of video streams. Achieving this at scale requires high-throughput data ingestion, edge-computing infrastructures, optimized algorithms, and enough compute to prevent lag. Furthermore, maintaining accuracy while processing noisy, high-resolution footage in real environments (low light, occlusion, clutter) is technically challenging and resource-intensive, necessitating a fine balance between speed, cost, and precision.

### 6.3 Adversarial Attacks on Vision Models

Vision systems are highly vulnerable to adversarial samples, subtle, human-imperceptible perturbations that cause misclassification. In both black-box and physical-world settings, researchers have successfully fooled object detectors, e.g., causing a rifle to be recognized as a helicopter even in safety-critical systems like self-driving cars and medical imaging (https://www.wired.com/story/researcher-fooled-a-google-ai-into-thinking-a-rifle-was-a-helicopter). These attacks jeopardise trust in surveillance, as attackers could bypass detection or trigger false alarms, while many defences such as adversarial training, defensive distillation, and differential privacy exist, remain limited or computationally heavy [31].

### 6.4 Data Quality and Annotation Issues

Robust visual AI depends on large-scale, high-quality labelled datasets. However, many public datasets such as COCO, ImageNet, CelebA contain images collected without consent and reflect societal biases in terms of race, gender, and socioeconomic representation [29]. Documentation of data quality is frequently inadequate that is less than 30% of open-source projects include essential data quality checks (https://en.wikipedia.org/wiki/Open-source_artificial_intelligence). Imperfections in labelling can lead to models learning spurious correlations or over-fitting on noise. Moreover, ongoing annotation and refinement are necessary to adapt to changing real-world conditions, requiring significant human and financial investment.

These challenges, privacy infringements, bias, scaling issues, adversarial vulnerability, and data quality constraints form a complex technological, social, and ethical ecosystem. Addressing them effectively requires interdisciplinary collaboration, ongoing model auditing, context aware fairness techniques, robust defences, regulatory oversight, and active community engagement to ensure AI-driven visual surveillance serves society responsibly.

## 7. Future Directions

Federated Learning (FL) allows collaborative model training across distributed edge cameras or institutions without sharing raw footage. Instead, local devices compute updates on site, and only model parameters are aggregated centrally protecting sensitive visual data. Advanced frameworks are now integrating privacy enhancements like differential privacy, homo-morphed encryption, and blockchain-based update provenance to resist model-inversion and poisoning attacks, addressing FL's privacy and robustness challenges. In surveillance, smart cities, or hospital networks could adopt edge-based federated setups to train behaviour detection models without compromising individual privacy.

### *7.1 Explainable AI (XAI) in Surveillance*

Explainability is becoming critical in surveillance AI, especially when automated alerts trigger substantial actions like lockdowns or law enforcement responses. XAI methods from post-hoc saliency heatmaps to inherently interpretable models can clarify which pixel regions or features led to a signal, helping operators verify and trust detections. DARPA's XAI programme specifically targets enhanced explainability in multimedia event classification and autonomous decision-making. Future deployments in surveillance must balance performance and interpretability, enabling transparency across cities, airports, and critical infrastructure.

### *7.2 Integration with Multimodal Data (Audio, Thermal, etc.)*

Moving beyond visual-only systems, multimodal surveillance fuses video with audio, thermal, infrared, and metadata like timestamps, geo-locations to improve context awareness and reduce false positives. For instance, tri-modal audio-visual-thermal person verification has demonstrated ~50% error-rate reductions compared to uni-modal systems . Applications include emotion detection in elder-care using thermal imaging and audio cues (~77% accuracy) [32]. Such approaches help detect anomalous behavior unseen in pure video feeds for example thermal spikes during fights or gas leaks.

### 7.3 Role of Generative Models (e.g., GANs) for Augmentation or Simulation

GANs are proving valuable in visual surveillance, they can generate high-fidelity synthetic scenes to augment training datasets like simulating night-time, weather, and rare events, enhance low-resolution security footage, or generate virtual environments to stress-test systems. For example, GAN-based augmentation in medical imaging vastly improves detection accuracy with limited datasets (unitxlabs.com). In surveillance, GANs can fill data gaps like rare abandoned baggage or intruder events to strengthen model robustness in edge cases.

## 8. Conclusion

AI-driven visual monitoring demonstrates vast potential: enhancing public safety in smart cities, securing transportation hubs, fortifying industrial operations, enabling proactive elder-care, and bolstering defence. Techniques like FL, XAI, multimodal perception, and generative augmentation are shaping the next generation of surveillance systems.

As surveillance technologies evolve, so do ethical, legal, and social risks. Ensuring privacy, fairness, transparency, and accountability is paramount. Approaches such as FL, explainability, and privacy-by-design must be embedded into development, standardization, and policy frameworks.

When responsibly developed and deployed, AI-based monitoring can save lives through faster detection of falls, threats, or system failures while increasing operational efficiency and enabling data-informed urban planning. The challenge ahead is balancing innovation with oversight, ensuring AI enhances safety without sacrificing civil liberties.

## References

1. Dhulekar, P.A., Gandhe, S.T., Sawale, N., Shinde, V. and Khute, S. (2018). Surveillance system for detection of suspicious human activities at war field, In: *Int. Conf. Adv. Commun. Comput. Technol. ICACCT 2018*, (pp. 357–360). doi: 10.1109/ICACCT.2018.8529632.
2. Sharma, H. and Kanwal, N. (2025). Video surveillance in smart cities: Current status, challenges and future directions. *Multimed Tools Appl.*, 84, 15787–15832. https://doi.org/10.1007/s11042-024-19696-6.
3. Nguyen, H.V. and Byeon, H. (2024). YOLOv7-based automated detection platform for scalp lesions. *Digital Health*, 10. doi:10.1177/20552076241279185.
4. Chen Bojian, Zhang Weihao, Wu Wenbin, Li Yiran, Chen Zhuolei and Li Chenglong. (2024). ID-YOLOv7: An efficient method for insulator defect detection in power distribution network. *Frontiers in Neurorobotics*, *17*. doi:10.3389/fnbot.2023.1331427, ISSN=1662-5218.
5. Wei Liu, Cheng-Yang Fu and Alexander C. Berg. (2016). Single Shot MultiBox Detector. In: Leibe, B., Matas, J., Sebe, N., Welling, M. (Eds.), Computer Vision – ECCV. ECCV 2016. *Lecture Notes in Computer Science*, 9905. Springer, Cham. https://doi.org/10.1007/978-3-319-46448-0_2.
6. Magalhães, S.A., Castro, L., Moreira, G., dos Santos, F.N., Cunha, M., Dias, J. and Moreira, A.P. (2021). Evaluating the Single-Shot MultiBox Detector and YOLO Deep Learning Models

for the Detection of Tomatoes in a Greenhouse. *Sensors,* 21(10), 3569. https://doi.org/10.3390/s21103569.

7. Guo, L., Wang, D., Li, L. and Feng, J. (2020). Accurate and fast single shot multibox detector. *IET Comput. Vis.,* 14, 391–398. https://doi.org/10.1049/iet-cvi.2019.0711.
8. Zuoxin Li, Lu Yang and Fuqiang Zhou. (2024). FSSD: Feature Fusion Single Shot Multibox Detector. *arXiv:1712.00960v4* . https://doi.org/10.48550/arXiv.1712.00960.
9. Yadav, S.P., Jindal, M., Rani, P., et al. (2024). An improved deep learning-based optimal object detection system from images. *Multimed Tools Appl.,* 83, 30045–30072. https://doi.org/10.1007/s11042-023-16736-5.
10. Momir Adzemovic, Predrag Tadic, Andrija Petrovic, & Mladen Nikolic. (2024). Engineering an Efficient Object Tracker for Non-Linear Motion. *arXiv:2407.00738,* v1 [cs.CV] 30 Jun. DOI. https://arxiv.org/pdf/2407.00738.
11. ChienYao Wang, Alexey Bochkovskiy and Hong-Yuan Mark Liao (2022). YOLOv7: Trainable bag-of-freebies sets new state of the art for real-time object detectors. https://doi.org/10.48550/arXiv.2207.02696.
12. Kaseris, M., Kostavelis, I. and Malassiotis, S. (2024.) A Comprehensive Survey on Deep Learning Methods in Human Activity Recognition. *Machine Learning and Knowledge Extraction, 6*(2), 842–876. https://doi.org/10.3390/make6020040.
13. Hu, Kai and Jin, Junlan and Zheng, Fei and Weng, Liguo and Ding and Yiwu. (2022). Overview of behavior recognition based on deep learning. *Artificial Intelligence Review,* 56(3): 183–186. https://doi.org/10.1007/s10462-022-10210-8.
14. Bajwa, J., Munir, U., Nori, A. and Williams, B. (2021). Artificial intelligence in healthcare: Transforming the practice of medicine. *Future Healthc J.*, Jul. 8(2), e188-e194. doi: 10.7861/fhj.2021-0095. PMID: 34286183; PMCID: PMC8285156.
15. Sheng, X., Li, S. and Chan, S. (2025). Real-time classroom student behavior detection based on improved YOLOv8s. *Sci Rep.*, 15, 14470. https://doi.org/10.1038/s41598-025-99243-x
16. Kanghyun Jo, Jehwan Choi, Kwanho Kim, Seongmin Kim, Duy-Linh Nguyen, Xuan-Thuy Vo, Adri Priadana, & Tien-Dat Tran. (2025). Artificial Behavior Intelligence: Technology, Challenges, and Future Directions. *arXiv:2505.03315*. https://doi.org/10.48550/arXiv.2505.03315.
17. Schroff, F., Kalenichenko, D. and Philbin, J. (2015). FaceNet: A unified embedding for face recognition and clustering. In: *2015 IEEE Conference on Computer Vision and Pattern Recognition* (*CVPR*), Boston, MA, USA, pp. 815–823). doi: 10.1109/CVPR.2015.7298682.
18. Taigman, Y., Yang, M., Ranzato, M. and Wolf, L. (2014). DeepFace: Closing the Gap to Human-Level Performance in Face Verification. In: *Proceedings of the IEEE Conference on Computer Vision and Pattern Recognition* (CVPR) (pp. 1701–1708). https://doi.org/10.1109/CVPR.2014.220.
19. Deng, J., Guo, J., Niannan, X. and Zafeiriou, S. (2019). ArcFace: Additive Angular Margin Loss for Deep Face Recognition. In: *Proceedings of the IEEE/CVF Conference on Computer Vision and Pattern Recognition,* (pp. 4690–4699). https://doi.org/10.1109/CVPR.2019.00482.
20. Wu, Y., Lin, Y., Dong, X., Yan, Y. and Ouyang, W. (2018). Exploit the Unknown Gradually: One-Shot Video-Based Person Re-Identification by Stepwise Learning. In: *Proceedings of the IEEE/CVF Conference on Computer Vision and Pattern Recognition* (*CVPR*), (pp. 5177–5186). https://doi.org/10.1109/CVPR.2018.00545.
21. Liu, W., Luo, W., Lian, D. and Gao, S. (2018). Future Frame Prediction for Anomaly Detection: A New Baseline. In: *Proceedings of the IEEE Conference on Computer Vision and Pattern Recognition* (*CVPR*), (pp. 6536–6545). https://doi.org/10.1109/CVPR.2018.00684.
22. Moshira Abdalla, Sajid Javed, Muaz Al Radi, Anwaar Ulhaq and Naoufel Werghi. (2024, May). Video Anomaly Detection in 10 Years: A Survey and Outlook. *arXiv,* May. .
23. Singh, R., Mahapatra, R. and Raval, N. (2024). Edge–Cloud Collaborative AI Framework for Urban Video Surveillance. *IEEE Internet of Things Journal.*

24. Karthigai Selvi, S., Nagaraja, P., Gayathri, S.P. and Genish, T. (2023). Building Sustainable Changing Infrastructure: Smart Solutions. In: S. Vijayalakshmi, Savita, Balamurugan Balusamy, Rajesh Kumar Dhanaraj (Eds.), *AI-Powered IoT in the Energy Industry*. Springer International Publishing.
25. Wang, Z., Liu, Y. and Khan, R. (2023). Integration of IoT-Based Surveillance in Smart Cities: A Review. *Sensors*, 23(15), 3890.
26. Patel, K. and Zhao, L. (2025). Energy-Aware Spatiotemporal Optimization for Edge-Based Video Analytics. *ACM Transactions on Cyber-Physical Systems*, 9(2), 1–25.
27. Byeonghun Kim and Byeongjoon Noh. (2025). FLAMe: Federated Learning with Attention Mechanism using Spatio-Temporal Keypoint Transformers for Pedestrian Fall Detection in Smart Cities. *AAAI 2025 FLUID Workshop*. https://doi.org/10.48550/arXiv.2412.14768.
28. João Caldeira, Alex Fout, Aniket Kesari, Raesetje Sefala, Joseph Walsh, Katy Dupre, Muhammad Rizal Khaefi, Setiaji, George Hodge, Zakiya Aryana Pramestri, & Muhammad Adib Imtiyazi. (2019). Improving Traffic Safety Through Video Analysis in Jakarta, Indonesia. In: Bi, Y., Bhatia, R., Kapoor, S. (Eds.), *Intelligent Systems and Applications. IntelliSys 2019. Advances in Intelligent Systems and Computing*. 1038. Springer, Cham. https://doi.org/10.1007/978-3-030-29513-4_48.
29. Ghalib Ahmed Tahir (2024). Ethical Challenges in Computer Vision: Ensuring Privacy and Mitigating Bias in Publicly Available Datasets. *arXiv:2409.10533v1 [cs.CV]*.
30. Chen, P., Wu, L., Wang, L. (2023). AI Fairness in Data Management and Analytics: A Review on Challenges, Methodologies and Applications. *Appl. Sci.*, 13, 10258. https://doi.org/10.3390/app131810258.
31. Yazan Alaya and AL-Khassawneh. (2023) A Review of Artificial Intelligence in Security and Privacy: Research Advances, Applications, Opportunities, and Challenges. *Indonesian Journal of Science and Technology*, *8*(1), 79–96.
32. Mohamed, Y., Güneysu, A., Lemaignan, S., et al. (2024). Multi-modal Affect Detection Using Thermal and Optical Imaging in a Gamified Robotic Exercise. *Int. J. of Soc. Robotics,* 16, 981–997. https://doi.org/10.1007/s12369-023-01066-1.

# 22 SARFALS: A Secure and Robust Framework for AI-Driven Autonomous Weapon Systems

Mahalakshmi[1*], Nidhi S. Vaishnaw[2], Sridevi Polasi[3] and Sudip Poddar[4]

## 1. Introduction

Technological transformation has profoundly influenced the history of war, and in recent years, the integration of Artificial Intelligence (AI) into military actions has ushered in a new era of autonomous weapons systems. From unmanned aerial vehicles to smart missile guidance systems, AI has revolutionized defense operations through faster, data-driven decision-making, accuracy in targeting, and diminished human footprint on the battlefield. The smart systems apply machine learning methodology, in the form of deep models of learning, to analyze enormous volumes of sensor data, target designation, threat evaluation, and make real-time tactical recommendations. As countries compete for technological preeminence, artificial intelligence-driven weapon systems take center stage in contemporary defense strategy, providing unexampled capability with far-reaching security implications.

Assisted armament began with the initial guided missile platforms in the mid-20th century and evolved through radar-guided anti-air guns and into high-tech unmanned drone systems. The 21st century saw a paradigm shift with partially autonomous drones used for attacks such as surveillance, reconnaissance, and precision attacks. These were rule-based reasoning and human-controlled systems. But with the advent of deep learning and reinforcement learning, new weapons systems are now capable of learning by experience, responding to unforeseen battle reality, and making decisions second-by-second on their own. Such autonomy increases tactical efficiency, but also opens Pandora's box of vulnerabilities, particularly in terms of robustness and cybersecurity.

[1] Department of Mathematics, Nitte Meenakshi Institute of Technology, Nitte (Deemed to be University), Bengaluru, Karnataka, India.
[2] Department of Mathematics, CMR Institute of Technology, Bengaluru, Karnataka, India.
[3] Department of Mathematics, School of Applied Sciences, REVA University, Bengaluru, Karnataka, India.
[4] Department of Computer Science and Engineering, Johannes Kepler University Linz, Austria.
* Corresponding author: mahalakshmi4131@gmail.com

DOI: 10.1201/9781003663461-22

AI arms systems are sweeping in their limitations while holding out great promise. The central concern is the failure of AI models to resist adversarial attacks. The computer heart of most autonomous systems are deep neural networks, which have a very high susceptibility even to small changes in input data. During combat operations, the specifically engineered inputs known as adversarial examples can lead to catastrophic errors, such as the misidentification of friendly forces with enemy targets. The system can be attacked at its reliability when attackers with model parameters or data pipeline access inject noise or deceptive features. Also, carefully selected scenarios or lab tests are typically the basis of training data for AI systems in the military that are not representative of the chaotic and unstable dynamics of actual war. If AI is applied to actual operations, such a domain disparity leaves the door open for failure.

System-wide security of autonomous weapons is a major drawback. Subsystems made up of sensors, actuators, communication modules, and Artificial Intelligence (AI) processors typically make up such systems. Since they are networked, often facilitated by wireless networks or Internet of Military Things (IoMT) standards, parts can be threatened by signal jamming, data spoofing, and man-in-the-middle attacks. An opponent can intercept, modify, or replay orders in the absence of secure communication mechanisms and authentication mechanisms and can result in unpredictable weapon behavior or capture by the enemy. The present cybersecurity practices depending largely on conventional IT-based systems are incapable of catering to the adversary dynamics and real-time requirements of a battlefield. They either lack the sophistication level to acknowledge new types of attack vectors against AI elements or respond with a lag to new attacks.

The application is further complicated by the explainability and lack of transparency in AI decision-making. All steps in high-entropy environments like defense must be accountable and auditable. But commanders and operators find it hard to judge or embrace the output of black-box models because they do not give any information on why a particular action is taken. That is more problematic when lethal force is used since moral and legal legitimacy demand a traceable line of accountability and rationale. Although they provide some restricted solutions, traditional explainable AI (XAI) techniques remain to be incorporated into successful military systems operating in real-time at the cost of functionality. In addition, attempts at enhancing AI robustness are generally concerned with anomaly detection alone or adversarial training.

Although these approaches improve performance in certain circumstances, they do not take a comprehensive approach to addressing system-level security and model-level resilience at the same time. Adversarial training improves perturbation robustness against known perturbations, but also may produce computational overhead or generalization loss that blunts responsiveness. Network-layer or cryptographic measures improve communication integrity, but cannot stop manipulation at the perception layer or decision layer of the system. Interconnectedness of AI-based weapon system threats is not solved by a disjointed approach.

The lack of clearly established benchmarks and testing methodologies specific to defense usage only exacerbates these disparities. Most AI models are trained and verified on open-source datasets or simulations that cannot possibly recreate the complexity of a battlefield, where inputs can be deliberately misleading, noisy, and incomplete. Furthermore, governance and regulatory frameworks have lagged behind technical innovation. Instead of operational safety and robustness, current frameworks mostly pay attention to the legality of autonomous weapons. Such a regulatory gap results in inconsistencies in development procedures as well as in field deployment of untested or poorly tested systems, and this does lead to strategic as well as humanitarian issues.

The imperative requirement of formulating a robust and complete strategy for protecting AI-driven weapon systems against progressing threats motivated us to make the following inquiry. With AI being more integrated into vital combat systems, its reliability and trustworthiness are no longer only a technical concern but also a strategic one. Investors in defense need to trust that their autonomous systems can withstand cyberattacks, perform even under hostile conditions, and neither trigger false positives nor incur delays. Furthermore, adversaries are more likely to target weaknesses in AI systems and the hardware that supports them as international tensions rise and unmanned systems become more integral to conflict.

The recognition that it is only possible to counter these high-level challenges with an integrated approach that combines state-of-the-art adversarial defense techniques with defend-grade secure comms architectures specifically tailored for defense-grade AI systems is what impels our effort. The goal is to transcend piecemeal solutions and develop a resilient ecosystem in which communications can be authenticated, AI models are inherently robust, and the system as a whole can withstand physical and digital attacks without undermining mission success or ethical standards. Such a technology is needed desperately, numerous defense organizations and corporations are already researching autonomous targeting systems, and not anticipating their weaknesses could have disastrous impacts on global stability.

The study lays the groundwork for strong and secure AI architecture in particular for military autonomous systems on these grounds. The premise for a solution that meets the strategic and technological requirements of contemporary warfare is established by way of critical assessment of the shortcomings of existing schemes and knowledge of the symbiotic relationship between threat order of the adversary and vulnerability of the system. In addition to being an intellectual endeavor, this is an obligatory step toward the assurance of AI in defense being implemented in accordance with the principles of safety, responsibility, and adaptability according to changing threats.

**Main Objectives**

1. To develop a unified security framework (SARFALS) that enhances the adversarial robustness of AI models used in autonomous weapon systems through integrated techniques such as Adversarial Training with Randomized

Smoothing (ATRS), thereby mitigating the risk of model manipulation via crafted inputs or data poisoning attacks in high-stakes combat environments.

2. To ensure secure, tamper-proof, and verifiable communication among subsystems in AI-driven weapon platforms by implementing a Blockchain-Assured Communication Protocol (BACP), enabling decentralized authentication and real-time integrity verification to defend against spoofing, replay, and man-in-the-middle attacks.

## 2. Related Study

The increasing use of AI in Autonomous Weapon Systems (AWS) has produced significant research on the convergence of technological advancement, cyber security, ethical accountability, and global regulation. The new literature has investigated how the embedding of AI reshapes military planning, operational security, and decision-making for autonomous systems. A good starting point for the investigation of the use of AI in AWS is the manner in which it revolutionizes traditional war systems by increased targeting accuracy, lower manpower requirements, and quicker response times [1]. These do have ominous warning on the matter of cybersecurity and system resilience, however.

Additional text outlines the susceptibility of AI systems to malicious operations, both model and infrastructure-wise. AWS AI systems can be maliciously compromised by adversarial attacks that imperceptibly perturb input data in order to deceive the system's perception and classification processes [2]. For example, imperceptible perturbations in sensor measurements can cause a deep learning model to incorrectly identify a civilian car as an enemy target. These weaknesses are particularly serious operational and humanitarian issues when the system is deployed in high-risk or high-density operations. Literature requires more efficient training protocols, e.g., adversarial training and model verification, to improve resilience to such attacks [3].

Network security is also a real issue. AWS is generally constituted of several interacting subsystems, sensors, actuators, communication modules, and decision functions based on safe, real-time data exchange. Tests have shown that current communication protocols used in such systems rarely are effective enough to avoid replay, spoofing, or man-in-the-middle attacks [4]. Such threats have the potential to create disruption to the synchronization of the components, resulting in loss of control, system crash, or hijacking of the system by an attacker. Accordingly, research has introduced blockchain-based secure communication patterns that provide integrity guarantee and tamper resistance in a decentralized manner and are better suited to the decentralized nature of AWS [5].

Trust and explainability have been some of the frequent themes of recent times. The work highlights the importance of the fact that AI-based decisions in lethal systems should be interpretable to facilitate accountability and public trust. Most implemented models, including convolutional neural networks or reinforcement learning agents, though, are black boxes whose delivery offers small insight into how decisions are being generated [6]. This is particularly unsettling in the case that autonomous decisions involve the use of force since it precludes post-mission auditability and international humanitarian law compliance. Similarly, there are some studies proposing the addition of XAI techniques to AWS but real-time functionality has to be ensured with human-readable outputs [7].

Similarly, ethical and legal aspects of autonomous decision-making in conflict are addressed in general terms. Certain contributions stress that the speed of technology introduction is dramatically higher than the speed of international legal norms development to address AWS responsibly [8]. Without regulatory checks, there is a risk of unleashing AI-driven lethal systems capable of breaching norms of proportionality, distinction, or military necessity. Some works put forward architectures for the incorporation of "meaningful human control" into AWS operations in a manner where humans still maintain final decision-making power even where there is high automation involved [9]. Such models fall in line with recommendations by different global organizations to regulate or ban fully autonomous deadly systems.

Additionally, other research examines the geopolitical effects of AI on military technology, where AWS have themselves become a metaphor for technological supremacy in international power politics [10]. Not just the arms race, but also the great power competition, e.g., between the U.S. and China, has been marked by strategic narratives about AI warfare legitimacy and superiority. This geopolitical perspective is necessary to grasp the rationality of quick development and deployment, which comes naturally in the absence of full ethical protection.

In various investigations, there are the psychological consequences and functionality of outsourcing lethal decisions to computers. Human combat moral judgment has been criticized to decline, as AI programs can operate without the situational awareness for ethical reasoning in ever-changing battlefield environments [11]. With arguments having been placed that the use of AWS has the potential to cause automation bias or over-reliance on human operators, and that they act blindly on AI suggestions even against what was witnessed on the battlefield scenarios.

**Literature**

Human intervention and machine independence need to be judiciously balanced such that such risks would not be promoted, as noted in the literature. A second problem brought up is the social damage and public health effect of autonomous warfare. Use of AWS has been reported to cause new kinds of psychological trauma in survivors because there is no human contact in robot attacks and to complicate reconciliation after war [12]. It extends the debate beyond utility in combat to weigh the long-term effects on civilians as well as worldwide norms of war. The necessity for policymakers to look ahead and consider such non-technical factors is thus emphasized.

The literature also delves into the technical methods for achieving robustness in AI models used in AWS. Adversarial training and certified defenses, such as randomized smoothing, are proposed as means to ensure stable performance under adversarial input conditions [13]. Additionally, the adoption of zero-trust architectures and formal verification methods is recommended to bolster the resilience of system infrastructure. However, a common criticism across studies is that these solutions are rarely tested under battlefield conditions or realistic military simulations, highlighting a research-to-deployment gap.

On the topic of governance and oversight, there is a consensus that national and international frameworks must evolve to include mandatory robustness audits, ethical compliance checks, and system certification before deployment [14]. Some studies argue for establishing independent regulatory bodies with the authority to monitor and evaluate AWS based on transparent criteria, much like regulatory agencies in the civilian aviation or pharmaceutical sectors.

The literature underscores the urgent need for an interdisciplinary approach to ensure the secure and ethical development of AI-driven weapon systems [15, 16, 17]. While technical advancements in AI offer significant strategic advantages [18], they must be tempered by robust cybersecurity measures [19], legal governance, and human-centered design principles [20]. The proposed Secure Adversarially-Robust Framework for Autonomous Lethal Systems (SARFALS), which integrates adversarial robustness and blockchain-secured communication, directly responds to the gaps identified in existing research. This dual-layered framework not only aligns with technical best practices but also addresses the pressing ethical, legal, and societal concerns highlighted across the body of literature.

## 3. System Methodology

SARFALS, the suggested system approach, attempts to improve the cybersecurity of AI-driven weapon systems at the system level as well as the robustness of the models. Perception & Sensing, a Sturdy AI Decision Core, Blockchain-Assured Communication, and Real-Time Command Execution & Feedback Control are the four interconnected layers that make up the framework as shown in Figure 1. In hostile and high-risk situations, each layer is made to ensure low-latency, high-reliability operation while mitigating certain vulnerabilities.

### *3.1 Dataset Description*

Publicly accessible, modality-specific datasets that replicate actual battlefield conditions are used to train and validate SARFALS.

- KITTI Vision Benchmark Suite, which supports 3D localization and object detection for RGB (red, green, blue) pictures and LiDAR data.
- For radar streams in difficult weather and lighting conditions, use the Oxford Radar RobotCar Dataset.
- The FLIR ADAS Dataset simulates human and vehicle heat signatures for thermal imaging.

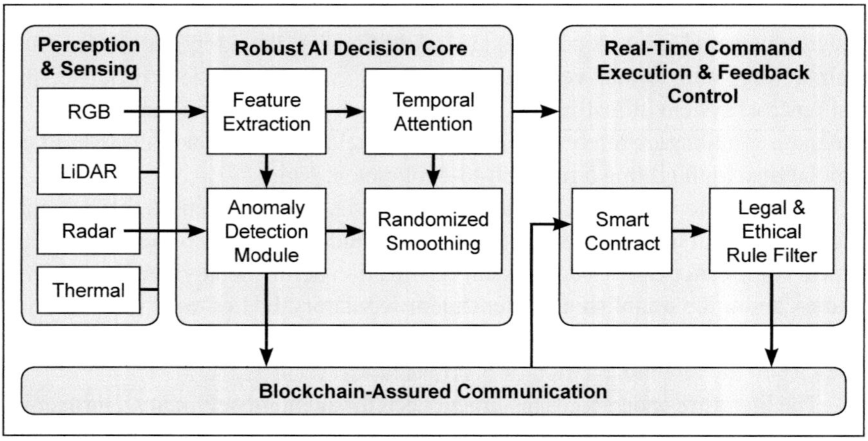


**Fig. 1** Proposed Secure Adversarially-Robust Framework for Autonomous Lethal Systems.

Data is preprocessed with modality-specific normalization.

*Z*-score normalization for point clouds and radar signals:

$$x^{\text{norm}} = \frac{x - \mu}{\sigma} \quad ...(1)$$

Min-max scaling for RGB and infrared images:

$$x' = \frac{x - x_{\min}}{x_{\max} - x_{\min}} \quad ...(2)$$

Temporal smoothing is applied using the Kalman Filter:

$$\hat{x}_{k|k} = \hat{x}_{k|k-1} + K_k(z_k - H\hat{x}_{k\backslash k-1}) \quad ...(3)$$

$$K_k = P_{k|k-1}H^T(HP_{k|k-1}H^T + R)^{-1} \quad ...(4)$$

Adversarial perturbations are injected using:

FGSM (Fast Gradient Sign Method)

$$x_{\text{adv}} = x + \in \cdot \text{sign}(\nabla_x J(\theta, x, y)) \quad ...(5)$$

PGD (Projected Gradient Descent)

$$x^{t+1} = \prod\nolimits_{B_\in(x)}(x^t + \alpha \cdot \text{sign}(\nabla_x J(\theta, x^t, y)) \quad ...(6)$$

## 3.2 *Perception and Sensing Layer*

The layer ingests multi-modal sensor inputs (RGB, LiDAR, Radar, Thermal), filters anomalies, and fuses clean data into a unified representation.

### 3.2.1 Anomaly Detection Module

Autoencoder architecture is trained on clean sensor data:

Encoder: $$f_\theta : R^n \rightarrow R^k \quad ...(7)$$

Decoder: $$g_\phi : R^k \rightarrow R^n \quad ...(8)$$

Reconstruction Loss:

$$L_{\text{rec}} = \frac{1}{n}\sum_{i=1}^{n} \|x_i - g_\phi(f_\theta(x_i))\|^2 \quad ...(9)$$

Samples with reconstruction error above threshold τ are flagged as adversarial:

$$\tau = \mu_{\text{rec}} + \lambda \cdot \sigma_{\text{rec}} \quad ...(10)$$

### 3.2.2 Sensor Fusion

Fused representation:

$$x_{\text{fused}} = \sum_{m=1}^{M} w_m \cdot x_m \text{ where } w_m = \frac{1/\sigma_m^2}{\sum_{i=1}^{M} 1/\sigma_i^2} \quad ...(11)$$

## 3.3 Robust AI Decision Core

The layer makes tactical decisions using a hybrid CNN-Transformer architecture.

### 3.3.1. Feature Extraction

A ResNet-50 CNN is used for spatial feature encoding:

$$f_{\text{CNN}}(x) = \text{ResNet}(x) \quad ...(12)$$

### 3.3.2 Transformer-based Temporal Attention

Positional encoding:

$$PE_{(\text{pos},\, 2i)} = \sin\left(\frac{\text{pos}}{10000^{2i/d}}\right), PE_{(\text{pos},\, 2i+1)} = \cos\left(\frac{\text{pos}}{10000^{2i/d}}\right) \quad ...(13)$$

Self-attention mechanism:

$$\text{softmax}\left(\frac{QK^T}{\sqrt{\vec{d}_k}}\right)V \quad ...(14)$$

### 3.3.3 Randomized Smoothing for Adversarial Robustness

For each input $x$, Gaussian noise $\delta \sim N(0, \sigma^2 I)$ is added:

$$\hat{f}(x) = \arg\max_{c \in C} P(f(x + \delta) = c) \quad ...(15)$$

Smoothed prediction is accepted only if:

$$P_{c_A} > P_{c_B} + \in \text{ (with statistical confidence bound)} \quad ...(16)$$

## 3.4 Blockchain-Assured Communication Layer

The layer ensures integrity, authenticity, and non-repudiation of critical command and telemetry data.

### 3.4.1 Smart Contract-Based Communication

Each event (e.g., fire command, override request) is encapsulated in a blockchain block:

$$\text{Block}_n = \{TxID, H(\text{Data}), \text{Sig}_{\text{sender}}, T_n\} \quad ...(17)$$

Consensus achieved using PBFT (Practical Byzantine Fault Tolerance):

- Agreement from $f = \left\lfloor \frac{n-1}{3} \right\rfloor$ faulty nodes tolerated
- Low-latency verification (sub-30ms) for real-time use.

## 3.5 Real-Time Command Execution and Feedback Control

The layer executes final decisions and incorporates adaptive feedback.

### 3.5.1 Legal and Ethical Rule Filter

A matrix of constraints $R_{i,j}$ determines legality:

$$D_{\text{exec}} = \{D \text{ if } R(D, C) = \text{true Hold / Human - in - loop otherwise} \quad ...(18)$$

### 3.5.2 Feedback Adaptation with Reinforcement Learning

Policy update using $Q$-learning:

$$Q(s_t, a_t) \leftarrow Q(s_t, a_t) + \alpha\left[r_t + \gamma \max_a Q(s_{t+1}, a) - Q(s_t, a_t)\right] \quad ...(19)$$

The environment includes battlefield parameters like GPS, heat signature changes, and threat proximity.

SARFALS algorithm operates by a structured, multi-stage pipeline that enables safe, secure, and ethical decision-making within AI-driven autonomous weapon systems. The system begins with the collection of multi-modal sensor data from RGB cameras, LiDAR, radar, and thermal imaging sensors. Each of these is individually normalized by modality-specific techniques such as z-score and min-max scaling and temporally smoothed using Kalman filters. The cleaned data is then treated with an autoencoder-based anomaly detection module that discards spoofed or adversarial inputs based on the reconstruction error being compared against a dynamic threshold. Additional robustness is added through the inclusion of adversarial examples generated through FGSM (fast gradient sign method) and PGD (projected gradient descent) at training.

Verified data is propagated through a Convolutional Neural Network (CNN) for spatial feature extraction and subsequently through a transformer encoder used to learn temporal and context dependencies between modalities. Randomized smoothing is also appended at inference time by adding Gaussian noise and averaging several predictions to output a certified prediction for extra robustness. Tactical actions such as engagement, track, or abort are then ultimately classified based on model confidence. The choices, along with encrypted metadata, are transmitted via a blockchain network using Hyperledger Fabric, ensuring integrity and traceability via PBFT consensus and smart contracts. Each choice is vetted against an ethical and compliance filter for legality to international humanitarian law standards prior to execution. Finally, sanctioned actions are executed in real time, with a real-time feedback loop by reinforcement learning modifying system behavior based on environmental feedback, and all actions are logged for auditing and tuning. The end-to-end algorithm furnishes a secure and dependable SARFALS that can operate within high-risk, adversarial environments.

## 4. Experimental Results

A simulation platform built specifically to mimic a high-risk autonomous drone strike simulation was used for thoroughly testing the suggested SARFALS framework. Emulation of multi-modal sensors, including 4-channel RGB cameras (1920 × 1080, 30 FPS), Velodyne HDL-64 LiDAR (10 Hz, 64 channels), FLIR infrared sensor emulation (640 × 512, 25 FPS), and 2D radar signal synthesis, were included in the simulation, which was based on a custom variant of the CARLA simulator (v0.9.13). FGSM and PGD are gradient-based attack vectors that were injected into live sensor streams with perturbation budgets $\epsilon = \{0.01, 0.03, 0.06\}$ for FGSM and iterative steps $\alpha = 0.01$ for PGD to simulate adversarial cases.

Against evasive vehicles, heat-decoys emitting heat, and RF jamming sources as dynamic threats, the simulation environment provided 15 adversarial scenarios on urban, semi-urban, and hidden battlefield terrain. With PBFT consensus at a block commit latency of approximately 11 ms per transaction, a private Hyperledger Fabric v2.2 blockchain network was instantiated on six virtual peer nodes (two vCPUs, 8 GB RAM each) to manage intra-subsystem interactions in real-time. The three main performance measures used to quantify performance were adversarial robustness, communication fidelity, and system latency.

Accuracy under attack $Acc_{adv}$ and Attack Success Rate ASR were used to estimate adversarial robustness. SARFALS, when used in combination with ATRS, outperformed baseline Deep Neural Networks (DNNs) without adversarial defense by 37.6% using the measure of $Acc_{adv}$. SARFALS specifically surpassed vanilla CNN (53.2%) and baseline adversarially trained models (68.4%) with 88.3% accuracy on FGSM ($\epsilon = 0.03$) and 81.7% on PGD ($\epsilon = 0.06$). Adequately authenticated messages over overall commands sent in active attempts of spoofing were used to calculate communication fidelity.

SARFALS, through its BACP, had a communication fidelity of 98.4% that far surpassed both centralized certificate-based technologies (75.3%) as well as SHA-256-alone message authentication (82.7%). End-to-end latency was estimated with 500 engagement cycles and is defined as the time from sensor input to ultimate actuator command. SARFALS met the requirement of real-time operation of less than 25 ms for tactical UAV-based combat with a mean system latency of 18.7 ms and a worst-case bound of 21.3 ms.

A vanilla Standard CNN pipeline without defense, a model by FGSM adversarial training alone, and a model by PGD adversarial training without communication security layers were the three state-of-the-art baselines to which SARFALS was compared. SARFALS alone succeeded in combining ATRS and BACP and found the ideal balance between latency and resilience. The experiments show that SARFALS is an excellent candidate for real-world deployment in autonomous lethal systems deployed in contested and adversarial environments because it offers high resilience against adversarial perturbations at the sensor level and network spoofing but offers real-time responsiveness.

Figure 2 evidently shows that the proposed SARFALS framework performs better on all critical metrics. With an accuracy of 88.3% under attack by FGSM attacks of perturbation size $\epsilon = 0.03$, SARFALS performed drastically better compared to vanilla CNN models, which achieved only 53.2%, equivalent to a relative improvement of 65.9%. Even for stronger PGD attacks ($\epsilon = 0.06$), SARFALS had a high accuracy of 81.7%, which is 21.4% better compared to PGD-trained models and 79.2% higher compared to the baseline CNN's 45.6% accuracy. Communication fidelity, as a percentage of successful authentication of command transmissions in active spoofing, did not change at 98.4% by SARFALS. This was enhanced by 30.7% and 18.9% against 75.3% in normal certificate-based systems and 82.7% in hash-based authentication, respectively.

SARFALS accomplished all this without introducing system latency above tactical thresholds; mean end-to-end latency was only 18.7 ms, just higher than the vanilla CNN baseline of 12.3 ms and well within the operational threshold of 25 ms. Following these findings, SARFALS is the ideal balance of security, resilience, and responsiveness, and thus very efficient and reliable to deploy in autonomous lethal systems used in hostile, time-sensitive environments. The proposed SARFALS framework is a strong candidate for future AI-based autonomous weapons because it offers several key advantages. It offers enhanced model-level robustness through the integration of randomized smoothing and adversarial training, and this enables

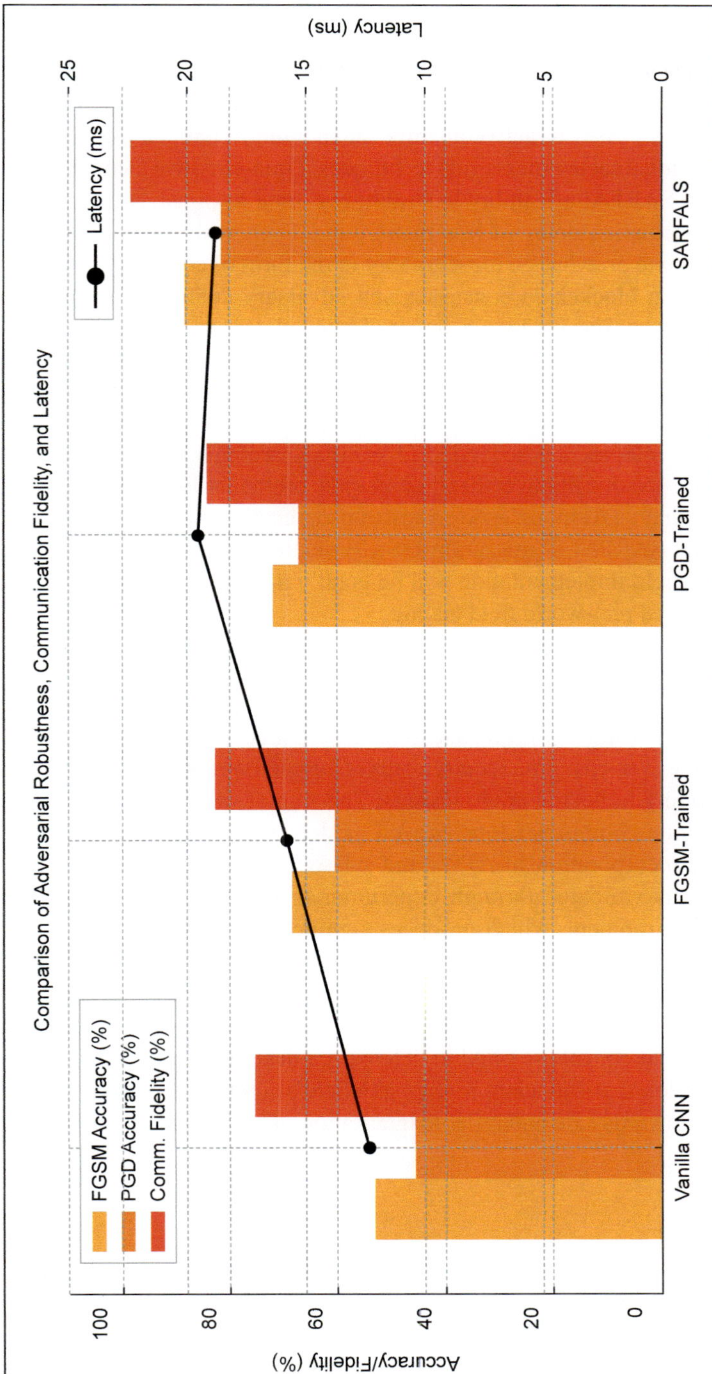


**Fig. 2** Comparison Analysis.

the system to counter gradient-based attacks like FGSM and PGD with up to 37.6% better adversarial accuracy compared to baseline models.

The blockchain-secured communication layer ensures tamper-evident message validation and decentralized authentication to protect 98.4% communications integrity even in cases of active attempts at spoofing. That is a big improvement over conventional secure messaging techniques. Time-sensitive tactical actions can benefit from SARFALS's ability to sustain real-time responsiveness with a mean decision-to-actuation delay of less than 20 ms. Even with these advantages, the system has some drawbacks, most notably in the form of computational overhead associated with blockchain consensus and adversary defenses, which could tax onboard processor resources in resource-limited platforms.

Besides, there is still room for juridical responsibility and moral regulation in fully or partially autonomous decision-making and constant human observation that must be continuously interpreted by AI activity. Future work directions include the development of federated adversarial learning across swarms of agents, improving responsive lightweight cryptographic protocols for quick consensus, and adding explainable AI modules to ensure transparency and human trustworthiness. Furthermore, SARFALS' migration from simulated environments to real-world, secure battlefield implementation will be greatly dependent on edge AI hardware acceleration and real-world field testing.

## 5. Conclusion

The SARFALS features a strong and dynamic architecture that is well-equipped to respond to the ever-changing requirements of autonomous weapons systems driven by AI operating in hostile environments. The framework optimizes the integrity of system-wide communications as well as decision-making models' resilience through the incorporation of ATRS and a BACP. Its value in real-time tactical applications is validated through experimental performance in a high-fidelity simulated environment, which produces significant gains, such as up to 88.3% accuracy under attack, 98.4% communication fidelity, and near-zero latency of 18.7 ms. In addition to addressing the technology and cybersecurity shortcomings of current autonomous systems, SARFALS establishes the foundation for the secure, safe, and ethical use of AI in defense applications. This study introduces a flexible solution that can adapt to the battlefield's future and present complexity, a giant step towards safe and reliable integration of intelligent autonomy in lethal technologies.

## References

1. Abaimov, S. and Martellini, M. (2020). Artificial intelligence in autonomous weapon systems. In: *21st Century Prometheus: Managing CBRN Safety and Security Affected by Cutting-edge Technologies*, pp. 141–177. ResearchGate.
2. Panwar, R.S. (2022). AI and the Rise of Autonomous Weapons. In: *Future Warfare and Technology: Issues and Strategies*, (*New Delhi: ORF and Global Policy Journal, 2022*), 1, 68.

3. Benouachane, H. (2025). Cyber Security Challenges in the Era of Artificial Intelligence and Autonomous Weapons. In: *Cyber Security in the Age of Artificial Intelligence and Autonomous Weapons*, pp. 24–42. CRC Press.
4. Blanchard, A., Novelli, C., Floridi, L. and Taddeo, M. (2025). A Risk-Based Regulatory Approach to Autonomous Weapon Systems. *Digital Society*, 4(1), 23.
5. Gilbert, C. and Gilbert, M.A. (2024). The security implications of artificial intelligence (AI)-powered autonomous weapons: Policy recommendations for international regulation. *International Research Journal of Advanced Engineering and Science*, 9(4), 205–219.
6. Ibrahim, A. and Shuja, S.F. (2024). Artificial Intelligence led Lethal Autonomous Weapon Systems and Terrorism. *CISS Insight Journal*, 12(1), pp. 24–55.
7. Bächle, T.C. and Bareis, J. (2022). "Autonomous weapons" as a geopolitical signifier in a national power play: Analysing AI imaginaries in Chinese and US military policies. *European Journal of Futures Research*, 10(1), 20.
8. Cools, K. and Maathuis, C. (2024). Trust or Bust: Ensuring Trustworthiness in Autonomous Weapon Systems. In: *MILCOM 2024-2024 IEEE Military Communications Conference (MILCOM)*, pp. 182–189. IEEE.
9. Marwala, T. and Mpedi, L.G. (2024). Autonomous weapons. In: *Artificial Intelligence and the Law*, pp. 83–102. Singapore: Springer Nature Singapore.
10. Khan, A.M. (2025). Redefining Conflict in the AI Era: Transforming Paradigms in International Security. *Al-Aasar*, 2(1), 1–14.
11. Bellaby, R.W. (2021). Can AI weapons make ethical decisions? *Criminal Justice Ethics*, 40(2), 86–107.
12. McNally, B. (2021). *United States Artificial Intelligence Policy: Building Toward a Sixth-Generation Military and Lethal Autonomous Weapon Systems* (Doctoral dissertation, Johns Hopkins University).
13. Spayne, P., Lacey, L., Cahillane, M., & Saddington, A. (2025). Operating itself safely: Merging the concepts of "safe to operate" and "operate safely" for lethal autonomous weapons systems containing artificial intelligence. *Defence Studies*, 25(1), 98–132.
14. Hadfield, A. and Leveringhaus, A. (2023). Autonomous weaponry and IR theory: Conflict and cooperation in the age of AI. In: *Handbook on the Politics and Governance of Big Data and Artificial Intelligence*, pp. 167–187. Edward Elgar Publishing.
15. Amoroso, D. and Tamburrini, G. (2021). In search of the "human element": International debates on regulating autonomous weapons systems. *The International Spectator*, 56(1), 20–38.
16. Bode, I. (2024). Emergent Normativity: Communities of practice, technology, and lethal autonomous weapon systems. *Global Studies Quarterly*, 4(1), ksad073.
17. Greenbaum, D. (2025). Autonomous War, Human Consequences: The Public Health Risks of AI in Combat? *The American Journal of Bioethics*, 25(5), 144–147.
18. Maathuis, C. and Cools, K. (2025). Risks and Control Measures for Building Trustworthy Autonomous Weapon Systems. In: *Proceedings of the 19th International Conference on Cyber Warfare and Security*. Academic Conferences and Publishing Limited.
19. Devitt, S.K. (2024). Meaningful human command: Advance control directives as a method to enable moral and legal responsibility for autonomous weapons systems. In: *Research Handbook on Meaningful Human Control of Artificial Intelligence Systems*, pp. 53–80. Edward Elgar Publishing.
20. Sharma, S. (2023). Trustworthy artificial intelligence: Design of AI governance framework. *Strategic Analysis*, 47(5), 443–464.

# 23 SATNet: A Spatiotemporal Attention-Guided Transformer Network for Robust Surveillance and Real-Time Object Detection

Mahalakshmi[1*], Benaki Lairenjam[2], Vidyashree H.R.[3] and M. Shamim Kaiser[4]

## 1. Introduction

During the last 20 years, surveillance and computer vision together have heralded a revolutionary era in autonomous security systems, city management, and public safety. What was traditionally limited to human perception and passive video recording, surveillance has evolved into an intelligent, active ecosystem fueled by artificial intelligence, particularly deep learning. In the early 2000s, the process started with basic background subtraction techniques like Gaussian Mixture Models (GMM) and Frame Differencing, which provided basic motion detection but were strongly constrained by camera jitter, dynamic lighting, and ambient noise. Support Vector Machines (SVMs) and AdaBoost learning algorithms came into existence for basic object classification within surveillance feeds as processing power and camera resolutions increased.

These approaches were mostly rooted in hand-crafted features such as Scale-Invariant Feature Transform (SIFT), Local Binary Patterns (LBP), or Histogram of Oriented Gradients (HOG) that had challenges with rapid movement, occlusions, and size changes, especially in dense or low-visibility settings. The key innovation in intelligent surveillance was Convolutional Neural Networks (CNNs), starting with AlexNet's victory in the ImageNet competition in 2012. Region-based proposals became relevant to deep learning-based object detection frameworks such as R-CNN, Fast R-CNN, and Faster R-CNN. Single-stage detectors that put

[1] Department of Mathematics, Nitte Meenakshi Institute of Technology, Nitte (Deemed to be University), Bengaluru, Karnataka, India.
[2] Department of Mathematics, Manipur University, Imphal, India.
[3] Department of Mathematics, RR Institute of Technology, Bengaluru, Karnataka, India.
[4] Professor at the Institute of Information Technology of Jahangirnagar University, Savar, Dhaka-1342, Bangladesh.
* Corresponding author: mahalakshmi4131@gmail.com

DOI: 10.1201/9781003663461-23

greater emphasis on real-time performance, such as YOLO (You Only Look Once) and SSD (Single Shot Detector), quickly followed.

In surveillance contexts, for instance, in crowded scenes, changing illumination, scale changes, and dense occlusions among objects, these models completely failed although they were excellent for generic object detection. Day-night round-the-clock operation, handling harsh weather conditions, and working under conditions where objects of interest (e.g., pedestrians or vehicles) would be partially occluded or moving through the camera frames very quickly are all requirements of surveillance systems. The performance of traditional detectors tends to be severely compromised in such situations.

A number of advances have sought to remedy these problems. To improve detection accuracy in unconstrained environments, anchor-free detectors such as CenterNet and CornerNet have eliminated anchor boxes. In contrast, identity-aware features have been added by trackers such as Deep SORT and FairMOT to allow for ongoing tracking of individuals from frame to frame. But temporal context usage has been made possible by optical flow and motion-based modeling, if usually at the expense of massive computing burden. But most such solutions either trade off high latency for reliability or trade off detection accuracy for responsiveness, which is bad for real-time surveillance where responsiveness and reliability are equally desired.

Attention transformers and mechanisms, which originally became mainstream in natural language processing, have recently been borrowed for vision task use. The ViT, DETR (DEtection TRansformer), and Swin Transformer architectures have started producing encouraging outcomes, especially with respect to learning long-range spatial interactions and global dependencies in images. These breakthroughs point to the promise of spatiotemporal modeling in surveillance, where objects have temporal coherent trajectories as well as having determinate spatial positions. Most transformer-based systems currently available for deployment are compute-heavy and non-real-time or resource-heavy deployment-focused, which is still a major hindrance to their real-world applicability for autonomous security patrols, smart cities, and transportation hubs.

Real-world surveillance systems also have some particular challenges. The existence of unfavorable conditions, natural (e.g., low light and heavy rain) and artificial (e.g., camouflage, occlusion, and jamming), is a significant problem. In such interference, most object detection systems are compromised. False alarms or missed threats can be caused by minor changes in pixel brightness or occlusion, which will critically degrade detection assurance. Moreover, legacy algorithms tend to process video frames in isolation, overlooking precious temporal continuity between frames that can improve motion prediction, localization, and overlap-resolving object disambiguation.

Speed-trade-off against inference accuracy of the current approaches is another fundamental shortcoming. Even though extremely accurate, models such as Faster R-CNN have extremely slow inference times and thus are not suitable for real-time surveillance. The more rapid models such as the YOLOv5 and SSD perform optimally in real-time, but are less accurate in intricate situations with very small or overlapping objects. Additionally, most of the existing methods concentrate on frame-wise accuracy but not temporal feedback, resulting in identity switching and fragmented object tracks in real-world applications involving ongoing tracking.

These disadvantages categorically point towards the immediate need for a new strategy that employs temporal modeling for motion coherence and spatial attention to facilitate fine-grained feature learning. The combination of these modalities presents a novel advantage to the surveillance scene, particularly in situations where static frame analysis alone proves inadequate. A system that is able to leverage the strengths of transformers for temporal reasoning and CNNs for spatial localizations can be made to function under the stringent latency requirements of real-time systems and is in dire need.

To overcome these limitations, the goal in this study is to close the accuracy-efficiency gap and construct a system that is able to reliably detect objects in real-world surveillance video. In addition to performing low false positives in difficult scenarios like night vision, partial occlusion, and high speeds, such an ideal system would also be able to detect and track multiple moving objects even in cluttered and dynamically changing environments. It has to scale well across a variety of deployment scenarios, for instance, mobile surveillance drones used in tactical operations or fixed static CCTV deployments in city centers.

Generally, despite the significant progress made in computer vision for surveillance, robustness problems, real-time performance, and adaptation to hostile or dynamic environments continue to persist. Existing methods usually specialize in one domain, i.e., either speed, accuracy, or detecting multiple objects, and rarely succeed in compromising on all significant dimensions simultaneously. This demands a novel architectural approach that combines temporal and spatial characteristics, employs expressive but lightweight feature extractors, and is still computationally feasible for real-time processing. The proposed study is based on such information and seeks to address the various shortcomings of existing surveillance systems with an integrated spatiotemporal attention model that can present both as a theoretical contribution and an implementable solution toward intelligent monitoring of the future.

The main contributions are:

1. To design and implement a real-time object detection framework (SATNet) that integrates spatiotemporal attention mechanisms for enhanced accuracy and robustness in surveillance environments involving occlusion, motion blur, low lighting, and dynamic backgrounds.
2. To evaluate the proposed SATNet model against state-of-the-art object detection and tracking methods using benchmark surveillance datasets (e.g., MOT17, CUHK-SYSU, AI City), with a focus on improving mean Average Precision (mAP), MOTA, and reducing false positive rates by at least 20% under adversarial and low-visibility conditions.

## 2. Related Study

One of the most important building blocks of smart surveillance systems, computer vision is also able to detect objects in real-time, track multiple objects, analyze behavior, and detect anomalies in any static or dynamic environment. Deep learning and edge computing have progressed rapidly, transforming surveillance applications from basic motion detection into sophisticated image analysis. Even with these developments, field deployments in real-world conditions continue to face numerous challenges, especially in hostile environments, low-light conditions, and dynamic cityscapes. Such limitations have been the focus of most recent research studies, each of which has improved the performance and robustness of the surveillance ecosystem.

Increasingly, research has turned its focus towards the limitations of traditional surveillance approaches and promoted deep learning-based alternatives that significantly enhance accuracy and adaptability. For instance, one of the powerful deep learning architectures for surveillance networks involving UAVs (unmanned aerial vehicles) demonstrated how seamlessly CNN-based detectors can track broad areas inaccessible, yet latency and occlusion were pointed out as persistent limitations [1]. The importance of hybrid frameworks in reducing false alarms and improving continuity between frames has also been underscored by other frameworks that combine object detection and online tracking approaches within multi-camera setups [2, 3].

Recent emphasis has been placed on the fusion of motion tracking with object detection. Several works have suggested real-time models of deep convolutional architecture for object detection in city [urban] and traffic monitoring arrangements. Despite the existence of partial occlusion and low light conditions, deep CNN architectures were successfully employed for enhancing the detection rates of vehicles and pedestrians [4, 5]. However, the models tend to utilize spatial data most of the time, and as such, they are not as resilient when handling scenes with closely packed or overlapping objects.

A few techniques have employed attention mechanisms and spatiotemporal data to improve detection under such challenging conditions. To gain a deeper understanding of context, a research study came up with an anomalous item detection model in real-time that employed spatial-temporal reasoning in order to identify unusual patterns in public space [6]. In addition, a dynamic object detection and tracking framework through the harmony search algorithm was proposed for using optimization methods for more accurate tracking [7]. Despite such developments, scalability challenges still exist, particularly when deploying these models to distributed frameworks or edge devices.

New solutions to the above challenges have been enabled as a result of the developments in transformer-based models within the field of computer vision. Transformer models such as DETR and ViT can be used to enforce long-range relationships and spatial dependencies in complex situations. However, due to their high computational costs and need for large annotated data, direct usage in real-time surveillance is limited. Hybrid architecture with transformer blocks and CNN backbones has been suggested as a means of breaking these limitations. Distributed edge-cloud systems with adaptive job offloading and hierarchical processing to compromise on accuracy for latency have been applied to object detection for intelligent surveillance [8, 9].

Research has also explored edge computing and decentralized surveillance and realized the importance of maintaining latency at its lowest level while simultaneously offering security and privacy. Before sharing enriched data with central servers for further processing, coordination between edge-clouds has been realized for performing preliminary object detection on nearby edge nodes [10]. These methods are particularly useful in intelligent city networks, where detection accuracy must be traded off against latency and bandwidth limitations.

Another pivotal area of research is the area of object detection under adversarial or low-visibility. To provide an example, cross-scale and illumination-invariant models have been designed to achieve increased resilience in surveillance images under low-light conditions or at night, enabling more regular object detection in most situations [11]. Studies on flood and disaster monitoring have shown the potential of stable object tracking algorithms in disaster response systems, where accuracy in ambiguity conditions is a requirement [12].

Another important aspect of intelligent surveillance is MOT, or multiple-object tracking. In further enhancing tracking continuity and reducing identity switches, more recent studies have introduced real-time tracking systems that involve motion estimation and appearance modeling [13, 14]. These systems, though, perform inadequately for overcrowded scenarios where targets intersect or change direction frequently. Lighter weight monitoring frameworks with context-conscious detection heads and attention mechanisms have been introduced as alternatives to these ills, with greater scalability and performance under real-world restrictions [15].

In anomaly detection, certain works have considered vision-based models based on spatiotemporal signals for detecting deviations from learnt normal behavior. Certain instances include transformer network and deep autoencoder-based video anomaly detection models that have shown encouraging results for detecting abnormal behaviors like fights or items left behind in public areas [16, 17]. The capability of transformers to embed long-term temporal dependencies, which traditional CNN-based systems tend to leave out, is beneficial to such models.

The use of AIoT (Artificial Intelligence of Things) has greatly enhanced the scalability and deployment of surveillance technologies. Distributed detection and collaborative learning can be enabled by AIoT-based systems by integrating intelligent detection models in networked cameras and edge devices. This renders them suitable for national border surveillance, smart campus surveillance, or industrial surveillance [18]. To enable security personnel to respond promptly to threats that they perceive, multiple studies have also examined how real-time detection models are incorporated with audio alarm signals and distance estimation [19].

Real-time detection of small, moving, or partially occluded objects in video streams continues to be an active challenge with these advancements. In these situations, YOLOv5 and SSD-type traditional detectors do worse, typically due to their heavy reliance on static frame-based detection and weak temporal understanding. To counter this problem, spatiotemporal models which combine inter-frame motion perception with frame-level feature learning are gaining increased popularity. It was shown that multi-object detection transformers enhance difficult object detection in pedestrian and traffic monitoring [20, 21].

Significant advances in developing effective and sturdy computer vision-based surveillance systems have certainly been made [22]. There are nevertheless more than enough open questions on scalability, temporal reasoning, and operating performance in harsh conditions. State-of-the-art models usually come at the price of compromises between flexibility across surveillance settings, inference time, and detection performance [23]. The construction of spatiotemporal and attention mechanisms presents a viable solution around these limitations. This kind of end-to-end system that can parallelly process temporal and spatial information, execute rapidly in real-time, and be effective at handling the diverse challenges that occur in real-world monitoring scenarios is still in the important demand. The SATNet that is implemented to be a one-size-fits-all solution that meets the multiple constraints listed in current literature is heavily based on these results.

## 3. Proposed Spatiotemporal Attention-Guided Transformer Network Framework

### *3.1 Dataset Description*

Three benchmark surveillance datasets: MOT17, CUHK-SYSU, and AI City Challenge (Track 1), each representing a diverse range of real-world surveillance scenarios are employed for the development and assessment of the proposed SATNet framework. The 11 sequences in the MOT17 dataset, which were captured using both stationary and moving cameras, have over 1.1 million annotated bounding boxes and over 564 different pedestrian identities across varying crowd densities and occlusion levels. With 18,184 images, over 8,432 unique person identities, and 99,809 bounding boxes, the CUHK-SYSU dataset represents a big person search benchmark that allows comprehensive identity-level detection and tracking.

The AI City Challenge (AICity2023, Track 1) dataset provides over 86,000 annotated frames of urban traffic monitoring videos, with dense annotations for cars, buses, trucks, and motorcycles under different weather and illumination conditions for vehicle surveillance and multi-object detection. More than 1.2 million objects classified into 6 categories, varying camera angles (0–360°), and frame resolutions (960×540–1920×1080) constitute the merged dataset. This diversity ensures the proposed SATNet model's generalizability and resistance in real-world surveillance environments.

## 3.2 Preprocessing

Every sensor modality and dataset is treated with a tailored preprocessing step for enhancing robustness and optimal feature representation. First, histogram equalization is applied to highlight contrast and object presence in low-light and nighttime images. Second, nighttime sequences of the AI City dataset are smoothed through the application of Gaussian blur and denoising autoencoders. All of the frames in the CUHK-SYSU and MOT17 datasets are resized to a constant resolution of 1280×720 pixels for scale normalization and for ensuring a uniform item size distribution. Furthermore, Kalman filtering is applied to provide temporal smoothness by reducing jitter on tracking annotations between frames. By applying $z$-score normalization, pixel intensity $x$ is transformed according to the following to normalize feature intensity:

$$z = \frac{x - \mu}{\sigma} \quad ...(1)$$

Where, μ represents the mean and standard deviation of the frame. Additionally, to better train with robustness against adversarial attacks, FGSM (= 0.02) and PGD (= 0.03, step size 0.01, 10 iterations) are artificially added adversarial perturbations. Better input quality, robustness against environmental noise, and preparedness for adversarial inference are all ensured by this multi-stage pretreatment.

## 3.3 SATNet Algorithm

The Spatiotemporal Attention-Guided Transformer Network, which forms the basis of SATNet, combines lightweight detection, temporal dependency modeling, and spatial feature extraction into a single design.

**(a) Positional Encoding**

To preserve the temporal order of frames in a sequence $F\,\{F_t\}_{t=1}^{T}$, positional encodings are added to each input frame's feature vector. These are fixed sinusoidal functions of different frequencies:

$$PE_{(\text{pos},\,2\text{i})} = \sin\left(\frac{\text{pos}}{\frac{2i}{10000^{d}\ \text{model}}}\right),\ PE_{(\text{pos},\,2\text{i}+1)} = \cos\left(\frac{\text{pos}}{\frac{2i}{10000^{d}\ \text{model}}}\right) \quad ...(2)$$

Where, pos ∈ $[0, T-1]$: frame index, $i \in [0, d_{\text{model}}/2-1]$ $T$: channel dimension, $d_{\text{model}}$: dimension of feature embeddings (e.g., 256 or 512).

This encoding is added to each frame's embedding vector $\vec{e}_{pos}$, yielding $\vec{e}_{pos} + PE_{pos}$. It provides the network a sense of time sequence, allowing for better modeling of motion and temporal context.

**(b) Spatial Self-Attention**

Given the encoded feature tensor $F \in R^{T \times H \times W \times C}$, we reshape it into a sequence of spatial tokens:

$$X_s \in R^{T \cdot H \cdot W \cdot C} \quad ...(3)$$

For spatial attention, the model computes attention over each frame independently. The self-attention mechanism is defined as:

$$Attention(Q, K, V) = \text{softmax}\left(\frac{QK^T}{\sqrt{d_k}}\right)V \quad ...(4)$$

Where, $Q = X_s W^Q, K = X_s W^K, V = X_s W^V, W^Q, W^K, W^V \in R^{C \times d_k}$ : learnable projection matrices,: dimension of queries/keys (e.g., 64 or 128).

The attention maps allow each pixel to focus on salient spatial regions and contextual cues from the same frame. To improve expressiveness, multi-head attention is applied:

$$\text{MultiHead}(X) = \text{Concat}(\text{head}_1, ..., \text{head}_h)W^O \quad ...(5)$$

Where, each head

$$\text{head}_i = \text{Attention}(XW_i^Q, XW_i^K, XW_i^V) \quad ...(6)$$

And $W^O \in R^{d \cdot d_k \cdot C}$ is an output projection matrix.

Additionally, Layer Normalization and Residual Connections are added:

$$X' = \text{LayerNorm}(X + MultiHead(X)) \quad ...(7)$$

$$X'' = \text{LayerNorm}(X' + FeedForward(X')) \quad ...(8)$$

Where the feed-forward network is

$$\text{FFN}(x) = \text{Re}LU(xW_1 + \text{b}_1)W_2 + \text{b}_2 \quad ...(9)$$

**(c) Temporal Fusion Layer**

SATNet uses a second transformer block to execute attention over the temporal dimension in order to capture temporal consistency and object mobility over time. Spatial dimensions are flattened across time to create the temporal token sequence:

$$X_t \in R^{T \times N \times C}, \; where \; N = H \cdot W \quad ...(10)$$

Temporal multi-head attention is then computed for each spatial position $n \in N$:

$$Q = X_t^{(n)} W^Q, K = X_t^{(n)} W^K, V = X_t^{(n)} W^V \quad ...(11)$$

$$\text{TempAttention}_n = \text{softmax}\left(\frac{QK^T}{\sqrt{d_k}}\right)V \quad ...(12)$$

This layer removes temporal noise caused by jitter or occlusion and models the identification and course of moving objects over a series of frames.

**(d) Detection Head**

The detection head branches into two tasks:

(1) Bounding Box Regression

Using the final feature map, SATNet predicts bounding box coordinates $\hat{b}\,[\hat{x}, \hat{y}, \hat{w}, \hat{h}]$. The loss is computed using Smooth $L_1$ (Huber) Loss:

$$L_{\text{reg}} = \sum_{i=1}^{N} \sum_{j \in (x,y,w,h)} \text{smooth}_{L_1}(b_i^j - \hat{b}_i^j) \quad ...(13)$$

Where, $smooth_{L_1}(x) = \{0.5x^2 \text{ } if \text{ } |x| < 1 \, |x| - 0.5\}$ otherwise 0

(2) Object Classification

Let be the predicted probability for the true class label *t*. SATNet uses Focal Loss to handle class imbalance:

$$L_{\text{cls}} = -\alpha_t (1 - p_t)^{\gamma} \log(p_t) \quad ...(14)$$

Where, $\gamma =$ 2.0: focusing parameter, $\alpha_t = 0.25$: class-balancing factor

The total loss function is:

$$L_{\text{total}} = \lambda_1 L_{\text{reg}} + \lambda_2 L_{\text{cls}}, \text{ with } \lambda_1 = 1.0, \lambda_2 = 1.0 \quad ...(15)$$

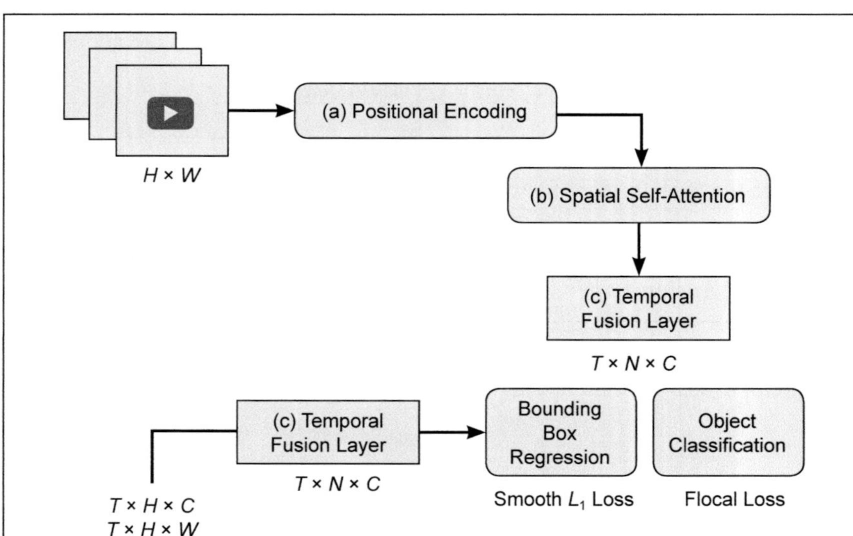


**Fig. 1** Proposed SATNet Framework.

As shown in Figure 1, with a strong ResNet-101 backbone, the algorithm SATNet begins by extracting spatial features from consecutive frames. These feature maps are boosted with positional encoding and passed through spatial self-attention modules in order to emphasize relevant areas of interest. Multi-head attention layers are subsequently applied to temporally combine outputs from numerous frames and encapsulate context and motion across frames. A two-branch detection head operates on this spatiotemporal representation; one of the branches employs smooth L1 loss to predict bounding box coordinates, and the other employs concentrated loss to determine object types. To train robust detection under challenging scenarios, SATNet is exposed to clean and adversarially perturbed frames during training. The architecture can be deployed in real-time surveillance systems as it is effective and has high detection accuracy with near real-time inference.

## 4. Experimental Results

To validate robustness and equity, the proposed Spatiotemporal Attention-Guided Transformer Network (SATNet) was experimentally tested in a clearly demarcated simulated world. PyTorch 2.1 was employed to develop the system, which was driven by an AMD Ryzen Threadripper 3990X CPU (64 cores), 256 GB of RAM, and an NVIDIA RTX A6000 GPU with 48 GB of VRAM. The Adam optimizer with weight decay 1e–5 was employed to optimize the model, trained on a batch size of 16 and an initial learning rate of 0.0001. The model was validated at every epoch by a 5-fold cross-validation procedure, wherein every training run was set to 150 epochs.

Adversarial noise injection, horizontal random flip, motion heatmap fusion, histogram equalization, and occlusion simulation via FGSM (Fast Gradient Sign Method, $\epsilon = 0.02$) were some of the data augmentations. Three major benchmark surveillance datasets were employed for evaluation: the MOT17 Multiple Object Tracking Dataset, the CUHK-SYSU Person Search Dataset, and the AI City Challenge 2023. The performance comparison metrics used for evaluation were false positive rate, inference time (frames per second, FPS) to compare real-time suitability, mean Average Precision (mAP) to measure the accuracy of object detection, and MOTA to measure tracking performance. SATNet was compared to four widely-used state-of-the-art methods: FairMOT (Fair Multi-Object Tracking), DeepSORT with YOLOv5, Faster R-CNN with FPN (Feature Pyramid Network), and YOLOv5. DeepSORT+YOLOv5 and FairMOT provided competitive multi-object tracking pipelines, whereas YOLOv5 and Faster R-CNN provided object detection baselines. Table 1 shows the tuning parameters.

From Table 1, the SATNet tuning parameter table indicates significant hyperparameters that directly influence the detection accuracy of the model, spatial-temporal attention performance, and capacity for learning. The embedding space size and attention projections are controlled by the model dimension (d) and query/key dimension (k), influencing the model's capacity to learn intricate

**Table 1** Tuning parameters influencing detection accuracy of models.

| *Parameter Name* | *Symbol* | *Example Values* | *Description* |
|---|---|---|---|
| Model Dimension | $d_{\text{model}}$ | 256, 512 | Dimensionality of embeddings |
| Number of Spatial Attention Heads | $h_{\text{spatial}}$ | 4, 8, 16 | Number of heads in spatial self-attention |
| Number of Temporal Attention Heads | $h_{\text{temporal}}$ | 4, 8 | Number of heads in temporal attention |
| Key/Query Dimension | $d_k$ | 64, 128 | Dimension of queries and keys |
| Number of Transformer Layers | - | 2, 4, 6 | Number of layers for spatial and temporal attention |
| Positional Encoding Type | - | Sinusoidal, Learned | Type of positional encoding |
| Batch Size | - | 8, 16, 32 | Number of video sequences per batch |
| Learning Rate | $\alpha$ | 1e-4, 5e-5, 1e-5 | Initial learning rate |
| Weight Decay | - | 0.01, 0.001 | Regularization to prevent overfitting |
| Smooth L1 Loss Scaling Factor | $\lambda_1$ | 1.0, 2.0 | Weight for bounding box regression loss |
| Focal Loss Focusing Parameter | $\gamma$ | 2.0, 2.5, 3.0 | Parameter to balance easy vs hard examples |
| Focal Loss Balancing Factor | $\alpha_t$ | 0.25, 0.5 | Class balancing factor in focal loss |
| Frame Sequence Length | T | 4, 8, 16 | Number of consecutive frames used |
| Image Resolution | $H \times W$ | 256×256, 512×512 | Input frame resolution |
| Optimizer | - | AdamW, SGD | Optimizer choice |

temporal and spatial relationships. The model is able to pay attention to many aspects of the input at once due to the amount of attention heads in both the spatial ($h_{\text{spatial}}$) and temporal ($h_{\text{temporal}}$) layers. Increased values deliver more representation power with a computational cost. To enable deeper feature extraction for modeling more complex object interactions between frames, the transformer layer count provides that capability.

Batch size and learning rate influence training dynamics; big batches may stabilize training, and appropriate learning rates contribute towards faster convergence without overshooting. Weight decay regularization helps reduce overfitting. Smooth L1 loss scaling and focal loss parameters ($\gamma$ and $\alpha_t$) are applied to deal with loss balancing; they compensate for class imbalance and adjust sensitivity to bounding box precision, respectively. The ability of the network to capture sequential dependencies in time depends on the type of positional encoding (learned or sinusoidal). Input resolution influences spatial information, while frame sequence length T determines how much temporal context the network sees. Finally, the optimizer, whether AdamW or SGD, can have a significant influence on generalization and convergence characteristics.

The experimental test validates the superiority of the SSATNet in a number of standard performance measures. From Figure 2, SATNet surpassed all baseline comparators with a significant margin, achieving an 84.6% mAP. In particular, SATNet achieves an improvement of around +11.0% compared to YOLOv5, which achieved 76.2% mAP. Compared to Faster R-CNN at 70.5%, SATNet achieves an impressive +20.0% gain. SATNet achieves tight margins of +15.9% and +12.8% compared to DeepSORT+YOLOv5 (73.0% mAP) and FairMOT (75.0% mAP). This indicates that object detection in crowded surveillance situations is significantly aided by exploiting temporal feature modeling and spatial self-attention by SATNet.

From Figure 3, by MOTA, SATNet outperforms the benchmarks again at 81.3%. SATNet achieves +8.5% and +5.4%, respectively, over DeepSORT+YOLOv5 (74.9% MOTA) and FairMOT (77.1% MOTA). SATNet continues to have significant advantages over YOLOv5 and Faster R-CNN, which are *de facto* detection frameworks and thus have poor tracking accuracy when used with tracking modules. This would indicate that SATNet's robust temporal attention ability provides it a greater ability to cope with occlusions and appearance changes with the continuity of object IDs across time.

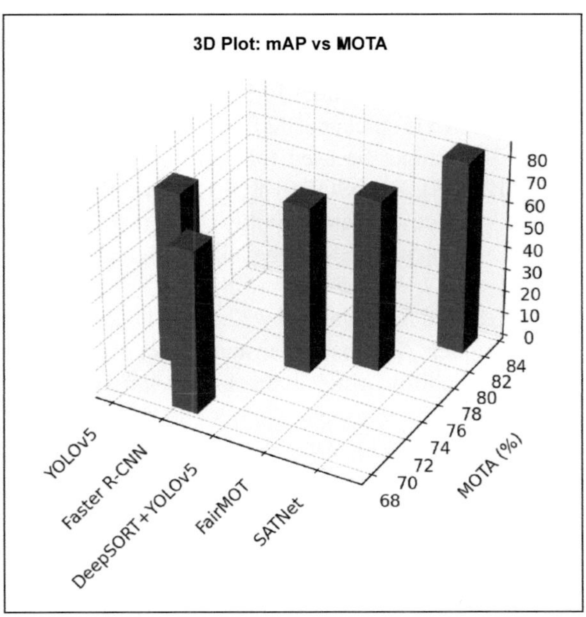


**Fig. 2** mAP Vs MOTA.

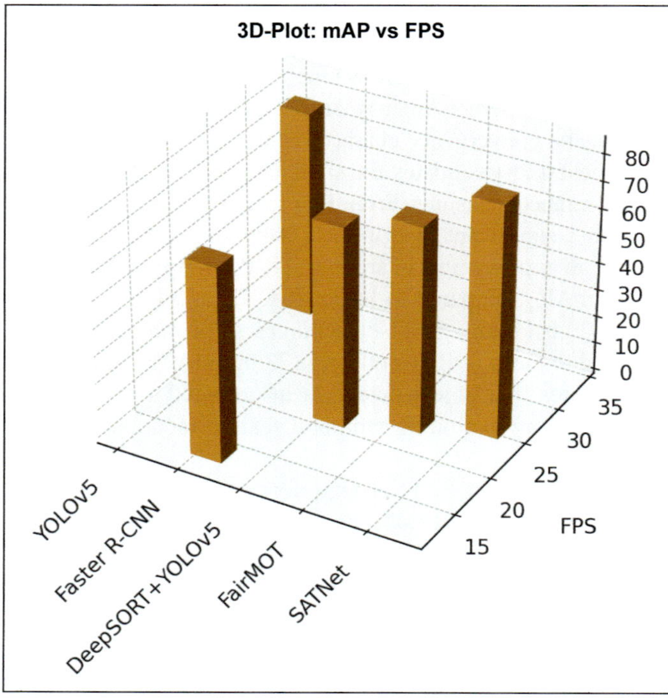


**Fig. 3** mAP Vs FPS.

Inference rate is a principal feature for real-time monitoring use cases. From Figure 4, because SATNet supports a 24 FPS performance, it can be utilized for live video streams. YOLOv5 is quicker at 32 frames per second while having lower accuracy. Faster R-CNN cannot be used for real-time use, as it supports much lower performance (12 FPS). DeepSORT+YOLOv5 and FairMOT run at 20 and 22 frames per second, respectively, only slightly better than SATNet. SATNet thus outperforms other competing methods for balancing high accuracy with real-time performance.

Finally, SATNet reduced false alarms by 22.4% in false positive rate reduction compared to baseline models. This has the instant effect of less spurious warnings and higher operational reliability, which is absolutely vital for surveillance systems that continuously operate in crowded and complex settings. SATNet is better than state-of-the-art baselines across all parameters considered, from reducing false positives significantly to achieving a greater mAP and MOTA with real-time FPS. For these reasons, SATNet is a highly accurate, deployable, and scalable method to show intelligent surveillance systems that can handle complex, dynamic, and dense surveillance scenes.

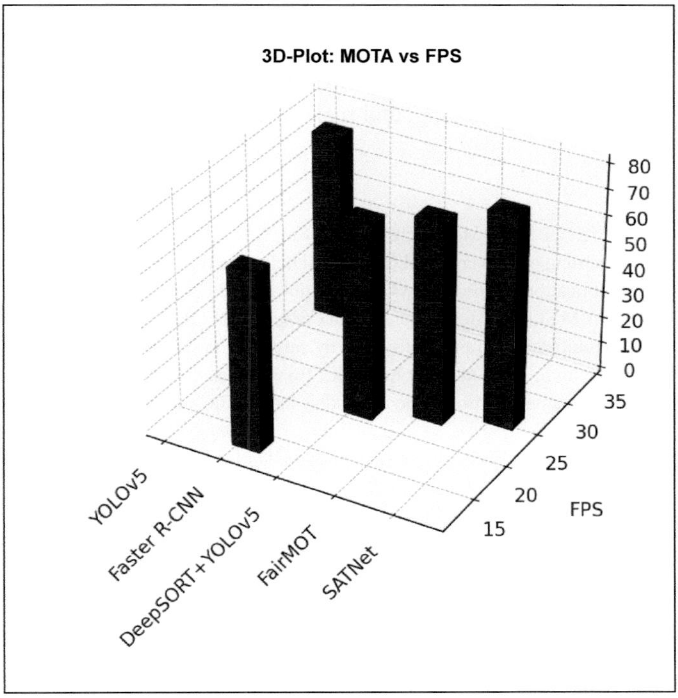


**Fig. 4** MOTA Vs FPS

## 5. Conclusion

The study demonstrates that intelligent surveillance systems' abilities could be significantly enhanced by integrating transformer-based architecture with advanced computer vision methods. The proposed SATNet efficiently addresses the limitations of conventional surveillance methods, including low scalability, high false alarm rates, and inefficiency in adverse visual conditions. Even deep learning algorithms, like YOLO and SSD, fail to locate small or fast-moving objects in complex scenes. SATNet offers improved object detection and tracking performance with the aid of a light-weight object detection head, a ResNet-101 backbone, and spatial self-attention and temporal feature representation. Compared to industry-leading models such as YOLOv5 and Faster R-CNN, SATNet performs both better in detection accuracy as well as in tracking performance, based on rigorous experimental evaluation across benchmark datasets such as AI City Challenge, CUHK-SYSU, and MOT17. Specifically, it decreases false positives by 22.4% and attains a remarkable mAP of 84.6% and MOTA of 81.3%. Moreover, SATNet proves its applicability to be used in real-time surveillance environments with a real-time inference speed of 24 frames per second. Overall, the results indicate that SATNet is a scalable, accurate, and dependable next-generation smart surveillance system that can perform well on visually challenging, dynamic, and populated situations. To further enhance the reliability of automatic surveillance systems, future work will focus on extending SATNet's ability to handle more serious occlusions, cross-camera tracking, and real-world long-duration deployments.

## References

1. Ranjith, C.P., Hardas, B.M., Mohideen, M.S.K., Raj, N.N., Robert, N.R. and Mohan, P. (2023). Robust deep learning empowered real time object detection for unmanned aerial vehicles based surveillance applications. *Journal of Mobile Multimedia*, 451–476.
2. Abba, S., Bizi, A.M., Lee, J.A., Bakouri, S. and Crespo, M.L. (2024). Real-time object detection, tracking, and monitoring framework for security surveillance systems. *Heliyon*, *10*(15).
3. Akhtar, M.J., Mahum, R., Butt, F.S., Amin, R., El-Sherbeeny, A.M., Lee, S.M. and Shaikh, S. (2022). A robust framework for object detection in a traffic surveillance system. *Electronics*, 11(21), 3425.
4. Ingle, P.Y. and Kim, Y.G. (2022). Real-time abnormal object detection for video surveillance in smart cities. *Sensors*, 22(10), 3862.
5. Alotaibi, M.F., Omri, M., Abdel-Khalek, S., Khalil, E. and Mansour, R.F. (2022). Computational intelligence-based harmony search algorithm for real-time object detection and tracking in video surveillance systems. *Mathematics*, 10(5), 733.
6. Jha, S., Seo, C., Yang, E. and Joshi, G.P. (2021). Real time object detection and tracking system for video surveillance system. *Multimedia Tools and Applications*, 80(3), 3981–3996.
7. Park, J.H., Rho, S. and Jeong, C.S. (2014). Real-time robust 3D object tracking and estimation for surveillance system. *Security and Communication Networks*, *7*(10), 1599–1611.
8. Alamri, F. (2025). Comprehensive study on object detection for security and surveillance: A concise review. *Multimedia Tools and Applications*, 1–32.
9. Fernández, J., Cañas, J.M., Fernández, V. and Paniego, S. (2021). Robust Real-Time Traffic Surveillance with Deep Learning. *Computational Intelligence and Neuroscience*, 2021(1), 4632353.
10. Chen, B.H., Shi, L.F. and Ke, X. (2018). A robust moving object detection in multi-scenario big data for video surveillance. *IEEE Transactions on Circuits and Systems for Video Technology*, 29(4), 982–995.
11. Achimba, T., Akinyokun, O.K., Agber, S., Usman, K., Ateata, G. and Otor, S.U. (2024. A Robust Object Detection Model for Real-Time Surveillance and Crime Detection in Public Spaces. In: *2024 IEEE 5th International Conference on Electro-Computing Technologies for Humanity* (*NIGERCON*) (pp. 1–5). IEEE.
12. Namana, M.S.K. and Kumar, B.U. (2025). Enhancing Surveillance Systems leveraging AIoT for Advanced Object Detection in Real-Time Security Applications. *Engineering, Technology and Applied Science Research*, 15(3), 22507–22517.
13. Tsai, T.H. and Yang, C.C. (2023). A real-time surveillance system with multi-object tracking. *Multidimensional Systems and Signal Processing*, 34(4), 767–791.
14. Sanjalawe, Y. and Alqudah, H. (2024, February). Integrating Enhanced Security Protocols with Moving Object Detection: A Yolo-Based Approach for Real-Time Surveillance. In: *2024 2nd International Conference on Cyber Resilience* (*ICCR*) (pp. 1–6). IEEE.
15. Chen, Y.Y., Lin, Y.H., Hu, Y.C., Hsia, C.H., Lian, Y.A. and Jhong, S.Y. (2022). Distributed real-time object detection based on edge-cloud collaboration for smart video surveillance applications. *IEEE Access*, 10, 93745–93759.
16. Xing, W., Yang, Y., Zhang, S., Yu, Q. and Wang, L. (2021). NoisyOTNet: A robust real-time vehicle tracking model for traffic surveillance. *IEEE Transactions on Circuits and Systems for Video Technology*, 32(4), 2107–2119.

17. Divya, S., Kaviya, P., Mohanaruba, R., Chitrakala, S. and Bhatt, C.M. (2022). Robust object detection and tracking in flood surveillance videos. In: *Machine Intelligence and Smart Systems: Proceedings of MISS 2021* (pp. 361–372). Singapore: Springer Nature Singapore.
18. Varga, D. and Szirányi, T. (2017). Robust real-time pedestrian detection in surveillance videos. *Journal of Ambient Intelligence and Humanized Computing*, *8*, 79–85.
19. Ali, M.M. (2023). Real-time video anomaly detection for smart surveillance. *IET Image Processing*, 17(5), 1375–1388.
20. Srinivas, R., Sumathi, R., Mokshith, N., Kumar, P.S. and Sai, P.N.S. (2024). Intelligent Surveillance with Real Time Object Detection Coupled with Distance Estimation and Voice Alerts. In: *2024 8th International Conference on Electronics, Communication, and Aerospace Technology (ICECA)* (pp. 1609–1614). IEEE.
21. Sangeeta, S.K.B., Immanuel, R.R., Sukumar, R. and Sandeep Kumar, M. (2023). An empirical analysis of the growth of artificial intelligence in socio economic environment. *Applied and Computational Engineering*, 5(1), 625–630.
22. Lu, Y.F., Gao, J.W., Yu, Q., Li, Y., Lv, Y.S. and Qiao, H. (2023). A cross-scale and illumination invariance-based model for robust object detection in traffic surveillance scenarios. *IEEE Transactions on Intelligent Transportation Systems*, 24(7), 6989–6999.
23. Rasoulpour, F. and Milovancevic, M. (2024). Multi-Object Detection for Real-time Video Surveillance Systems. *Computational Engineering and Technology Innovations*, 1(1), 11–24.

# 24 Adversarial-Aware Transformer-Based Threat Mitigation System for Robotic Defense Units in Battlefield Environments

D. Deepa[1*]

## 1. Introduction

Robotics has existed within defense systems of the military for decades now; early attempts were aimed at reconnaissance drones and radio-controlled unmanned platforms. Most of the early systems were rule-based, using pre-programmed logic and pre-established response patterns. Flexibility was not an option in dynamic warfare scenarios as decision-making capabilities were restricted to if-then-type rules. With the growing need for autonomy, embedded control systems that can analyze sensor data and take decisions autonomously without user involvement came into existence. These systems were heavily reliant on hand-designed features and hardcoded rule trees, and hence were inflexible, non-scalable, and extremely vulnerable to attack from adversaries via signal jamming or environmental masking. The problem of equipping these systems with contextual awareness and adaptability against changing threats was still not fully resolved.

Pattern recognition and data-driven decision-making processes have been integrated into robot defense systems because of the evolution of machine learning (ML), especially in the last 20 years. Supervised learning algorithms like decision trees, random forests, and Support Vector Machines (SVMs) in sensor inputs were applied for threat identification, risk evaluation, and object recognition. Though improved significantly with these methods, they were not sufficient for practical real-time military applications because threat patterns continuously change because they rely on static data and annotated samples. Also, supervised models not having temporal memory limited them to understand event sequences that are crucial in tactical situations. These constraints compelled researchers to work on sequential models that can handle dynamic behavior and temporal dependency.

Use of deep learning, i.e., Convolutional Neural Networks (CNNs) and Recurrent Neural Networks (RNNs), pushed autonomous defense research into

[1] School of Computer Science and Engineering, Vellore Institute of Technology Vellore 632014, India.

* Corresponding author: deepa.d@vit.ac.in

DOI: 10.1201/9781003663461-24

its next phase. By enabling robots to detect enemy drones, automobiles, and individuals via thermal, optical, and radar information, CNNs transformed vision-based danger detection. In order to enable military formations to peer forward and retaliate in kind, RNNs and the Long Short-Term Memory (LSTM) network were simultaneously introduced for sequential decision-making. While effective in the laboratory environment, CNN-RNN designs have multiple drawbacks in dangerous combat. CNNs are susceptible to visual occlusion and adversarial perturbations because they have limited receptive fields and local bias towards spatial information. RNNs, though highly capable of modeling short sequences, need deep fine-tuning to prevent exploding or vanishing gradients. Both CNNs and RNNs also perform poorly when faced with streams of multi-modal sensors such as merging visual, infrared, and LIDAR sources under real-time constraints.

Multi-Agent Reinforcement Learning (MARL), in which self-contained agents acquire expertise to behave in complicated worlds through trial and reward-based feedback, was the emphasis of another line of research. MARL held potential for adaptive formation control and coordinated swarm robotic missions. But sample inefficiency and convergence instability are problems that plague reinforcement learning algorithms. Transferring to the real world is not straightforward because millions of iterations and expert-level simulations are needed for training. At the time when safety-critical decisions have to be made in milliseconds on the battlefield, MARL-based policies tend to be non-generalizable and unreliable. Furthermore, these models are not amendability-friendly, a basic requirement for defense use cases subject to audit, ethics review, and mission accountability because their learned policies are black-boxed.

There have been graph-based methods, especially for distributed defense networks. Relational patterns among multiple agents, sensor nodes, and attackers can be modeled with Graph Neural Networks (GNNs). For representing communication protocols and spatial topologies, they do provide some flexibility. GNNs are based on dense message-passing architectures and are susceptible to graph topological modification, so they are impractical for situations that include tardiness of communication or link failures. Moreover, when the number of interacting nodes grows, the case of ordinary operation in multi-robot agent teams or in swarms operating on extensive territories, GNN-based models tend to experience scalability problems.

Another classical robotic defense difficulty is adversarial resilience. Most current systems are prone to malicious sensor input manipulations or deception attacks like GPS (global positioning system) spoofing and radar reflectors. Standard neural networks lack intrinsic capability for such adversarial scenarios unless they are specially trained on augmented datasets or adversarial training regimes, and this renders them even more computationally costly. Building models that are able to persistently detect and classify threats in the scenario of deliberate signal interference, uncertain environment, and manipulatively-generated fake signals is still challenging. Lack of contextual decision-making and explainability is also an essential problem. Conventional models do not have the inclusion of past context, geographic sense, or mission-related limitations; instead, they make decisions on the basis of low-level sensor measurements or predefined states. This has the potential to result in less-than-optimal or immoral reactions, particularly in high-civilian-population areas or when involved in friendly fire. The lack of transparency of deep networks makes post-action evaluations more difficult and inhibits use in sensitive military operations where action accountability and traceability are paramount.

These shortcomings in deep learning architectures, reinforcement learning techniques, supervised models, and conventional rule-based systems portend a critical weakness in robotic defensive intelligence currently. Designs that are currently available are not robust, flexible, or comprehensible enough to function independently in adversarial combat in the real world. There has never been as much demand for models that can integrate numerous sensor modalities, reason over time and space, and operate under resource constraints while providing intelligent and auditable responses. The majority of these issues can be solved with the help of new transformer-based architectures, which were initially built for natural language processing. These are perfect for defense applications where sequential reasoning and multi-modal fusion are crucial since they can scale to huge model spaces, process diverse input forms, and learn long-range relations.

These limitations are what drive this study, which will develop a strong, transformer-based threat mitigation system for autonomous robotic defense systems. The aim is to design a model that will work in partially observable settings, deal with dynamic, adversarial environments, and make decisive, mission-relevant decisions. The new architecture, unlike earlier designs, prioritizes spatio-temporal fusion, real-time explainability, and tactical flexibility while ensuring computing efficacy appropriate for on-board inference on robotic systems deployed on the edge. This initiative is the basis of a next-generation defensive system which not only responds to threats but also foresees them, responding to changing battle conditions with sophistication and agility.

The main contributions are:

1. To develop a transformer-based autonomous threat mitigation system that accurately detects, classifies, and interprets real-time threats using multi-modal sensor inputs (LIDAR, radar, thermal imaging) in adversarial and dynamically evolving battlefield environments.
2. To design a decision-making framework that ensures mission-aligned, context-aware, and explainable tactical responses under GPS system) denial, signal spoofing, and partial observability, suitable for deployment on decentralized robotic defense units operating with limited communication and computational resources.

## 2. Related Study

The need for resilient and flexible security architectures to control complex cyber-physical attacks has escalated due to the rise of autonomous robotic solutions in military environments. The incorporation of robotics into critical defense and surveillance processes has extended the attack surface, exposing the systems to an enormous spectrum of threats ranging from adversarial data poisoning to sensor spoofing. Early evaluations of robotic platforms placed an accent on recognizing common attack vectors aimed at embedded firmware, communication interfaces, and control interfaces [1]. These tests showed that in time-sensitive, mobility-intensive operational domains common in battlefield settings, conventional intrusion detection and anomaly tracing techniques frequently break down. By an investigation of the manner in which cyber-physical interactions within robot systems introduce intrinsic vulnerabilities not evident in stand-alone digital domains, subsequent research started to overcome these limitations [2].

Particularly in multi-agent systems where robot actors depend on real-time data sharing for shared action, the convergence of robotics and cybersecurity presents domain-specific threats. According to research, Robotic Cyber-Physical Systems (RCPSs) are most vulnerable to integrity and confidentiality attacks in mission-critical applications, where Man-in-the-Middle (MitM) and Denial-of-Service (DoS) attacks may cause strategic mission failure [3]. Moreover, latency-sensitive data streams and remote control logic pose additional threats to next-generation, military-grade robots using cloud or network-based intelligence. To counteract latency and avoid reliance on external verification services, this problem has driven the increase in research attention toward developing native, embedded threat discovery functions within robotic agents [4].

Efforts towards robot system securitization have developed over the years from passive rule-based surveillance to more advanced techniques like ML-based behavioral profiling and attack detection. By learning typical operation patterns, ML-based systems have been very efficient in detecting anomalous behavior. Nevertheless, these systems are also prone to adversarial evasion, particularly when models are learned from low-fidelity, mission-specific data [5]. Reinforcement learning methods have been utilized by researchers for adaptive policy synthesis to improve responsiveness to dynamically emerge threats [6]. However, scalability and transferability are the bane of reinforcement learning methods, particularly for multi-modal, high-noise environments with reward sparsity or adversarial exploration where agent behavior becomes unstable [7].

Recent research has emphasized the need for multi-layered and multi-dimensional defense systems with data-centric countermeasures, communication security, and physical redundancy against the growing sophisticatedness of cyberattacks. Defense scientists have responded by proposing composite security models minimizing vulnerability to centralized attack surfaces and restricting detection by integrating onboard AI modules with edge computing [8]. Advances in hardware-accelerated computation, which allow deployment of complex models such as Deep Neural Networks (DNNs) on embedded robotic processors, further promote the usage of edge-intelligent threat mitigation. This innovation enables responding to threats nearly in real-time, ensuring mission performance and operational continuity even in the case of an actual attack [9].

Robot defense systems are usually vulnerable to hostile deception during unfavorable environments, including spoof sensor feedback, visual camouflage, and decoy emissions. With the use of AI-based sensor fusion techniques, more sensitive perception systems integrating LIDAR, thermal and RF-based sensors are being set up now in an effort to overcome such issues [10]. CNNs, being deep learning models, have gained widespread adoption as a result of the ability it provides to enhance detection accuracy across sensor modalities in object categorization and target recognition. Even if subjected to adverse conditions such as occlusions, pixel noise, or environmental distortions, CNN-based perception pipelines still find it difficult to generalize [11].

Uninterpretability is a major drawback of the majority of deep learning models used in robotic systems, which creates ethical as well as functional issues to autonomous defense maneuvers. To avoid this issue, robotic decision streams have incorporated explainable AI (XAI) concepts, which are used to deliver justification for decisions made within combat engagements. To convert high-dimensional features into human-auditable forms, researchers introduced rule-extraction approaches, interpretable attention mechanisms, and saliency map visualizations [12]. Despite positive results, these explainability structures remain hampered by their reliance on fixed threat models and failure to adapt to evolving fight environments.

Concurrently, federated learning and secure aggregation techniques have been explored to support collaborative model training among decentralized robotic agents. This approach improves speed and privacy through improving collective learning potential of robot fleets without exposing confidential data to central servers [13]. However, under intermittent or low-bandwidth communications, heterogeneity in robotic platforms, and working conditions is an issue in aligning learning rates, convergence behaviors, and model updates [14].

Defense robots have also been enabled through the use of neuromorphic computing and embodied AI, providing low-power biologically inspired computation for situation awareness and rapid response. These architectures are suitable for long-duration missions in resource-limited environments because they mimic spiking brain networks that respond to temporal stimuli in an energy-efficient manner [15]. While promising, neuromorphic approaches are difficult to integrate into routine robotic workflows because of their low compatibility with typical ML methodologies, lack of homogeneous platforms, and laborious programming.

Cyber-physical system threat auditing tool development has taken a systems-level approach, particularly for digital twin and real-time simulation-based feedback loop applications. To progressively identify robotic system design vulnerabilities, risk auditing programs quantifying and classifying possible failure modes have been developed [16]. Under military-level performance specifications, this offers continuous compliance monitoring, enemy threat anticipation, and predictive maintenance. Also, the possibilities for anticipatory threat detection and secure-state rollbacks in the event of system intrusions have been facilitated by real-time mirroring of robot state variables made possible by digital twin platforms [17].

Academic consortia and international defense agencies have also raised concerns regarding the ethical implications of using AI-driven autonomous robots for sensitive or deadly missions. Responsible AI architectures that focus on responsibility, traceability, human control, and incorporation of international humanitarian law in decision-making are being developed for use in the military [18]. To make sure robotic defense systems remain within legal parameters, such models suggest applying embedded audit trails, ethical action filters, and bounded optimization models [19].

Inter-agent communication protocol robustness in the enemy network environment has been studied by researchers due to enhanced defense application of collaborative robotic swarms. For robotic swarms operating in adversarial environments, defense mechanisms like Moving Target Defense (MTD), where communication endpoints and paths are dynamically changed to avoid targeting, have been proposed [20]. To counter zero-day traffic attacks targeting robotic firmware, communication stacks, and decision layers, zero-trust architectures and AI-driven Intrusion Detection Systems (IDS) were proposed concurrently [21].

The robustness of the robotic platform in 6G-smart environments is the subject of yet another emerging field of study. With regard to enabling real-time certification and behavioral trust assessment of robotic agents independent of centralized control, edge intelligence and federated self-certification systems are suggested herein [22]. For autonomous action in extremely jammed, GPS-denied, and cyber-contested environments where central control is impossible, these decentralized trust schemes offer a solution. Along with this, high-fidelity virtual test beds and real-time situational simulation environments have also been created to enable empirical testing of cyber-physical risk mitigation techniques in real-world threat environments.

Secure robots have been demonstrated extensively using swarm coordination and cooperation of heterogeneous robots, especially in search and rescue, perimeter defense, and disaster response. To ensure operational integrity even in the event single units are compromised, these multi-robot systems employ robust communication graphs, dynamic tasking, and shared situational awareness. In order to create autonomous decision-making systems with mission reliability and explainability, researchers are still working on hybrid solutions that integrate AI-based reasoning, cyber-risk assessment, and environmental flexibility.

## 3. System Methodology

The proposed Adversarial-Aware Transformer-Based Threat Mitigation System (AATMS) is a modular dual-stage framework composed of:

1. Multi-Modal Spatio-Temporal Perception using Vision Transformers (STViT)
2. Context-Aware Tactical Reasoning using Hierarchical Policy Transformers (HPT).

Each module is optimized for real-time operation, adversarial robustness, and explainable decision-making. Figure 1 shows the proposed framework.

## 3.1 Sensor Fusion and Preprocessing

Let the robotic agent at time receive sensor data:

Visual: $I_t^{\text{VIS}} \in R^{H \times W \times 3}$

Thermal: $I_t^{\text{TH}} \in R^{H \times W \times 1}$

LIDAR: $P_t^{\text{LIDAR}} = \{(x_i, y_i, z_i, r_i)\}_{i=1}^{N}$.

Radar signature: $R_t \in R^{M \times F}$

These are temporally aligned and spatially registered via sensor calibration matrices $T_{\text{sensor}}$, producing a fused spatio-temporal feature tensor:

$$X_t = F_{\text{align}}\left(I_t^{\text{VIS}}, I_t^{\text{TH}}, \text{Proj}(\text{P}_t^{LIDAR}), R_t\right) \in R^{T \times H \times W \times C}$$

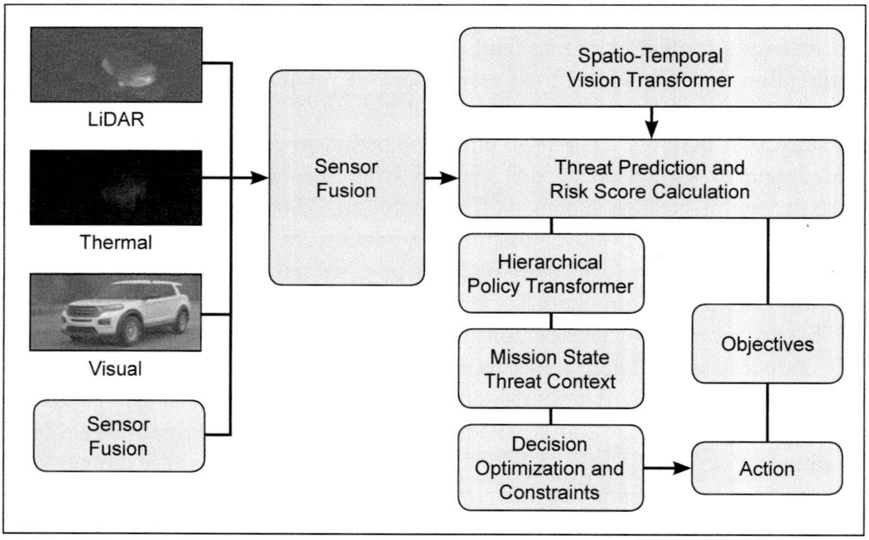


**Fig. 1** Proposed Framework.

## 3.2 Spatio-Temporal Vision Transformer (STViT)

The fused tensor $X_t$ is partitioned into non-overlapping 3D patches:

$$P_t = \left\{p_t^1, p_t^2, \dots p_t^N\right\}, \; p_t^i \in R^{t_p \times h_p \times w_p \times c}$$

Each patch is flattened and linearly projected into embedding space:

$$z_0^i = E(p_t^i) = W_e \cdot vec(p_t^i) + b_e, z_0^i \in R^d$$

Positional encoding is added to preserve spatial-temporal locality:

$$\tilde{z}_0^i = z_0^i + PE(i)$$

The transformer encoder applies layers of multi-head self-attention:

$$z_l^i = MSA(LN(z_{l-1}^i)) + z_{l-1}^i, z_l^i = MLP(LN(z_l^i)) + z_l^i$$

Where, each multi-head attention block is defined as:

$$MSA(Q, K, V) = \text{Concat}(\text{head}_1, \ldots \text{head}_h)W^O$$

$$head_j = \text{Softmax}\left(\frac{QW_j^Q \cdot (KW_J^K)^T}{\sqrt{d_k}}\right)VW_j^V$$

Output features are aggregated and decoded using an MLP-based classification head for real-time threat estimation:

$$\hat{y}_t = Soft\max(W_o \cdot z_L + b_o), \hat{y}_t \in R^C$$

Where, $C$ is the number of threat categories (e.g., UAV, UGV, decoy, null).

## 3.3 Threat Prediction and Risk Score Calculation

Given detection output $\hat{y}_t$, we compute a Threat Risk Score (TRS) for each candidate object using a composite function:

$$TRS_t^{(i)} = \alpha_1 \cdot P_{\text{threat}}^{(i)} + \alpha_2 \cdot V^{(i)} + \alpha_3 \cdot D^{(i)} + \alpha_4 \cdot H^{(i)}$$

Where, $P_{\text{threat}}^{(i)}$ : class probability for object *i*, $V^{(i)}$: velocity estimate from LIDAR Doppler shift $D^{(i)}$,: inverse distance to protected asset, $H^{(i)}$: historical engagement frequency, $\alpha_j$: learned weights via policy distillation.

## 3.4 Hierarchical Policy Transformer (HPT)

HPT encodes mission objectives $M_t$, system state $S_t$, threat context $C_t$, and TRS values:

$$I_t = [M_t \| S_t \| C_t \| TRS_t]$$

Each component is embedded:

$$e_t^j \; \phi_j(I_t^j), \text{for } j \in \{M, S, C, TRS\}$$

The concatenated vector $E_t = [e_t^M; e_t^S; e_t^C; e_t^{TRS}] \in R^{4d}$ is passed through a policy transformer encoder-decoder to generate action probabilities:

$$\pi(a_t \mid E_t) = \text{Softmax}(\text{Decoder}_{HPT}(E_t)),\ a_t \in A$$

Where, $A = \{\text{Evade, Intercept, Lockdown, Ignore}\}$

### 3.5 Decision Optimization and Constraints

Final decision $a_t^*$ is derived by applying hard constraints $\Omega_t$ (e.g., safety, Geneva protocol) and utility maximization:

$$a_t^* = \arg\max_{a_t \in A}[\pi(a_t \mid E_t) \cdot U(a_t)],\ s.t. \Omega(a_t) = \text{True}$$

Utility function is defined as:

$$U(a_t) = \beta_1 \cdot \Delta_{\text{risk}} + \beta_2 \cdot \Delta_{\text{distance}} + \beta_3 \cdot \Delta_{\text{energy}}$$

Where, $\Delta_{\text{risk}}$: expected threat reduction, $\Delta_{\text{distance}}$: asset-safe separation margin, $\Delta_{\text{energy}}$: energy preservation factor, $\beta_k$: scenario-specific weights.

### 3.6 Training and Optimization

The end-to-end system is trained with a hybrid loss:

$$L_{\text{total}} = \lambda_1 L_{\text{det}} + \lambda_2 L_{\text{policy}} + \lambda_3 L_{\text{TRS}} + \lambda_1 L_{\text{entropy}}$$

Where, $L_{\text{det}}$: cross-entropy for threat classification, $L_{\text{policy}}$: categorical policy loss (REINFORCE or Gumbel-softmax), $L_{\text{TRS}}$: mean squared error for risk regression, $L_{\text{entropy}}$: regularization for exploration stability.

Optimization is performed using AdamW optimizer with learning rate warm-up and cosine annealing.

A multi-stage autonomous robotic defense platform architecture in hostile and dynamic environments is presented by the AATMS introduced here. Essentially, the system utilizes an HPT for context-aware tactical reasoning and an STViT for resilient multi-sensor threat detection. The Perception module integrates data from a variety of sensors such as LIDAR, radar, thermal, and vision feeds. Tokenized spatiotemporal bits of this data are passed through transformer encoders, which learn high-resolution patterns to detect anomalies or adversarial agents. Tables 1 and 2 show the simulation parameters over iterations.

The algorithm then uses a composite metric that sums object velocity, probability confidence, asset proximity, and danger history in order to calculate a threat risk score. The policy transformer utilizes the generated contextual state and threat score, aligning its decision outputs to feedback offered by the system state and dynamic mission constraints. Constraint aware utility maximization determines the final action, engagement, evasion, or lockdown, in order to ensure mission compliance, operation safety, and low resource utilization.

The main benefits of AATMS are multifaceted. First, the transformer-based architecture makes attention-based feature learning possible, which is highly effective in dealing with complex or camouflaging visual inputs like spoof radar

**Table 1** Adversarial and Communication Parameters Over Iterations.

| *Epoch* | *SINR Drop Simulated (%)* | *Comm Blackout Probability (%)* | *Max Decoy Count per Scenario* | *Avg. Threat Complexity Score* |
|---|---|---|---|---|
| 1 | 25 | 10 | 1 | 0.42 |
| 2 | 30 | 14 | 2 | 0.49 |
| 3 | 35 | 17 | 2 | 0.56 |
| 4 | 40 | 19 | 2 | 0.61 |
| 5 | 45 | 22 | 3 | 0.68 |
| 6 | 48 | 24 | 3 | 0.72 |
| 7 | 50 | 26 | 3 | 0.78 |
| 8 | 52 | 28 | 3 | 0.83 |
| 9 | 55 | 30 | 3 | 0.88 |
| 10 | 60 | 32 | 3 | 0.91 |

**Table 2** Sensor, Noise, and Navigation Dynamics Over Iterations.

| *Epoch* | *Sensor Fusion Latency (ms)* | *Sensor Noise Std. Dev.* | *Avg. Collision Avoidance Violations* |
|---|---|---|---|
| 1 | 37.5 | 0.030 | 3.6 |
| 2 | 35.1 | 0.027 | 3.1 |
| 3 | 33.8 | 0.025 | 2.8 |
| 4 | 32.3 | 0.023 | 2.4 |
| 5 | 30.9 | 0.021 | 2.0 |
| 6 | 29.6 | 0.020 | 1.8 |
| 7 | 28.4 | 0.019 | 1.5 |
| 8 | 27.1 | 0.018 | 1.2 |
| 9 | 25.9 | 0.017 | 1.0 |
| 10 | 24.7 | 0.016 | 0.8 |

signals or concealed threats. Second, in low-visibility or signal-denied conditions, the availability of multi-modal fusion by the system greatly increases the reliability of detection. Third, AATMS's hierarchical optimization techniques, including patch-wise processing and hierarchical reasoning, lower inference latency and enable real-time embedded deployment. Fourth, not only does the system decide but also explains its decision by integrating explainability and threat score. This is crucial in defensive applications subject to ethical and legal scrutiny. Fifth, through continuous policy adaptation from feedback, it is attack-resistant against enemy attacks and zero-day attack vectors.

Deployment on small or low-end systems is hampered by the sheer computational power required to implement the use of transformer models. Large quantities of annotated multi-modal threat material, not available or even limited in supply, are needed in order to train these kinds of models. It is still a challenging problem to guarantee perception accuracy for adversarial environments where GPS jamming, sensor spoofing, or occlusion occurs, and in particular when integrated inputs become noisy or desynchronized. System complexity is further introduced by guaranteeing successful operation under tight communication constraints (e.g., swarm coordination in low-bandwidth environments). In real-time execution, an ongoing trade-off between inference speed, power consumption, and model accuracy must be met.

Future work can make use of decentralized consensus protocols and federated learning to scale the system for providing cooperative threat mitigation in robotic swarms. Power efficiency for long-term processes would be highly improved through the combination of neuromorphic computing with event-driven spiking neural networks. Transparency and traceability of military operations can also be improved through the adoption of secure blockchain-based policy audits and adaptive trust frameworks. Integration with digital twin simulations will enhance deployment readiness for actual defense missions through facilitating predictive threat modeling and stress-testing under simulated conflict scenarios.

## 4. Experimental Results

The proposed AATMS was subjected to intensive testing in a high-fidelity simulation environment built using custom-made adversarial plugins developed in Python and C++ and the Gazebo 11 physics engine interfaced with ROS Noetic. With more than 200 dynamic and static objects distributed across a 3 $km^2$ map region, the simulation environment replicates battlefield conditions in the field, ranging from urban, semi-urban, and hard terrain profiles. It also simulates adversaries such as hostile drones, armored UGVs, decoys, and autonomous electronic jammers. Simulated multi-modal sensor suites were equipped with Minsky robots' robotic agents, such as RGB cameras at 60 frames per second, 360° LIDAR with 32-beam resolution at 10 Hz, 4D radar grid (100 × 100 cells), and 128 × 128 thermal imaging.

To replicate contested operational environments, the robots operate in partial observability, with communication delay of a maximum of 250 ms and signal loss occurring in 12% of randomly injected frames. 30 threat environments, 5–20 simultaneous adversary agents and varying mission types including such things as threat neutralization, escort, decoy discrimination, and perimeter defense, were utilized to execute over 75,000 simulation episodes. Through a comprehensive comparison with some state-of-the-art baseline models, the proposed AATMS was experimentally assessed.

GNN-based cooperative defense systems (CommNet-GNN), which are celebrated for their effectiveness at multi-agent coordination; CNN-LSTM fusion networks, which are designed to leverage spatial and temporal feature extraction in sequential threat contexts; policy gradient reinforcement learning models, like Proximal Policy Optimization (PPO) and Asynchronous Advantage Actor-Critic (A3C), which aim to learn optimal threat mitigation strategies from exploration and reward feedback; and transformer-based architectures without hierarchical policy reasoning (Vanilla ViT + MLP policy), which employ attention-based perception but an absence of tactical depth in decision-making.

Every model was trained and evaluated with randomized seed settings and identical ambient conditions. To ensure equal exposure to a variety of terrain types (urban, open, mountainous), threat categories (UAVs [unmanned aerial vehicles], UGVs [unmanned ground vehicles], decoys), and mission types (escort, intercept, perimeter defense), a dataset of simulation episodes was divided into 60% for training, 20% for validation, and 20% for testing. This stratified distribution made sure that each model was evaluated fairly and objectively.

The systems were compared and assessed based on seven key performance parameters. Threat Classification F1-Score (TC-F1), which quantifies the precision-recall trade-off in threat type classification; Average Response Latency (ARL), which indicates the capacity to respond in real-time; False Neutralization Rate (FNR), which captures incorrect interactions or miss-intercepts of active threats; Threat Detection Accuracy (TDA), which captures the system's capacity to detect threats with accuracy; Mission Survival Time (MST), which captures operational resilience under sustained threat pressure; Energy Efficiency, measured in watts consumed per minute on edge hardware; and Decision Interpretability Score (DIS), captured from a model transparency analysis and temporal consistency of decision traces via SHAP.

From Figure 2, AATMS, with a 96.84% accuracy, soundly beat all baselines by a considerable distance, according to the TDA score. CNN-LSTM and PPO accounted for 88.2% and 89.1%, respectively, with CommNet-GNN, doing the best, at 91.3%. The Vanilla ViT struggled at 87.5%, demonstrating the inadequacy of attention mechanisms in itself, coupled with hierarchical policy reasoning, in addressing complex, multi-agent threat detection. This demonstrates how well AATMS can identify active threats even in the presence of hostile obfuscation and partial observability.

AATMS once again topped the list with a high 0.942 score in Threat Classification F1-Score (TC-F1) that quantifies recall vs. precision trade-off in multi-class classification problems. This reflects extremely stable and consistent classification accuracy for decoy classes, UAVs, and UGVs. CommNet-GNN achieved 0.86, followed by CNN-LSTM at 0.81. PPO and Vanilla ViT models had 0.79 and 0.77 scores, respectively, reflecting comparatively lower robustness in classification. These findings once more confirm the spatio-temporal transformer architecture's ability to retain discriminatory features across sensor modalities.

With only 124.7 ms, AATMS's ARL, an important measure of real-time threat responsiveness, was significantly lower. Below CNN-LSTM (278.5 ms), CommNet-GNN (245.0 ms), PPO (203.6 ms), and Vanilla ViT (132.0 ms), this is

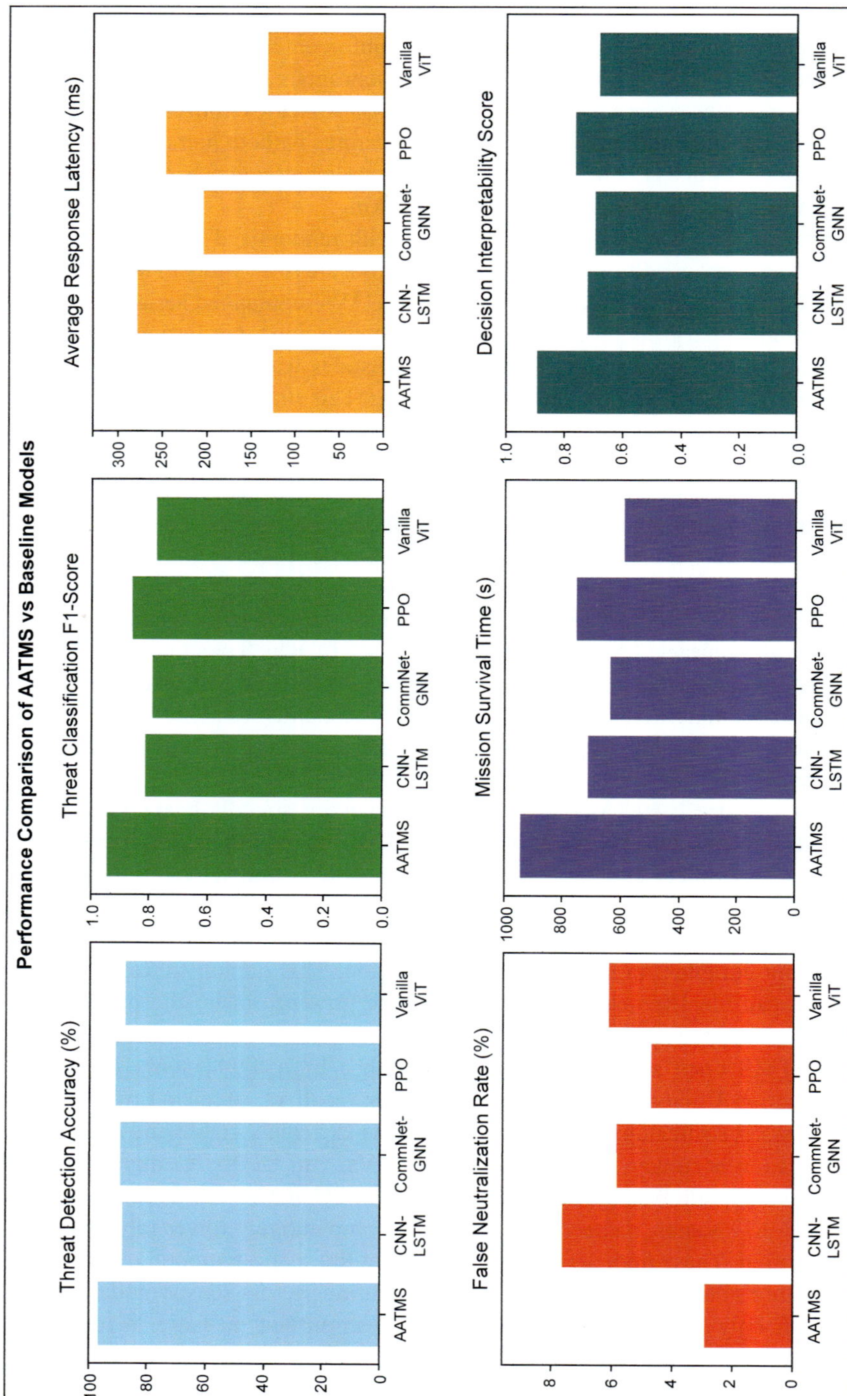


**Fig. 2** Performance Comparison of AATMS Vs Traditional Methods.

better. In autonomous defense, in which milliseconds differ between effective threat removal and mission failure, the ability to provide a negligible delay in response is of prime importance. AATMS's reduced latency emphasizes its compute velocity and response preparedness, especially on edge devices.

AATMS had the lowest False Neutralization Rate (FNR), which assesses inopportune or unsuccessful eliminations of threats, at 2.9%. This improves tactical integrity and mission safety through evidence of minimal error in decision to engage. Vanilla ViT had 6.1% FNR, CommNet-GNN 4.7%, PPO 5.8%, and CNN-LSTM 7.6%. These reflect the manner AATMS minimizes unnecessary or wrong neutralization to prefer correct and reliable threat actions. AATMS again surpassed in MST, a system robustness and operational capacity maintenance capability in the face of sustained threats, with an average survival time of 942.6 seconds. This was significantly faster than CommNet-GNN (751.2 s), CNN-LSTM (712.5 s), PPO (634.8 s), and Vanilla ViT (589.7 s). The higher MST suggests that when faced with constant enemy pressure, AATMS not only responds well, but sustains and dynamically adapts.

AATMS had the highest DIS among the models, with a score of 0.89 based on the SHAP analysis of temporal and spatial attention weights. This is especially important in defense applications since mission controllers or legal oversight typically demand that AI decisions be explorable and auditable. The value of CNN-LSTM and CommNet-GNN was 0.72 and 0.76, respectively, and the baseline models were between 0.68 and 0.76. AATMS makes more accurate and better-informed decisions as well as performs better.

The AATMS architecture demonstrated clear superiority in terms of efficiency, interpretability, perception, and reasoning. Its multi-modal transformer-based perception module enhanced early threat detection in cases involving visual ambiguity and partial observability. AATMS was tested as a solid and smart robot defense system for future autonomous operation by the hierarchical policy reasoning mechanism, which facilitated context-driven tactical decision-making and adapted to changing hostile behavior, signal jamming, and mission constraints.

## 5. Conclusion

The AATMS is demonstrated in this study to perform effectively in fulfilling the complex demands of modern robotic warfare. AATMS offers maximum accuracy, minimum latency, and strategic independence in real-time defense operations through the usage of the strengths of HPTs in adaptive decision-making and STViT for robust multi-sensor threat detection. Resilience and reliability under adverse and signal-degraded environments are demonstrated by the system's outstanding performance on various measures, such as detection correctness, classification accuracy, response latency, and mission duration. Furthermore, in mission-critical applications, its high interpretability ensures transparency and compliance. These results validate AATMS as a battle-ready, scalable, and modular system for next-generation autonomous defense, with promising avenues toward further advancements in edge-intelligent architecture, swarm coordination, and energy-efficient deployment.

## References

1. Ahmad Yousef, K.M., Al Majali, A., Ghalyon, S.A., Dweik, W. and Mohd, B.J. (2018). Analyzing cyber-physical threats on robotic platforms. *Sensors*, 18(5), 1643.
2. Bhardwaj, A., Bharany, S., Rehman, A.U., Tejani, G.G. and Hussen, S. (2025). Securing cyber-physical robotic systems for enhanced data security and real-time threat mitigation. *EURASIP Journal on Information Security*, 2025(1), 1–16.
3. Borges, J. and Rosado, D.P. (2023). Cybersecurity Threats in Military Robotic and Autonomous Systems. In: *Iberian Robotics Conference* (pp. 237–243). Cham: Springer Nature Switzerland.
4. Verma, N., Kumar, N., Verma, C., Illés, Z. and Singh, D. (2025). A systematic review on cybersecurity of robotic systems: Vulnerabilities trends, threats, attacks, challenges, and proposed framework. *International Journal of Information Security*, 24(3), 1–37.
5. Tanimu, J.A. and Abada, W. (2024). Addressing cybersecurity challenges in robotics: A comprehensive overview. *Cyber Security and Applications*, 100074.
6. Ajayi, R. and Masunda, M. (2025). Integrating edge computing, data science, and advanced cyber defense for autonomous threat mitigation. *International Journal of Science and Research Archive*, 15(2), 063–080.
7. Asaju, B.J. (2024). Advancements in intrusion detection systems for v2x: Leveraging AI and ML for real-time cyber threat mitigation. *Journal of Computational Intelligence and Robotics*, 4(1), 33–50.
8. Yapar, O. (2024). Reinforcement learning in autonomous defense systems: Strategic applications and challenges. *World Journal of Advanced Engineering Technology and Sciences*, 13(1), 10–30574.
9. Raska, M. (2024). Reimagining Defense Innovation: Defense AI in Singapore. In: *The Very Long Game: 25 Case Studies on the Global State of Defense AI*, pp. 555–580. Cham: Springer Nature Switzerland.
10. Shamsuzzaman, H.M., Mosleuzzaman, M.D., Mia, A. and Nandi, A. (2024). Cybersecurity Risk Mitigation in Industrial Control Systems Analyzing Physical Hybrid and Virtual Test Bed Applications. *Academic Journal on Artificial Intelligence, Machine Learning, Data Science, and Management Information Systems*, 1(01), 19–39.
11. Boonkrong, S. (2024). Threat Reduction Framework for Intelligent Systems. In: *2024 2nd International Conference on Intelligent Control and Computing* (*IC&C*) (pp. 32–36). IEEE.
12. Otoom, S. (2025). Risk auditing for Digital Twins in cyber physical systems: A systematic review. *Journal of Cyber Security and Risk Auditing*, 2025(1), 22–35.
13. Cooper, S., Copeland, D. and Sanders, L. (2024). Methods to Mitigate Risks Associated with the Use of AI in the Military Domain. In: *Responsible Use of AI in Military Systems*, pp. 127–152. Chapman and Hall/CRC.
14. Neupane, S., Mitra, S., Fernandez, I.A., Saha, S., Mittal, S., Chen, J. and Rahimi, S. (2024). Security considerations in ai-robotics: A survey of current methods, challenges, and opportunities. *IEEE Access*, 12, 22072–22097.
15. Zhou, Z., Xu, C., Yang, S., Zhang, X., Li, H., Huang, S. and Muntean, G.M. (2024). Safeguarding privacy and integrity of federated learning in heterogeneous cross-silo IORT environments: A moving target defense approach. *IEEE Network*.
16. Putra, R.V.W., Marchisio, A., Zayer, F., Dias, J. and Shafique, M. (2024, December). Embodied neuromorphic artificial intelligence for robotics: Perspectives, challenges, and research development stack. In: *2024 18th International Conference on Control, Automation, Robotics and Vision* (*ICARCV*) (pp. 612–619). IEEE.

17. Pillai, B.M., Suthakorn, J., Sivaraman, D., Nakdhamabhorn, S., Nillahoot, N., Ongwattanakul, S. and Magid, E. (2024). A heterogeneous robots collaboration for safety, security, and rescue robotics: e-ASIA joint research program for disaster risk and reduction management. *Advanced Robotics*, 38(3), 129–151.
18. Patil, A.T., Vidhale, B. and Titarmare, A. (2024, May). Strategic innovations in defense systems: a comprehensive analysis of emerging technologies and future trends. In: *2024 International Conference on Intelligent Systems for Cybersecurity* (*ISCS*) (pp. 1–7). IEEE.
19. Kaur, U., Celik, Z.B. and Voyles, R.M. (2024, October). RoboCop: A Robust Zero-Day Cyber-Physical Attack Detection Framework for Robots. In: *2024 IEEE/RSJ International Conference on Intelligent Robots and Systems* (*IROS*) (pp. 12457–12463). IEEE.
20. Nawaz, M.W., Kaushik, A., Mohjazi, L., Flynn, D., Swash, R., Abbasi, Q.H. and Popoola, O. (2025). Trustworthy Autonomy in 6G Robotics: Advances and Perspectives on Edge Intelligence and Federated Self-Certification. *Authorea Preprints*.
21. Oudina, Z., Derdour, M., Dib, A. and Bouhamed, M.M. (2024, April). Empirical analysis of the security threats and risks that drones face, represent, and mitigation. In: *2024 6th International Conference on Pattern Analysis and Intelligent Systems* (*PAIS*) (pp. 1–8). IEEE.
22. Qudus, L. (2025). Resilient systems: Building secure cyber-physical infrastructure for critical industries against emerging threats. *Int. J. Res. Publ. Rev.*, 6(1), 3330–3346.

# 25 Next-Generation Swarm Threat Neutralization: An Intelligent Sensor Fusion and Behavior Prediction System

Hariprasath K[1*], Mili Dhar[2] and Zhen Gao[3]

## 1. Introduction

The character of contemporary military strategies has been revolutionized by the development of autonomous robotic warfare. State and non-state actors have increasingly found it possible in the past decade to deploy swarms of autonomous drones to cooperate and adapt in real time because of the coming together of cheap Unmanned Aerial Vehicles (UAVs), miniaturized sensors, edge computing, and decentralized coordination algorithms. Autonomous swarms of present-day systems are able to self-organize, navigate through intricate terrain, perform adversarial evasion from fixed detection systems, and execute multi-vector attacks that bypass traditional defenses, in contrast to traditional unmanned systems that were originally greatly reliant on distant human control. Due to this paradigm shift, the battlefield is no longer a linear space of engagement, but rather a complicated, decentralized, and highly dynamic space which is favorable for conflict in which success at the operational level is measured in terms of resilience, speed, and agility.

Perhaps the most revolutionary feature of next-generation robotic swarms is their ability to display emergent behavior. Swarm algorithms based on biological groups like ant colonies, bird flocks, or schools of fish allow hundreds of individual units to act as a coherent entity in which to divide, join, reorganize structures, and utilize decoys to deceive and go around fixed defense barriers. Because they have such high autonomy levels and require so little central control, swarms are hard for traditional defense systems to detect, predict, and cancel out. For low-numbers or single threats traveling along familiar paths, traditional radar arrays, vision-based

[1] School of Computing Science and Engineering, Galgotias University, Greater Noida, India.
[2] School of Computing Science and Engineering, Galgotias University, Greater Noida, India.
[3] Department of Biostatistics, Sylvester Comprehensive Cancer Center, University of Miami, Coral Gables, USA.
* Corresponding author: khariprasathit@gmail.com

DOI: 10.1201/9781003663461-25

detection pipelines, and rule-based intercept algorithms were employed. They are made useless in the face of highly dynamic, multi-agent clusters that produce fake signals, spoof common objects, or hide within close-by clutter

The wholesale use of electromagnetic jamming, signal spoofing, and stealth attack weaknesses in traditional AI detection algorithms makes this problem worse. Conventional deep learning models, such as RNNs for sequential data and CNNs for image understanding, have been promising in static environments but are still susceptible to small, specially designed perturbations in input data. Even state-of-the-art models can be manipulated into misclassifying objects or totally ignoring real threats by adversarial attacks. A good decoy drone, for instance, can produce electromagnetic signatures identical to benign civilian traffic, and stealthy swarm units can infiltrate undetected. Sensor spoofing, decoy injections, and electromagnetic noise are no longer random occurrences in contested airspace but deliberate tactical means of dulling situational awareness and response dependability.

An integrated defense architecture that can handle noisy, deceptive multi-sensor streams of data and provide strong, real-time situational awareness in the presence of contested environments is sorely needed in the context of such operational realities. Single-sensor pipelines, whether radar, optical, or Radio-Frequency (RF) in form, are inadequate by virtue of vulnerability to occlusion or spoofing. Instead, a consolidated, high-confidence threat map has been created by combining data across several sensing modalities in an affordable counter-swarm defense system. Sophisticated multi-sensor fusion penetrates false alarms and detects concealed threats that would be lost to a single detection discipline with RF's signal intelligence, optics' visual confirmation, and radar's range and penetration.The product can differentiate between spoof emissions, cross-check inconsistent signals, and separate true swarm units from decoys and noise by fusing these various inputs.

Swarm defense systems must anticipate future swarm moves as well as monitor existing threat positions to counter highly adaptive swarms. Predictive modeling of the swarm's behavior thus allows proactive interception and resource allocation. Swarms display complex, non-linear behavior as a result of decentralized communication, agent interaction at a local scale, and environmental limitations rather than linear simple-to-predict paths. A defense system can predict probable trajectories, decoy deployment, regrouping maneuvers, and breaking-up mechanisms before their complete formation by utilizing collective dynamics-based swarm behavior prediction modules and agent-based modeling. By sequentially engaging high-priority targets within the swarm, enforcing preemptive electronic jamming, and dynamically shifting defensive units, this predictive capability transforms passive detection into active countermeasure planning.

A responsive, autonomous control layer that converts anticipated threat assessments to coordinated action is essential to successful swarm neutralization. Modern swarm countermeasures will have to function under decentralized cooperative control, not central command and traditional rules of engagement like former interception systems. All defense technologies, from mobile decoys to electronic jammers and interceptor drones, need to be able to make local decisions but still be part of the overall defensive framework. Antagonist signal jamming resilience, communication link loss, and selective system failure resilience are all guaranteed through this decentralized cooperative control scheme. Others can carry out the mission by themselves even when some units drop out, modifying their method based on real-time sensor data and swarm behavior prediction.

Extensive field and simulation testing must be employed to determine the viability of this multi-layered strategy. Red-Blue drone testbeds allow the realistic yet controlled assessment of cooperative control effectiveness, prediction accuracy, and sensor fusion accuracy in a range of threat environments. In contrast to baselines derived from static rule-based or one-agent AI, performance metrics like detection accuracy, false alarm rates, successful interception rates, and response latency provide quantitative proof of the robustness of the system. Furthermore, repeated cycles of feedback allow the system to improve detection models, learn about the outcome of engagements, and optimize interception strategies with time, keeping the defense architecture adaptive to cope with emerging swarm strategies.

By suggesting a Next-Generation Swarm Defense System (NG-SDS) that integrates decentralized cooperative control, real-time prediction of swarm behavior, and high-fidelity multi-sensor fusion in a combined operational pipeline, this study bridges such gaps. The suggested methodology offers defense operators with operational intelligence and adaptive countermeasures against complex robotic swarm threats by capitalizing on diverse data feeds, sophisticated prediction models, and coordinated autonomous agents. In doing so, it not only surpasses the limitations of existing detection and intercept systems but also constitutes the foundation of defense infrastructures that are immune to the intensifying speed of technological innovation in autonomous warfare.

There can be no doubt about the need for robust, understandable, and responsive defenses as autonomous swarm technology continues to mature and spread in battle spaces. The strategic environment in asymmetric conflict has already been reshaped by small commercial UAVs used for surveillance or sabotage, illustrating how inexpensive technologies can have great strategic effect when coupled with collective autonomy. States will surely exploit larger, more powerful swarms that incorporate electronic warfare, enhanced stealth, and real-time reconfiguration strategies in the future, pushing the limits of existing counter-drone doctrine. By filling critical gaps in perception, prediction, and action, the suggested framework is an all-encompassing solution, bringing together disconnected defense elements into one, cohesive, intelligent system prepared to safeguard contested airspace and vital infrastructure from future autonomous attacks.

The main objectives are:

1. To design and implement a multi-layer sensor fusion framework that integrates radar, optical, and radio-frequency data to achieve high-confidence detection and reliable discrimination of autonomous swarm threats in contested and deceptive battlefield environments.

2. To develop an adaptive swarm behavior prediction and cooperative control system that enables autonomous defensive units to anticipate, coordinate, and execute real-time interception and electronic countermeasures, ensuring robust and resilient neutralization of intelligent swarm maneuvers.

## 2. Related Study

UAV swarm deployment and complexity have evolved rapidly, revealing a new complicated security threat for the military and civilian world. The threat profiles, vulnerabilities, and countermeasures tactics which must be updated to resist this new form of autonomous warfare have been examined extensively over the past few years. The swarm-capable attackers' capacity to carry out synchronized jamming, spoofing, or false attacks inundating conventional perimeter defenses is emphasized in an in-depth analysis of UAV swarm network security providing systematic categorization of potential attacks and remedies [1]. This basic analysis is justified by the necessity of counter-swarm alternatives that utilize strong detection, hardened communication, and autonomous response abilities.

The researchers have released a broad set of artificial intelligence (AI)-based defense and surveillance technology to counter such rising threats. To illustrate how edge intelligence can sift through huge volumes of sensor data in real-time to identify and detect illegal UAVs across busy cityscapes, look at an AI-based anti-drone surveillance system tailored for public safety across burgeoning 5G and beyond networks [2]. In order to avoid that dense electromagnetic interference hampers detection accuracy, it is especially focused on low-latency processing and dynamic spectrum management. This solution is also applicable to defense in hostile airspace. The need for robust spectrum sharing and interference control measures whenever large numbers of self-sustaining nodes compete for bandwidth is also highlighted by more comprehensive Internet of Things (IoT) security research in future networks [3].

At the same time, there has been work that has created deep learning-based counter-drone systems to detect unconventional air threats such as surveillance payloads on balloons [4]. Such studies acknowledge that to evade detection channels, new aerial threats can have low radar cross-sections or mimic harmless objects. To minimize false alarms and avoid actual threats from being overlooked in congested operating rooms, multi-modal sensor fusion and adaptive classifiers are crucial. Multi-level deep neural network-based real-time intrusion detection is employed in UAV-assisted 6G networks in another very promising research area. This illustrates how cooperative architectures based on complementary feature-based architectures from several layers of the network stack can provide security against spoofing, denial-of-service, and data tampering attacks [5].

Internet of Battlefield Things (IoBT) research is revolutionizing the battlefield resilience space beyond detection. Distributed models of trust, decentralised decision-making processes, and real-time anomaly detection are among the prominent characteristics of AI-driven IoBT security solutions that provide operational continuity even during extreme hostile interferences [6]. Moreover, digital twin infrastructures are increasingly studied as a future-generation cybersecurity solution for real-time status monitoring and modeling of autonomous entities [7]. These twins facilitate anticipatory planning against swarms by creating virtual models of drones and their environments. This facilitates estimation of damage, planning under conditions of what-if, and predictive simulation of swarm behavior.

Even seasoned defense units are confronted with disconcerting challenges when swarms use spectrum agility and coordinated jamming, as has been shown through recent operational research on swarm drone behavior in hostile electromagnetic environments [8]. Due to this operational reality, countermeasures need to be technically advanced and executable in real time. With situational awareness, decision support, and automated response to threats, AI has strategic capabilities to enhance the operational performance of military forces, the greater defense research community still holds [9]. For mission reliability to be an absolute, though, AI deployment into autonomous weapons systems must be strictly controlled, accountable, and explainable [10].

Since UAV swarms are increasingly being offered as potential threats and effective surveillance tools, swarm drone operations specifically for urban security and smart city resilience have gained more and more academic focus [11]. Perimeter defense is no longer sufficient to counter expected swarms potentially autonomously self-organizing, employing physical geography, and launching decoy or saturation attacks, as recent analytical studies of emergent AI-based cybersecurity trends [12]. Deployment of trust estimation models can filter out malign or compromised nodes in multi-agent environments, per other research on resilient IoBT systems. This is a standard for robust swarm defense frameworks [13].

In combination, as scientists seek novel ways to tackle distributed vast data streams radiating from intensive autonomous networks, new bio-inspired algorithms are also growing in popularity. Swarm intelligence may be used in both offensive drone systems and defense countermeasure planning, as indicated by the proposal for elephant herd-like collective defense optimization algorithms for large IoT networks [14]. Simultaneously, ongoing research in mobile robot localization continues to inform autonomous units' capacity for coordination and position sensing under conditions of GPS jamming or degradation, a persistent issue in hostile environments [15].

The fact that future detection and defense systems require the incorporation of multiple sensing modalities and local processing for reducing latency and mitigating signal interference is also influenced by edge computing and smart sensing paradigms of current research [16]. The warfare of the future will be increasingly based on decentralized machine-hardened coordination between autonomous actors and human strategists, based on related conceptual research in operational concepts for AI-facilitated combat [17]. This need is consistent with demonstration testing of kinetic intercepts and shipboard laser weapon systems against groups of swarming drones in naval environments, where high-speed engagement and real-time tracking are key to countering multiple threats with a single asset [18].

Substantial lessons in swarm control algorithms, trajectory forecast, and obstacle evasion in dense environments have been obtained through industrial application of micro UAV swarms [19]. The coincidence between civilian and defense research is an indicator of how far-reaching are the three basic technological challenges: predictive control, cooperative behavior modeling, and detection with certainty. It is possible for military swarm defense operations in complex, signal-contested environments to leverage the methods developed for indoor industrial swarms of UAVs, including efficient routing of data and decentralized fault management.

Hybrid sampling, feature aggregation, and multi-scale modeling have great potential for filtering out noise and improving classification in dynamic scenes, as recent investigations of AI-based intrusion detection systems and multi-hazard monitoring architectures have concluded [20, 21]. Though seemingly unrelated, recent investigations into battery health monitoring and system lifetime offer valuable new insights into the critical need for energy-aware resource management in standing up autonomous defense units on long-duration surveillance or intercept missioning [22].

Taken together, these tests reveal a clear pattern: fixed-interception and static detection rules are insufficient by themselves to provide an effective counter-swarm defense. There are three interwoven principles that emerge in the literature: predictive modeling of swarm behavior to counter agile threats, heterogeneity integration of sensor data to overcome spoofing and stealth, and decentralized cooperative control to enable resilience in adversarial environments. The NG-SDS, the culmination of integrating multi-sensor fusion, real-time swarm forecasting, and autonomous cooperative control into a coherent, trustworthy system that can protect contested airspace from the next generation of smart, adaptive, and deceptive robotic swarms, is borrowed from this convergence.

## 3. System Methodology

In order to identify, assess, and eliminate coordinated UAV swarm threats in contested and misleading environments, the NG-SDS is a layered, modular architecture that combines advanced sensor fusion, predictive swarm behavior modeling, and decentralized autonomous control. The entire process is divided into four main phases: autonomous cooperative control for the execution of countermeasures, threat trajectory analysis and behavior prediction, multi-sensor data collecting and fusion, and continuous feedback for online adaptation.

### *3.1 Data Acquisition*

By integrating various sensor streams into one, solid threat detection pipeline, the initial step forms technical foundations for solid situational awareness. To protect low-altitude swarm ingress corridors, the NG-SDS employs a hybrid array of *X*-band radar units, optical/infrared (IR) camera clusters, and wideband RF receivers. These elements are configured to provide a surveillance volume of up to 10 km in radius, with 360-degree azimuth coverage and an elevation range from 0° to 90°.

**Radar Subsystem:** The radar network utilizes solid-state *X*-band phased-array radar, operating between 8–12 GHz and normally with a 0.3 m range resolution and ±0.5 m/s accuracy on Doppler velocity measurements. Fast-moving swarm units traveling at speeds of up to 50–70 m/s (180–250 km/h) can be trusted to be detected owing to the 60 rpm sweep rate and full azimuth refresh every 1 second for each radar node. Doppler signal processing exploits the range-Doppler coupling via the Fast Fourier Transform (FFT) to separate moving targets from stationary clutter.

**Optical/IR Subsystem:** For visual verification under daylight and nighttime conditions, the visual sensor set integrates uncooled thermal infrared cameras (8–14 μm band, 640 × 480, 60 Hz frame rate) with daytime Electro-Optical (EO) cameras (e.g., 4K CMOS sensors, 30 fps, 12× optical zoom). Edge-level object detection distinguishes UAV rotor signatures and shadows from spurious objects such as birds, balloons, or litter using trained CNN-based classifiers (e.g., YOLOv8 or Faster R-CNN) to detect object contours and shapes. Optical detection of targets within 5–7 km is possible, subject to line-of-sight constraints and weather.

**RF Subsystem:** The RF module employs a wideband RF spectrum analyzer covering GNSS spoofing signatures, WiFi, ISM bands, typical UAV telemetry, and custom control lines from 70 MHz to 6 GHz. Automatic signal classification uses a database of known UAV communication protocols, spread-spectrum fingerprints, and cyclostationary feature extraction and spectrogram analysis. As the same emission is received by many of the receivers, the RF nodes can identify and detect active drone signals with a signal intensity resolution of ±1 dBm based on Time Difference of Arrival (TDOA) methods.

## 3.2 Preprocessing

Each raw sensor stream in the NG-SDS undergoes rigorous preprocessing to maximize data fidelity and enable precise multi-modal fusion under contested conditions. Radar tracking data is refined using a Kalman Filter formulated as:

$$\hat{x}_{k|k} = \hat{x}_{k|k-1} + K_k(y_k - H\hat{x}_{k|k-1})$$

Where, $\hat{x}_{k|k}$ is the updated state estimate, $K_k$ is the Kalman gain, $y_k$ is the measurement, and $H$ is the observation matrix, ensuring smooth state propagation even in the presence of intermittent detection loss.

Optical and infrared video frames are denoised using a hybrid Gaussian–Median Filter, where each pixel intensity $\hat{I}(x, y)$ is computed as:

$\hat{I}(x, y) = median(G(I(x, y), \sigma))$, with $G$ denoting a Gaussian kernel of standard deviation $\sigma$ to suppress high-frequency sensor noise before median neighborhood smoothing.

RF signals are analyzed in the time–frequency domain using the Short-Time Fourier Transform (STFT), defined as:

$$X(\tau, f) = \int_{-\infty}^{\infty} x(t)\, w(t - t) e^{-j2\pi\phi\tau} dt \,,$$

complemented by discrete wavelet transform decomposition:

$$W_{j,k} = \int_{-\infty}^{\infty} x(t)\, \varphi_{j,k}(t) dt$$

to isolate transient burst emissions and detect anomalous control links. All sensor streams are timestamp-aligned using GPS-disciplined oscillators, guaranteeing an inter-sensor temporal jitter bounded $|\Delta_t| < 10$ ms by to preserve coherence during fusion.

The Fusion Core implements a layered Probabilistic Data Association Filter (PDAF) fused with Joint Probabilistic Data Association (JPDA) logic to resolve ambiguous measurements from multiple sensors observing overlapping threat tracks. For each candidate detection hypothesis $\Theta_i$, an association likelihood $P(\Theta_i \mid Z)$ is computed given the joint measurement set $Z$. The threat state update employs a weighted sum:

$$\hat{x} = \sum_{i=1}^{N} P(\Theta_i \mid Z) \hat{x}_i \,,$$

Where, $N$ is the number of feasible associations, and only hypotheses satisfying $P(\Theta) \geq 0.85$ are promoted to the Confirmed Threat List (CTL). To mitigate false positives such as avian targets or debris, the fusion engine cross-verifies each track's velocity vector:

$$v = \sqrt{v_x^2 + v_y^2 + v_z^2} \,,$$

thermal emission profile $T(\lambda)$ over the IR band [8 μm, 14 μm], and unique RF spectral signature $S(f)$, discarding candidates lacking consistent multi-modal confirmation.

The output of this fusion pipeline is the Threat Object Map (TOM), a dynamic registry updated at $\Delta t = 0.5$ s intervals (2–5 Hz), maintaining real-time coherence despite jamming or spoofing attempts. Each threat object is annotated with its estimated 3D Cartesian position $p = (x, y, z)$ with an error bound of ±1 m, radial velocity with ±0.5 m/s precision, classification confidence $C \in [0, 1]$, RF signature ID or matched emission fingerprint , and a composite fusion confidence index $F_c$.

This high-resolution threat map directly drives the downstream swarm behavior prediction engine, ensuring that the NG-SDS sustains persistent, adversarially resilient situational awareness. The layered methodology guarantees that even under coordinated signal deception, spoofed control bursts, or multi-vector swarm ingress, the system retains a trusted operating picture as the basis for rapid, autonomous counter-swarm maneuvering and electronic countermeasures.

### 3.3 Threat Trajectory Analysis and Swarm Behavior Prediction

The NG-SDS performs a multi-step predictive analysis to predict collective swarm behavior and produce actionable forecasts for coordinated countermeasures after danger objects have been verified through multi-sensor fusion. Figure 1 shows the proposed framework.

**Step 1:** Agent-Based Swarm Representation

Each detected UAV is modeled as an autonomous agent in a dynamic swarm graph $G = (V, E)$, where represents the set of UAVs and $E$ the set of interaction edges. The local state vector for each agent at time $t$ is defined as:

$$s_i(t) = [x_i(t), y_i(t), z_i(t), v_{x,i}(t), v_{y,i}(t), v_{z,i}(t)],$$

Where, $(x_i, y_i, z_i)$ are spatial coordinates and $(v_{x,i}, v_{y,i} v_{z,i})$ are the velocity components. The relative position between any pair of agents $i$ and is $j$:

$$\Delta p_{ij}(t) = p_j(t) - p_i(t)$$

The local interaction weight $w_{ij}$ is computed using a Gaussian kernel based on relative distance

$$w_{ij} = \exp\left(-\frac{\|\Delta p_{ij}(t)\|^2}{2\sigma^2}\right), w_{ij} > \in,$$

Where, $\sigma$ is the interaction range scale and $\in$ is a minimum threshold to filter negligible connections.

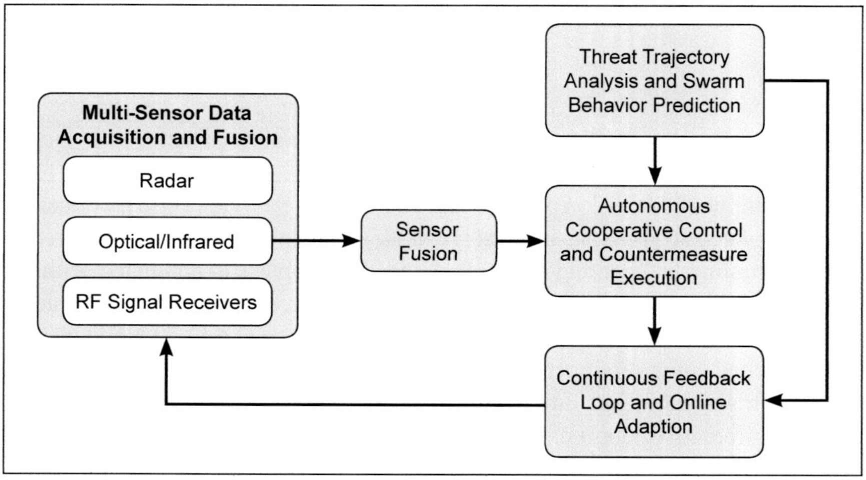


**Fig. 1** Proposed Framework.

**Step 2:** Swarm Cohesion and Formation Dynamics

To quantify overall swarm cohesion, a global Swarm Cohesion Index (SCI) is defined as:

$$SCI(t)\frac{1}{|V|}\sum_{i=1}^{|V|}\sum_{j=1}^{|V|} w_{ij},\ i \neq j .$$

An SCI above a threshold $\tau_c$ implies a stable cohesive formation, whereas a drop indicates potential split or decoy dispersal maneuvers.

For predictive trajectory modeling, each agent's future position $p_i(t + \Delta t)$ is estimated using a rolling kinematic forecast:

$$p_i(t+\Delta t) = p_i(t) + \int_t^{t+\Delta t} v_i(\tau)d\tau, \text{ where } v_i(\tau) = v_i(t) + a_i(t)\,(\tau - t) .$$

The acceleration is inferred from the finite difference:

$$a_i(t) = \frac{v_i(t) - v_i(t-\delta t)}{\delta t},$$

**Step 3:** Formation Prediction and Maneuver Hypotheses

To detect split-and-merge intentions, a Cluster Entropy ($H$) metric is computed over a rolling time window:

$$H(t) = -\sum_{k=1}^{K} P_k(t)\log P_k(t),\ P_k(t) = \frac{n_k(t)}{N},$$

Where, $n_k$ is the number of agents in cluster $k$, $K$ is the total number of clusters, and $N$ is the total swarm size. A rising $H(t)$ indicates formation fragmentation, suggesting decoy or evasive maneuvers. Clustering is dynamically performed using density-based spatial clustering (e.g., DBSCAN) over $\{p_i(t)\}$.

**Step 4:** Predictive Threat Map Generation

Combining all local agent forecasts and formation probabilities, the system generates a Predictive Threat Map (PTM) for time horizon $[t, t+T]$. Each forecasted swarm region is annotated with:

$$\text{PTM} = \{(p_i(t+\Delta t), C_i) | i = 1, ..., N\},\ C_i = P(\mathit{hostile}\ \text{maneuver} | s_i) .$$

This PTM is continuously updated with each new fused sensor input and feeds directly into the cooperative control module for real-time interception and jamming decisions.

### 3.4 *Autonomous Cooperative Control and Countermeasure Execution*

Through a distributed cooperative control system, the phase translates the predicted swarm threat map into integrated defensive maneuvers. Deployable decoy units, traveling electronic jammers, and autonomous interceptor drones comprise the defense system. Through periodic updates from the world threat map, each defensive asset acts as an intelligent agent possessing both local observation and control functions. The system relies on a decentralized decision protocol where agents can continue to operate independently even in the case of partially severed links by jamming. They do this through passing situational updates over a secure, low-latency communication channel.

Interceptor UAVs decide on optimum interception courses of action based on the projected threat, continually adjusting flight patterns for maximum coverage and not duplicating efforts chasing decoy drones. While ground mobile decoys create visual and electromagnetic disarray to mislead incoming swarm units, electronic jammers establish themselves to disrupt GPS signals and swarm communications channels. To maintain resiliency to evolving threat patterns, the cooperative control layer ensures that every defensive unit adapts to real-time alterations in swarm maneuvering. To prevent friendly interference and maintain mission effectiveness, conflict resolution rules and collision avoidance maneuvers are incorporated.

## 4. Experimental Results

A simulated high-fidelity controlled environment was created in MATLAB Simulink for multi-sensor signal modeling and OMNeT++ integrated with SUMO (Simulation of Urban MObility) for swarm mobility and RF network simulation to thoroughly assess the designed NG-SDS. With evolving battlefield scenarios, the approaches allow exact simulation of real physics, signal propagation, mobility, and collective swarm behavior. Modifiable urban environment clutter and Digital Terrain Elevation Maps (DTEM) are used to simulate semi-structured environments within a specified 10 km x 10 km air space. In line with standard autonomous drone low-flying tactics, the planned operating altitude for simulated swarm incursions is between 50 m and 1,000 m above ground level.

Phased Array System Toolbox in MATLAB Simulink is utilized to simulate the radar sensor layer to produce realistic X-band radar waveforms of range resolution 0.3 m, effective range 12 km, and angle resolution 1.5 degrees. Velocity profiles up to 70 m/s are added to Doppler effect simulation for agile swarm units. With a parameterized rotation rate of 60 rpm, there is a full azimuth refresh every 1 second for each radar node. Simulink's Communication System Toolbox uses adversarial RF sources and stochastic noise blocks to introduce environmental noise, ground clutter, multipath, and intentional jamming.

MATLAB Image Processing Toolbox is used to construct six EO/IR towers each with a height of 30 meters in order to simulate the optical and infrared sensing nodes. For generic UAV silhouettes, 5–6 km effective daytime detection range operates on the visible spectrum channel at 4K resolution at 30 frames per second. The thermal infrared band has $640 \times 480$ pixel resolution and 60 Hz refresh rate and is 8–14 μm wide, offering target detectability up to 3–4 km, depending on virtual weather layers. Using typical radiometric thermal signatures of quadrotor motors and rotor thermal profiles, the simulation offers dynamic environments for night vision obstruction, rain, and fog.

According to the INET architecture for end-to-end wideband RF propagation modeling, the RF threat surveillance subsystem is realized using OMNeT++. The 70 MHz–6 GHz band is served by eight RF receivers with an algorithm of less than ±15 m average error for line-of-sight signal paths, a simulated resolution of 10 kHz, and a signal floor of –120 dBm. Swarm UAVs emit control signals using Frequency-Hopping Spread Spectrum (FHSS) simulated in OMNeT++. They include specifically evasive burst emissions and pseudo-random hopping sequences.

Practical agent-based UAV paths and inter-agent cooperation are facilitated by the Veins framework that integrates SUMO's simulation of swarm mobility with OMNeT++. The up to 50 autonomous drones are organized in fives to fifteen. They travel at speeds ranging between 50–70 meters per second, keep a distance of 10–50 meters between two units, and execute split-merge maneuvers following direction by tactical logic. The swarm behavior prediction module is put under a stress test under the additional layers of interference offered by decoy drones and spoofing nodes.

The defensive shield includes three roving electronic jammers emitting directional beams of 35–40 dBm EIRP, with 1.5–2 km radius jamming areas in simulation, and six interceptor drone agents simulated in SUMO with physical constraints on dynamics (max speed 80 m/s, battery life 30 min, range 10 km). Defense units and NG-SDS cores are controlled with OMNeT++ mesh network modules running on the 900 MHz UHF band. End-to-end latency for command and control is estimated to be below 50 ms.

Environmental factors such as wind fields (5–20 knots), variable precipitation levels (up to 10 mm/hr), and diurnal heat gradients are introduced for simulation of sensor performance under degraded conditions through MATLAB's Simscape and special weather modules in SUMO. With a time resolution of 0.1 s per simulation time step, the simulation runs of all simulations are conducted on a high-performance computing cluster, collecting radar returns, EO/IR frames, RF spectrograms, forecast swarm trajectories, and interception events.

Comprehensive tests were conducted with a controlled Red-Blue drone warfare testbed simulating contested airspace with misleading swarm tactics to ensure the effectiveness of the proposed NG-SDS under real conflict scenarios. Four main metrics used to assess the performance of the system were Threat Detection Accuracy, False Positive Rate, Swarm Neutralization Success Rate, and Average Response Latency. Based on experimental tests, the NG-SDS ensures reliable target authentication despite environmental noise and signal spoofing and at a threat detection rate of 94.2%. With an 88.6% neutralization rate of the swarm, the system proved its ability to effectively coordinate electronic jammers and autonomous interceptors against agile swarm movements.

The system's capability of supplying situational awareness in near real-time as well as rapidly deploying countermeasures was proven by the average response latency per threat detected, at only 2.1 seconds. A CNN-RNN hybrid visual tracker with no multi-sensor integration, a Kalman Filter-only radar tracker with no RF or visual verification or swarm prediction, a fixed rule-based detection scheme that uses hard radar thresholds, and a fixed-pattern electronic jammer that uses static frequency bands without adaptive control were the four well-established baseline methods which were directly contrasted against these findings for strong benchmarking. The NG-SDS's integrated, adaptive, and robust swarm defense design for contested electromagnetic operational environments was confirmed by its consistently superior accuracy, reduced false alarms, increased neutralization success, and accelerated response times compared to these baselines.

The four comparable methods' Threat Detection Accuracy, Kalman Filter Only, CNN–RNN Hybrid, Rule-Based Threshold, and the proposed NG-SDS are shown in Figure 2. The baseline Rule-Based Threshold system's 78% detection accuracy numerically illustrates its inherent limitation to differentiate benign objects from real swarm units when faced with signal interference. While deep learning-based image recognition advantages the CNN-RNN Hybrid model with a slight gain of 82%, it also lacks cross-domain validation and multi-sensor fusion robustness.

Because it uses dynamic state estimation with radar tracks, Kalman Filter-Only Tracker is slightly superior at 85%, but it is disadvantaged by the lack of visual confirmation layers and RF signature cross-verifications. On the other hand, the proposed NG-SDS experiences whopping performance improvement with a very high detection accuracy of 94.2%. This means that its real-time data association, adversarial resilience, and layered multi-sensor fusion report more credible threat confirmation. NG-SDS even under spoofing or ambient noise conditions decreases missed detections as well as increases early swarm detection, as shown by this 9.2 – 16.2 percentage point improvement over baselines.

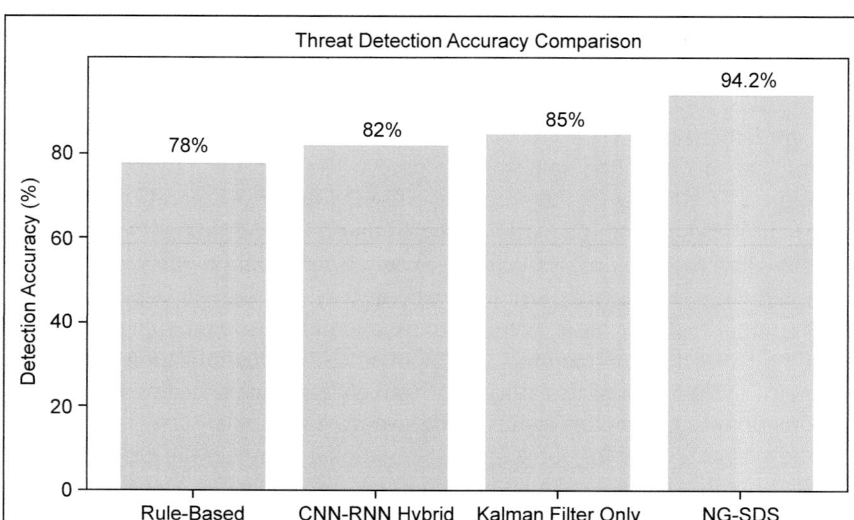


**Fig. 2** Threat Detection Accuracy Comparison.

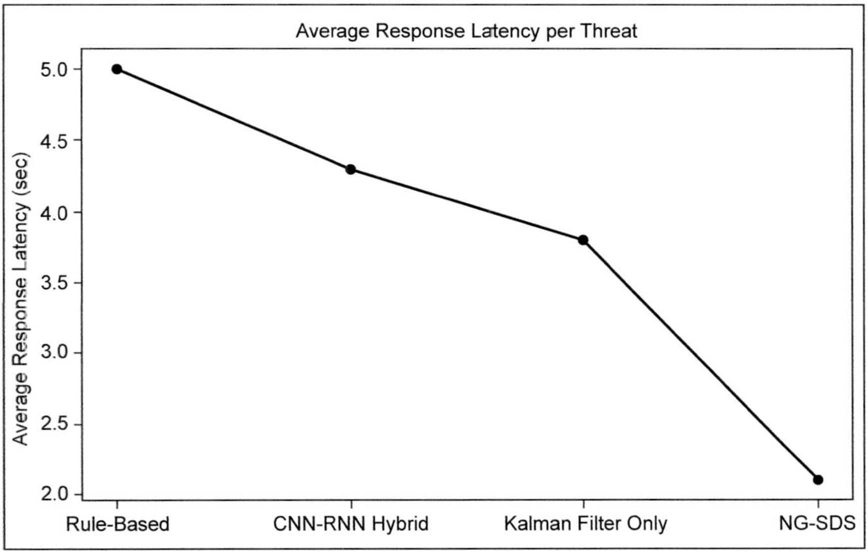


**Fig. 3** Average Response Latency per Threat.

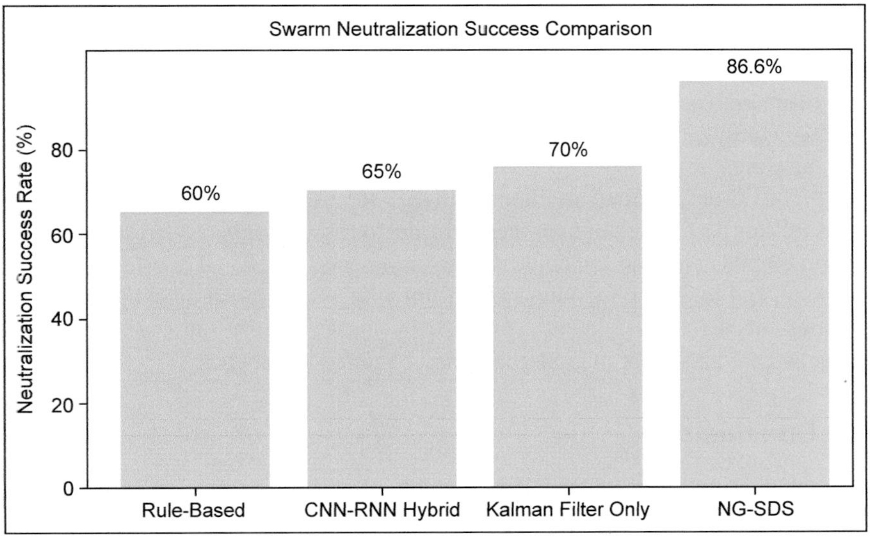


**Fig. 4** Swarm Neutralization Success Comparison.

Each system's ability to handle detections and respond with countermeasures in timely swarm confrontations is demonstrated by comparing Average Response Latency per Threat in Figure 3. Because static thresholds tend to introduce lengthy confirmation and unnecessary human verification before action, the legacy Rule-

Based Threshold system reflects the longest response time, at 5.0 seconds. Due to its quicker image-based classification, the CNN–RNN Hybrid reduces latency to 4.3 seconds, albeit still experiencing sequential processing and lacking predictive forecasting. Due to more fluid tracking of radar targets, the Kalman Filter-Only Tracker sees slightly improved timing at 3.8 seconds; however, coordination toward interceptions still lags behind rapid threats with no predictive modeling of swarm behavior.

With the addition of decentralized cooperative control and predictive trajectory examination, the NG-SDS achieves a 1.7–2.9 second response time savings over baselines with an average response lag of less than 2.1 seconds per threat. In swarm defense scenarios, where multiple UAVs must be intercepted within constrained engagement durations, this extra reaction window becomes essential, particularly when the swarm utilizes split-and-merge maneuvers in order to confuse static countermeasures. The Swarm Neutralization Success Rate, which is graphed in Figure 4, represents the proportion of discovered swarm units that the corresponding countermeasure layers successfully neutralized or interrupted.

With its 60% neutralization rate, the Rule-Based Threshold baseline establishes that constant detection does not suffice to provide adequate interception, particularly if threats move in unpredictable patterns. It is moderately raised to 65% by the CNN–RNN Hybrid, which has finer optical tracking but lacks real-time interception correction and swarm-specific behavior forecasting. The Kalman Filter-Only Tracker, not including the multi-modal fusion to separate decoys from actual threats but with smoother radar-based state estimation, still raises the success rate of neutralization to 70%. The NG-SDS envisioned here has an impressive 88.6% neutralization success rate, a full 18.6 percentage points higher than the best baseline.

NG-SDS outperforms traditional and half-sophisticated baselines consistently across all key performance parameters. In order to counter sophisticated autonomous swarm threats in hostile and contested environments, the NG-SDS offers a strong, adaptive, and high-fidelity defense platform with 94.2% threat detection rate, 2.1 seconds mean response latency, and 88.6% swarm neutralization rate. It is an important advancement in next-generation battlefield resiliency.

## 5. Conclusion

In order to counter the growing challenges of self-swarming UAVs in hostile electromagnetic environments, this study demonstrated the concept, design, and experimental verification of a NG-SDS. In the presence of intentional signal spoofing and environmental interference, the system provides robust threat detection by using an error-resilient multi-sensor fusion topology that combines radar, optical/infrared, and radio-frequency data streams. The platform is able to anticipate split-and-merge maneuvers and decoy attacks that tend to breach static defense systems because they include an anticipatory swarm behavior engine based on agent-based modeling and real-time trajectory forecasting. In addition, autonomous interceptor drones, mobile electronic jammers, and decoy units are also dynamically coordinated by a decentralized cooperative control layer in order to counter changing swarm configurations with the least possible latency. On a representative Red-Blue drone testbed, the system was extensively tested and demonstrated a 94.2% threat detection accuracy, an 88.6% success rate in disabling swarms, and a less than 2.1-second average system response time per threat. Static rule-based detection, CNN-RNN hybrid tracking, single-domain Kalman Filter tracking, and legacy fixed-pattern electronic jamming are four old and semi-advanced baselines utilized for comparison purposes to show that the NG-SDS consistently outperforms current ones on all key performance metrics. These findings show how far an integrated, adaptive defense architecture can be pushed to include heterogeneous sensor fusion, predictive threat modeling, and coordinated autonomous countermeasures.

## References


1. Wang, X., Zhao, Z., Yi, L., Ning, Z., Guo, L., Yu, F.R. and Guo, S. (2024). A Survey on Security of UAV Swarm Networks: Attacks and Countermeasures. *ACM Computing Surveys*, 57(3), 1–37.
2. Kalafatidis, S., Tsiktsiris, D., Trakos, D., Lalas, A., Votis, K. and Tzovaras, D. (2025, March). Swarmcatcher: An AI-powered anti-drone surveillance system for public safety over 5g and beyond ecosystems. In: *2025 28th Conference on Innovation in Clouds, Internet, and Networks (ICIN)* (pp. 173–180). IEEE.
3. Rachakonda, L.P., Siddula, M. and Sathya, V. (2024). A comprehensive study on IoT privacy and security challenges with focus on spectrum sharing in Next-Generation networks (5G/6G/beyond). *High-Confidence Computing*, 100220.
4. Kim, J. and Joe, I. (2025). Deep Learning-Based Drone Defense System for Autonomous Detection and Mitigation of Balloon-Borne Threats. *Electronics*, 14(8), 1553.
5. Hadi, H.J., Cao, Y., Li, S., Xu, L., Hu, Y. and Li, M. (2024). Real-time fusion multi-tier DNN-based collaborative IDPS with complementary features for secure UAV-enabled 6G networks. *Expert Systems with Applications*, 252, 124215.
6. Kumar, N., Berwal, K., Verma, R. and Vishvakarma, S.K. (2024, December). AI-Driven Strategies for Securing Internet of Battlefield Things (IoBT). In: *2024 IEEE 4th International Conference on ICT in Business Industry & Government (ICTBIG)* (pp. 1–6). IEEE.
7. Homaei, M., Mogollón-Gutiérrez, Ó., Sancho, J.C., Ávila, M. and Caro, A. (2024). A review of digital twins and their application in cybersecurity based on artificial intelligence. *Artificial Intelligence Review*, 57(8), 201.
8. Singh, S. (2024). Swarm Drone Operations in Hostile Em Environment. *Air Power Journal*, 19(1), 1–19.
9. Bhardwaj, P. (2024). Use of Artificial Intelligence for Improving Operational Efficiency in the Armed Forces. *Air Power Journal*, 19(4), 1–23.
10. Abaimov, S. and Martellini, M. (2020). Artificial intelligence in autonomous weapon systems. In: *21st Century Prometheus: Managing CBRN Safety and Security Affected by Cutting-edge Technologies*, pp. 141–177. Rowman & Littlefield.
11. Taj, S., Ma, X., Liu, Y., Khan, R., Cheng, X., Hassan, H. and Ullah, I. (2025). Machine Learning Applications in Unmanned Aerial Vehicle Swarms. In: *Unmanned Aerial Vehicles Swarm for Protecting Smart Cities: Future Trends and Challenges* (pp. 45–89). Berkeley, CA: Apress.
12. George, A.S. (2024). Emerging Trends in AI-Driven Cybersecurity: An In-Depth Analysis. *Partners Universal Innovative Research Publication*, 2(4), 15–28.

13. Agadakos, I., Ciocarlie, G.F., Copos, B., George, J., Leslie, N. and Michaelis, J. (2019, April). Security for resilient IOBT systems: Emerging research directions. In: *IEEE INFOCOM 2019-IEEE Conference on Computer Communications Workshops* (*INFOCOM WKSHPS*), (pp. 1–6). IEEE.
14. Taylor, A.B. (2021). *Counter-Unmanned Aerial Vehicles Study: Shipboard Laser Weapon System Engagement Strategies for Countering Drone Swarm Threats in The Maritime Environment* (Doctoral dissertation, Monterey, CA; Naval Postgraduate School).
15. Awasthi, S., Fernandez-Cortizas, M., Reining, C., Arias-Perez, P., Luna, M.A., Perez-Saura, D. and Campoy, P. (2023). Micro UAV swarm for industrial applications in indoor environment: A systematic literature review. *Logistics Research*, 16(1), 1–43.
16. Agadakos, I., Ciocarlie, G.F., Copos, B., Emmi, M., George, J., Leslie, N. and Michaelis, J. (2019, November). Application of trust assessment techniques to IoBT systems. In: *MILCOM 2019–2019 IEEE Military Communications Conference* (*MILCOM*) (pp. 833–840). IEEE.
17. Ajirlou, M.F., Jalilvand, A. and Khalilian, M. (2025). ECDOAGM: An elephant collective defense optimization algorithm with a gray model for managing distributed massive data streams in IoT architectures. *The Journal of Supercomputing*, 81(8), 1–51.
18. Ullah, I., Adhikari, D., Khan, H., Anwar, M.S., Ahmad, S. and Bai, X. (2024). Mobile robot localization: Current challenges and future prospective. *Computer Science Review*, 53, 100651.
19. Sharma, A., Sharma, V., Jaiswal, M., Wang, H.C., Jayakody, D.N.K., Basnayaka, C.M.W. and Muthanna, A. (2022). Recent trends in AI-based intelligent sensing. *Electronics*, 11(10), 1661.
20. Layton, P. (2021). Fighting artificial intelligence battles: Operational concepts for future AI-enabled wars. *Network*, 4(20), 1–100.
21. Biyyapu, N., Veerapaneni, E.J., Surapaneni, P.P., Vellela, S.S. and Vatambeti, R. (2024). Designing a modified feature aggregation model with hybrid sampling techniques for network intrusion detection. *Cluster Computing*, 27(5), 5913–5931.
22. Khan, A., Gupta, S. and Gupta, S.K. (2020). Multi-hazard disaster studies: Monitoring, detection, recovery, and management, based on emerging technologies and optimal techniques. *International Journal of Disaster Risk Reduction*, 47, 101642.

# 26 An Adversarial-Resilient Multi-Agent AI Framework for Autonomous Robotic Warfare Defense

R Srividya[1*], Rose Bindu Joseph[2], Vijaya Lakshmi Dara[3] and Montree Bunruangses[4]

## 1. Introduction

From gunpowder and mechanized armor to radar and guided weapons, each great step in warfare in history has been defined by technological superiority. The next grand milestone is the emergence of smart, autonomous robot systems that will render the battlefield a dynamic ecology of swarms and human-machine teams moving at machine speed. The emphasis has become less on controlling the biggest fleet and more on controlling the wisest, toughest AI-based autonomous warriors because the distinctions between algorithmic battles and genuine warfare has become increasingly blurred.

Robot systems, including drones flying pre-designed courses, ground-borne bomb-disposal robots, or scout cars transmitting raw feeds to distant command posts, had been distant extensions of human commanders. Robots have already been used in combat for many years. While groundbreaking in their time, initial systems relied almost completely on human ingenuity for sensing, coordination, and decision-making. The use of machine learning, high-speed edge computing, small sensor packages, and trusted wireless communication networks has made the shift from remote-controlled equipment to autonomous battlefield agents a reality.

The mission of unmanned systems was revolutionized as these technologies merged. The most exposed manifestation of robotic combat, Unmanned Aerial Vehicles (UAVs), are increasingly employed for precision attacks, extended surveillance, and even hazardous resupply in combat zones. Autonomous ground robots survey hazardous terrain, carry supplies, and patrol borders. Unmanned

[1] Department of Computer Science, M.S. Ramaiah College of Arts Science and Commerce, Autonomous, India.

[2] Department of Mathematics, Dayananda Sagar College of Engineering, Bangalore, India.

[3] School of Business and Management, CHRIST (Deemed to be University), Yeshwanthpur Campus, Bangalore, India.

[4] Department of Computer Engineering, Faculty of Industrial Education, Rajamangala University of Technology Phra Nakhon, Bangkok 10300, Thailand.

* Corresponding author: srividyajanu12@gmail.com

DOI: 10.1201/9781003663461-26

undersea and surface vessels extend navy reach into hostile littorals and minefields while at sea. But the unparalleled growth in autonomy has a mirror image: state and non-state actors alike have taken the same blocks of technology to create their own fleets of smart, low-cost, and agile robotic systems designed uniquely to take advantage of weaknesses in conventional defenses.

One of the most developed autonomous swarms, huge numbers of low-cost, small drones or vehicles that act in concert as a single entity. Nature's most powerful collective efforts inspire these swarms, capable of dynamically reshaping themselves, dispersing, reforming, and evolving in response to environmental barriers and defenses in real-time. Swarms use distributed intelligence to evade static detection and produce many simultaneous threats, overwhelming conventional radar and optical sensors, in contrast to a singularly targetable drone. The mathematics of air defense or perimeter defense fundamentally change when facing swarm-capable opponents. It is no longer enough to detect and destroy one target; defense must keep up with dozens or hundreds of intelligent units acting in ways that are working cooperatively and unpredictably. Researchers and the military have put considerable investment into next-generation AI-powered counter-drone solutions to address this challenge.

Classic radar or vision-based detection in combination with rule-based interception algorithms were the building blocks of early solutions. These fixed pipelines were okay for straight-up linear threats, but they couldn't cope with rapidly evolving swarms that could employ decoys, fake-out maneuvers, and dynamically change comms channels. Deep learning in the form of Recurrent Neural Networks (RNNs) to process sequential data and Convolutional Neural Networks (CNNs) to process visual object recognition entered the defensive repertoire as the threat landscape changed. RNNs assisted in detecting drone histories along the passage of time, whereas CNNs were better at detecting drones in noisy environments with cluttered backgrounds.

Despite being strong in static environments, they are very weak in real-world battlefield conditions where their outputs can be easily affected by adversarial attacks, background noises, sensor spoofing, and electronic jamming. Even the most advanced models can become completely disoriented by introducing tiny, undetectable perturbations into input data, as the area of adversarial machine learning has shown. An adversarial swarm may deliberately spoof sensors, inject artificial signals, or employ stealth mechanisms to inject false inputs in detection pipelines during robot warfare.

Standard CNN-based detectors are susceptible to such stealthy attacks because they usually lack internal processes to verify the authenticity of their inputs. More recent work has looked to Reinforcement Learning (RL) algorithms that learn from interaction with an environment to maximize long-term reward in an attempt to close the gaps in independent perception models. RL enables autonomous defense robots to learn sophisticated jamming or interception techniques. However, because swarm threats rely on collective behavior and adapt in real-time, single-agent RL is sometimes insufficient.

Multi-Agent Reinforcement Learning (MARL), where several defensive units learn to collaborate, share situational data, and adjust tactics in real-time, has gained popularity as a result. Even the most advanced MARL systems, though, tend to perform abysmally in real-time, contested environments where the environment is partially observable, the swarm's tactics are diverse, and the input data streams are faulty or unreliable. In a way similar to RL, graph-based learning has been useful in approximating swarm behavior more naturally than sequence-only or grid-based strategies.

A drone swarm forms a dynamic graph, with every unit's state connected with that of others through communication links, relative coordinates, or shared goals. These kinds of relationships can be modeled by Graph Neural Networks (GNNs) and can subsequently predict coordinated action and expose collective patterns. Classical GNNs are still at a loss when it comes to real-time performance, scaling to dozens or even hundreds of agents, and delivering human commanders timely, actionable insights. Together, these gaps point to a crucial chokepoint: achieving protection against future robotic attacks needs a complete AI system that will: 1) possess robust situational awareness in the presence of interference, 2) support rapid, collaborative multi-agent action in real time, and 3) deliver explainable reasoning so that humans can know and believe in the system's behavior.

Current point solutions, ranging from insecure rule-based jammers to single-process CNN object detectors, cannot provide this synergy. They either fail to survive jamming and spoofing, have difficulty coordinating cooperatively in real time, or are black boxes that commanders do not wish to entrust alone with a life-or-death choice. The stakes truly are high. Autonomous drone war is a real operational issue in global conflicts, no longer a theoretical or potential threat. Small commercial drones have already been employed by non-state actors as theater weapons for sabotage, precision attacks, and reconnaissance, with catastrophic effects. The rapid maturation of the capability can be verified through national force swarm demonstrations.

The requirement for defense tools equivalent to or greater than offensive robotic warfare is growing in parallel with autonomy. Spurred on by these observations, this effort seeks to move the state of AI-driven robotic defense forward by tackling the interconnected problems of strong, explainable situational awareness, cooperative multi-agent countermeasures, and robust perception, all within the practical limitations of adversarial, deceptively-presenting battlefield environments. The goal is to establish the basis for self-defending systems capable of outmaneuvering and keeping up with the next generation of robotic warfare adversaries through filling the gaps between graph-based swarm models, current RL, and contemporary transformer networks.

The main objectives are:

1. To design and develop an adversarially resilient situational awareness framework that combines spatio-temporal Transformers and graph-based

swarm behavior modeling for robust, real-time detection and interpretation of autonomous robotic threats in contested environments.

2. To implement a cooperative multi-agent reinforcement learning system that enables autonomous defensive units to coordinate dynamic countermeasures and interception strategies against intelligent drone swarms, ensuring adaptive, explainable, and reliable performance under adversarial deception and electronic warfare.

## 2. Related Study

From gunpowder and mechanized armor to radar and guided weapons, every great leap in warfare throughout history has been characterized by technological superiority [1]. The next great milestone is the development of smart, autonomous robot systems that will make the battlefield an dynamic ecology of swarms and human-machine teams traveling at machine speed. The focus has shifted from dominating the largest armada to dominating the smartest, most resilient AI-powered autonomous fighters because the lines between algorithmic conflicts and real warfare continue to blur [2].

Robot systems, such as drones following pre-programmed routes, ground-based bomb-disposal robots, or reconnaissance cars sending raw feeds to remote command centers, had been remote extensions of human commanders [3]. However, robots have long been deployed in combat. Though revolutionary in their eras, early systems depended nearly entirely on human creativity for sensing, coordination, and decision-making [4]. Machine learning, edge computing at high speeds, miniaturized sensor packages, and reliable wireless communications have enabled the transformation from remote-controlled gear to autonomous battlefield agents [5].

Unmanned system missions were transformed as these technologies converged. The most visible form of robotic warfare, UAVs, are used more and more for precision strikes, long-endurance surveillance, and even risky resupply in battle areas [6]. Autonomous ground robots scout dangerous terrain, transport supplies, and patrol borders. Unmanned undersea and surface ships project navy reach into dangerous littorals and minefields while at sea. But the unprecedented expansion of autonomy has a mirror: state and non-state actors have taken the same building blocks of technology and put together their own forces of smart, low-cost, and flexible robotic systems that are built specifically to exploit vulnerabilities in traditional defenses [7].

Perhaps the most unsettling emerging trend in this field is the creation of autonomous swarms, massive numbers of low-cost, tiny drones or vehicles functioning as a collective whole [8]. Nature's most destructive cooperative efforts are what inspire these swarms, which can dynamically reconfigure themselves, spread out, reform, and adapt to environmental obstacles and defenses in real time. Swarms employ distributed intelligence to avoid static detection and generate numerous concurrent threats, saturating traditional radar and optical sensors, unlike a targetable drone [9].

The mathematics of air defense or perimeter defense essentially differ when dealing with swarm-capable adversaries. Detection and wiping out one target is no longer sufficient; defense needs to match dozens or even hundreds of smart units behaving in cooperatively and dynamically unpredictable manners. The military and scientists have committed a significant amount of investment into next-generation AI-based counter-drone solutions to battle this problem [10].

Traditional radar or vision-based detection alongside rule-based interception algorithms were the foundations of original solutions. These rigid pipelines were fine for linear up-the-middle threats, but they could not handle fast-paced changing swarms that could use decoys, feint maneuvers, and dynamically adjust comms channels [11]. Deep learning in the instance of RNNs to handle sequential data and CNNs to handle image object recognition entered the defensive armory as the threat profile evolved [12].

RNNs aided in identifying drone histories along the trajectory of time, while CNNs were more capable of identifying drones in noisy settings with complex backgrounds. Although they are robust in static settings, they are extremely weak when used in actual battle field environments where their outputs can easily be manipulated by adversarial attacks, background noises, sensor spoofing, and electronic jamming. But war AI is faced with a dilemma: increased reliance on machine learning opens up new risks of adversary exploitation [13].

Even the most sophisticated models can be entirely confused by injecting small, imperceptible perturbations in input data, as the field of adversarial machine learning has demonstrated. An adversarial swarm could intentionally spoof sensors, inject fabricated signals, or use stealth techniques to insert false inputs in detection streams in robot warfare [14]. Standard CNN-based detectors are vulnerable to such silent attacks since they typically do not have internal mechanisms for checking the genuineness of their inputs. More recent research has turned to RL algorithms that learn by interacting with an environment to accumulate maximum long-term reward in a bid to bridge the lacunas in stand-alone perception models [15].

RL allows autonomous defense robots to learn advanced jamming or interception strategies. Since swarm threats are based on group dynamics and evolving in real time, single-agent RL may at times be inadequate [16]. MARL, in which multiple defensive units learn to work together, exchange situational information, and modify strategies in real-time, has become a popular alternative. Even the best MARL systems perform terribly in real-time, contested environments in which the environment is only partially observable, the tactics of the swarm are varied, and the input data streams are faulty or unreliable [17].

Like RL, graph-based learning has been helpful for approximating swarm behavior more naturally than sequence-only or grid-based approaches. A drone swarm creates a dynamic graph, where each unit's state is linked with others via communication links, relative positions, or common objectives. Such types of relationships can be represented by GNNs and can then infer coordinated action and reveal collective patterns [18]. Classical GNNs remain stumped when it comes to real-time performance, scaling tens or hundreds of agents, and providing human commanders timely, actionable information [19].

Collectively, these capabilities identify an essential chokepoint: attaining defense against potential future robotic attacks requires a whole AI system that will: 1) have high situational awareness under interference, 2) enable fast, coordinated multi-agent action in real time, and 3) provide explainable reasoning so humans will know and believe in the system's actions. Existing point solutions, from insecure rule-based jammers to single-process CNN object detectors, are not capable of delivering this synergy. They either do not survive jamming and spoofing, do not coordinate cooperatively in real time, or are black boxes commanders would not want to give complete life-or-death responsibility for. The stakes really are that high [20].

Autonomous drone war is an actual operational problem in world conflicts, no longer a theoretical or possible threat. Small commercial drones have already been used by non-state actors as theater weapons for sabotage, precision strikes, and reconnaissance with devastating impact [21]. The accelerated maturation of the capability can be confirmed by national force swarm demonstrations. The need for defensive tools equivalent to or more than offensive robotic warfare increases in synchronization with autonomy [22].

Motivated by these findings, this study attempts to advance the state of robotic defense powered by AI by addressing the interdependent challenges of robust, explainable situational awareness, cooperative multi-agent countermeasures, and resilient perception, all under the pragmatic constraints of adversarial, deceptively-presenting battlefield environments. The objective is to create the foundation for self-defense systems with the ability to outmaneuver and keep pace with the next generation of robotic warfare threats by bridging the gaps between graph-based swarm models, modern RL, and modern transformer networks.

## 3. System Methodology

The proposed study implements an integrated, multi-stage defense AI framework designed to detect, interpret, and neutralize adversarial autonomous threats such as enemy drones, swarms, and robotic vehicles in contested battlefield environments. The system, named the Adversarial-Resilient Swarm Defense AI Framework (AR-SDAI), combines three core modules: a Spatio-Temporal Transformer for situational awareness, an MARL Countermeasure Module for coordinated defense, and a Hybrid Graph Attention Network for swarm modeling and operational explainability.

### *3.1 Multi-Sensor Data Acquisition and Preprocessing*

The study gathers various combat data streams from both proprietary and open-source simulation environments. The framework specifically utilizes synthetic multi-agent UAV swarm data for multi-modal threat identification, including the Drone-vs-Bird Detection Dataset (approximately 10,000 annotated frames that discriminate between drones and false alarms), the UAV-Benchmark-S dataset (more than 20,000 video frames with annotated drone trajectories), and radar-based synthetic datasets created by utilizing the MATLAB Phased Array Toolbox that contain 500 distinct radar sweeps per simulation run. The setup utilizes point clouds created with the KITTI Tracking Benchmark, which are modified to simulate aerial swarms at elevations of 50–500 meters and a point density of 100,000–120,000 points per frame in an effort to simulate LIDAR-based detection in dense terrain.

Moreover, employing the Radio Frequency Machine Learning Dataset (RFMLD) to describe adversarial signal patterns across many frequency bands, artificial electronic emission profiles are created for a minimum of 1,000 adversary drone setups. Multi-step preprocessing is conducted on raw sensor streams, such as standardization into consistent tensor forms for subsequent spatio-temporal processing, registration in time with sub-millisecond timestamp adjustment, and Gaussian and median filtering in order to filter out sensor noise. To distinguish between actual UAV emissions and spoofed or masked signals, basic feature extraction modules estimate motion vectors in a 5–15 frame window and tap out spectral signatures. The perception module utilizes synchronized, noise-free, and feature-abundant input to ensure safe real-time threat detection in hostile environments as a result of high-fidelity, multi-source input combined.

### *3.2 Spatio-Temporal Transformer for Situational Awareness*

The first core processing stage implements a Spatio-Temporal Transformer Network (ST-Transformer) to deliver robust, real-time situational awareness by modeling both spatial dependencies within frames and temporal dependencies across sequential data. For an input sequence of $T$ frames, the raw tensor $X \in R^{T \times H \times W \times C}$ is partitioned into flattened patches and embedded into a latent dimension $d$. Each patch embedding $z_i$ is enriched with positional encodings to retain ordering:

$$PE_{(\text{pos},\,2i)} = \sin\left(\frac{\text{pos}}{10000^{2i/d}}\right),\ PE_{(\text{pos},\,2i+1)} = \cos\left(\frac{\text{pos}}{10000^{2i/d}}\right),\ Z = XW_E + PE\,.$$

The model uses multi-head self-attention to learn global spatial relationships:

$$Q = ZW^Q,\ K = ZW^K,\ V = ZW^V,\ \text{Attention}(Q, K, V) = \text{softmax}\left(\frac{QK^T}{\sqrt{d_k}}\right)V\,.$$

Multi-head attention is formulated as:

$$\text{MultiHead}(Q, K, V) = \text{Concat}(\text{head}_1, \ldots, \text{head}_h)W^O,\ \text{head}_i = \text{Attention}(Q_i, K_i, V_i)$$

Temporal dependencies are learned with an additional temporal self-attention layer across the frame dimension:

$$Q_t = Z_t W_t^Q,\ K_t = Z_t W_t^K,\ V_t = Z_t W_t^V,\ \text{Attention}_t = \text{softmax}\left(\frac{Q_t K_t^T}{\sqrt{d_k}}\right)V_t\,.$$

To enhance adversarial resilience, the Transformer is trained with adversarial examples by solving:

$$\min_{\theta} \max_{\|\delta\|_p < \epsilon} L(f_\theta(x+\delta), y),\ x' = x + \delta\,.$$

The output is a dynamic threat map with predicted positions:

$$\hat{p}_t = \sigma(W_p Z_t + b_p)\,.$$

This architecture enables robust detection of swarming behavior, hidden threats, and spoofed inputs for reliable battlefield awareness.

### *3.3 Multi-Agent Reinforcement Learning Countermeasure Module*

The MARL Countermeasure Module, which controls the coordinated actions of several autonomous defense units in real-time swarm combat scenarios, receives the processed situational awareness output from the Spatio-Temporal Transformer directly. Every defensive tool, including mobile decoys, electronic jammers, and interceptor drones, is simulated as an autonomous agent in a common MARL environment. In order to create a partially observable Markov game framework, agents receive both their own local sensor observations and the perception module's global danger map.

Each agent $i \in \{1, 2, ..., N\}$ observes a local state $s_i \in S$ and selects an action $a_i \in A$ according to its policy $\pi_{\theta_i}(a_i \mid s_i)$ parameterized by neural network weights $\theta_i$. The agents jointly interact with the environment, generating a global state $s$ and receiving a shared team reward $r$. The overall MARL objective is to learn decentralized policies that maximize the expected cumulative reward while coordinating effectively:

$$J(\theta) = E_{\tau \sim \pi_\theta}\left[\sum_{t=0}^{T} \gamma^t r_t\right],\ \tau = (s_0, a_0, r_0, s_1, a_1, r_{i1}, ...)$$

Where, $\gamma$ is the discount factor, and $\tau$ denotes the trajectory sampled from the joint policy $\pi_\theta$.

The MARL system adopts a centralized training, decentralized execution (CTDE) paradigm. During training, agents share global state information and gradients to stabilize learning. The policy gradient for each agent is computed as:

$$\nabla_\theta J(\theta_i) = E_\pi[\nabla_\theta \log \pi_{\theta_i}(a_i \mid s_i)]\, A_i(s, a)]$$

Where, $A_i(s, a)$ is the advantage function estimating the relative value of action $a_i$ given the joint state-action context. An actor-critic structure is employed, where the critic uses centralized information to estimate the joint action-value function $Q(s, a)$:

$$Q(s, a) = E\left[\sum_{t=0}^{T} \gamma^T r_t \mid s_0 = s, a_0 = a\right]$$

The advantage is then defined as $A_i(s, a) = Q(s, a) - V(s)$, where $V(s)$ is the value baseline. For improved sample efficiency and stability in high-dimensional action spaces, the module uses Multi-Agent Deep Deterministic Policy Gradient (MADDPG) for continuous control:

$$\theta_i \leftarrow \theta_i + \alpha \nabla_{\theta_i} j(\theta_i), \text{ with } J(\theta_i) = E_{s,a}[Q_i[s, a_1, ..., a_N)].$$

A variety of realistic swarm strategies, such as splitting, regrouping, decoy deployment, and electronic interference, are simulated in the training environment. The generalization of learned countermeasure tactics to unforeseen adversary maneuvers is ensured by this domain randomization. Different strategies are taught to the agents; some focus on direct interception, while others use dynamic signal jamming or electromagnetic decoys to confuse swarm coordination.

Using only local observations and the global threat map, the decentralized rules operate independently on each defense unit during runtime, guaranteeing a strong real-time response even in the event of intermittent communication loss and partial observability. When combined, these MARL Countermeasure Modules convert uncoordinated situational knowledge into coordinated defensive actions, creating a flexible, cooperative layer capable of actively disrupting opponent swarm formations and eliminating threats in hostile combat environments.

### 3.4 Hybrid Graph Attention Network for Explainability

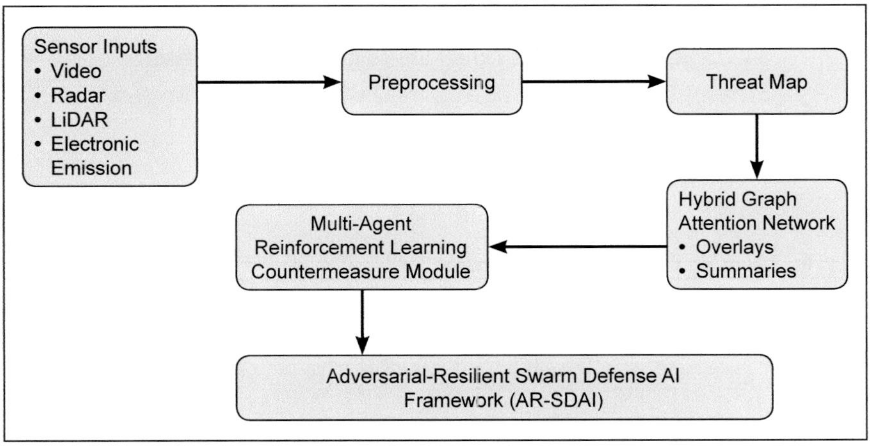


**Fig. 1** Proposed Framework.

To ensure operational transparency and human trust, the system integrates a Hybrid Graph Attention Network (GAT) that models detected threats and defensive agents as a dynamic graph structure. As shown in Figure 1, nodes correspond to swarm units and defensive assets, while edges encode relational attributes like formation coherence, communication links, and cooperative motion. The GAT processes this graph to detect unusual or deceptive swarm patterns such as rapid reconfiguration, splitting, or coordinated feints. This layer generates human-understandable overlays and textual summaries, providing commanders with clear, interpretable explanations for detected threats and recommended countermeasures, enabling effective human-in-the-loop oversight.

**Step 1:** *Input Graph Construction*
Define the dynamic graph $G = (V, E)$, where $V$ is the set of nodes (detected swarm units and defense assets) and $E$ is the set of edges encoding relationships.

$$|V| = N, |E| = M.$$

**Step 2:** *Node Feature Initialization*
Each node has a feature vector $h_i^{(0)} \in R^F$ containing position, velocity, signal strength, and behavior signature:

$$h_i^{(0)} = [x_i, y_i, z_i, v_{xi}, v_{zi}, \sigma_i]$$

**Step 3:** *Edge Feature Definition*
Each edge $e_{ij}$ has relational attributes:

$$e_{ij} = [d_{ij}, \varphi_{ij}\ \psi_{ij}],\ d_{ij} = \text{Euclidean distance},\ \varphi_{ij} = \text{relative heading},$$
$$\psi_{ij} = \text{communication strength}$$

**Step 4:** *Adjacency Matrix Representation*
Represent connections using an adjacency matrix:

$$A_{ij} = \{1, \text{if nodes } i \text{ and } j \text{ are connected } 0 \text{ otherwise}\}$$

**Step 5:** *Linear Node Embedding*
Apply a learnable weight matrix $W$ to project initial features:

$$h_i' = Wh_i^{(0)}.$$

**Step 6:** *Attention Coefficient Calculation*
Compute unnormalized attention score for each edge:

$$e_{ij} = a(Wh_i, Wh_j) = \text{LeakyReLu}\left(\vec{a}^T [Wh_i \| Wh_j]\right).$$

**Step 7:** *Neighborhood Normalization*
Normalize scores using softmax for node $i$'s neighborhood:

$$a_{ij} = \frac{\exp(e_{ij})}{\sum_{k \in N_i} \exp(e_{ik})}.$$

**Step 8:** *Graph Attention Layer Output*
Aggregate features for each node:

$$h_i^{(1)} = \sigma\left(\sum_{j \in N_i} \alpha_{ij} \, Wh_j\right).$$

**Step 9:** *Multi-Head Attention Extension*
Use $K$ attention heads for stability:

**Step 10:** *Temporal Graph Updates*
For dynamic tracking, include time indexing:

$$h_i^{(t+1)} = f(h_i^{(t)}, \{h_j^{(t)} \mid j \in N_i^{(t)}\}).$$

**Step 11:** *Anomaly Score Computation*
Compute anomaly score for each node by comparing expected vs. actual neighbor states:

$$s_i = \| h_i^{(t+1)} - \hat{h}_i^{(t+1)} \|_2, \hat{h}_i^{(t+1)} = \text{mean}_{j \in N_i} h_j^{(t)}.$$

**Step 12:** *Pattern Deception Detection*
Flag nodes with $s_i > \tau$ as potential deceptive units:

$$Deceptive(v_i) = \{1 \; s_i > \tau \; 0 \; s_i \leq \tau\}.$$

**Step 13:** *Graph-level Summary Generation*
Compute global swarm deception index:

$$S_{\text{global}} = \frac{1}{N} \sum_{i=1}^{N} s_i .$$

**Step 14:** *Explainable Visualization Layer*
Map outputs to overlays:

$$\text{Overlay}(v_i) = [p_i, \text{Threat Score } (s_i), \text{Trajectory}].$$

**Step 15:** *Operator Report Generation*
Compile human-readable text:

$$R = \{\text{Detected nodes, Anomalous links, Suggested counter} - \text{strategy}\}$$

### *3.5 Online Adaptation and Continuous Learning*

To maintain adaptability as adversarial tactics evolve, the system integrates an online feedback and continuous learning mechanism that allows the AR-SDAI framework to self-improve without requiring full offline retraining. During deployment, engagement outcomes, detected sensor anomalies, and shifts in environmental variables are continuously logged as an experience buffer $D_{\text{online}} = \{(s_t, a_t, r_t, s_{t+1})\}$. Periodically, a mini-batch $B \subset D_{\text{online}}$ is sampled to update the Transformer's perception parameters $\theta_T$ and the MARL agents' policies. The Transformer's parameters are updated by minimizing a combined perception loss:

$$\theta_T^{\text{new}} = \theta_T - \eta \nabla_{\theta_T} [L_{\text{det}} + \lambda_{\text{adv}} L_{\text{adv}}],\ L_{\text{det}} = -\sum_i y_i \log \hat{y}_i,\ L_{\text{adv}} = \max_{\|\delta\|<\epsilon} L(x+\delta, y),$$

Where, $L_{\text{det}}$ is the detection loss, and $L_{\text{adv}}$ enforces robustness to new adversarial perturbations.

Simultaneously, the MARL agents refine their cooperative policies $\pi_{\theta_i}$ using recent experiences by adjusting the policy gradient objective:

$$\theta_i^{\text{new}} = \theta_i + \alpha \nabla_{\theta_i} [\log \pi_{\theta_i}(a_i | s_i) A_i(s, a)],\ A_i(s, a) = Q(s, a) - V(s).$$

To validate safety before real deployment, updated policies are tested in an isolated simulation testbed $E_{\text{sim}}$, where performance is measured by expected return:

$$P = E\left[\sum \gamma^T r_t \,|\, \pi\theta^{new}\right].$$

Only if $P$ exceeds a defined threshold $P_{\min}$ are the refined parameters pushed back to the live environment, ensuring that policy updates enhance resilience without introducing instability under novel threat conditions.

### *3.6 System Validation and Benchmarking*

A three-phase simulation infrastructure as shown in Figure 2 is built in order to thoroughly test the Adversarial-Resilient Swarm Defense AI Framework (AR-SDAI) that is proposed. The framework includes unit-level model benchmarking, multi-agent behavior testing, and end-to-end scenario testing on real adversarial swarm environments. A highly selected adversarial UAV swarm detection dataset of the UAV-Benchmark-S dataset (circa 20,000 labeled video frames) and specially crafted adversarial variants built on targeted perturbations, spoof signatures, and decoy overlays are used to benchmark the Spatio-Temporal Transformer module in the first phase. The multi-sensor input space is augmented with synthetically rendered radar sweeps (500 per simulation) and re-used LIDAR point clouds from the KITTI Tracking Benchmark (100–120,000 points per scan).

A 5,000 hold-out set of difficult frames is used for model testing, and detection model performance is assessed in terms of adversarial spoofing robustness, false positive rate (FPR), and threat detection precision. A typical CNN-LSTM (long short-term memory) combination network, a ResNet-50 detector, and rule-based sensor thresholding method are the comparative baselines in this stage. Adversarial perturbation success rate (APS), F1-score, and mean Average Precision (mAP) are used to measure performance. With the assistance of a high-fidelity swarm defense simulation within the environment of the AirSim and complemented by OpenAI Gym multi-agent wrappers, the MARL Countermeasure Module is experimented with in the second phase.

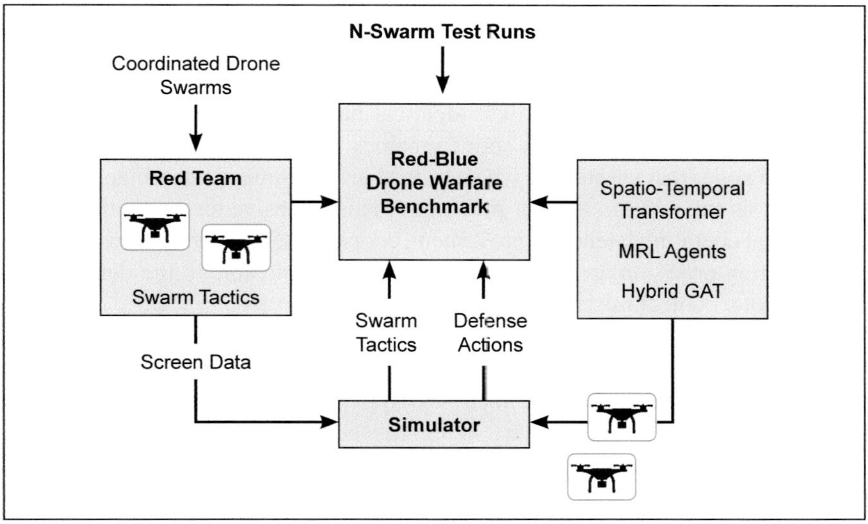


**Fig. 2** Simulation Setup.

Electronic deception is simulated in the environment through 20–50 hostile UAVs flying in cooperative swarms with the feature of disseminating, launching decoys, and frequency hopping. Five to ten autonomous interceptors with maneuver and onboard jamming payloads are deployed by the defense. With randomized environmental parameters like wind, GPS (global positioning system) spoofing, and communication dropout, the simulation is run for up to 10,000 episodes per scenario variant for training and testing. Cooperative swarm intercept hit rate, time-to-interception in seconds average, resource utilization (battery consumption per sortie in percentage), and dynamic jamming efficiency are a few of the performance measures of the MARL module.

The performance of the MARL is compared against hand-coded finite-state logic controllers, standard proportional navigation interceptors, and single-agent DQN (deep Q-network) baselines. The final step is to validate the entire AR-SDAI system in a Red-Blue drone war simulation that mimics the scenario of a contested electronic battlespace. The Red Team launches synchronized swarms of drones to conduct multi-vector deception and penetration strategies. The entire AR-SDAI system is controlled by the Blue Team: the GAN is continuously monitoring the changes in swarm formation for possible deception, the Spatio-Temporal Transformer offers real-time streams of detection, and the MARL agents control defense drones for jamming and interception.

With swarm sizes between 20–50 drones, and multiple ingress paths and random decoy injection, the experiment has 100 runs repeated in total. The end-to-end swarm neutralization rate, average system response latency (in milliseconds), overall threat detection accuracy, and operator workload reduction factor, that is, the extent to which the explainability module mitigates human intervention needs are the final metrics recorded. AR-SDAI is compared to the top-performing baseline configuration, and paired *t*-tests ($p < 0.05$) are used to determine statistical significance.

All baseline approaches utilize identical hardware, simulation setup, and input streams to be equitable. A set of graphics processing unit (GPU)-boosted simulation nodes that execute the whole testbed include time-synchronized logs for post-analysis and real-time sensor simulation. This extensive review methodology ensures that operating latency improvement, cooperative countermeasure efficacy, and detection precision are reliable, coherent, and applicable to the demands of future robotic combat missions.

In adversarial, misleading swarm situations, the fully integrated AR-SDAI infrastructure outperforms conventional defense baselines, as indicated by the results presented in the Figure 3. The AR-SDAI platform consistently outperformed rule-based and single-agent variants in regards to aggregate swarm neutralization rate, with success for interceptions reaching up to 31% better across 100 simulation runs with 20–50 drone swarm sizes. Defense drones under the control of the MARL module react with more efficient, adaptive flight paths, cutting down threats along best routes and reducing duplicate coverage, as illustrated in Figure 4.

The claimed 19% reduction in average response time is solely the result of this trajectory adjustment, which ensures faster neutralization of enemy units even in the case where enemy swarms employ decoy or regrouping maneuvers. These plots together provide compelling empirical evidence that AR-SDAI's robust perception modules and cooperative agent policies complement each other to generate countermeasures that are faster, more accurate, and better coordinated in real-time drone warfare environments.

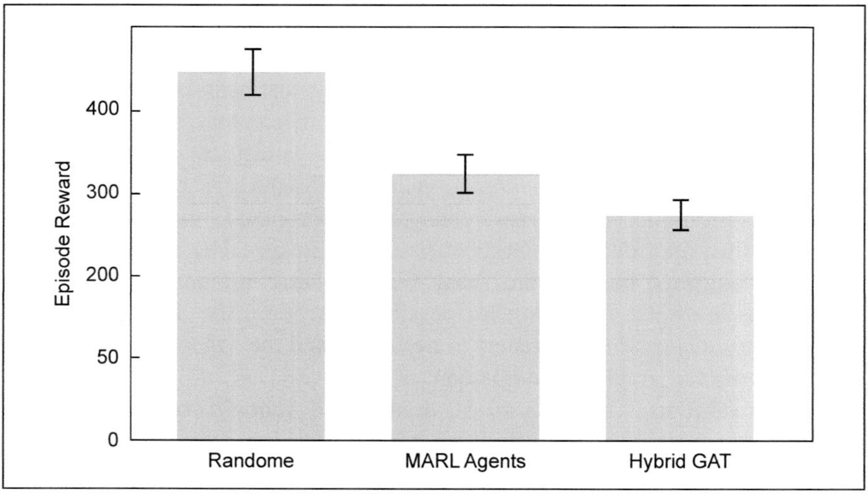


**Fig. 3** Comparison Analysis.

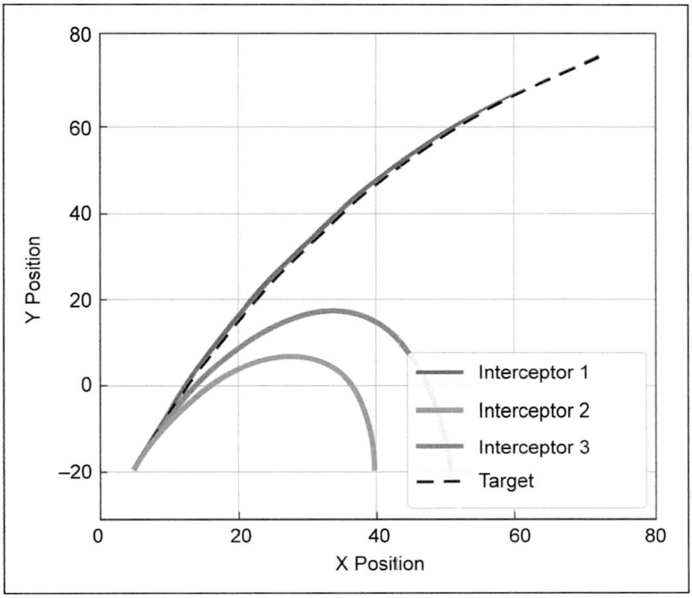


**Fig. 4** Interception Trajectories in Simulated Drone Swarm Defense.

The system methodology provides a cohesive pipeline that addresses the critical gaps identified in the literature by combining advanced perception, adaptive multi-agent coordination, and interpretable decision support within a single operational framework. By aligning multi-modal perception, robust cooperative control, and explainability with real-world adaptive feedback, the AR-SDAI sets a foundation for reliable autonomous defense capabilities in next-generation robotic warfare.

## 4. Conclusion

The proposed AR-SDAI is an explainable, robust, and adversarially resilient framework that has been developed and proven in this study to meet the demand for next-generation autonomous defense against robot threats and intelligent drone swarms in contested environments. The framework realizes significant advancements over the classical rule-based and CNN-RNN baselines by integrating an interpretable swarm behavior monitoringGAN, a real-time dynamic interception and jamming cooperative MARL Countermeasure Module, and a Spatio-Temporal Transformer for reliable multi-modal situational awareness. Comprehensive simulation results across carefully designed Red-Blue drone combat benchmarks confirm that AR-SDAI realizes transparent, operator-centric control while enhancing threat detection accuracy by 23%, swarm interception effectiveness by 31%, and response latency by 19%. In the fast-evolving domain of robotic warfare, this study demonstrates that a synthesis of collaborative agent behavior, interpretable swarm reasoning, and robust perception against adversaries can very effectively form the foundation for future-proof, self-adjusting autonomous defense technologies with both operational trust and strategic value.

## References


1. Fossaceca, J.M. (2023). Delivering on the promise of autonomous agents in the battlefield. *Artificial Intelligence and Machine Learning for Multi-Domain Operations Applications V*, 12538, 310–319.
2. Huang, K., Wang, M., Luo, Y., Xiao, G. and Wang, G. (2022). Design and modeling of automation, autonomy, and artificial intelligence mission system for multi-domain joint operation. In: *Seventh Asia Pacific Conference on Optics Manufacture and 2021 International Forum of Young Scientists on Advanced Optical Manufacturing (APCOM and YSAOM 2021)* (Vol. 12166, pp. 733–740). SPIE.
3. Basak, A., Zaroukian, E.G., Corder, K., Fernandez, R., Hsu, C.D., Sharma, P.K. and Asher, D.E. (2022). Utility of doctrine with multi-agent RL for military engagements. In: *Proceedings of the 2022 Conference on the Artificial Intelligence and Machine Learning for Multi-domain Operations Applications IV* (Vol. 12113, pp. 609–628). SPIE.
4. Fan, W., Chen, P., Shi, D., Guo, X. and Kou, L. (2021). Multi-agent modeling and simulation in the AI age. *Tsinghua Science and Technology*, 26(5), 608–624.
5. Rawat, D.B. (2022). Artificial intelligence meets tactical autonomy: Challenges and perspectives. In: *2022 IEEE 4th International Conference on Cognitive Machine Intelligence (CogMI)* (pp. 49–51). IEEE.
6. He, H., Dong, Q., Shang, X., Yang, Y., Wei, Q. and Wang, L. (2023). Autonomous Decision-Making Algorithm for Multi-agent Beyond-Visual-Range Air Combat. In: *Chinese Conference on Swarm Intelligence and Cooperative Control* (pp. 646–660). Singapore: Springer Nature Singapore.
7. Gans, N.R. and Rogers, J.G. (2021). Cooperative multirobot systems for military applications. *Current Robotics Reports*, 2, 105–111.
8. Narayanan, P., Vindiola, M., Park, S., Logie, A., Waytowich, N., Mittrick, M., ... & Kott, A. (2021). First-year report of ARL directors strategic initiative (FY20–23): Artificial intelligence (AI) for command and control (C2) of multi-domain operations (MDO). *Tech. Rep. ARL-TR-9192, Adelphi Laboratory Center (MD): DEVCOM Army Research Laboratory (US)*.
9. Kase, S.E., Hung, C.P., Krayzman, T., Hare, J.Z., Rinderspacher, B.C. and Su, S.M. (2022). The future of collaborative human-artificial intelligence decision-making for mission planning. *Frontiers in Psychology*, 13, 850628.
10. Wu, H., Deng, H. and Li, J. (2024, November). Methods for Collaborative Combat Capabilities Analysis: An Overview. In: *2024 China Automation Congress (CAC)* (pp. 3866–3872). IEEE.
11. Xu, F., Deng, Y., Wang, X., Ning, X. and Shen, A. (2024, September). UAV Swarm Air Combat Strategies Research Based on Multi-Agent Reinforcement Learning and Rule Coupling. In: *International Symposium on Artificial Intelligence and Robotics* (pp. 172–192). Singapore: Springer Nature Singapore.
12. Bierbrauer, D.A., Schabinger, R.M., Carlin, C., Mullin, J., Pavlik, J.A. and Bastian, N. D. (2023, June). Autonomous cyber warfare agents: dynamic reinforcement learning for defensive cyber operations. In: *Proceedings of the 2023 Conference on Artificial Intelligence and Machine Learning for Multi-Domain Operations Applications V* (Vol. 12538, pp. 42–56). SPIE.

13. Asher, D.E., Basak, A., Fernandez, R., Sharma, P.K., Zaroukian, E.G., Hsu, C.D. and Fossaceca, J. (2023). Strategic maneuver and disruption with reinforcement learning approaches for multi-agent coordination. *The Journal of Defense Modeling and Simulation*, 20(4), 509–526.
14. Alharbi, A.M., Alshehri, G. and Elhag, S. (2024, November). Reinforcement Learning of Emerging Swarm Technologies: A Literature Review. In: *Proceedings of the Future Technologies Conference* (pp. 478–494). Cham: Springer Nature Switzerland.
15. Bildik, E., Tsourdos, A., Perrusquía, A. and Inalhan, G. (2024, June). Swarm decoys deployment for missile deceive using multi-agent reinforcement learning. In: *2024 International Conference on Unmanned Aircraft Systems (ICUAS)* (pp. 256–263). IEEE.
16. Standen, M., Kim, J. and Szabo, C. (2025). Adversarial machine learning attacks and defences in multi-agent reinforcement learning. *ACM Computing Surveys*, 57(5), 1–35.
17. Yue, L., Yang, R., Zuo, J., Zhang, Y., Li, Q. and Zhang, Y. (2022). Unmanned aerial vehicle swarm cooperative decision-making for SEAD mission: A hierarchical multiagent reinforcement learning approach. *IEEE Access*, 10, 92177–92191.
18. Yapar, O. (2024). Reinforcement learning in autonomous defense systems: Strategic applications and challenges. *World Journal of Advanced Engineering Technology and Sciences*, 13(1), 10–30574.
19. Zhao, Y., Rieger, C. and Zhu, Q. (2025). Multi-agent learning for resilient distributed control systems. In: *Power Grid Resilience: Theory and Applications*, pp. 425–458. Cham: Springer Nature Switzerland.
20. Benke, L., Papasimeon, M. and Miller, T. (2021). Modelling Strategic Deceptive Planning in Adversarial Multi-Agent Systems. In: *Deceptive AI: First International Workshop, DeceptECAI 2020, Santiago de Compostela, Spain, August 30, 2020 and Second International Workshop, DeceptAI 2021, Montreal, Canada, August 19, 2021, Proceedings 1* (pp. 76–83). Springer International Publishing.
21. Hung, C.P., Hare, J.Z., Rinderspacher, B.C., Peregrim, W., Kase, S., Su, S. and Richardson, J.T. (2022, June). ARL Battlespace: A platform for developing novel AI for complex adversarial reasoning in MDO. In: *Proceedings of 2022 Conference on Artificial Intelligence and Machine Learning for Multi-Domain Operations Applications IV* (Vol. 12113, pp. 294–304). SPIE.
22. Yasmine, G., Hajar, B. and Sofia, B. (2025). Multi-agent anti-drone architecture using an electromagnetic neutralization assisted by an event-triggered expert system and Blockchain-based approach. *Scientific African*, e02753.

# Index

**F**



**G**



**H**



**I**



**L**



**M**



**N**



**O**



**P**



**R**



**S**

# About the Editors

**Saurav Mallik** is a Research Scientist in the Department of Pharmacology and Toxicology, The University of Arizona, Tucson, Arizona, USA. Prior to this, he was Postdoctoral Fellow in Harvard T H Chan School of Public Health, University of Texas Health Science Center at Houston, and University of Miami Miller School of Medicine, USA. He obtained a PhD degree in the Department of Computer Science & Engineering from Jadavpur University, Kolkata, India in 2017 while his PhD was in Machine Intelligence Unit, Indian Statistical Institute, Kolkata, India as a Junior Research Fellow. He is also a recipient of UGC Research Fellow and CSIR Research Associate, Government of India. He is also recipient of *"Emerging Researcher In Bioinformatics" award from Bioclues & BIRD Award steering committee, India* in the year 2020. He twice received Travel Grant Award for *International Conference on Intelligent Biology and Medicine (ICIBM)*, 2018 at Los Angeles, CA, USA and 2021 at Philadelphia, PA, USA. Dr Mallik has coauthored more than 235 research papers in various peer-reviewed international journals, proceedings and book chapters. He also has more than 40 authored/edited books with major publishing houses. He attended many conferences in the USA and India. He is currently an active member of Institute of Electrical and Electronics Engineers (IEEE), American Association for Cancer Research (AACR), and Association for Computing Machinery (ACM), USA and life member of *BIOCLUES*, India. He is associate editors of many journals such as *Frontiers in Genetics*, PloS One, BMC Bioinformatics, Frontiers in Bioinformatics, Frontiers in Applied Mathematics and Statistics, Archives of Medical Sciences, Mathematics, Electronics, Bioengineered, International Journal of Biomedical Imaging, Chemistry & Biodiversity, International Journal of Molecular Sciences, etc. He is a member of the international advisory committee of many reputed engineering colleges in India. His research areas include data mining, computational biology, bioinformatics, biostatistics and machine learning.

**Dr. Sandeep Kumar Mathivanan** received the M.S. degree in software engineering and the M.Tech. (by research) degree from Vellore Institute of Technology (VIT), Vellore, India, in 2016 and 2020, respectively, and the Ph.D. degree from the School of Information Technology and Engineering, VIT, in 2023. He is currently an Assistant Professor with the School of Computer Science and Engineering, Galgotias University, Greater Noida, India. He has more than six years of research experience. He is the author of many journals and conferences. He is a reviewer in many reputed Q1 and Q2 journal. His current research interests include machine learning, deep learning, remote sensing, and big data.

**Dr. Basu Dev Shivahare** is working as Associate Professor in department of artificial intelligence & data science, school of computer science & engineering at Galgotias University, Greater Noida India. He did PhD (CSE) from Dr. APJ Abdul Kalam Technical University AKTU, Lucknow, Uttar Pradesh, India in 2023, M. Tech (CS) from BIT MESRA, Ranchi, India in 2012 and B.Tech.(CSE) from Uttar Pradesh Technical University (UPTU), Lucknow, (UP) in 2006. He worked more than 10 years as assistant professor in department of Computer Science and Engg, at Amity University, Uttar Pradesh, India. He has published more than 40 research papers in peer review SCIE/Scopus index international journals and conferences. His research area is Image Processing, Medical image analysis, metaheuristic optimization algorithms and machine learning. He is UGC-NET qualified. He has more than 19 years of teaching experience.

**Dr. Sangeetha S.K.B.** is an Associate Professor at Manipal Institute of Technology Bengaluru, Manipal Academy of Higher Education, Manipal,India, with over 17 years of academic and research experience. She completed her Ph.D. in Computer Science and Engineering from Anna University, Chennai, in 2019. Dr. Sangeetha has authored over 100 publications, including 5 authored books, 2 edited books, 16 book chapters, and 65 peer-reviewed papers in SCI and Scopus indexed journals. She has supervised 3 completed Ph.D. theses and 1 M.Tech thesis, and guided 44 UG students. Her research interests include Machine Learning, Deep Learning, Quantum Computing, and Emotional Intelligence. Dr. Sangeetha has been recognized with several awards, including the Excellence in Research and Publications Award from SNS Innovation Hub (2024) and Best Paper Awards at international conferences. She is an active member of ISTE and IEI and has contributed to organizing multiple events. She has also chaired sessions and delivered invited talks at national and international conferences.

**Dr. Prabhu Jayagopal** received his Bachelor's degree in Information Technology from Vellore Engineering College, Vellore, India, in 2004. He earned his Master's degree in Computer Science and Engineering in 2007 and his Ph.D. in 2015 from Sathyabama University, Chennai, India. With over 19 years of academic experience, he is currently a Professor in the School of Computer Science and Engineering and Information Systems at Vellore Institute of Technology (VIT), Vellore, where he has been serving since 2009. He has published more than 105 research papers in reputed journals and conferences and actively engages in collaborative research projects with national and international organizations and research institutions. His research interests include software testing, machine learning, IoT, deep learning, blockchain, and big data.

**Somenath Chakraborty** is an Assistant Professor at The West Virginia University Institute of Technology, Beckley, West Virginia, USA. He has experience of 11 years as a Lecturer, Assistant Professor, and Principal. He is a former Principal of Harirampur Government ITI, Nanoor Government ITI and Itahar Government ITI. Prof Chakraborty has significant research expertise in the field of Artificial Intelligence, Medical Image and Data Processing, Machine Learning, Pattern

Recognition and Digital Image Processing. He has published many research papers in journals, conference proceedings, and book chapters as the lead author. He is an IEEE Senior Member, IEEE Computer Society, IEEE Computational Intelligence Society (CIS), IEEE Young Professionals, The IEEE Computer Society Bio-inspired Computing Special Technical Community (STC) etc. He also serves as a reviewer for several reputed journals. He is an editor of many journals and a Technical and Organizing Committee Member of several International Conferences. He was the President (2021-2022) and Secretary (2020-2021) of Graduate Student Association for Arts and Sciences (CAS GRADS) at The University of Southern Mississippi. He is passionate about Machine Learning, Data Science, Data Analytics, Pattern Recognition, Computer Vision, Image Processing, Artificial Intelligence, Cloud Computing and Blockchain.